LONDON MATHEMATICAL SOCIETY LECTURE NOTE SERIES

Managing Editor: Professor M. Reid, Mathematics Institute, U[r ...] m

The titles below are available from booksellers, or from Cambr

216 Stochastic partial differential equations, A. ETHERID[
217 Quadratic forms with applications to algebraic geometry
218 Surveys in combinatorics, 1995, P. ROWLINSON (ed)
220 Algebraic set theory, A. JOYAL & I. MOERDIJK
221 Harmonic approximation, S.J. GARDINER
222 Advances in linear logic, J.-Y. GIRARD, Y. LAFONT & [...]
223 Analytic semigroups and semilinear initial boundary value problems, K. TAIRA
224 Computability, enumerability, unsolvability, S.B. COOPER, T.A. SLAMAN & S.S. WAINER (eds)
225 A mathematical introduction to string theory, S. ALBEVERIO et al
226 Novikov conjectures, index theorems and rigidity I, S.C. FERRY, A. RANICKI & J. ROSENBERG (eds)
227 Novikov conjectures, index theorems and rigidity II, S.C. FERRY, A. RANICKI & J. ROSENBERG (eds)
228 Ergodic theory of Z^d actions, M. POLLICOTT & K. SCHMIDT (eds)
229 Ergodicity for infinite dimensional systems, G. DA PRATO & J. ZABCZYK
230 Prolegomena to a middlebrow arithmetic of curves of genus 2, J.W.S. CASSELS & E.V. FLYNN
231 Semigroup theory and its applications, K.H. HOFMANN & M.W. MISLOVE (eds)
232 The descriptive set theory of Polish group actions, H. BECKER & A.S. KECHRIS
233 Finite fields and applications, S. COHEN & H. NIEDERREITER (eds)
234 Introduction to subfactors, V. JONES & V.S. SUNDER
235 Number theory: Séminaire de théorie des nombres de Paris 1993–94, S. DAVID (ed)
236 The James forest, H. FETTER & B.G. DE BUEN
237 Sieve methods, exponential sums, and their applications in number theory, G.R.H. GREAVES et al (eds)
238 Representation theory and algebraic geometry, A. MARTSINKOVSKY & G. TODOROV (eds)
240 Stable groups, F.O. WAGNER
241 Surveys in combinatorics, 1997, R.A. BAILEY (ed)
242 Geometric Galois actions I, L. SCHNEPS & P. LOCHAK (eds)
243 Geometric Galois actions II, L. SCHNEPS & P. LOCHAK (eds)
244 Model theory of groups and automorphism groups, D.M. EVANS (ed)
245 Geometry, combinatorial designs and related structures, J.W.P. HIRSCHFELD et al (eds)
246 p-Automorphisms of finite p-groups, E.I. KHUKHRO
247 Analytic number theory, Y. MOTOHASHI (ed)
248 Tame topology and O-minimal structures, L. VAN DEN DRIES
249 The atlas of finite groups - Ten years on, R.T. CURTIS & R.A. WILSON (eds)
250 Characters and blocks of finite groups, G. NAVARRO
251 Gröbner bases and applications, B. BUCHBERGER & F. WINKLER (eds)
252 Geometry and cohomology in group theory, P.H. KROPHOLLER, G.A. NIBLO & R. STÖHR (eds)
253 The q-Schur algebra, S. DONKIN
254 Galois representations in arithmetic algebraic geometry, A.J. SCHOLL & R.L. TAYLOR (eds)
255 Symmetries and integrability of difference equations, P.A. CLARKSON & F.W. NIJHOFF (eds)
256 Aspects of Galois theory, H. VÖLKLEIN, J.G. THOMPSON, D. HARBATER & P. MÜLLER (eds)
257 An introduction to noncommutative differential geometry and its physical applications (2nd Edition), J. MADORE
258 Sets and proofs, S.B. COOPER & J.K. TRUSS (eds)
259 Models and computability, S.B. COOPER & J. TRUSS (eds)
260 Groups St Andrews 1997 in Bath I, C.M. CAMPBELL et al (eds)
261 Groups St Andrews 1997 in Bath II, C.M. CAMPBELL et al (eds)
262 Analysis and logic, C.W. HENSON, J. IOVINO, A.S. KECHRIS & E. ODELL
263 Singularity theory, W. BRUCE & D. MOND (eds)
264 New trends in algebraic geometry, K. HULEK, F. CATANESE, C. PETERS & M. REID (eds)
265 Elliptic curves in cryptography, I. BLAKE, G. SEROUSSI & N. SMART
267 Surveys in combinatorics, 1999, J.D. LAMB & D.A. PREECE (eds)
268 Spectral asymptotics in the semi-classical limit, M. DIMASSI & J. SJÖSTRAND
269 Ergodic theory and topological dynamics of group actions on homogeneous spaces, M.B. BEKKA & M. MAYER
271 Singular perturbations of differential operators, S. ALBEVERIO & P. KURASOV
272 Character theory for the odd order theorem, T. PETERFALVI. Translated by R. SANDLING
273 Spectral theory and geometry, E.B. DAVIES & Y. SAFAROV (eds)
274 The Mandelbrot set, theme and variations, T. LEI (ed)
275 Descriptive set theory and dynamical systems, M. FOREMAN, A.S. KECHRIS, A. LOUVEAU & B. WEISS (eds)
276 Singularities of plane curves, E. CASAS-ALVERO
277 Computational and geometric aspects of modern algebra, M. ATKINSON et al (eds)
278 Global attractors in abstract parabolic problems, J.W. CHOLEWA & T. DLOTKO
279 Topics in symbolic dynamics and applications, F. BLANCHARD, A. MAASS & A. NOGUEIRA (eds)
280 Characters and automorphism groups of compact Riemann surfaces, T. BREUER
281 Explicit birational geometry of 3-folds, A. CORTI & M. REID (eds)
282 Auslander–Buchweitz approximations of equivariant modules, M. HASHIMOTO
283 Nonlinear elasticity, Y.B. FU & R.W. OGDEN (eds)
284 Foundations of computational mathematics, R. DEVORE, A. ISERLES & E. SÜLI (eds)
285 Rational points on curves over finite fields, H. NIEDERREITER & C. XING
286 Clifford algebras and spinors (2nd Edition), P. LOUNESTO
287 Topics on Riemann surfaces and Fuchsian groups, E. BUJALANCE, A.F. COSTA & E. MARTÍNEZ (eds)
288 Surveys in combinatorics, 2001, J.W.P. HIRSCHFELD (ed)
289 Aspects of Sobolev-type inequalities, L. SALOFF-COSTE
290 Quantum groups and Lie theory, A. PRESSLEY (ed)
291 Tits buildings and the model theory of groups, K. TENT (ed)
292 A quantum groups primer, S. MAJID

London Mathematical Society Lecture Note Series: 368

Geometry of Riemann Surfaces

Proceedings of the Anogia Conference to celebrate
the 65th birthday of William J. Harvey

Edited by

FREDERICK P. GARDINER

Brooklyn College, CUNY

GABINO GONZÁLEZ-DIEZ

Universidad Autónoma de Madrid

CHRISTOS KOUROUNIOTIS

University of Crete

CAMBRIDGE
UNIVERSITY PRESS

CAMBRIDGE UNIVERSITY PRESS
Cambridge, New York, Melbourne, Madrid, Cape Town,
Singapore, São Paulo, Delhi, Tokyo, Mexico City

Cambridge University Press
The Edinburgh Building, Cambridge CB2 8RU, UK

Published in the United States of America by Cambridge University Press, New York

www.cambridge.org
Information on this title: www.cambridge.org/9780521733076

© Cambridge University Press 2010

First published 2010

A catalogue record for this publication is available from the British Library

ISBN 978-0-521-73307-6 Paperback

Contents

v

Contents

Preface

This conference on the Geometry of Riemann Surfaces and related topics was held in the beautiful hill town of Anogia at the Conference Centre of the University of Crete, spanning four days in June and July 2007. The pretext was the celebration of Bill Harvey's 65th birthday and retirement from teaching. About 50 mathematicians and friends came, many from far-flung points of the globe, to enjoy this opportunity to refresh mind, body and spirit.

We invited all participants to contribute articles based on their talks or related material; the response was wholehearted and expert, with the result that you see before you. The broad range of topics addressed by the articles reflects the pervasive influence of the theory of Riemann surfaces and the remarkable variety of geometric ideas and methods which flow from it; this expansive aspect of the field will be discussed by Bill Harvey in the introduction which follows.

We take this opportunity also to thank Professor Harvey himself, whose supportive and knowledgeable comments provided foil and inspiration for all the participants.

On behalf of those lucky enough to be at Anogia for this conference, we thank all the sponsors, the Universidad Autónoma de Madrid and Comunidad Autónoma de Madrid (Grant C-101), the Spanish Government Ministerio Español de Educación y Ciencia (Grants MTM2006-01859 & MTM2006-28257-E), the Department of Mathematics of the University of Crete and the Anogia Academic Village, Crete, for generous financial support which made the meeting possible; in particular, the funding provided subsistence and travel expenses for graduate students and others lacking support.

We are also very grateful to the local organisers and staff at the University of Crete, in Iraklio and at the Conference Centre, and especially the conference secretary Marina Vasilaki, for their work in planning and preparation and for friendly assistance during the meeting. Significant editorial assistance with the Proceedings was given by David Torres Teigell (UAM) and is gratefully acknowledged. Finally, the appearance of this volume is thanks to the unselfish hard work of many people including all the anonymous referees, testament in itself to the continuing good health of this section of the mathematical community.

Fred Gardiner
Gabino González-Diez
Christos Kourouniotis

Foreword: Riemann surfaces and a little history

William Harvey

King's College London

A Riemann surface is a thing of beauty, possessing geometric shape as well as analytic or algebraic structure. From its introduction in 1851 in Riemann's inaugural dissertation, his first great work establishing the foundations of geometric complex analysis, the concept has exerted an unusual influence as a powerful clarifying mental tool.

Today, the pervasive role of complex analysis in the mathematical and physical sciences has brought these ideas into a significance wider than even their founder could have predicted. In the present book, the reader will find a selection of results which can only indicate the part currently played by surfaces and their spaces of deformations: just as a single convergent power series is enough to generate by continuation an entire Riemann surface structure, so the foundational ideas of our discipline extend and evolve beyond our present view of them.

Central to the contemporary study of Riemann surfaces is the interplay between different aspects, geometric ideas and algebraic or analytical calculations, leading to insights into the deeper properties these objects possess. The basic notion provides a topological base for deploying the most powerful ideas of algebra, geometry and analysis: indeed it establishes a central role for topology in bringing about a unique mathematical synthesis. A single accessible theory serves to interconnect complex analysis and the various algebraic invariants, the fundamental group, field of functions, homology and period lattices. In the reassuring familiarity of a two dimensional framework, we have a global base for complex analytic and covering space methods, interacting with Galois-theoretic properties of the function field. And awaiting new developments, there are so many areas where Riemann surfaces are direct contributors: hyperbolic manifolds and kleinian groups, iteration of polynomial or holomorphic functions, crystallographic groups, geometric group theory, conformal quantum field theory and symplectic geometry. Not least, there is the burgeoning field of complex geometry, which relies on the theory of Riemann surfaces for insights and testing of ideas as well as for essential tools in the study of complex or symplectic manifolds, via deformation theory and the existence of embedded curves. All these areas are represented among the articles included here, a blend of original research, broad surveys and applications, as indicated in the next section.

1 Background and development.

The brief historical narrative which follows will (it is hoped) set the scene and draw together all the diverse themes to be found in this volume. Much of it relies, directly or indirectly, on the mapping theorem of Riemann and its successor, the uniformisation theorem of Poincaré and Koebe, which deliver an intrinsic geometric structure by covering projection from the unit disc to any Riemann surface or complex algebraic curve, casting new light on purely analytic or algebraic matters. Thus, as an instance of how important insights follow, any closed loop not bounding a disc on a hyperbolic surface S determines (by the calculus of variations) a unique closed geodesic in the same free homotopy class, whose length is an important geometric invariant, an element of the *length spectrum* of that surface. This discrete collection of positive numbers encodes the conformal shape in an extremely subtle way, related via famous work by A. Selberg to the analytic study of the Laplace-Beltrami operator, a global elliptic operator on $\mathcal{L}_2(S)$, whose spectral properties are of wide interest. Taking a different line from the same beginning, such a loop in S determines a smooth real *length function* on the moduli space of conformal classes of hyperbolic surfaces diffeomorphic to S, which is part of a Morse-theoretic topological decomposition of the moduli space. Three articles in the book pursue this theme. In Peter Buser's article, an algorithmic approach is given to the question of computing the set of shortest geodesics from a suitable specification of the (marked) surface's moduli and carried out for a specific example, a breakthrough in explicit geometric computation on a surface, with roots in work by Poincaré and Dehn on algorithms for simple loops. The paper by Greg McShane and Hugo Parlier addresses the issue of multiplicities for the length spectrum for simple loops on a (punctured) torus, a classic test case for hyperbolic surface phenomena. Robert Silhol's paper addresses a quite different issue in this geometric arena, which has bearing on the coefficient field of the uniformising fuchsian group.

Over the past sixty years, Riemann surface theory has seen a progression of new ideas and methods, none more influential than the development during the 1950s, by L. V. Ahlfors, L. Bers and H. E. Rauch, of a theory of moduli for surface deformations, holomorphic parameters for a family of surfaces of specified topological type. The case of the flat torus (genus 1) is classical, but for higher genus it was not until the foundational ideas of O. Teichmüller were digested and sharpened that Ahlfors and Bers could establish a comprehensive analytic theory of Riemann surface families based on *quasiconformal homeomorphisms*, or q-c maps. This type of controlled deformation has special flexibility and plays a prominent part in the study of holomorphic families of discrete groups and dynamical systems in complex dimension 1. The crucial first step was the construction of a complex analytic Teichmüller space of moduli for a given compact Riemann surface or orbifold, a parameter space for all holomorphic deformations of fixed type carrying a given topological marking; this was established using q-c maps soon after 1960. The complex analytic and Kähler metric structure of the moduli spaces $\mathcal{M}_g$ of conformal classes for surfaces of genus $g \geq 2$ soon ensued and attention then focussed on the construction of a suitable compactification within the parallel development of q-c deformations for all finitely generated kleinian groups – these are discrete groups of Möbius transformations which occur as cover transformations in uniformisation of surfaces when the covering region is not necessarily a Euclidean disc.

A very different approach to moduli of curves was completed by D. Mumford around 1965, using geometric invariant theory, the algebraic geometer's approach to classification problems; later with P. Deligne he extended his earlier results with A. Mayer on stable degeneration of curves by acquisition of double points to produce a projective completion $\widehat{\mathcal{M}_g}$, the algebraic variety of moduli of stable curves of fixed genus $g \geq 2$. The noded 'stable curves' which fill up the missing boundary divisor represent cusps of the corresponding mapping class group action within a complex analytic compactification, analogous to the point at infinity, and its orbit under the modular group $SL_2(\mathbb{Z})$, in the space of marked tori. This chimed with the detailed classification by Bers, by B. Maskit and others of regular b-groups, a particular kind of kleinian group which represent noded curves within a kind of local completion of the familar spaces of fuchsian and quasifuchsian groups; this too gives a compactification of complex moduli space, isomorphic in an appropriate sense to that of Deligne and Mumford. A theory of deformations for all projective complex manifolds had been developed earlier by K. Kodaira and D.C. Spencer and by M. Kuranishi, but the special nature of geometry in one complex dimension has created a vastly different, more detailed landscape. This was enhanced after 1980 by S. A. Wolpert's penetrating work on the Weil-Petersson metric structure, linking it to the length spectrum and hyperbolic surface geometry.

Around 1970, H. E. Rauch and his student H. M. Farkas were considering the *Schottky problem*, which asks for a precise characterisation of the place of curves and their associated Jacobi varieties – these are complex tori of dimension g which emerge by integrating a basis of holomorphic 1-forms along a set of generating loops for the homology on a curve – within the general theory of (principally polarised) abelian varieties and their periods. This question received a remarkable reformulation as part of the theory of completely integrable systems of differential equations, resulting in T. Shiota's resolution of the problem in 1985; however, because this approach is very inexplicit, further study continues today, and the article here by Victor Gonzalez contributes to it for the case of curves with symmetry. The projective geometry of linear systems of divisors on p-gonal curves, analysed in the article by Gabino González-Díez, is also significant for the study of these special curves and their Jacobi varieties.

By a theorem of Hurwitz, any compact Riemann surface of genus g (at least 2) has finite automorphism group with order at most $84(g-1)$. These groups lift to discrete (infinite) covering group actions on the universal covering disc as fractional linear automorphisms, thus providing a valuable way to study surface automorphism groups first discovered by Klein and Fricke and exploited by A.M. Macbeath and his students in the 1960s. In work pursued later by David Singerman and co-workers, this approach was extended to the case of all symmetry groups by expanding the framework to involve Macbeath's classification of the nonorientable hyperbolic crystallographic groups and their Teichmüller spaces together with symmetries of real algebraic curves. Articles presenting results on this aspect of the theory include a foundational one by Clifford Earle and two joint papers, by Emilio Bujalance, Javier Cirré & Gregor Gromadski and by Antonio Costa & Milagros Izquierda. Also, the uniformisation of surfaces using Schottky groups – another classical technique, which uses free kleinian groups operating on an intermediate planar covering of the Riemann surface – is extended here by Rubén Hidalgo & Bernard Maskit to include reflection symmetry.

Low dimensional topology and complex analysis are old friends; mapping class groups and braid groups have been part of this relationship from the beginning with Hurwitz, Klein and Poincaré, followed by Max Dehn's foundational work on abstract discrete groups and J. Nielsen's geometric analysis of surface mapping classes in the 1930s, with later contributions by W. Fenchel and by a long German tradition of low-dimensional topologists. The link was greatly strengthened in the 1960s by the school established under Wilhelm Magnus at New York University, who developed a detailed theory of finitely presentable groups with strong geometric ingredients. There is a direct connection between kleinian groups and hyperbolic 3-manifold structures stemming from the fact that the Riemann sphere, on which these groups act, forms the boundary of a standard model of hyperbolic space: this was observed by Poincaré, but thereafter lay dormant until Ahlfors reignited the subject after 1960 with his finiteness theorem. Several preliminary developments using hyperbolic space began after 1970; A. Marden established some key properties of kleinian 3-manifold quotients and their deformation spaces, and with C.J. Earle applied this to develop a distinctive new approach to completion of Teichmüller moduli spaces. Meanwhile R. Riley was studying the group theory of hyperbolic knot complements and T. Jorgensen began a deeper exploration of the special geometric crystallography which underlies the structure of discrete hyperbolic groups in dimensions 2 and 3. This topic expanded greatly, bringing further geometric insight into the general area of discrete group actions in real and complex hyperbolic spaces. Several articles in this book belong to this tradition, for instance the one by Jane Gilman and Linda Keen on the combinatorial structure of generating sets for free Möbius groups and the article by John Parker and Yiannis Platis describing recent progress on a higher dimensional complex hyperbolic analogue of quasifuchsian groups.

These aspects of Riemann surfaces were surveyed in an Instructional Conference on Discrete Groups and Automorphic Functions at Cambridge (UK) in 1975, which aimed to promote the field more widely among a new generation of students. A year later, W.P. Thurston brought about a revolution in the theory of kleinian groups and 3-manifolds by the introduction of new ideas and methods from dynamics, differential geometry and foliation theory, formulating his vision of geometrisation for 3-manifolds, the 3D analogue of surface uniformisation, and setting out an ambitious new agenda for research in low-dimensional geometry. The effect was electric. The existing framework of researchers expanded to meet the challenge, with concentrations across the USA and Europe, and all signed up with enthusiasm to Thurston's geometric approach, seeking a synthesis with the complex analytic one. A decade of intense activity ensued, with an explosion of results and fresh initiatives. Dynamical properties of real quadratic maps were analysed by J. Milnor and Thurston; a fundamental dictionary between the twin theories of kleinian groups and conformal dynamics was established by D. Sullivan, bringing quasiconformal methods and renormalisation into the dynamics of iteration for rational maps of the Riemann sphere; the classification of critically finite rational maps by Thurston was followed by the theory of polynomial-like mappings due to A. Douady and J.H. Hubbard and their group at Paris-Sud (Orsay), who worked out a remarkable detailed analysis of the Mandelbrot family of quadratic polynomials. The insights thus gained into the role of quasiconformal deformations in conformal geometry, rational maps and fractal structures in the plane fed into C.T. McMullen's work, which emphatically expanded Sullivan's dictionary and established major aspects of the Thurston programme. They continue to bear fruit today. In particular, the concept of *holomorphic motion* of a closed planar set (such as the limit set of a kleinian group or Julia set of a rational map) introduced in a seminal paper of Mañé, Sad & Sullivan, and developed by Sullivan and Thurston and, independently, by Bers and Royden, has been very influential; the fundamental theorem that any holomorphic motion of a closed set extends to one of the whole Riemann sphere emerged in definitive form in later work by Slodkowski and by Chirka. There are two papers on this topic in the present volume: the first, by Frederick Gardiner, Yunping Jiang & Zhe Wang presents a new proof of the Slodkowski Theorem and a second, by Sudeb Mitra, surveys the applications to deformation theory. In addition, the article by Shaun Bullett describes a special blend of rational maps and kleinian groups to which these methods apply.

As if the arrival of Thurston on the scene were not enough upheaval, the end of the 1970s also marked the beginning of M. Gromov's comprehensive deconstruction and redesign of discrete group theory and hyperbolic differential geometry. This has seen the creation of fresh ways to distinguish types of group action on metric spaces, involving a flexible approach to curvature. Notably, Gromov introduced the concept of hyperbolic group, providing a unifying theme for a broad swath of results in topology involving groups which, like fuchsian and kleinian groups, act on a space of negative curvature. A new landscape emerged, now known as geometric group theory, subsuming the work of Dehn and

Magnus on combinatorial group theory. This is a broad-brush weakening of the standard approach to rigidity (which uses isometric conjugacy to compare group actions on spaces): familiar groups arising from geometric origins are taken as benchmarks for classification, while any discrete group is viewed as a geometric object in its own right by its action on a Cayley graph, and the classification of group actions on spaces is carried out using a new measure of distortion, called quasi-isometry. Martin Bridson's paper here focusses on the position of the mapping class group within this environment, bringing out in the process the significance of its action as the modular group of isometries of the Teichmüller space.

As a final comment concerning these articles, for me much of the excitement of mathematics lies in gaining access to new results and relating them to already familiar facts and ideas; here the part played by topology has a special power because of the brevity, flexibility and universality of its language and methods. For instance, the variational principles underlying Morse theory, first seen in relation to the structure of compact surfaces, pervade the theory of manifolds and Lie groups as well as Teichmüller theory and play an essential part in the applications of all this to theoretical physics: the same interaction between topology and optimisation seen in the article by Paolo Teofilatto & Mauro Pontani on orbital trajectories may one day, perhaps by way of string theory, produce a practical application of our developing comprehension of moduli spaces to the real world.

2 Some personal history.

This account of developments in Riemann surfaces drew on my own experience within this field which has fascinated me ever since I first encountered it as a graduate student in 1962. I have tried to make clear what lies behind the mathematics considered in this volume, how such a fusion of topics and combination of methods happened and why it continues to attract wide attention. This falls short of the comprehensive historical treatment the subject merits.

2.1 *Mathematics at Birmingham after 1960.*

My doctoral adviser A.M. Macbeath, now retired but still active, was unfortunately unable to attend the Anogia conference. Instead he has written the following brief comments on the early days of my postgraduate career.

Bill was a new graduate student (in 1962) when I was a new kid on the block too, coming from Dundee to succeed Peter Hilton (as chairman of the Department of Mathematics) at the University of Birmingham. It was a landmark in my life as well as his. Bill had planned to work with Bill Parry in Ergodic Theory, but Parry took leave that year and Bill began his research with me.

I had the challenge of adapting to the way of life in a new environment. Birmingham was very different from Dundee, where the mathematics department belonged to a college of the venerable University of St. Andrews. At that time my interest had changed from number theory and convex sets to automorphisms of Riemann surfaces. Like Ergodic Theory, this is a branch of Analysis, but our approach, using Fuchsian groups and uniformization, was quite algebraic. With this method Bill solved the problem of finding the lowest genus on which a cyclic group of given order acts. His result had as an immediate corollary the classical result of Wiman giving the maximum order $2(2g+1)$ for automorphisms of a curve of genus g, but the answer to Bill's question depended on the prime factorisation of the prescribed order and was more intricate. This was included in his doctoral thesis.

I had to cope too with administrative and other matters, so Bill and his fellow graduate students (Hugh Wilkie and Colin Maclachlan arrived at the same time) were often left to their own devices, but they saw problems for themselves, talked to one another as well as to me. They experimented with various types of finite group acting on surfaces of low genus, and several further results soon emerged, for instance with Colin finding the maximum order of a group (rather than an element) acting on a surface of given genus. (The classical Hurwitz maximum $84(g-1)$ is attained only for a sparsely distributed set of values of g.)

When Bill submitted his work for publication, the manuscript got mislaid somewhere. This is no fun at any stage and can be a devastating matter for a first-time author. After a time we sorted things out and Bill's work was published in the Oxford Quarterly Journal of Mathematics. By this time he had graduated and moved across the Atlantic (in 1966) to work with Lipman Bers. Riemann surfaces with non-trivial automorphisms are the branch points of the Teichmüller orbifold, so Bill's previous work fitted in well.

In New York Bill met Michele Linch, also a disciple of Lipman Bers, and they have been together ever since. They returned to the UK (in 1972) and have both held positions in London since then (Bill at Kings College London and Michele at the LSE). Sometimes on one side of the Atlantic, sometimes on the other, we have met at conferences and meetings. Bill edited the proceedings of the Cambridge conference in 1976. His direct research contributions include the invention of the (so-called) curve complex.

Perhaps a few comments (from me) are in order here. Colin and I did indeed start as graduate students together and have remained firm friends to this day. Lacking any real knowledge of German, we painfully translated together Wiman's classic papers of 1892 (on automorphism groups of low genus curves) into a form of quasi-English, and spent much time looking into the Todd-Coxeter enumerative technique for cosets in discrete groups, a topic on which I was not at all keen. Fortunately it was soon possible for me to concentrate on something simpler: finite cyclic groups are a comfort. Colin's result is that the maximal automorphism group in genus g has order at least $8g + 8$; it was obtained by different methods at around the same time by Bob Accola at Brown. Murray was the first European to appreciate the true significance of the Ahlfors-Bers theory; he introduced us to it in a seminar on Ahlfors's work in 1964, and applied Teichmüller's extremal mapping theorem to extend the classic Nielsen theorem in surface topology. In the next few years Murray directed many students; both mathematical and social activity were lively in Birmingham.

The reference to the conference volume which I edited, *Discrete Groups and Automorphic Functions* (Academic Press 1976), prompts me to recall just how thinly documented this field was in those days. The Ahlfors book *Lectures on Quasiconformal Mappings* (recently reissued by the American Math. Society) and the Bers Zürich Notes *Moduli of Riemann surfaces* (ETH, 1964) were very influential and so was Irwin Kra's *Automorphic forms and Kleinian groups* (W.A. Benjamin, 1972), but no comprehensive text appeared on Teichmüller spaces until Fred Gardiner's *Teichmüller Theory and Quadratic Differentials* (John Wiley-Interscience, 1987), followed by Subha Nag's *Complex Analytic Theory of Teichmüller Spaces* (John Wiley, 1988) and Bernard Maskit's treatise *Kleinian Groups* (in Springer's Grundlehren series, 1988). In the interim there was little background reading, apart from that volume and Bill Abikoff's valuable Springer Lecture Notes (*Real analytic theory of Teichmüller spaces*). To fill the demand created by the Thurston revolution we had to wait until the Warwick group under David Epstein and Caroline Series began to fill in the geometric background to Thurston's

Princeton Lectures with several important volumes in the LMS Lecture Note (Cambridge University Press) series.

As to the curve complex – there are infinitely many of them in fact, one for each value of the genus $g \geq 2$ – this abstract simplicial structure and the intricate part it plays in low dimensional manifold theory epitomises an aspect of mathematics which has always fascinated me, the interplay between geometrically defined groups (such as reflection groups, braid groups and mapping class groups) and more primitive simplicial objects such as buildings in algebraic topology. It has been in the spotlight recently largely because of the part it plays in the monumental resolution of Thurston's ending lamination conjecture by Yair Minsky, but it occurs in work on abstract presentations and other actions of the mapping class group and appears to carry some kind of special status as a base of operations for geometric actions of this group. It also stands in as a surrogate Tits building to provide a key ingredient in N.V. Ivanov's remarkable proof in the style of Mostow rigidity of Royden's theorem for the Teichmüller modular group.

2.2   The New York research community: seminars and conferences.

An intrinsic part of the explosive growth of research in mathematics and the sciences generally post-Sputnik has been a steady supply of new blood in the field from all parts of the globe, thanks to the growth in funding for graduate study in the USA and Western Europe, and the hub of research activity within the mathematical community has been the system of weekly seminars. The importance of this for the mental well-being of the research community cannot be overstated: in my case, a first postdoctoral position at Columbia University and the Bers weekly seminar provided intellectual mother's milk, an intrinsic part of my life as an immigrant, feeding my educational and social needs, a model of its kind that I appreciated immensely and long sought to duplicate in London without much success.

In addition, the NSF in America and also corresponding Science Research Councils in Britain and Europe, have a tradition of supporting specialised conferences, in recognition of their cost effectiveness in sustaining progress; in addition, the research community supports this itself through the various national societies. Beginning from the early 1960s, there was a pattern of summer conferences featuring Teichmüller theory every two or three years, lasting for a week or more, enabling workers in the field to renew contacts, develop friendships across all frontiers, exchange ideas and establish longer term research collaborations; this has grown in significance over the years along with the research community. The original conference series on Riemann surfaces began with a famous Princeton conference on Analytic Functions in 1957, passing through a period of concentration on Riemann surfaces to the broader agenda now in place, with coordinated organising committees such as the Ahlfors-Bers Colloquium in the USA, the Ibero-American Congress of Geometry in Latin America, symposia at the Warwick Mathematics Institute and the Rolf Nevanlinna Colloquium in Europe. In this way, research in Riemann surfaces and associated theories has thrived, gaining a much more varied program, too broad now for any single meeting to encompass. At the same time, the growth of cooperation between national societies and formation of international groups for regional congresses in America, Europe and Asia has changed the mould of larger meetings.

3 Final comments.

The proliferation of conferences, enhanced by the freedom brought by e-mail and wide access to the Internet, has led to a blossoming of cooperative small group research worldwide, representing a potential for development of the global scientific community that is unprecedented. What we do when we do our research may be attributable to personal curiosity or a thirst for fame, but it is fuelled by an urge to communicate, clarified by all those conversations which go on within each circle of interested friends and then at conferences. Lipman Bers liked to say jokingly that the reason why we spend the time and effort we do on our research is "for the grudging respect of a few close friends"; but he always emphasized the importance of the seminar as an essential part of academic life, especially for those too burdened with other commitments to deliver a talk, and at conferences he was both an enthusiastic audience member and the most inspirational of speakers. The legacy of the Bers seminar lives on in all the research groups around the world run by his students and their students, which continue to nurture young talent and provide a forum for research in complex analysis and geometry.

For my part, nothing in my career has given me more pleasure than to help my students find their feet in this corner of mathematics. It is a further source of satisfaction to see my former students, Christos Kourouniotis, Gabino González and Paolo Teofilatto, take on the mantle of researcher and pass on the flame of mathematical excitement to their own students. Long may it continue.

The organisation of conferences like the one at Anogia is always carried out by a small self-selecting set of members of our community whose work often goes unrecognised beyond those who are direct beneficiaries of their efforts. Their true reward is the continuing health of our field, but for this event we are all very grateful to our organisers, Christos Kourouniotis, on whom all the local arrangements depended, Gabino González-Diez and Fred Gardiner.

King's College, London WC2R 2LS   bill.harvey@kcl.ac.uk

Semisimple actions of mapping class groups on CAT(0) spaces

Martin R. Bridson [1]

Mathematical Institute, University of Oxford.
bridson@maths.ox.ac.uk

Abstract

Let Σ be an orientable surface of finite type and let $\mathrm{Mod}(\Sigma)$ be its mapping class group. We consider actions of $\mathrm{Mod}(\Sigma)$ by semisimple isometries on complete CAT(0) spaces. If the genus of Σ is at least 3, then in any such action all Dehn twists act as elliptic isometries. The action of $\mathrm{Mod}(\Sigma)$ on the completion of Teichmüller space with the Weil-Petersson metric shows that there are interesting actions of this type. Whenever the mapping class group of a closed orientable surface of genus g acts by semisimple isometries on a complete CAT(0) space of dimension less than g it must fix a point. The mapping class group of a closed surface of genus 2 acts properly by semisimple isometries on a complete CAT(0) space of dimension 18.

1 Introduction

This article concerns actions of mapping class groups by isometries on complete CAT(0) spaces. It records the contents of the lecture that I gave at Bill Harvey's 65th birthday conference at Anogia, Crete in July 2007.

A CAT(0) space is a geodesic metric space in which each geodesic triangle is no fatter than a triangle in the Euclidean plane that has the same edge lengths (see Definition 2.1). Classical examples include complete 1-connected Riemannian manifolds with non-positive sectional curvature and metric trees. The isometries of a CAT(0) space X divide naturally into two classes: the *semisimple* isometries are those for which there exists $x_0 \in X$ such that $d(\gamma.x_0, x_0) = |\gamma|$ where $|\gamma| := \inf\{d(\gamma.y, y) \mid y \in X\}$; the remaining isometries are said to

1

be *parabolic*. Semisimple isometries are further divided into *hyperbolics*, for which $|\gamma| > 0$, and *elliptics*, which have fixed points. Parabolics can be divided into *neutral parabolics*, for which $|\gamma| = 0$, and non-neutral parabolics. If X is a polyhedral space with only finitely many isometry types of cells, then all isometries of X are semisimple [Bri1].

E. Cartan [Ca] proved that the natural metric on a symmetric space of non-compact type has non-positive sectional curvature. This gives an action of the mapping class group on a complete CAT(0) space: the morphism $\mathrm{Mod}(\Sigma_g) \rightarrow \mathrm{Sp}(2g, \mathbb{Z})$ induced by the action of $\mathrm{Mod}(\Sigma_g)$ on the first homology of Σ_g, the closed orientable surface of genus g, gives an action of $\mathrm{Mod}(\Sigma_g)$ by isometries on the symmetric space for the symplectic group $\mathrm{Sp}(2g, \mathbb{R})$. In this action, the Dehn twists in non-separating curves act as neutral parabolics.

In the fruitful analogy between mapping class groups and lattices in semisimple Lie groups, Teichmüller space takes the role of the symmetric space. Unfortunately, the Teichmüller metric does not have non-positive curvature [Mas1]. On the other hand, the Weil-Petersson metric, although not complete, does have non-positive curvature [Wol1, Wol2]. Since the completion of a CAT(0) space is again a CAT(0) space [BriH] p. 187, it seems natural to complete the Weil-Petersson metric and to examine the action of the mapping class group on the completion $\overline{\mathcal{T}}$ in order to elucidate the structure of the group.

Theorem A *Let Σ be an orientable surface of finite type with negative euler characteristic and empty boundary. The action of $\mathrm{Mod}(\Sigma)$ on the completion of Teichmüller space in the Weil-Petersson metric is by semisimple isometries. All Dehn twists act as elliptic isometries (i.e. have fixed points).*

This theorem is a restatement of results in the literature but I wanted to highlight it as a motivating example. The essential points in the proof are described in Section 3. The fact that the Dehn twists have fixed points in the action on $\overline{\mathcal{T}}$ is a manifestation of a general phenomenon:

Theorem B *Let Σ be an orientable surface of finite type with genus $g \geq 3$. Whenever $\mathrm{Mod}(\Sigma)$ acts by semisimple isometries on a complete CAT(0) space, all Dehn twists in $\mathrm{Mod}(\Sigma)$ act as elliptic isometries (i.e. have fixed points).*

One proves this theorem by comparing the centralizers of Dehn twists in mapping class groups with the centralizers of hyperbolic elements in isometry groups of CAT(0) spaces — see Section 2. The action of Mod(Σ) on the symmetric space for Sp($2g$, $\mathbb{R}$) shows that one must weaken the conclusion of Theorem B if one wants to drop the hypothesis that Mod(Σ) is acting by semisimple isometries. The appropriate conclusion is that the Dehn twists act either as elliptics or as neutral parabolics (see Theorem 2.6).

Theorem B provides information about actions of finite-index subgroups of mapping class groups. For if H is a subgroup of index n acting by semisimple isometries on a CAT(0) space X, then the induced action of Mod(Σ) on X^n is again by semisimple isometries: $\gamma \in$ Mod(Σ) will be elliptic (resp. hyperbolic) in the induced action if and only if any power of γ that lies in H was elliptic (resp. hyperbolic) in the original action (cf. remark 2.5).

The situation for genus 2 surfaces is quite different, as I shall explain in Section 6.

Theorem C *The mapping class group of a closed orientable surface of genus 2 acts properly by semisimple isometries on a complete* CAT(0) *space of dimension* 18.

The properness of the action in Theorem C contrasts sharply with the nature of the actions in Theorems A, B and D.

I do not know if the mapping class group of a surface of genus $g > 1$ can admit an action by semisimple isometries, without a global fixed point, on a complete CAT(0) space whose dimension is less than that of the Teichmüller space. However, one can give a lower bound on this dimension that is linear in g (cf. Questions 13 and 14 of [BriV] and Problem 6.1 in [Farb1]). To avoid complications, I shall state this only in the closed case.

Theorem D *Whenever the mapping class group of a closed orientable surface of genus g acts by semisimple isometries on a complete* CAT(0) *space of dimension less than g it fixes a point.*

Here, "dimension" means topological covering dimension. An outline of the proof is given in Section 5; the details are given in [Bri3]. The strategy of proof is based on the "ample duplication" criterion in [Bri2]. The semisimple hypothesis can be weakened: it is sufficient to assume that there are no non-neutral parabolics, cf. Theorem 2.6.

I make no claim regarding the sharpness of the dimension bounds in Theorems C and D.

The lecture on which this paper is based was a refinement of a lecture that I gave on 8 December 2000 in Bill Harvey's seminar at King's College London. Bill worked tirelessly over many years to maintain a geometry and topology seminar in London. Throughout that time he shared many insights with visiting researchers and always entertained them generously. His mathematical writings display the same generosity of spirit: he has written clearly and openly about his ideas rather than hoarding them until some arcane goal was achieved. The benefits of this openness are most clear in his highly prescient and influential papers introducing the curve complex [Harv3, Harv2].

The book [Harv1] on discrete groups and automorphic forms that Bill edited in 1977 had a great influence on me when I was a graduate student. At a more personal level, he and his wife Michele have been immensely kind to me and my family over many years. It is therefore with the greatest pleasure that I dedicate these observations about the mapping class group to him on the occasion of his sixty fifth birthday.

2 Centralizers, fixed points and Theorem B

Definition 2.1 *Let X be a geodesic metric space. A geodesic triangle Δ in X consists of three points $a, b, c \in X$ and three geodesics $[a, b]$, $[b, c]$, $[c, a]$. Let $\overline{\Delta} \subset \mathbb{E}^2$ be a triangle in the Euclidean plane with the same edge lengths as Δ and let $\overline{x} \mapsto x$ denote the map $\overline{\Delta} \to \Delta$ that sends each side of $\overline{\Delta}$ isometrically onto the corresponding side of Δ. One says that X is a CAT(0) space if for all Δ and all $\overline{x}, \overline{y} \in \overline{\Delta}$ the inequality $d_X(x, y) \leq d_{\mathbb{E}^2}(\overline{x}, \overline{y})$ holds.*

Note that in a CAT(0) space there is a unique geodesic $[x, y]$ joining each pair of points $x, y \in X$. A subspace $Y \subset X$ is said to be convex *if $[y, y'] \subset Y$ whenever $y, y' \in Y$.*

Lemma 2.2 *If X is a complete CAT(0) space then an isometry γ of X is hyperbolic if and only if $|\gamma| > 0$ and there is a γ-invariant convex subspace of X isometric to $\mathbb{R}$. (Each such subspace is called an axis for γ.)*

Proposition 2.3 *Let Γ be a group acting by isometries on a complete CAT(0) space X. If $\gamma \in \Gamma$ acts as a hyperbolic isometry then γ has infinite order in the abelianization of its centralizer $Z_\Gamma(\gamma)$.*

Proof This is proved on page 234 of [BriH] (remark 6.13). The main points are these: if γ is hyperbolic then the union of the axes of γ splits isometrically as $Y \times \mathbb{R}$; this subspace and its splitting are preserved by $Z_\Gamma(\gamma)$; the action on the second factor gives a homomorphism from $Z_\Gamma(\gamma)$ to the abelian group $\mathrm{Isom}_+(\mathbb{R})$, in which the image of γ has infinite order.   □

In the light of this proposition, in order to prove Theorem B it suffices to show that if Σ is a surface of finite type with genus $g \geq 3$, then the Dehn twist T about any simple closed curve c in Σ does not have infinite order in the abelianization of its centralizer.

Proposition 2.4 *If Σ is an orientable surface of finite type that has genus at least 3 (with any number of boundary components and punctures) and if T is the Dehn twist about any simple closed curve c in Σ, then the abelianization of the centralizer of T in* Mod(Σ) *is finite.*

Proof The centralizer of T in Mod(Σ) consists of mapping classes of homomorphisms that leave c invariant. This is a homomorphic image of the mapping class group of the surface obtained by cutting Σ along c. (This surface has two boundary components corresponding to c and hence two Dehn twists mapping to T.) Since Σ has genus at least 3, at least one component of the cut-open surface has genus $g \geq 2$. The mapping class group of such a surface has finite abelianization — see [Kor] for a concise survey and references.   □

Remark 2.5 *As we remarked in the introduction, Theorem B gives restrictions on how subgroups of finite index in* Mod(Σ) *can act on* CAT(0) *spaces. For example, given an orientable surface of genus $g \geq 3$ and a homomorphism ϕ from a subgroup $H <$ Mod(Σ) of index n to a group G that acts by hyperbolic isometries on a complete* CAT(0) *space X, we can apply Theorem B to the induced action† of* Mod(Σ) *on X^n and hence deduce that any power of a Dehn twist that lies in H must lie in the kernel of ϕ. Taking $G = \mathbb{Z}$ tells us that powers of Dehn twists cannot have infinite image in the abelianization of H. A more explicit proof of this last fact was given recently by Andrew Putman [Put].*

† One can regard this as "multiplicative induction" in the sense of [Di] p. 35: if $H < G$ has index n and acts on X then one identifies X^n with the space of H-equivariant maps $f : G \to X$ and considers the (right) action $(g.f)(\gamma) := f(\gamma g)$; a power of each $g \in G$ preserves the factors of X^n, so it follows from [BriH] pp. 231–232 that the action of G is by semisimple isometries if the action of H is.

I am grateful to Pierre-Emmanuel Caprace, Dawid Kielak, Anders Karlsson and Nicolas Monod for their comments concerning the following extension of Theorem B.

Theorem 2.6 *Let* Σ *be an orientable surface of finite type with genus* $g \geq 3$. *Whenever* $\mathrm{Mod}(\Sigma)$ *acts by isometries on a complete* $\mathrm{CAT}(0)$ *space, each Dehn twist* $T \in \mathrm{Mod}(\Sigma)$ *acts either as an elliptic isometry or as a neutral parabolic (i.e.* $|T| = 0$ *).*

Proof The proof of Theorem B will apply provided we can extend Proposition 2.3 to cover non-neutral parabolics. In order to appreciate this extension, the reader should be familiar with the basic theory of Busemann functions in CAT(0) spaces, [BriH] Chap. II.8.

If γ is a parabolic isometry with $|\gamma| > 0$, then a special case of a result of Karlsson and Margulis [KaMa] (cf. [Ka], p. 285) shows that γ has a unique fixed point at infinity $\xi \in \partial X$ with the property that $\frac{1}{n} d(\gamma^n . x, c(n|\gamma|)) \to 0$ as $n \to \infty$ for every $x \in X$ and every geodesic ray $c : [0, \infty) \to X$ with $c(\infty) = \xi$. Now, $Z_\Gamma(\gamma)$ fixes ξ and acts on any Busemann function centred at ξ by the formula $z.\beta(t) = \beta(t) + \phi(z)$, where $\phi : Z_\Gamma(\gamma) \to \mathbb{R}$ is a homomorphism. Since $\phi(\gamma) = -|\gamma|$, this is only possible if γ has infinite order in the abelianization of $Z_\Gamma(\gamma)$. $\square$

M. Kapovich and B. Leeb [KL] were the first to prove that if $g \geq 3$, then $\mathrm{Mod}(\Sigma_g)$ cannot act properly by semisimple isometries on a complete CAT(0) space; cf. [BriH] p. 257.

3 Augmented Teichmüller space and Theorem A

Let Σ be an orientable hyperbolic surface of finite type with empty boundary and let $\overline{\mathcal{T}}$ denote the completion of its Teichmüller space equipped with the Weil-Petersson metric. $\overline{\mathcal{T}}$ is equivariantly homeomorphic to the augmented Teichmüller space defined by Abikoff [Abi]. Wolpert's concise survey [Wol3] provides a clear introduction and ample references to the facts that we need here.

Masur [Mas2] describes the metric structure of $\overline{\mathcal{T}} \setminus \mathcal{T}$ as follows. It is a union of strata $\mathcal{T}_C$ corresponding to the homotopy classes of systems C of disjoint, non-parallel, simple closed curves on Σ. The stratum corresponding to C is the Teichmüller space of the nodal surface obtained by shrinking each loop $c \in C$ to a pair of cusps (punctures); more explicitly, it is the product of Teichmüller spaces for the components of

$\Sigma \smallsetminus \bigcup C$ (with punctures in place of the pinched curves), each equipped with its Weil-Petersson metric. The identification of $\mathcal{T}_C$ with this product of Teichmüller spaces is equivariant with respect to the natural map from the subgroup of $\mathrm{Mod}(\Sigma)$ that preserves C and its components to the mapping class group of the nodal surface. Importantly, $\mathcal{T}_C \subset \overline{\mathcal{T}}$ is a convex subspace [DaWe].

The Dehn twists (and hence multitwists μ) in the curves of C act trivially on $\mathcal{T}_C$ and hence are elliptic isometries of $\overline{\mathcal{T}}$.

Daskalopoulos and Wentworth [DaWe] proved that every pseudo-Anosov element ψ of $\mathrm{Mod}(\Sigma)$ acts as a hyperbolic isometry of $\overline{\mathcal{T}}$; indeed each has an axis contained in $\mathcal{T}$. If $\rho \in \mathrm{Mod}(\Sigma)$ leaves invariant a curve system C and each of the components of $\Sigma \smallsetminus C$, then either it is a multitwist (and hence acts as an elliptic isometry of $\overline{\mathcal{T}}$) or else it acts as a pseudo-Anosov on one of the components of $\Sigma \smallsetminus C$. In the latter case ρ will act as a hyperbolic isometry of $\mathcal{T}_C$, and hence of $\overline{\mathcal{T}}$, since $\mathcal{T}_C \subset \overline{\mathcal{T}}$ is convex. An isometry of a complete CAT(0) space is hyperbolic (resp. elliptic) if and only if every proper power of it is hyperbolic (resp. elliptic) [BriH], pp. 231–232. Every element of the mapping class group has a proper power that is one of the three types μ, ψ, ρ considered above [Thu]. Thus Theorem A is proved.

Remarks 3.1 *(1) I want to emphasize once again that I stated Theorem A only to provide context: nothing in the proof is original. Moreover, since I first wrote this note, Ursula Hamenstädt has given essentially the same proof in [Ham].*

(2) Masur-Wolf [MasW] and Brock-Margalit [BroM] have shown that $\mathrm{Mod}(\Sigma)$ is the full isometry group of $\overline{\mathcal{T}}$. I am grateful to Jeff Brock for a helpful correspondence on this point.

(3) In the proof of Theorem A, the roots of multitwists emerged as the only elliptic isometries (provided that we regard the identity as a multitwist). Combining this observation with Theorem B, we see that any homomorphism from the mapping class group of a surface of genus $g \geq 3$ to another mapping class group must send roots of multitwists to roots of multitwists. This contrasts with the fact that there are injective homomorphisms between mapping class groups of once-punctured surfaces of higher genus that send pseudo-Anosov elements to multitwists [ALS].

4 Criteria for common fixed points

The classical theorem of Helly concerns the combinatorics of families of convex subsets in $\mathbb{R}^n$. There are many variations on this theorem in the literature. For our purposes the following will be sufficient (see [Bri2], [Farb2] and references therein).

Proposition 4.1 *Let X be a complete* CAT(0) *space of (topological covering) dimension at most n, and let $C_1, \ldots, C_N \subset X$ be closed convex subsets. If every $(n + 1)$ of the C_i have a point of intersection, then $\bigcap_{i=1}^N C_i \neq \emptyset$.*

When applied to the fixed point sets $C_i = \mathrm{Fix}(s_i)$ with $s_i \in S$, this implies:

Corollary 4.2 *Let Γ be a group acting by isometries on a complete* CAT(0) *space X of dimension at most n and suppose that Γ is generated by the finite set S. If every $(n+1)$-element subset of S fixes a point of X, then Γ has a fixed point in X.*

We shall need a refinement of this result that relies on the following well-known proposition, [BriH] p. 179. We write $\mathrm{Isom}(X)$ for the group of isometries of a metric space X and $\mathrm{Ball}_r(x)$ for the closed ball of radius $r > 0$ about $x \in X$. Given a subspace $Y \subseteq X$, let $r_Y := \inf\{r \mid Y \subseteq \mathrm{Ball}_r(x), \text{ some } x \in X\}$.

Proposition 4.3 *If X is a complete* CAT(0) *space and Y is a bounded subset, then there is a unique point $c_Y \in X$ such that $Y \subseteq \mathrm{Ball}_{r_Y}(c_Y)$.*

Corollary 4.4 *Let X be a complete* CAT(0) *space. If $H < \mathrm{Isom}(X)$ has a bounded orbit then H has a fixed point.*

Proof The centre c_O of any H-orbit O will be a fixed point.   □

Corollary 4.5 *Let X be a complete* CAT(0) *space. If the groups $H_1, \ldots, H_\ell < \mathrm{Isom}(X)$ commute and $\mathrm{Fix}(H_i)$ is non-empty for $i = 1, \ldots, \ell$, then $\bigcap_{i=1}^\ell \mathrm{Fix}(H_i)$ is non-empty.*

Proof A simple induction reduces us to the case $\ell = 2$. Since $\mathrm{Fix}(H_2)$ is non-empty, each H_2-orbit is bounded. As H_1 and H_2 commute, $\mathrm{Fix}(H_1)$ is H_2-invariant and therefore contains an H_2-orbit. As $\mathrm{Fix}(H_1)$ is convex, the centre of this (bounded) orbit is also in Fix(H_1), and there-fore is fixed by $H_1 \cup H_2$.   □

Building on these elementary observations, one can prove the follow-ing; see [Bri2].

Proposition 4.6 (Bootstrap Lemma) *Let* $k_1, \ldots, k_n$ *be positive in-tegers and let* X *be a complete* CAT(0) *space of dimension less than* $k_1 + \cdots + k_n$. *Let* $S_1, \ldots, S_n \subset \mathrm{Isom}(X)$ *be subsets with* $[s_i, s_j] = 1$ *for all* $s_i \in S_i$ *and* $s_j \in S_j$ $(i \neq j)$.

If, for $i = 1, \ldots, n$, *each* k_i-*element subset of* S_i *has a fixed point in* X, *then for some* i *every finite subset of* S_i *has a fixed point.*

When applying the Bootstrap Lemma one has to overcome the fact that the conclusion only applies to *some* S_i. A convenient way of gaining more control is to restrict attention to conjugate sets.

Corollary 4.7 (Conjugate Bootstrap) *Let* k *and* n *be positive inte-gers and let* X *be a complete* CAT(0) *space of dimension less than* nk. *Let* $S_1, \ldots, S_n$ *be conjugates of a subset* $S \subset \mathrm{Isom}(X)$ *with* $[s_i, s_j] = 1$ *for all* $s_i \in S_i$ *and* $s_j \in S_j$ $(i \neq j)$.

If each k-*element subset of* S *has a fixed point in* X, *then so does each finite subset of* S.

5 Some surface topology

The reader will recall that, given a closed orientable surface Σ and two compact homeomorphic sub-surfaces with boundary $T, T' \subset \Sigma$, there exists an automorphism of Σ taking T to T' if and only if $\Sigma \smallsetminus T$ and $\Sigma \smallsetminus T'$ are homeomorphic. In particular, two homeomorphic sub-surfaces are in the same orbit under the action of $\mathrm{Homeo}(\Sigma)$ if the complement of each is connected.

The relevance of this observation to our purposes is explained by the following lemma, which will be used in tandem with the Conjugate Boot-strap.

Lemma 5.1 *Let* H *be the subgroup of* $\mathrm{Mod}(\Sigma)$ *generated by the Dehn twists in a set* C *of loops all of which are contained in a compact sub-surface* $T \subset \Sigma$ *with connected complement. If* Σ *contains* m *mutu-ally disjoint sub-surfaces* T_i *homeomorphic to* T, *each with connected* *complement, then* $\mathrm{Mod}(\Sigma)$ *contains* m *mutually-commuting conjugates* H_i *of* H.

Proof Since the complement of each T_i is connected, there is a homeomorphism ϕ_i of Σ carrying T to T_i. Define H_i to be the subgroup of $\mathrm{Mod}(\Sigma)$ generated by the Dehn twists in the loops $\phi_i(C)$. Since the various H_i are supported in disjoint sub-surfaces, they commute. $\quad\square$

5.1 The Lickorish generators

Raymond Lickorish [Lic] proved that the mapping class group of a closed orientable surface of genus g is generated by the Dehn twists in $3g - 1$ non-separating loops, each pair of which intersects in at most one point. Let Lick denote this set of loops.

We say that a subset $S \subset$ Lick is *connected* if the union $U(S)$ of the loops in S is connected. An analysis of Lick reveals the following fact, whose proof is deferred to [Bri3]. In this statement all sub-surfaces are assumed to be compact.

Proposition 5.2 *Let* $S \subset$ Lick *be a connected subset.*

   (i) *If* $|S| = 2\ell$ *is even, then* $U(S)$ *is either contained in a sub-surface of genus* ℓ *with* 1 *boundary component, or else in a non-separating sub-surface of genus at most* $\ell - 1$ *with* 3 *boundary components.*
  (ii) *If* $|S| = 2\ell + 1$ *is odd, then* $U(S)$ *is either contained in a non-separating sub-surface of genus* ℓ *with at most* 2 *boundary components, or else in a non-separating sub-surface of genus at most* $(\ell - 1)$ *that has at most* 3 *boundary components.*

5.2 The Proof of Theorem D: an outline

We must argue that when the mapping class group of a closed orientable surface of genus g acts without neutral parabolics on a complete CAT(0) space X of dimension less than g it must fix a point.

The case $g = 1$ is trivial. A complete CAT(0) space of dimension 1 is an $\mathbb{R}$-tree, so for $g = 2$ the assertion of the theorem is that the mapping class group of a genus 2 surface has property FR. This was proved by Culler and Vogtmann [CuVo].

Assume $g \geq 3$. According to Corollary 4.2, we will be done if we can show that each subset $S \subset$ Lick with $|S| \leq g$ has a fixed point in X. We proceed by induction on $|S|$. Theorem 2.6 covers the base case $|S| = 1$.

If S is not connected, say $S = S_1 \cup S_2$ with $U(S_1) \cap U(S_2) = \emptyset$, then the subgroups $\langle S_1 \rangle$ and $\langle S_2 \rangle$ commute. Each has a fixed point since $|S_i| < |S|$, so Corollary 4.5 tells us that S has a fixed point.

Suppose now that S is connected. If $|S| = 2\ell$ is even then Proposition 5.2 tells us that $U(S)$ is contained either in a sub-surface of genus ℓ with 1 boundary component or else in a non-separating sub-surface of genus $\ell - 1$ with 3 boundary components. In either case one can fit $\lfloor g/\ell \rfloor$ disjoint copies of this sub-surface into Σ_g. Lemma 5.1 then provides us with $\lfloor g/\ell \rfloor$ mutually-commuting conjugates of $\langle S \rangle$. As all proper subsets of S are assumed to have a fixed point and the dimension of X is less than $(2\ell - 1)\lfloor g/\ell \rfloor$, the Conjugate Bootstrap (Corollary 4.7) tells us that S will have a fixed point.

The argument for $|S|$ odd is similar. For a detailed proof, see [Bri3].
□

6 A proper semisimple action of $\mathrm{Mod}(\Sigma_2)$

Theorem 6.1 *The mapping class group of a closed surface of genus* 2 *acts properly by semisimple isometries on a complete* CAT(0) *space of dimension* 18.

Proof The hyperelliptic involution τ is central in $\mathrm{Mod}(\Sigma_2)$. The quotient orbifold $\Sigma_2/\langle \tau \rangle$ is a sphere with 6 marked points and the action of $\mathrm{Mod}(\Sigma_2)$ on this quotient induces a homomorphism $\mathrm{Mod}(\Sigma_2) \to \mathrm{Mod}(\Sigma_{0,6})$. This is onto [BiHi] and the kernel is $\langle \tau \rangle$. Thus we have a short exact sequence

$$1 \to \mathbb{Z}_2 \to \mathrm{Mod}(\Sigma_2) \to \mathrm{Mod}(\Sigma_{0,6}) \to 1.$$

For each positive integer $n \geq 2$ there is a natural homomorphism from the braid group B_n to $\mathrm{Mod}(\Sigma_{0,n+1})$; the image of this map is of index $(n+1)$ and the kernel is the centre of B_n, which is infinite cyclic. This map admits the following geometric interpretation. Regard B_n as the mapping class group of the n punctured disc D with 1 boundary component. One maps B_n to $\mathrm{Mod}(\Sigma_{0,n+1})$ by attaching the boundary of a once-punctured disc to ∂D and extending homeomorphisms of D by the identity on the attached disc. The image of B_n is the subgroup of $\mathrm{Mod}(\Sigma_{0,n+1})$ that stabilizes the puncture in the added disc; this has index $n + 1$. The centre of B_n is the mapping class of the Dehn twist ζ in a loop parallel to the boundary ∂D. This twist becomes trivial in

$\mathrm{Mod}(\Sigma_{0,n+1})$, and it generates the kernel of $B_n \to \mathrm{Mod}(\Sigma_{0,n+1})$. Thus we have a second short exact sequence

$$1 \to \mathbb{Z} \to B_n \to \Gamma \to 1$$

where $\Gamma \subset \mathrm{Mod}(\Sigma_{0,n+1})$ is of index $n+1$.

Brady and McCammond [BrMc] and independently Krammer (unpublished), showed that B_5 is the fundamental group of a compact nonpositively curved piecewise-Euclidean complex X of dimension 4 that has no free faces. It follows from [BriH] II.6.15(1) and II.6.16 that the universal cover of X splits isometrically as a product $Y \times \mathbb{R}$ and that the quotient of B_5 by its centre acts properly on Y (II.6.10(4) *loc. cit.*) by semisimple isometries (II.6.9 *loc. cit.*). By inducing, as in remark 2.5, we obtain an action of $\mathrm{Mod}(\Sigma_{0,6})$ on Y^6 that is again proper and semisimple. And since the kernel of $\mathrm{Mod}(\Sigma_2) \to \mathrm{Mod}(\Sigma_{0,6})$ is finite, the resulting action of $\mathrm{Mod}(\Sigma_2)$ on Y^6 is also proper. $\qquad\square$

Notes

1 This research was supported by a Senior Fellowship from the EPSRC of the UK and a Royal Society Wolfson Research Merit Award.

References

[Abi] Abikoff, W. Degenerating families of Riemann surfaces. *Ann. of Math.* **105** (1977), 29–44.

[ALS] Aramayona, J., Leininger, C. J. and Souto, J. Injections of mapping class groups. Preprint, arXiv:0811.0841.

[BiHi] Birman, J. S. and Hilden, H. M. On the mapping class groups of closed surfaces as covering spaces, in *"Advances in the Theory of Riemann Surfaces"* (Stony Brook 1969), Ann. of Math. Studies **66**, Princeton Univ. Press, Princeton NJ, 1971, pp. 81–115.

[BrMc] Brady, T. and McCammond, J. Braids; posets and orthoschemes. Preprint; Santa Barbara, 2009.

[Bri1] Bridson, M. R. On the semisimplicity of polyhedral isometries. *Proc. Amer. Math. Soc.* **127** (1999), 2143–2146.

[Bri2] Bridson, M. R. Helly's theorem, CAT(0) spaces, and actions of automorphism groups of free groups. Preprint, Oxford, 2007.

[Bri3] Bridson, M. R. On the dimension of CAT(0) spaces where mapping class groups act. Preprint, Oxford, 2009, arXiv: 0908.0690.

[BriH] Bridson, M. R. and Haefliger, A. *Metric Spaces of Non-Positive Curvature*, Grundlehren der Math. Wiss. **319**, Springer-Verlag, Berlin, 1999.

[BriV] Bridson, M. R. and Vogtmann, K. Automorphism groups of free groups, surface groups and free abelian groups, in *Problems on Mapping Class Groups and Related Topics*, ed. B. Farb, Proc. Sympos. Pure Math. **74**, Amer. Math. Soc., Providence, RI, 2006, pp. 301–316.

[BroM] Brock, J. and Margalit, D. Weil-Petersson isometries via the pants complex. *Proc. Amer. Math. Soc.* **135** (2007), 795–803.

[Ca] Cartan, E. Sur une classe remarquable d'espaces de Riemann. *Bull. Soc. Math. France* **54** (1926), 214–264.

[CuVo] Culler, M. and Vogtmann, K. A group theoretic criterion for property FA. *Proc. Amer. Math. Soc.* **124** (1996), 677–683.

[DaWe] Daskalopoulos, G. and Wentworth, R. Classification of Weil-Petersson isometries. *Amer. J. Math.* **125** (2003), 941–975.

[Di] tom Dieck, T. *Transformation Groups*, Studies in Mathematics **8**, de Gruyter, Berlin, 1987.

[Farb1] Farb, B. Some problems on mapping class groups and moduli space, in *Problems on Mapping Class groups and Related Topics*, ed. B. Farb, Proc. Sympos. Pure Math. **74**, Amer. Math. Soc., Providence, RI, 2006, pp. 11–55.

[Farb2] Farb, B. Group actions and Helly's theorem. Preprint, arXiv:0806.1692.

[Ham] Hamenstadt, U. Dynamical properties of the Weil-Petersson metric. Preprint, arXiv:0901.4301.

[Harv1] Harvey, W. J. (ed.) *Discrete Groups and Automorphic Forms*, Academic Press, London, 1977.

[Harv2] Harvey, W. J. Geometric structure of surface mapping class groups, in *Homological Group Theory* (Durham 1977), London Math. Soc. Lecture Note Ser. **36**, Cambridge Univ. Press, Cambridge, 1979, pp. 255–269.

[Harv3] Harvey, W. J. Boundary structure of the modular group, in *Riemann Surfaces and Related Topics* (Stony Brook 1978), Ann. of Math. Stud. **97**, Princeton Univ. Press, Princeton, NJ, 1981, pp. 245–251.

[Iv] Ivanov, N. V. Mapping class groups, in *Handbook of Geometric Topology*, ed. R. Daverman and R.B. Sher, Elsevier Science, Amsterdam, 2002, pp. 523–633.

[Ka] Karlsson, A. Nonexpanding maps, Busemann functions, and multiplicative ergodic theory, in *Rigidity in Dynamics and Geometry* (Cambridge, 2000), Springer-Verlag, Berlin, 2002, pp. 283–294.

[KaMa] Karlsson, A. and Margulis, G. A multiplicative ergodic theorem and nonpositively curved spaces. *Comm. Math. Phys.* **208** (1999), 107–123.

[KL] Kapovich, M. and Leeb, B. Actions of discrete groups on nonpositively curved spaces. *Math. Ann.* **30** (1996), no. 2, 341–352.

[Kor] Korkmaz, M. Low-dimensional homology groups of mapping class groups: a survey. *Turk. J. Math.* **26** (2002), 101–114.

[Lic] Lickorish, W. B. R. A finite set of generators for the homeotopy group of a 2-manifold. *Proc. Cambridge Philos. Soc.* **60** (1964), 769–778.

[Lin] Linch, M. R. *On metrics in Teichmüller spaces*, PhD Thesis, Columbia Univ., New York, 1971.

[Mas1] Masur, H. On a class of geodesics in Teichmüller space. *Ann. of Math.* *(2)* **102** (1975), 205–221.

[Mas2] Masur, H. The extension of the Weil-Petersson metric to the boundary of Teichmüller space. *Duke Math. J.* **43** (1976), 623–635.

[MasW] Masur, H. and Wolf, M. The Weil-Petersson isometry group. *Geom. Dedicata* **93** (2002), 177–190.

[Put] Putman, A. A note on the abelianizations of finite-index subgroups of the mapping class group. Preprint arXiv:0812.0017.

[Thu] Thurston, W. P. On the geometry and dynamics of diffeomorphisms of surfaces. *Bull. Amer. Math. Soc.* **19** (1988), 417–431.

[Wol1] Wolpert, S.A. Noncompleteness of the Weil-Petersson metric for Teichmüller space. *Pacific J. Math.* **61** (1975), 573–577.

[Wol2] Wolpert, S. A. Geodesic length functions and the Nielsen problem. *J. Diff. Geom.* **25** (1987), 275–296.

[Wol3] Wolpert, S. A. Weil-Petersson perspectives, in *Problems on Mapping Class Groups and Related Topics*, ed. B. Farb, Proc. Sympos. Pure Math. **74**, Amer. Math. Soc., Providence, RI, 2006, pp. 301–316.

A survey of research inspired by Harvey's theorem on cyclic groups of automorphisms

Emilio Bujalance [1]

Departamento Matemáticas Fundamentales, UNED
eb@mat.uned.es

Francisco-Javier Cirre

Departamento Matemáticas Fundamentales, UNED
jcirre@mat.uned.es

Grzegorz Gromadzki

Institute of Mathematics, University of Gdańsk
greggrom@math.univ.gda.pl

Abstract

In 1966 Harvey found necessary and sufficient conditions on the signature of a Fuchsian group for it to admit a smooth epimorphism onto a cyclic group. As a consequence, he solved the *minimum genus* and the *maximum order* problems for the family of cyclic groups (the latter already solved by Wiman). This was a seminal work in the study of groups of automorphisms of compact Riemann surfaces. Since then, much research has been conducted to extend Harvey's theorem to other classes of finite groups and, in particular, to solve the minimum genus and maximum order problems for these classes. In this survey we present some results obtained so far.

Introduction

In 1966 Bill Harvey published *Cyclic groups of automorphisms of a compact Riemann surface* in the Quarterly Journal of Mathematics, [Har]. The main result in it gives necessary and sufficient conditions on the signature of a Fuchsian group for it to admit a smooth epimorphism onto a cyclic group. This allowed him to solve the *minimum genus* and the *maximum order* problems for the family of cyclic groups, although the latter had already been solved by Wiman [Wim]. Some years earlier,

15

at the beginning of the sixties, A. M. Macbeath [Mcb1] revisited the Poincaré uniformization theorem and settled the basis of the combinatorial study of compact Riemann surfaces and their groups of automorphisms. This combinatorial point of view opened the door to a fruitful research, subsequently developed by Macbeath's students Bill Harvey, Colin Maclachlan and David Singerman.

It is well known that every finite group G acts as a group of automorphisms of some compact Riemann surface of genus $g \geq 2$. We will briefly say that G acts on genus g. The same group may act on different genera and also the same surface may admit different actions of the same abstract group. Some questions, which we shall examine in this survey, arise naturally. The *minimum genus problem* is one of them; it asks for the minimum genus $g = g(G) \geq 2$ of a Riemann surface S on which a given finite group G acts as a group of conformal automorphisms. The minimum genus is a parameter which has been studied for more than a century. It was already considered by Burnside [Bur] with the restriction that the orbit space S/G be the Riemann sphere, and by Hurwitz [Hur]. In the modern terminology this parameter is called the *strong symmetric genus*, although in general the restriction $g \geq 2$ is not imposed. Another parameter closely related to this is the *symmetric genus* of a group G, which is the minimum genus of a surface on which G acts but now G is allowed to contain orientation-reversing automorphisms (recall that Riemann surfaces are orientable). This modern terminology, which we adopt here with the restriction of genus bigger than one, is due to Tucker in his significant paper [Tuc].

Closely related to the minimum genus problem is the *maximum order problem* for a family $\mathcal{F}$ of finite groups, which for a given integer $g \geq 2$, asks for the maximum order $N = N_{\mathcal{F}}(g)$ of the groups in $\mathcal{F}$ acting on genus g, where we allow $N = 0$ if such a group action does not exist. This problem makes sense in view of the classical result by Schwarz in [Sch], who showed that the order of a group acting on genus $g \geq 2$ is finite (in fact, Hurwitz in [Hur] showed that it is not bigger than $84(g-1)$ if the action preserves the orientation). However, to solve the maximum order problem is a rather difficult task in general. It is more feasible to look for *uniform upper bounds* for $N_{\mathcal{F}}(g)$ and study those values of g for which the bound is attained. By a uniform upper bound we mean a formula of $N_{\mathcal{F}}(g)$ valid for infinitely many values of g. For example, in the case of orientation-preserving automorphisms, the Hurwitz bound $84(g-1)$ is never attained by a soluble group and so it is natural to look for a uniform upper bound for the order of a soluble group acting on genus g. The answer in this case, see Section 3, is $N_{\mathcal{F}}(g) = 48(g-1)$ (valid by infinitely many values of g) which, in addition, is the largest uniform bound for soluble groups. However, it is not known the precise values of g for which this formula is valid.

In this survey we examine the above problems for some important families of finite groups. Most of the results deal with the classical case of groups consisting of orientation-preserving automorphisms, although groups with orientation-reversing automorphisms also have their own place in this survey. Observe that, for these last groups, the orientation-preserving automorphisms constitute a subgroup of index two. Consequently, if a group G has no such subgroups (*e.g.,* simple groups or odd-order groups) then G always acts preserving orientation; in particular, its strong symmetric genus coincides with its symmetric genus.

There is a large number of papers dealing with the above problems or with problems closely related to them. The purpose of this survey is to give a general overview of the state of art of this topic, and not to analyze in detail the techniques used in each paper. The interested reader is referred to the appropriate reference in the Bibliography, made to that end as complete as possible, although we have probably missed some important papers.

1 Preliminaries

The combinatorial study of compact surfaces is based on the theory of groups acting discontinuously on the hyperbolic plane $\mathcal{H}$. By a *Fuchsian group* we mean a discrete cocompact subgroup of the group of orientation-preserving isometries of $\mathcal{H}$. The algebraic structure of a Fuchsian group Λ and the geometric structure of the orbit space $\mathcal{H}/\Lambda$ are determined by the *signature* $\sigma(\Lambda)$ of Λ, which is a collection of non-negative integers of the form

$$\sigma(\Lambda) = (\gamma; m_1, \ldots, m_r)$$

with $m_i \geq 2$ for all i. The integers m_i are called the *periods* of Λ and γ is its *orbit genus*. The compact orbit space $\mathcal{H}/\Lambda$ has the structure of a hyperbolic 2-orbifold of topological genus γ with r conic points of orders $m_1, \ldots, m_r$. These are the branching orders of the branch points of the ramified covering projection $\mathcal{H} \to \mathcal{H}/\Lambda$.

A Fuchsian group with this signature has an abstract presentation in terms of 2γ hyperbolic generators $a_1, b_1, \ldots, a_\gamma, b_\gamma$ and r elliptic isometries $x_1, \ldots, x_r$ subject to the defining relations

$$x_1^{m_1} = \cdots = x_r^{m_r} = x_1 \cdots x_r [a_1, b_1] \cdots [a_\gamma, b_\gamma] = 1,$$

where $[a, b] = aba^{-1}b^{-1}$. The ordering in which proper periods are written is irrelevant since any permutation of them yields an isomorphic Fuchsian group, see [Mcb3].

The hyperbolic area $\mu(\Lambda)$ of any fundamental region of Λ equals

$$\mu(\Lambda) = 2\pi \left(2\gamma - 2 + \sum_{i=1}^{r} \left(1 - \frac{1}{m_i} \right) \right).$$

If Λ' is a subgroup of Λ of finite index $[\Lambda : \Lambda']$ then Λ' is also a Fuchsian group and the following Hurwitz-Riemann formula holds:

$$\mu(\Lambda')/\mu(\Lambda) = [\Lambda : \Lambda'].$$

By the Uniformization Theorem, a compact Riemann surface X of genus $g \geq 2$ can be represented as the orbit space $\mathcal{H}/\Gamma$ of the hyperbolic plane $\mathcal{H}$ under the action of a *surface Fuchsian group* Γ. This is a torsion free Fuchsian group, whose signature is therefore $(g; -)$. An abstract finite group G then acts as a group of automorphisms of a surface X so represented if and only if G is isomorphic to the factor group Λ/Γ for some Fuchsian group Λ containing Γ as a normal subgroup, or equivalently, if and only if there exists an epimorphism $\theta : \Lambda \to G$ with ker $\theta = \Gamma$. An epimorphism from a Fuchsian group Λ onto a finite group G whose kernel Γ is torsion free will be called a *smooth epimorphism* and the factor group Λ/Γ will be called a *smooth factor*.

Representing groups and surfaces in this way yields a combinatorial approach to the problem of finding conditions for a given group G to act on some compact Riemann surface of given genus. For instance, a direct application of the Hurwitz-Riemann formula to Λ and Γ produces the following arithmetic relation between the order $|G|$ of $G = \Lambda/\Gamma$ and the genus g of the surface $\mathcal{H}/\Gamma$ on which it acts:

$$2\pi(2g - 2) = |G|\,\mu(\Lambda).$$

If, for instance, we want to minimize the genus g on which a given group G acts, then we have to minimize the hyperbolic area $\mu(\Lambda)$.

Among the Fuchsian groups with least hyperbolic area we find the *triangle Fuchsian groups,* that is, Fuchsian groups whose signature is

of the form $(0; k, l, m)$ (this signature shall be abbreviated as (k, l, m)).
These groups are also called (k, l, m)-triangle groups and algebraically
they are generated by a pair of elements of orders k and l whose product
has order m. Such a pair is called a (k, l, m)-generating pair. Observe
that if G is a smooth factor of a (k, l, m)-triangle group then G also has
a (k, l, m)-generating pair. These pairs play a key role in the minimum
genus problem for the group G since one of the main tools to solve
it is to find minimal (k, l, m)-generating pairs for G. Minimal means
that there is no other (r, s, t)-generating pair with $1/r + 1/s + 1/t >
1/k+1/l+1/m$. Obviously, by the Hurwitz-Riemann formula, a minimal
(k, l, m)-generating pair yields an action on minimal genus.

The Fuchsian groups with least hyperbolic area are the $(2, 3, 7)$-
triangle groups, as is easy to see. Consequently, the Hurwitz bound
$84(g - 1)$ for the order of a group acting on genus g is attained just by
smooth factors of $(2, 3, 7)$-triangle groups. These factor groups are called
Hurwitz groups. Macbeath in [Mcb2] showed that there exist infinitely
many values of g for which the Hurwitz bound is attained, and infinitely
many values of g for which it is not attained. Hurwitz groups are a continuous source of research, not only for Riemann surface theory but also
for group theory, see the excellent survey by Conder [Con5]. Observe
that, as a consequence of the Hurwitz bound, we have $g \geq |G|/84 + 1$
for a group G acting on genus g, with equality holding only if G is a
Hurwitz group. Consequently, the minimum genus problem is trivial for
these groups.

In the sequel, by a Riemann surface we mean a compact Riemann
surface of genus bigger than one, unless otherwise stated.

2 Cyclic, abelian and dihedral groups

As mentioned above, the minimum genus and maximum order problems
are, in general, rather difficult. They are completely solved just for
special classes of finite groups. Here we shall present the solutions for
cyclic, abelian and dihedral groups.

The above problems for $G = Z_n$, the **cyclic group** of order n, were
completely solved by Harvey in [Har] using the following characterization
of the signature of a Fuchsian group Λ such that Z_n is a smooth factor
of Λ.

Let Λ be a Fuchsian group with signature $(\gamma; m_1, \ldots, m_r)$ and let $m =$

$lcm(m_1, \ldots, m_r)$. *Then there exists a smooth epimorphism from* Λ *onto* Z_n *if and only if*

a) $m = lcm(m_1, \ldots, m_{i-1}, m_{i+1}, \ldots, m_r)$ *for all* i;

b) m *divides* n, *and if* $\gamma = 0$ *then* $m = n$;

c) $r \neq 1$, *and if* $\gamma = 0$ *then* $r \geq 3$;

d) *if* m *is even then the number of periods* m_i *such that* m/m_i *is odd is also even.*

As a consequence, he obtained the solution of the minimum genus problem.

The minimum genus $g \geq 2$ *of a surface which admits an automorphism of order* $n = p_1^{r_1} \cdots p_k^{r_k}$, *where* $p_i < p_{i+1}$ *and* p_i *prime for all* i, *is given by*

a) $g = \max\left\{2, \left(\frac{p_1-1}{2}\right)\frac{n}{p_1}\right\}$ *if* $r_1 > 1$ *or if* n *is prime*,

b) $g = \max\left\{2, \left(\frac{p_1-1}{2}\right)\left(\frac{n}{p_1}-1\right)\right\}$ *if* $r_1 = 1$.

He also obtained Wiman's solution to the maximum order problem.

The maximum order for an automorphism of a surface of genus $g \geq 2$ *is* $2(2g+1)$.

Harvey also showed that this maximum is attained for each value of g, since triangle groups with signature $(2, 2g+1, 2(2g+1))$ admit a smooth epimorphism onto the cyclic group $Z_{2(2g+1)}$.

A natural extension of Harvey's results is to consider cyclic groups generated by orientation-reversing automorphisms. This was carried out by Etayo in [Eta]. The combinatorial theory of Fuchsian groups is no longer valid since these are generated by orientation-preserving automorphisms of the hyperbolic plane. Instead, the *non-Euclidean crystallographic groups*, NEC groups for short, provide the appropriate tool for dealing with orientation-reversing automorphisms of Riemann surfaces. An NEC group is a discrete cocompact subgroup of the group of all, including orientation-reversing, automorphisms of the hyperbolic plane. It is worth mentioning here that these groups provide also the appropriate tool for dealing with non-orientable surfaces or surfaces with boundary, see Chapters 0 and 1 in [BEGG] for an introduction to the combinatorial theory of NEC groups.

A single orientation-reversing automorphism of a Riemann surface must have even order n and, moreover, it generates orientation-reversing involutions if and only if $n/2$ is odd. Etayo in [Eta] characterized, according to the parity of $n/2$, the NEC signatures with which the group generated by such an automorphism may act. This characterization allows him to solve the minimum genus problem for this type of group actions, that is, in Tucker's terminology, he computes the symmetric genus of a cyclic group. It depends, as in Harvey's solution, on the least prime divisor of n. Moreover, Etayo also obtained partial results concerning the maximum order problem, showing that an orientation-reversing automorphism of a Riemann surface of genus g has order bounded by $3g + 6$ if $n/2$ is odd, and $4g + 4$ otherwise and that these bounds are sharp for infinitely many values of g.

In 1965 Maclachlan in [Mcl1] published the solution to the minimum genus problem for the family of non-cyclic **abelian groups**. He used no characterization of Fuchsian signatures with which such groups may act. Instead, a precise analysis of the abelianizing homomorphism, which factors out the derived group of a Fuchsian group, together with some technical results on abelian group theory led him to find "minimal" sets of generators of an abelian group. The sets are minimal in the sense that they allow for the definition of a smooth epimorphism from a Fuchsian group with minimal hyperbolic area onto the abelian group they generate. Let us write an abelian group G in its canonical form $Z_{m_1} \oplus \cdots \oplus Z_{m_s}$ with $m_i | m_{i+1}$. The formulae for the minimum genus of G given by Maclachlan depend on the parity of the number s of direct summands of G.

Much later, Breuer [Bre] found necessary and sufficient conditions on the signature of a Fuchsian group Γ for the existence of a smooth epimorphism from Γ onto a prescribed abelian group. As a consequence of either the Breuer or Maclachlan result, we have that the maximum order of an abelian group of automorphisms of a compact Riemann surface of genus $g \geq 2$ is $4g+4$. This maximum is attained for each value of g, since as Breuer showed, triangle groups with signature $(2, 2(g + 1), 2(g + 1))$ admit a smooth epimorphism onto the direct product $Z_2 \times Z_{2(g+1)}$. This solution to the maximum order problem can also be found without a proof in Accola's paper [Acc]. A proof using Sah's results in [Sah] is given by Weaver in his nice survey [Wea2].

The minimum genus problem for the actions of abelian groups which allow orientation-reversing automorphisms was solved by May and Zimmerman in [MZ1]. In other words, they computed the symmetric genus of abelian groups. Relabelling the invariants of an abelian group A, its canonical form can be rewritten as $A = (Z_2)^a \oplus Z_{m_1} \oplus \cdots \oplus Z_{m_d}$ with $m_i > 2$, so that the Z_2 factors are explicitly exhibited. May and Zimmerman showed that the symmetric genus of A depends on the number a of Z_2 factors, and calculated it according to the three cases $a = 0$, $1 \le a \le d+1$ and $a \ge d+2$.

Dihedral groups of automorphisms were studied by Maclachlan in his Ph. D. Thesis [Mcl2], where he solved the minimum genus problem. The same solution was obtained by Michael as a particular case of his study on metacyclic groups in his Ph. D. Thesis. The results were published in [Geo] some years later.

The minimum genus $g \ge 2$ of a Riemann surface which admits a dihedral group of automorphisms of order $2n > 4$ is given by

a) $g = (p_1 - 1)\left(\frac{n}{p_1} - 1\right)$ *if* $r_1 = 1$ *and* n *is not prime,*

b) $g = (p_1 - 1)\frac{n}{p_1}$ *if* $r_1 > 1$ *or* n *is prime,*

where $n = p_1^{r_1} \cdots p_k^{r_k}$, *with* $p_i < p_{i+1}$ *is the prime decomposition of* n.

The maximum order problem is solved by the authors and J.M. Gamboa in [BCGG], where the characterization of the signatures of Fuchsian groups Λ admitting smooth epimorphisms onto the dihedral group D_n was given. The characterization splits naturally into the cases n even and n odd. It is more involved than in the cases of cyclic or abelian groups, and too large to be included here. We just mention that the conditions are different according to whether the orbit genus γ of $\sigma(\Lambda)$ is equal to zero or not, and that in both cases, a key role is played by the number of periods of $\sigma(\Lambda)$ which are equal to two. As a consequence, the maximum order of a dihedral group of conformal automorphisms of a genus g Riemann surface was solved: it is $4g + 4$ if g is even and $4g$ if g is odd. The largest dihedral group is not a smooth factor or a triangle group but of a Fuchsian group with signature $(0; 2, 2, 2, g+1)$ if g is even or $(0; 2, 2, 2, 2g)$ if g is odd.

3 Soluble, metabelian and metacyclic groups

The largest uniform bound for the order of a **soluble group** of automorphisms of a genus g Riemann surface was proved by Chetiya in [Che2] to be $48(g - 1)$ (for infinitely many values of g). This is the second largest possible order of a group acting on genus g, and it is attained by smooth factors of a $(2, 3, 8)$-triangle group. Chetiya showed that for every positive integer n there is a Riemann surface of genus $g = 2n^6 + 1$ which admits a soluble group of automorphisms of order $96n^6$. However, the exact description of those g for which the bound is sharp is unknown.

The study of soluble groups of automorphisms was continued by the same author and Kalita in the subsequent paper [ChK1], where in particular they found soluble groups of derived length 4 attaining the above bound for infinitely many values of g. It turns out however that this bound is not sharp for groups of derived length ≤ 3 and the third author showed in [Gro1] that the largest uniform bound for the order of a soluble group of automorphisms of derived length 3 is $24(g - 1)$ unless $g = 3, 5, 6, 10$. Moreover, he also found infinitely many values of g for which this bound is attained.

We now consider **metabelian groups** *i.e.*, soluble groups of derived length 2. For $g \neq 3, 5$ the largest metabelian groups acting conformally on genus $g \geq 2$ were shown by Chetiya and Patra in [ChP1] to have order $16(g-1)$. For $g = 3$ or 5 the bound is $16g$. They showed that the $(2, 4, 8)$-triangle groups admit metabelian smooth factors of order $16(g - 1)$ for infinitely many values of g. In [Gro2] the third author improved these results by finding the precise values of $g \geq 4$ for which this bound is attained. These are the odd numbers of the form $g = k^2\beta + 1$ where k and β are integers such that -1 is a quadratic residue (mod β). In addition, he also found presentations by means of generators and defining relations of the corresponding metabelian groups.

Certain subclasses of metabelian groups, as **metacyclic groups,** have also been the focus of interest as groups of automorphisms on Riemann surfaces. Chetiya and Patra in [ChP2] considered K-metacyclic groups, that is, metacyclic groups G having commutator subgroup G' of prime order p, commutator quotient G/G' of order $p - 1$, and faithful action of the commutator quotient G/G' on the commutator subgroup G'. They found necessary and sufficient conditions for the existence of a smooth epimorphism from a Fuchsian group onto a K-metacyclic group, and used such conditions to solve, in several cases, the minimum genus and maximum order problems. In [DPCh] the same authors together with Dutta studied the same problem for the ZS-metacyclic groups of order pq where p and q are primes and $q|(p - 1)$. Michael in [Geo] solved the minimum genus problem for two special families of metacyclic groups, namely, those with presentations $G_s^0(n) = \langle a, b \mid a^n = 1 = b^2, b^{-1}ab = a^s \rangle$ and for n even, $G_s^1(n) = \langle a, b \mid a^n = 1, a^{n/2} = b^2, b^{-1}ab = a^s \rangle$.

The *genus spectrum* of a finite group G is the set of all integers $g \geq 2$ such that G acts on genus g. In [Wea1] Weaver determined the genus spectrum of split metacyclic groups of order pq where p and q are primes, that is, groups with presentation $\langle a, b \mid a^q = b^p = 1, bab^{-1} = a^r \rangle$, where $r^p \equiv 1 \pmod{q}$, $(r \neq 1)$. In particular, he solved the minimum genus problem for these groups, correcting a wrong statement on the minimum genus in [DPCh].

Let us consider finally soluble actions which allow orientation-reversing elements. May and Zimmerman in [MZ2] gave a lower bound for the symmetric genus of groups which have a cyclic factor group of order q, where $q = 5$ or $q \geq 7$. The lower bound depends on the highest power of 2 dividing q. They showed infinite families of groups, many of them metacyclic, for which the lower bound is attained. In particular, it is attained for the family of K-metacyclic groups and their non-abelian even-order subgroups. They also considered K-metacyclic groups of symmetric genus zero and one.

4 Supersoluble groups

The results concerning this class of groups are fairly complete. The problem of finding the largest uniform bound for the order of a super-soluble group of automorphisms of a compact Riemann surface of genus $g \geq 3$ was solved independently by Maclachlan and the third author in [GM] and Zomorrodian in [Zom3] (for $g = 2$ the bound was known to be 24). The largest bound turns out to be $18(g - 1)$, and it is attained by smooth factors of $(2, 3, 18)$-triangle groups. In the first paper the authors also showed that the values of $g \geq 3$ for which the bound is attained are exactly those for which 3^2 divides $g - 1$ and the only other prime factors of $g - 1$ are congruent to 1 (mod 3). This extends the class of genera for which the bound is attained given in [Zom3] where the author proved wrongly, see his own correction in [Zom4], that the divisibility condition $3^2 \mid (g - 1)$ is a necessary and sufficient one for the existence of such surfaces.

5 Nilpotent and *p*-groups

Zomorrodian in [Zom1] showed that the largest uniform bound for the order of a **nilpotent group** acting on genus $g \geq 2$ is $16(g - 1)$, and this upper bound is attained by smooth factors of the $(2, 4, 8)$-triangle groups. In addition, he also found the values of g for which this bound is attained, namely, those g such that $g - 1$ is a power of 2. In particular, the largest nilpotent groups of automorphisms are 2-groups.

Motivated by the above results on the largest nilpotent groups of automorphisms, Zomorrodian studied in [Zom2] the largest **p-groups** of automorphisms for any prime p. The case $p = 2$ follows from the results in the nilpotent case. The largest 3-groups have order $9(g - 1)$ and are smooth factors of a $(3, 3, 9)$-triangle group; moreover, the bound is attained for all values of g such that $g - 1$ is a power of 3. If $p \geq 5$ then the largest p-groups have order $2p(g - 1)/(p - 3)$ and are smooth factors of a (p, p, p)-triangle group; in this case, the bound is attained for all values of g such that $2(g - 1)/(p - 3)$ is a power of p.

The same bounds were independently obtained by Kulkarni in [Kul], where he also showed that the largest uniform bound for the order of a cyclic p-group of automorphisms is $2pg/(p - 1)$. For the non-cyclic p-groups, sharp upper bounds in terms of p and the minimum number of generators of the group were given.

A remarkable result in Kulkarni's paper [Kul] is that for every finite group G there exists an integer $n = n(G)$ with the following two properties:

   a) if G acts on a Riemann surface of genus g then $g \equiv 1 \pmod{n}$,
   b) for all but finitely many values of $g \equiv 1 \pmod{n}$, the group G acts on a Riemann surface of genus g.

The integer $n(G)$ depends on the structure of the p-subgroups of G and so in order to determine $n(G)$, it is natural to consider first the case of p-groups. Kulkarni and Maclachlan in [KM] dealt with the case of cyclic p-groups. For a fixed prime power p^e, they characterized in terms of its p-adic expansion the list of all genera g on which the cyclic group of order p^e acts. In the reverse direction, for a fixed genus g they determine all primes p such that the cyclic group of order p acts on genus g.

These results were extended to a wider class of p-groups with $p \neq 2$ by Maclachlan and Talu in [MT]. They considered p-groups of cyclic p-deficiency ≤ 2 and also a class of p-groups satisfying a condition known as the maximal exponent condition. The cyclic p-deficiency of a p-group is $n - e$ where p^n is the order of the group and p^e is its exponent, so cyclic p-groups have deficiency zero.

Other infinite families of p-groups whose strong symmetric genus has been calculated are the following ones, considered by May and Zimmerman. In [MZ5] they studied groups of strong symmetric genus two and three. Some of these groups belong to two infinite families of non-abelian 2-groups, namely the quasidihedral groups $\mathrm{QD}_n = \langle a, b \mid a^{2^{n-1}} = b^2 = 1, bab = a^{-1+2^{n-2}} \rangle$, and the quasiabelian groups $\mathrm{QA}_n = \langle a, b \mid a^{2^{n-1}} = b^2 = 1, bab = a^{1+2^{n-2}} \rangle$, both of order 2^n. They computed the strong symmetric genus of each of these groups: that of QD_n is 2^{n-3} and it is attained by smooth factors of the $(2, 4, 2^{n-1})$-triangle groups, while that of QA_n is $2^{n-2} - 1$ and it is attained by smooth factors of the $(2, 2^{n-1}, 2^{n-1})$-triangle groups. This is in contrast with their symmetric genus, which equals 1 for both families. In particular, this shows that the strong symmetric genus can be arbitrarily larger than the symmetric genus.

Later on, the same authors in [MZ7] succeeded in classifying the groups of strong symmetric genus 4, the more intricate case being the groups of order 2^5. Extending their above mentioned results on 2-groups, they considered p-groups in general. More precisely, they considered non-abelian p-groups that have an element of maximal possible order, that is, non-abelian p-groups with a cyclic subgroup of index p. For p odd these are the groups $\mathrm{QA}_n(p)$ of order p^n and presentation $\mathrm{QA}_n(p) = \langle a, b \mid a^{p^{n-1}} = b^p = 1, \ b^{-1}ab = a^{1+p^{n-2}} \rangle$. The strong symmetric genus (and also the symmetric genus) of $\mathrm{QA}_n(p)$ equals $(p-1)(p^{n-1} - 2)/2$ and it can be achieved by a smooth factor of a (p, p^{n-1}, p^{n-1})-triangle group.

Let us consider now p-group actions with orientation-reversing elements. This only makes sense for $p = 2$ since these elements have even order. Zimmerman in [Zim] showed that a lower bound for the symmetric genus of a non-abelian 2-group G is $1 + |G|/32$, provided that it is bigger than one. The main result in this paper is the explicit computation of the symmetric genus of all groups of order 32, where the possibility of symmetric genus 0 and 1 is included.

6 Alternating and symmetric groups

Following arguments of Higman on the use of coset diagrams for the $(2, 3, 7)$-triangle group and a technique of pasting diagrams, Conder in [Con1] showed that the **alternating group** A_n is a Hurwitz group for all $n \geq 168$, and for all but 64 integers n in the range $3 \leq n \leq 167$. Consequently, the strong symmetric genus of A_n is $n!/168 + 1$ for all but finitely many values of n. The exceptions were considered by Conder himself in [Con4], completing previous work by Patra in [Pat] where the cases $A_6, A_{10}, A_{11}, A_{12}, A_{13}, A_{14}$ and A_{15} were solved. Also in [EM], and as a corollary of their main result, Etayo and Martínez determined the strong symmetric genus of A_n for n in the range $5 \leq n \leq 19$.

Since Hurwitz groups are perfect, no non-trivial **symmetric group** S_n attains the maximal bound $84(g - 1)$ of the order of a group acting on genus g. The second and third largest bounds are $48(g - 1)$ and $40(g - 1)$ respectively. The first is attained by the smooth factors of the $(2, 3, 8)$-triangle groups and the second by smooth factors of the $(2, 4, 5)$-triangle groups. In his unpublished thesis [Che1] Chetiya showed that no symmetric group of degree ≤ 17 can attain the bound $48(g - 1)$, whilst Kalita [Kal] showed that S_5, S_{11}, S_{15} and S_{16} do attain the bound $40(g - 1)$. The definitive step towards the computation of the largest order of an S_n-action on genus g was given later on by Conder in [Con2], where he showed that for even k all but finitely many of the S_n are smooth factors of the $(2, 3, k)$-triangle groups. In particular, taking $k = 8$, all but finitely many of the S_n attain the bound $48(g - 1)$. Since none of them attains the Hurwitz bound, it follows that the strong symmetric genus of S_n is $n!/48 + 1$ for all but finitely many values of n. The exceptions are those n in the range $1 \leq n \leq 17$, and $n = 22, 23, 26$ and 29. These were handled also by Conder in [Con3] where, as a consequence of some results about two-element generation of certain permutation groups, he found minimal (k, l, m)-generating pairs for each of these exceptions.

If we allow orientation-reversing automorphisms then the Hurwitz bound extends to $168(g - 1)$. Conder in [Con1] showed that this bound is attained by the symmetric group S_n for all $n \geq 168$, and for all but 64 integers n in the range $3 \leq n \leq 167$. Consequently, the symmetric genus of S_n is $n!/168 + 1$ for all but finitely many values of n (compare this with the previous result on alternating groups, which, having no subgroup of index 2, always act without orientation-reversing automorphisms; that means that the strong symmetric and the symmetric genera of A_n

coincide). The symmetric genus of S_n for the finite number of exceptions was calculated by Conder in [Con4].

7 Linear fractional groups

Macbeath showed in [Mcb4] that the projective linear fractional group $PSL(2, q)$ is a Hurwitz group if and only if either $q = 7$, or q is a prime congruent to ± 1 (mod 7), or $q = p^3$ for some prime p congruent to ± 2 or ± 3 (mod 7). Consequently, the strong symmetric genus of $PSL(2, q)$ for such values of q is $q(q-1)(q+1)/168 + 1$ for q odd and 7 for $q = 8$.

For $q = p$ prime and $p \geq 5$, the strong symmetric genus for $PSL(2, p)$ was calculated by Glover and Sjerve in [GS1]. For $p = 5, 7$ and 11 the strong symmetric genus is 0, 3 (attained by the Klein's quartic) and 26 respectively. For $p \geq 13$ the study splits into four cases according to whether p is congruent to ± 1 (mod 5) and (mod 8). The same authors extended these results in [GS2], where they found the strong symmetric genus of $PSL(2, p^n)$ for all primes p and exponents n. In all cases with $p^n \geq 7$ the minimum genus is attained when $PSL(2, p^n)$ is represented as a smooth factor of the (r, s, t)-triangle groups (for $p^n = 2, 3, 4$ and 5 the minimum genus is 0). In most of the cases, $(r, s, t) = (2, 3, d)$ where $d \geq 7$ is arithmetically determined by p and n. The remaining triangle signatures are $(2, 4, 5)$, $(3, 3, 4)$ and $(2, 5, 7)$.

The problem of characterizing the signatures $(h; m_1, \ldots, m_r)$ which allow smooth epimorphisms of the corresponding Fuchsian group onto $PSL(2, p)$ was solved in case $p > 11$ and $h = 0$ by Özaydin, Simmons and Taback in [ÖST]. The two obvious necessary conditions, namely, that each period m_i is p or a divisor of $(p \pm 1)/2$ (since these are the orders of the elements in $PSL(2, p)$) and $\sum_{i=1}^{r}(1 - 1/m_i) > 2$ (for the Fuchsian group to have positive hyperbolic area) turn out to be also sufficient. In other words, they showed that $PSL(2, p)$ acts with any signature $(0; m_1, \ldots, m_r)$ satisfying some obvious conditions. Moreover, they also calculated the *spherical Hurwitz semigroup* of $PSL(2, p)$ for $p > 11$. The *Hurwitz semigroup* of a finite group G is the translation by $|G| - 1$ of the genus spectrum of G, and it is a subset of the positive integers with the remarkable property of being closed under addition, see [MM]. The spherical Hurwitz semigroup of $PSL(2, p)$ consists then of the translation by $p(p-1)(p+1)/2 - 1$ of those genera of surface on which $PSL(2, p)$ acts yielding a genus zero quotient. In particular, the first element of the spherical Hurwitz semigroup yields, up to translation, the strong symmetric genus of $PSL(2, p)$, since this is attained by means of triangle signatures, whose orbit genus is zero.

The projective linear fractional group $PGL(2, q)$ with $q \neq 2^n$ is not perfect and so it cannot be a Hurwitz group. Chetiya and Kalita in [ChK2] showed that $PGL(2, q)$ with $q = p^n$ and p an odd prime attains infinitely often both the second and third largest uniform upper bounds $48(g - 1)$ and $40(g - 1)$ for the order of a group acting on genus g. Moreover, they determined the (infinitely many) values of q for which each bound is attained.

The strong symmetric genus for the special linear group $SL(2, p^n)$ was calculated by Voon in [Voo] for every odd prime p and exponent n. Observe that for $p = 2$, the groups $SL(2, p^n)$ and $PSL(2, p^n)$ coincide. As in the case of $PSL(2, p^n)$, also the strong symmetric genus is attained when $SL(2, p^n)$ is represented as a smooth factor of a triangle group; this time, however, in most of the cases its signature is not $(2, 3, d)$ but $(3, 3, d)$ where $d \geq 7$ is the smallest integer which divides $p^n - 1$ or $p^n + 1$ but not $p^m \pm 1$ for any proper divisor m of n. The remaining triangle signatures are $(3, 3, 4)$ for $SL(2, 3)$, whose genus is 2, $(3, 4, 5)$ for $SL(2, 5)$, whose genus is 14, and $(3, 5, 7)$ for $SL(2, p^6)$ and $d \geq 105$. In addition, Voon also computed the full automorphism group of a surface of minimum genus on which $SL(2, p^n)$ acts. It turns out that in most of the cases, *e.g.*, for $(3, 3, d)$-actions, the full automorphism group is strictly larger than $SL(2, p^n)$. This gives a negative answer to the following interesting conjecture by Glover [Voo, p. 529]: let G be a finite simple group and let S be a surface of least genus on which G acts; then S/G is the Riemann sphere, the covering projection $S \to S/G$ ramifies over three points and S can be chosen as a Riemann surface such that its full group of automorphisms is G.

8 Other groups and results

The largest uniform upper bound for the order of an **odd-order group** acting on genus g was shown by Weaver in [Wea3] to be $3(3 + 2^{l+1})(g - 1)/2^l$, where 2^l is the highest power of 2 dividing $g - 1$. He first showed that if an odd-order group G acting on genus g is a smooth factor of a (k, l, m)-triangle group of least hyperbolic area then $(k, l, m) = (3, 3, N_l)$, where $N_l = 3 + 2^{l+1}$. The third member of the derived series of such a triangle group is a surface group, and this allowed Weaver to obtain smooth factors of a $(3, 3, N_l)$-triangle group. Moreover, he found an infinite sequence of values of g for which the bound is attained, namely, those of the form $g = 1+2^l N_l m^{2h}$ where $l \geq 0$, $h = 1+2^l N_l$ and m is odd. Weaver also showed that the bound fails to be attained infinitely often, as it happens for the values $g = 1 + 2^l d$ where $d > N_l$ is a square-free odd integer having no common factor with $3N_l$.

Weaver also considered the problem of lower uniform bounds for the order of the largest odd-order groups acting on genus g. (In the general case, Accola [Acc] and Maclachlan [Mcl3] showed independently that such bound is $8g + 8$, and that it is attained for every $g \geq 2$; in other words, for every $g \geq 2$ there exists a Riemann surface of genus g with at least $8g+8$ automorphisms.) Weaver showed that the order of the largest odd-order group acting on genus g is not smaller than $(1 + 2^{k+1})g/2^k$, where 2^k is the highest power of 2 dividing g. This bound is attained for cyclic groups whenever g is odd. If $g = 5$ then the bound is also valid for all odd-order groups, although it is not known whether this is true for other values of g.

In searching the groups of strong symmetric genus two and three, May and Zimmerman in [MZ5] had to deal with **groups of low order**. In their paper we can find the strong symmetric genus of each group of order ≤ 24. The list was extended to groups of order ≤ 36 by the same authors in [MZ7]. Concerning groups with orientation-reversing elements, the symmetric genus of groups of order < 48 with the exception of order 32 was computed also by the same authors in [MZ3]. Later on, Zimmerman in [Zim] solved the remaining case of groups of order 32. In both papers, the possibility of symmetric genus 0 and 1 is included.

Instead of fixing groups of low order and looking for their symmetric or strong symmetric genus, there is the interesting problem of fixing a low value of g and looking for the (finitely many) groups with symmetric or strong symmetric genus g. May and Zimmerman showed in [MZ5] that there are six groups of strong symmetric genus two and ten groups of strong symmetric genus three. (Actually, they claimed that there exist nine groups of strong symmetric genus three, but the mistake was corrected by themselves in [MZ7].) The same authors extended the classification in [MZ7] finding the ten groups with strong symmetric genus four. As to groups of small symmetric genus, a good reference for them is Section 6.3 in the book on graph theory [GT], by Gross and Tucker. It turns out that there are four groups with symmetric genus two (see, *e.g.*, [MZ3, Thm. 4] for an explicit description of them) and three with symmetric genus three, found by May and Zimmerman in

[MZ4]. In these two last papers, and in many others, the close relation between Riemann surfaces with large groups of automorphisms and the theory of regular maps on surfaces is quite useful (a classical reference for the basic definitions on maps is the book [CM] by Coxeter and Moser). Let us briefly explain this relation here.

Let G act preserving orientation on a Riemann surface S of genus g. If G is a smooth factor of a $(2, l, m)$-triangle group (this happens, for example, whenever $|G| > 24(g-1)$ by a result of Singerman in [Sin]) then there is a regular map of type $\{l, m\}$ on the topological surface S (a map is said to be of type $\{l, m\}$ if it is composed of l-gons, m meeting at each vertex). Moreover, G is isomorphic to the rotation group of the map. Conversely, if G is the rotation group of a regular map of type $\{l, m\}$ on a surface S, then G is a smooth factor of a $(2, l, m)$-triangle group and G acts as a group of orientation-preserving automorphisms on a Riemann surface homeomorphic to S. Hence the classification of regular maps is an important tool in the study of large groups acting (preserving orientation) on genus g, and conversely.

A general **lower bound for the symmetric genus** of a group G was obtained by May and Zimmerman in [MZ3]. Let S be a generating set of G and let $t_2(S)$, $t_3(S)$ and $t_h(S)$ denote the number of generators in S of order 2, 3 or higher, respectively. Let $\psi(G)$ be the minimum of $9t_h(S) + 8t_3(S) + 3t_2(S)$ when S runs over all generating sets of G. Then a lower bound for the symmetric genus of G is $1 + |G|(\psi(G) - 16)/24$, provided it is bigger than one. They exhibited two infinite families of groups for which this bound is attained. As an application, they calculated the symmetric genus of the Hamiltonian groups with no odd-order part. A Hamiltonian group is a non-abelian group in which every subgroup is normal. The finite Hamiltonian groups have the form $Q \times A \times B$ where Q is the quaternion group of order 8, A is an elementary abelian 2-group and B is an abelian group of odd order, [CM]. They showed that the symmetric genus of $Q \times (Z_2)^a$ equals $1 + 2^a(a + 1)$ if $1 \le a \le 3$, and $1 + 2^a(a + 2)$ if $a \ge 4$.

An interesting question, completely answered nowadays, is to determine which of the 26 **sporadic simple groups** is a Hurwitz group and, for each of the non-Hurwitz groups, to calculate its symmetric genus (recall that for simple groups, the symmetric genus and the strong symmetric genus coincide). In all cases, the minimum genus is attained when the group is a smooth factor of a (k, l, m)-triangle group, and the goal is to find minimal (k, l, m)-generating pairs of each sporadic group. To that end a fruitful method consists in using the group character table to calculate the appropriate structure constants, and the help of some information about maximal subgroups (known in almost all cases). For the sporadic groups of intractable size, the task is obviously impossible without the help of modern algorithms and fast computers.

The results are summarized in Table .1. In its first column we list the triples (k, l, m) for which there exists a minimal (k, l, m)-generating pair for the corresponding group. In all cases, the symmetric genus is given by the triple (k, l, m) and the Hurwitz-Riemman formula.

Table .1. *The symmetric genus of the sporadic simple groups.*

(k, l, m)	Symmetric genus	Group	Proof
$(2, 3, 7)$	$|G|/84 + 1$	J_1	[Sah]
	$|G|/84 + 1$	J_2	[FR]
	$|G|/84 + 1$	He, Ru, HN, Ly	[Wol1]
	$|G|/84 + 1$	Co_3	[Wor, Wol1]
	$|G|/84 + 1$	Fi_{22}, J_4	[Wol3]
	$|G|/84 + 1$	Th	[Lin]
	$|G|/84 + 1$	Fi'_{24}	[LW]
	$|G|/84 + 1$	M	[Wil3]
$(2, 3, 8)$	$|G|/48 + 1$	O'N, Co_1	[CWW]
	$|G|/48 + 1$	B	[Wil1]
	$|G|/48 + 1$	Fi_{23}	[Wil2]
$(2, 4, 5)$	$|G|/40 + 1$	J_3, Suz	[CWW]
$(2, 3, 11)$	$5|G|/132 + 1$	HS	[Wol4]
	$5|G|/132 + 1$	Co_2	[CWW]
$(3, 3, 4)$	$|G|/24 + 1$	M_{24}	[Con6]
$(2, 3, 10)$	$|G|/30 + 1$	M_{12}	[Wol2, Con6]
$(2, 5, 7)$	$11|G|/140 + 1$	M_{22}	[Wol2, Con6]
$(2, 4, 11)$	$7|G|/88 + 1$	M_{11}	[Wol2, Con6]
$(2, 5, 8)$	$|G|/10 + 1$	McL	[CWW]
$(2, 4, 23)$	$19|G|/184 + 1$	M_{23}	[Con6]

It is also interesting to mention that if G is a sporadic group and S is a surface of minimal genus on which G acts then G is the full group Aut(S) of all orientation-preserving automorphisms of S. This was shown by Woldar in [Wol1] for all sporadic groups other than McL, and by Conder, Wilson and Woldar himself in [CWW] for McL.

The determination of the strong symmetric genus of some finite Coxeter groups led Jackson in [Jac] to consider the **hyperoctahedral groups**. These are the finite Coxeter groups of type B_n, where the group B_n may be defined for $n \geq 3$ as the group of symmetries of the

n-dimensional cube. Jackson showed that for any pair a, b of generators of B_n, the orders of a, b and ab are all even. The main result is that for all $n \geq 3$ except $n = 5, 6$ and 8, the hyperoctahedral group is a smooth factor of the $(2, 4, 6)$-triangle groups. The groups B_5, B_6 and B_8 are smooth factors of $(2, 4, 10)$, $(2, 4, 12)$ and $(2, 4, 8)$-triangle groups respectively. As a consequence, the strong symmetric genus of B_n equals $(2^{n-3}n!)/3 + 1$ for $n \geq 3$, while B_5, B_6 and B_8 have strong symmetric genus 289, 3841 and 645 121 respectively.

For $n \geq 2$ the **dicyclic group** is the group of order $4n$ with presentation $DC_n = \langle a, b \mid a^{2n} = 1, a^n = b^2, b^{-1}ab = a^{-1} \rangle$. May and Zimmerman in [MZ5] showed that the strong symmetric genus of DC_n equals n if n is even and $n-1$ if n is odd. Consequently, they obtained the interesting result that for each even genus g there are at least two groups of strong symmetric genus g. A natural question arises: is there at least a group of strong symmetric genus g for odd values of g? The same authors answered this question in the affirmative in [MZ6]. They achieved this result by computing the strong symmetric genus of the **direct product** $Z_k \oplus D_n$ of a cyclic group of order k with a dihedral group of order $2n$, for every value of k and n. For example, if k is even, $k \geq 4$, n is odd and $n \geq 3$ then the strong symmetric genus of $Z_k \oplus D_n$ equals $1 + nk(1/2 - 1/k - 1/\mathrm{lcm}(k, n))$. As a consequence, taking $k = 4$, they obtained that the strong symmetric genus of $Z_4 \oplus D_n$ equals n for all odd $n \geq 3$. Together with their previous results, this shows that there is at least one group of strong symmetric genus g for each value of g. They conjectured that this is also true for the symmetric genus.

Jones and Silver in [JS] calculated the symmetric genus of the **Suzuki groups** $Sz(q) = {}^2B_2(q)$, where $q = 2^f$ for odd $f > 1$. These are non-abelian simple groups of order $q(q - 1)(q^2 + 1)$ which, particularly important here, have no order-three elements. So they cannot be a smooth factor of either a $(2, 3, 7)$-triangle group or a $(2, 3, 8)$-triangle group. The next triangle signature of least hyperbolic area is $(2, 4, 5)$, and Jones and Silver proved that the Suzuki groups are smooth factors of a $(2, 4, 5)$-triangle group. In particular, the Hurwitz-Riemann formula yields that the symmetric genus (and also the strong symmetric genus) of the Suzuki group $G = Sz(q)$ with $q = 2^f$ for odd $f > 1$ equals $1 + |G|/40 = 1 + q(q - 1)(q^2 + 1)/40$. The surfaces of minimum genus realizing this bound are not unique, as Jones and Silver showed by calculating the number of (conformal equivalence classes of) such surfaces. They achieved this result by first computing the number of smooth epimorphisms from a $(2,4,5)$-triangle group onto G, and studying then the number of orbits of Aut$G$ on the set of such epimorphisms. They also proved that G is the full group of conformal automorphisms of each of these surfaces.

To conclude this section, and as a curiosity, we mention Conder's study of the **group of Rubik's cube**. This is the group of all patterns achievable by natural manipulations of a standard six-coloured Rubik's cube. The group has $2^{10} \cdot 3^7 \cdot 12! \cdot 8!$ elements, but its structure is quite well-known. As a consequence of his results about two-element generation of certain permutation groups in [Con3], Conder found a minimal $(2,4,12)$-generating pair of the group of Rubik's cube. Hence, the minimum genus of surface on which it acts is $1 + 2^{10} \cdot 3^7 \cdot 11! \cdot 8!$

Notes

1 All authors partially supported by MTM 2008–00250. The third author was also partially supported by The Research Grant N N201 366436 of The Polish Ministry of Sciences and Higher Education.

Bibliography

[Acc] Accola, R. D. M. On the number of automorphisms of a closed Riemann surface. *Trans. Amer. Math. Soc.* **131** (1968), 398–408.

[Bre] Breuer, T. *Characters and Automorphism Groups of Compact Riemann Surfaces,* London Math. Soc. LNS **280**, Cambridge Univ. Press, Cambridge, 2000.

[BCGG] Bujalance, E., Cirre, F. J., Gamboa, J. M. and Gromadzki, G. On compact Riemann surfaces with dihedral groups of automorphisms. *Math., Proc. Cambridge Phil. Soc.* **134**, (2003), 465–477.

[BEGG] Bujalance, E., Etayo, J. J., Gamboa, J. M. and Gromadzki, G. *Automorphism Groups of Compact Bordered Klein Surfaces,* Lecture Notes in Math. **1439**, Springer-Verlag, Berlin, 1990.

[Bur] Burnside, W. *Theory of Groups of Finite Order,* Cambridge University Press, Cambridge, 1911.

[Che1] Chetiya, B. P. Groups of automorphisms of compact Riemann surfaces, PhD Thesis, Birmingham University, 1971.

[Che2] Chetiya, B. P. On genuses of compact Riemann surfaces admitting solvable automorphism groups. *Indian J. Pure Appl. Math.* **12**(11), (1981), 1312–1318.

[ChK1] Chetiya, B. P. and Kalita, S. C. On solvable automorphism groups of compact Riemann surfaces. *Indian J. Pure Appl. Math.* **15**(9) (1984), 978–983.

[ChK2] Chetiya, B. P. and Kalita, S. C. A classification of the projective lineal groups. *J. Indian Math. Soc. (N.S.)* **48**(1–4) (1984), (1986), 89–99.

[ChP1] Chetiya, B. P. and Patra, K. On metabelian groups of automorphisms of compact Riemann surfaces. *J. London Math. Soc.* (2) **33**(3) (1986), 467–472.

[ChP2] Chetiya, B. P. and Patra, K. *K*-metacyclic groups of automorphisms of compact Riemann surfaces. *Far East J. Math. Sci.* **2**(2) (1994), 127–136.

[Con1] Conder, M. D. E. Generators for alternating and symmetric groups. *J. London Math. Soc. (2)* **22**(1), (1980), 75–86.

[Con2] Conder, M. D. E. More on generators for alternating and symmetric groups. *Quart. J. Math. Oxford (2)* **32** (1981), 137–163.

[Con3] Conder, M. D. E. Some results on quotients of triangle groups. *Bull. Austral. Math. Soc.* **30**(1) (1984), 73–90.

[Con4] Conder, M. D. E. The symmetric genus of alternating and symmetric groups. *J. Combin. Theory Ser. B* **39**(2) (1985), 179–186.

[Con5] Conder, M. D. E. Hurwitz groups: a brief survey. *Bull. Amer. Math. Soc.* **23**(2) (1991), 359–370.

[Con6] Conder, M. D. E. The symmetric genus of the Mathieu groups. *Bull. London Math. Soc.* **23** (1991), 445–453.

[CWW] Conder, M. D. E., Wilson, R. A. and Woldar, A. J. The symmetric genus of sporadic groups. *Proc. Amer. Math. Soc.* **116**(3), (1992), 653–663.

[CM] Coxeter, H. S. M. and Moser, W. O. J. *Generators and Relations for Discrete Groups*, Springer-Verlag, Berlin, 1957, viii+155 pp.

[DPCh] Dutta, S. K., Patra, K. and Chetiya, B. P. *ZS*-metacyclic groups of automorphisms of compact Riemann surfaces. *Indian J. Pure Appl. Math.* **28**(1) (1997), 63–74.

[Eta] Etayo, J. J. Nonorientable automorphisms of Riemann surfaces. *Arch. Math. (Basel)* **45**(4) (1985), 374–384.

[EM] Etayo, J. J. and Martínez, E. Alternating groups as automorphism groups of Riemann surfaces. *Internat. J. Algebra Comput.* **16**(1) (2006), 91–98.

[FR] Finkelstein, L. and Rudvalis, A. Maximal subgroups of the Hall-Janko-Wales group. *J. Algebra,* **24** (1973), 486–493.

[Geo] George Michael, A. A. Metacyclic groups of automorphisms of compact Riemann surfaces. *Hiroshima Math. J.* **31**(1) (2001), 117–132.

[GS1] Glover, H. and Sjerve, D. Representing PSl₂(*p*) on a Riemann surface of least genus. *Enseign. Math. (2)* **31**(3–4) (1985), 305–325.

[GS2] Glover, H. and Sjerve, D. The genus of PSl₂(*q*). *J. reine angew. Math.,* **380** (1987), 59–86.

[Gro1] Gromadzki, G. On soluble groups of automorphisms of Riemann surfaces. *Canadian Math. Bull.* **34**(1) (1991), 67–73.

[Gro2] Gromadzki, G. Metabelian groups acting on compact Riemann surfaces. *Rev. Mat. Univ. Complut. Madrid* **8**(2) (1995), 293–305.

[GM] Gromadzki, G. and Maclachlan, C. Supersoluble groups of automorphisms of compact Riemann surfaces. *Glasgow Math. J.* **31**(3) (1989), 321–327.

[GT] Gross, J. L. and Tucker, T. W. *Topological Graph Theory*, Wiley-Interscience Series in Discrete Mathematics and Optimization, John Wiley & Sons, Inc., New York, 1987, xvi+351 pp.

[Har] Harvey, W. J. Cyclic groups of automorphisms of a compact Riemann surface. *Quart. J. Math.* **17** (1966), 86–97.

[Hur] Hurwitz, A. Über algebraische Gebilde mit eindeutigen Transformationen in sich. *Math. Ann.* **41** (1893), 403–442.

[Jac] Jackson, M. A. The strong symmetric genus of the hyperoctahedral groups. *J. Group Theory* **7**(4) (2004), 495–505.

[JS] Jones, G. A. and Silver, S. A. Suzuki groups and surfaces. *J. London Math. Soc. (2)* **48**(1) (1993), 117–125.

[Kal] Kalita, S. C. Some symmetric groups of small degrees as automorphism groups of compact Riemann surfaces. *Indian J. Pure Appl. Math.* **10**(9) (1979), 1157–1166.

[Kul] Kulkarni, R. Symmetries of surfaces. *Topology* **26**(2) (1987), 195–203.

[KM] Kulkarni, R. and Maclachlan, C. Cyclic p-groups of symmetries of surfaces. *Glasgow Math. J.* **33** (1991), 213–221.

[Lin] Linton, S. A. The maximal subgroups of the Thompson group. *J. London Math. Soc.* **39** (1989), 79–88.

[LW] Linton, S. A. and Wilson, R. A. The maximal subgroups of the Fischer groups Fi'_{24} and Fi_{24}. *Proc. London Math. Soc.* **63** (1991), 113–164.

[Mcb1] Macbeath, A. M. Discontinuous groups and birational transformations. Proc. Dundee Summer School, 1961.

[Mcb2] Macbeath, A. M. On a theorem of Hurwitz. *Proc. Glasgow Math. Assoc.* **5** (1961), 90–96.

[Mcb3] Macbeath, A. M. The classification of non-euclidean plane crystallographic groups. *Canad. J. Math.* **19** (1967), 1192–1205.

[Mcb4] Macbeath, A. M. Generators of the linear fractional groups, 1969 Number Theory (Proc. Sympos. Pure Math., Vol. XII, Houston, Texas, 1967), Amer. Math. Soc., Providence, RI, pp. 14–32.

[Mcl1] Maclachlan, C. Abelian groups of automorphisms of compact Riemann surfaces. *Proc. London Math. Soc. (3)* **15** (1965), 699–712.

[Mcl2] Maclachlan, C. Groups of automorphisms of compact Riemann surfaces. PhD, Univ. Birmingham, 1966.

[Mcl3] Maclachlan, C. A bound for the number of automorphisms of a compact Riemann surface. *J. London Math. Soc.* **44** (1969), 265–272.

[MT] Maclachlan, C. and Talu, Y. p-groups of symmetries of surfaces. *Michigan Math. J.* **45** (1998), 315–332.

[MZ1] May, C. L. and Zimmerman, J. The symmetric genus of finite abelian groups. *Illinois J. Math.* **37**(3) (1993), 400–423.

[MZ2] May, C. L. and Zimmerman, J. The symmetric genus of metacyclic groups. *Topology Appl.* **66**(2) (1995), 101–115.

[MZ3] May, C. L. and Zimmerman, J. Groups of small symmetric genus. *Glasgow Math. J.* **37** (1995), 115–129.

[MZ4] May, C. L. and Zimmerman, J. The groups of symmetric genus three. *Houston J. Math.* **23**(4) (1997), 573–590.

[MZ5] May, C. L. and Zimmerman, J. Groups of small strong symmetric genus. *J. Group Theory* **3**(3) (2000), 233–245.

[MZ6] May, C. L. and Zimmerman, J. There is a group of every strong symmetric genus. *Bull. London Math. Soc.* **35**(4) (2003), 433–439.

[MZ7] May, C. L. and Zimmerman, J. The groups of strong symmetric genus 4. *Houston J. Math.* **31**(1), (2005), 21–35.

[MM] McCullough, D. and Miller, A. The stable genus increment for group actions on closed 2-manifolds. *Topology* **31**(2) (1992), 367–397.

[ÖST] Özaydin, M., Simmons, C. and Taback, J. Surface symmetries and $\mathrm{PSL}_2(p)$. *Trans. Amer. Math. Soc.* **359**(5) (2007), 2243–2268.

[Pat] Patra, K. On determination of the marks of some alternating groups of small degree. *Indian J. Pure Appl. Math.* **14**(7) (1983), 864–872.

A survey on a theorem by Harvey

[Sah] Sah, C.-H. Groups related to compact Riemann surfaces. *Acta Math.* **123** (1969), 13–42.

[Sch] Schwarz, H. A. Über diejenigen algebraischen Gleichungen zwischen zwei Veränderlichen Grossen, welche eine Schaar rationaler eindeutig umkehrbarer Transformationen in sich selbst zulassen. *J. reine und angew. Math.* **87** (1879), 139–145.

[Sin] Singerman, D. Symmetries of Riemann surfaces with large automorphism group. *Math. Ann.* **210**, (1974), 17–32.

[Tuc] Tucker, T. W. Finite groups acting on surfaces and the genus of a group. *J. Combin. Theory Ser. B* **34**(1) (1983), 82–98.

[Voo] Voon, S.-N. The genus of $SL_2(F_q)$. *Michigan Math. J.* **40**(3) (1993), 527–544.

[Wea1] Weaver, A. Genus spectra for split metacyclic groups. *Glasgow Math. J.* **43**(2) (2001), 209–218.

[Wea2] Weaver, A. *Automorphisms of Surfaces, Combinatorial and Geometric Group Theory*, Contemp. Math., **296**, Amer. Math. Soc., Providence, RI, 2002, pp. 257–275.

[Wea3] Weaver, A. Odd-order group actions on surfaces. *J. Ramanujan Math. Soc.* **18**(3) (2003), 211–220.

[Wil1] Wilson, R. A. The symmetric genus of the Baby Monster. *Quart. J. Math. Oxford (2)* **44**(176) (1993), 513–516.

[Wil2] Wilson, R. A. The symmetric genus of the Fischer group Fi_{23}. *Topology*, **36**(2) (1997), 379–380.

[Wil3] Wilson, R. A. The Monster is a Hurwitz group. *J. Group Theory* **4**(4) (2001), 367–374.

[Wim] Wiman, A. Über die hyperelliptischen Kurven und diejenigen vom Geschlecht $p = 3$, welche eindeutige Transformationen in sich zulassen. *Bihang Till. Kungl. Svenska Vetenskaps-Akademiens Handlingar* **21**, 1(1), (1895), 23 pp.

[Wol1] Woldar, A. J. On Hurwitz generation and genus actions of sporadic groups. *Illinois J. Math.* **33**(3) (1989), 416–437.

[Wol2] Woldar, A. J. Representing M_{11}, M_{12}, M_{22} and M_{23} on surfaces of least genus. *Comm. Algebra* **18**(1) (1990), 15–86.

[Wol3] Woldar, A. J. Sporadic simple groups which are Hurwitz. *J. Algebra* **144** (1991), 443–450.

[Wol4] Woldar, A. J. The symmetric genus of the Higman-Sims group HS and bounds for Conway's groups Co_1, Co_2. *Illinois J. Math.* **36**(1), (1992), 47–52.

[Wor] Worboys, M. F. Generators for the sporadic group Co_3 as a $(2, 3, 7)$-group. *Proc. Edinburgh Math. Soc. (2)* **25** (1982), 85–68.

[Zim] Zimmerman, J. The symmetric genus of 2-groups. *Glasgow Math. J.* **41**(1) (1999), 115–124.

[Zom1] Zomorrodian, R. Nilpotent automorphism groups of Riemann surfaces. *Trans. Amer. Math. Soc.* **288**(1) (1985), 241–255.

[Zom2] Zomorrodian, R. Classification of p-groups of automorphisms of Riemann surfaces and their lower central series. *Glasgow Math. J.* **29**(2) (1987), 237–244.

[Zom3] Zomorrodian, R. Bounds for the order of supersoluble automorphism groups of Riemann surfaces. *Proc. Amer. Math. Soc.* **108**(3) (1990), 587–600.

[Zom4] Zomorrodian, R. On a theorem of supersoluble automorphism groups. *Proc. Amer. Math. Soc.* **131**(9) (2003), 2711–2713.

Algorithms for simple closed geodesics

Peter Buser

Institut de Géométrie, Algèbre et Topologie
SB-IGAT-GEOM, Station 8
École Polytechnique Fédérale de Lausanne
CH-1026 Lausanne, Switzerland

Abstract

We enumerate and represent graphically the simple closed geodesics on a compact hyperbolic surface S of genus $g \geq 2$, with the aim of drawing pictures of the Birman-Series set or parts of it. An algorithm is presented that enumerates the homotopy classes of simple closed curves on S based on, and adapted to, pairs of pants decompositions of S. The coding is similar to the Dehn-Thurston coding. The algorithm runs in polynomial time with respect to both code and geodesic length. The topological design is such that the curves constructed iteratively are by themselves pairwise non-homotopic so that no comparison with previously constructed curves is needed. This is achieved by grouping the homotopy classes into suitable families. As an illustration, some figures of sets of simple closed geodesics drawn on a fundamental domain are shown in Section 9. For these figures we used an implementation of the algorithm for genus 2.

1 Introduction

Closed geodesics on hyperbolic surfaces have been studied along with the development of the geometry of the surfaces themselves. One of the remarkable achievements obtained in the middle of the last century is certainly the relationship with the spectrum of the Laplace-operator and the resulting asymptotic formula

$$\mathcal{N}_S(L) \sim e^L/L, \quad L \to \infty, \tag{1.1}$$

38

where $\mathcal{N}_S(L)$ is the number of oriented closed geodesics of length $\leq L$ on a closed hyperbolic surface S of genus $g \geq 2$ (Huber, [H], Selberg [Sel]).

Among the closed geodesics, the simple ones, i.e. those without self-intersections, are of special interest; they are, for instance, used for the description of hyperbolic surfaces via building blocks. There are considerably fewer of them, of course, but it was not until quite recently in the work of McShane-Rivin [MR1, MR2], Rivin [Ri], and Mirzakhani [Mz1, Mz2] that precise statements about their growth rate were made, the strongest result being the asymptotic formula

$$\mathcal{NS}_S(L) \sim c_S L^{6g-6}, \quad L \to \infty, \tag{1.2}$$

in [Mz2], where $\mathcal{NS}_S(L)$ is the number of simple closed geodesics of length $\leq L$ on S, and c_S is a constant that depends on the geometry of S. The simple closed geodesics are thus quite sparse.

A sparseness of a different type was discovered by Birman and Series in [BS2]. It is well known that the union of all closed geodesics on S (in fact, on any closed negatively curved Riemannian manifold) is a dense subset. In contrast to this, the result in [BS2], in its simplest form, is that on any closed hyperbolic surface S the union of the simple closed geodesics or, more generally, the union of *all* complete simple geodesics is a nowhere dense subset of S; the latter is called the *Birman-Series set*. We shall denote it by $\mathcal{BS}(S)$. In [BuP] it is shown that for some constant r_g depending only on g, the complement of $\mathcal{BS}(S)$ always contains a disk of radius r_g. The enumeration algorithm for simple closed geodesics presented in this paper grew out of the wish to get insight into the size of r_g by looking at explicitly computed images of $\mathcal{BS}(S)$.

The paper is organized as follows. In Section 2, we introduce a coding for the homotopy classes of the simple closed curves that is adapted to a pants decomposition of S, and show that the coding is one-to-one. Section 3 provides some topological lemmas used in the proof that were difficult to find in the literature, at least in the form needed here. Section 4 shows ways to translate code words into words in terms of generators of the fundamental group of S and vice versa. Sections 5 and 6 contain the description of the algorithm, one of its features being that a curve (in the form of its code word) is never computed twice so that no pairwise comparison of curves is needed. Section 7 shows the metric equivalence between code length and geodesic length. Section 8 outlines a procedure to draw geodesics by depicting them as sequences of geodesic arcs on a convex geodesic fundamental domain of S. In Section 9 we collect all formulae and conventions that were used to produce the pictures of simple closed geodesics on surfaces of genus 2 shown towards the end of the paper. Finally, we finish with some complementary literature and remarks concerning the enumeration algorithm.

As for the topological background, we assume that the reader is familiar with the material covered, for example, in the chapters by Macbeath [Mb] and Birman [Bi] in Bill Harvey's book "Discrete groups and automorphic functions" [Ha]; for the geometry of discrete groups used in Sections 8 and 9 we make use of the material covered by Beardon's chapter [Bd1] in the same book.

2 Pasting schemes and codes

The algorithm in Section 5 is designed for surfaces that are given together with a decomposition into pairs of pants. We shall now fix the notation for such a decomposition, then introduce a system of arcs that will allow us to characterize the homotopy classes of simple closed curves by certain codes and, finally, show that codes and homotopy classes are in one-to-one correspondence.

A *pair of pants* is a topological sphere Y as shown in Fig. 2.1 with the interiors of three disjoint closed disks removed. By the classification theorem of surfaces (e.g. [St, sect. 1.3]) any compact orientable surface S of genus $g \geq 2$ may be obtained by pasting together $2g - 2$ copies $Y^1, \ldots, Y^{2g-2}$ of Y along their boundaries as illustrated in Fig. 2.2. For the coding we carry this out explicitly as follows.

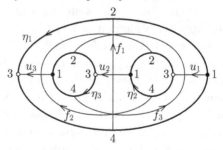

Fig. 2.1. A pair of pants with parameter strands.

Let η_1, η_2, η_3 be the boundary curves of Y parametrized in the form $t \mapsto \eta_i(t)$, $t \in [0,1]$, $i = 1,2,3$. The orientations are as indicated by the arrows in Fig. 2.1; these arrows are such that an observer on Y who approaches the boundary will always have to turn left if he wishes to continue in the direction of the arrows. On each boundary curve we introduce four points, $p_{ij} = \eta_i(\frac{1}{4}(j-1))$, $i = 1, 2, 3$, $j = 1, 2, 3, 4$, and connect them by simple arcs on Y in the following way. Arcs u_i go from p_{i1} to $p_{i+1,3}$, and arcs f_i from p_{i4} to p_{i2}, $i = 1, 2, 3$ (i-indices modulo 3). Furthermore, it is required that the u_i are pairwise disjoint and that each f_i intersects u_{i+1} in exactly one point. A selection of such arcs is illustrated in Fig. 2.1.

Pasting together two copies Y', Y'' of Y along boundary curves η_i' of Y' and η_j'' of Y'' amounts to the following procedure (see [Bu, sect. 1.3] for details). First, the parametrizations of the boundary curves are extended from the interval $[0, 1]$ to $\mathbb{R}$ periodically with period 1; then one declares that

$$\eta_i'(t) \text{ is identified with } \eta_j''(\frac{1}{2} - t), \quad t \in \mathbb{R}. \tag{2.1}$$

This establishes an equivalence relation on the union $Y' \cup Y''$: for each $\eta_i'(t)$, respectively $\eta_j''(\frac{1}{2} - t)$, the equivalence class is the set $\{\eta_i'(t), \eta_j''(\frac{1}{2} - t)\}$; for any other $p \in Y' \cup Y''$ the equivalence class is $\{p\}$. The quotient space for this equivalence relation together with the quotient topology will be called "$Y' \cup Y''$ modulo identification (2.1)". Here we allow that $Y' = Y''$ but in this case i must be different from j. Note that (2.1) pastes the copies $p_{i1}', p_{i2}', p_{i3}', p_{i4}'$ in Y' of the corresponding points in Y to the copies $p_{j3}'', p_{j2}'', p_{j1}'', p_{j4}''$ (in this order) in Y''.

In the same way we may paste together several copies of Y. Fig. 2.2 shows an example with four copies pasted together so as to form a topological sphere with six holes. If we paste the remaining boundary curves together as indicated by the arrows we get a closed orientable surface of genus 3.

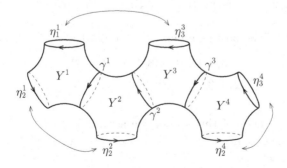

Fig. 2.2. This pasting yields a surface of genus 3.

To get surfaces of arbitrary genus $g \geq 2$ we use a *pasting scheme*: let $Y^1, \ldots, Y^{2g-2}$ be copies of Y with boundary curves η_i^k, points p_{ij}^k and arcs u_i^k, f_i^k on Y^k, named after the corresponding elements on Y, $i = 1, 2, 3$, $j = 1, 2, 3, 4$, $k = 1, \ldots, 2g - 2$. The pasting is now carried out in two steps. In the first step we construct surfaces $S^1, \ldots, S^{2g-2}$ iteratively in the following way. The first surface is $S^1 = Y^1$. On S^1 we select a boundary curve, rename this curve γ^1 and paste together S^1 and Y^2 along γ^1 and η_1^2 to get a four-holed sphere S^2. On S^2 we select a boundary curve γ^2 and paste together S^2 and Y^3 along γ^2 and η_1^3 to get S^3, and so on. The final surface S^{2g-2} is a sphere with $2g$ holes. By *convention* the curves $\gamma^1, \ldots, \gamma^{2g-3}$ keep their names and their parametrizations on S^{2g-2}.

In a second step the boundary curves of S^{2g-2} are grouped into pairs $\alpha^1, \tilde{\alpha}^1, \ldots, \alpha^g, \tilde{\alpha}^g$ and then each α^l is pasted to $\tilde{\alpha}^l$ so as to get a closed surface S of genus g. Here the curves $\alpha^1, \ldots, \alpha^g$ keep their names and their parametrizations on S. For practical reasons we also name these curves $\gamma^{2g-2}, \ldots, \gamma^{3g-3}$.

From the combinatorial point of view the pasting is uniquely determined by the ordered list of all pairs

$$\eta_{i_1}^{k_1}; \eta_{i_1'}^{k_1'}, \ldots, \eta_{i_{3g-3}}^{k_{3g-3}}; \eta_{i_{3g-3}'}^{k_{3g-3}'} \tag{2.2}$$

together with the convention that the resulting curves on S are named and parametrized in this way:

$$\gamma^l(t) = \eta_{i_l}^{k_l}(t) = \eta_{i_l'}^{k_l'}(\tfrac{1}{2} - t), \quad t \in \mathbb{R}, \quad l = 1, \ldots, 3g - 3. \tag{2.3}$$

Instead of (2.2) it suffices to write down the list

$$(k_1, i_1; k_1', i_1'), \ldots, (k_{3g-3}, i_{3g-3}; k_{3g-3}', i_{3g-3}'). \tag{2.4}$$

We call the list (2.2), respectively (2.4), a *pasting scheme*.

Example 2.1 The pasting scheme for the illustration in Fig. 2.2 is

$$(1, 3; 2, 1), \ (2, 3; 3, 1), \ (3, 2; 4, 1),$$
$$(1, 1; 3, 3), \ (1, 2; 2, 2), \ (4, 2; 4, 3).$$

For the remainder of this section we assume that S is a closed orientable surface of genus $g \geq 2$ constructed in the above way based on a given pasting scheme.

On each Y^k, $k = 1, \ldots, 2g - 2$, we have oriented arcs u_1^k, u_2^k, u_3^k,

f_1^k, f_2^k, f_3^k, with the configuration as shown in Fig. 2.1. In addition to this we subdivide each of the closed curves $\gamma^1, \ldots, \gamma^{3g-3}$ (along which the Y^k are pasted together, see (2.3)) into four oriented arcs c_j^l:

$$c_j^l = \gamma^l|_{[(j-1)/4, j/4]}, \quad j = 1, 2, 3, 4. \tag{2.5}$$

If, with respect to our pasting scheme, the boundary curve η_i^k of Y^k has obtained the name γ^l on S and if the parametrization has remained the same, i.e. η_i^k plays the role of $\eta_{i_l}^{k_l}$ in (2.3), then the $c_1^l, c_2^l, c_3^l, c_4^l$ show up on Y^k as the arcs on η_i^k that go successively from p_{i1}^k to p_{i2}^k, from p_{i2}^k to p_{i3}^k, etc. If the parametrization has changed, i.e. if η_i^k plays the role of $\eta_{i_{l'}}^{k_{l'}}$, then these arcs go successively from p_{i3}^k to p_{i2}^k, from p_{i2}^k to p_{i1}^k, etc. Our goal is to use the arcs u_i^k, f_i^k, c_j^l, to associate a code with each homotopy class of closed curves that identifies the homotopy class in a one-to-one way.

In what follows, if a is an oriented or parametrized arc, then $\bar{a}$ denotes the arc with the opposite orientation, respectively the parametrization in the inverse direction. (The inverse of an element δ in the fundamental group of S, however, will be denoted by δ^{-1}.)

Definition 2.2 A closed curve $\varkappa$ on S is called a *code polygon* if it can be written as a product $\varkappa = w_1 w_2 \cdots w_N$, where each w_ν is one of the arcs u_i^k, f_i^k, c_j^l, or the inverse of such an arc. Furthermore, the following two conditions have to be fulfilled:

(i) $w_\nu \neq \bar{w}_{\nu+1}$, $\nu = 1, \ldots, N$ (indices mod N);
(ii) all crossings of $\varkappa$ with $\gamma^1, \ldots, \gamma^{3g-3}$ are transversal (see below).

The cyclic word $\langle w_1, w_2, \ldots, w_N \rangle$ is called the *code* of $\varkappa$. A cyclic word is called a *code word* if it is the code of a code polygon.

In this definition the *cyclic word* $\langle w_1, w_2, \ldots, w_N \rangle$ is the equivalence class of the sequence of letters $w_1, \ldots, w_N$ modulo cyclic permutation, i.e. $\langle w_1, w_2, \ldots, w_N \rangle = \langle w_2, \ldots, w_N, w_1 \rangle$, etc. The terminology "transversal" is shorthand for the following property: if $w_\nu \cdots w_\mu$ is part of the code polygon (possibly after cyclic permutation), where w_ν, w_μ are among the u_i^k, f_i^k or their inverses, while $w_{\nu+1}, \ldots, w_{\mu-1}$ all lie on the same γ^l, then the approach to γ^l of w_ν and the departure from γ^l of w_μ do not take place on the same side of γ^l.

Example 2.3 $\langle c_1^l, c_2^l, c_3^l, c_4^l \rangle$ and $\langle \bar{c}_3^l, \bar{c}_2^l, \bar{c}_1^l, \bar{c}_4^l \rangle$ are code words. If Y^k and $Y^{k'}$ are pasted together along η_i^k and $\eta_{i'}^{k'}$, then $\langle f_i^k, \bar{f}_{i'}^{k'} \rangle$ is a code word.

The coding is based on the following.

Theorem 2.4 *Any simple closed curve α on S which is not homotopic to a point is homotopic to a uniquely determined code polygon.*

We denote the code polygon of a simple closed curve α on S by $\mathcal{P}(\alpha)$. The proof of the theorem will follow in the second part of the next section.

3 Some curve topology

For the proof of Theorem 2.4 as well as the description of the algorithm in Section 5 we introduce some terminology. The surface S with the pairs of pants decomposition $Y^1, \ldots, Y^{2g-2}$, the system of curves $\gamma^1, \ldots, \gamma^{3g-3}$, etc., are as in the preceding section.

A simple arc (i.e. a simple curve with endpoints) s on S is called a *Y-strand* (relatively to the given pants decomposition) or simply a *strand* if it is contained in one of the pairs of pants Y^k of the decomposition and satisfies the following conditions:

– the endpoints of s lie on the boundary of Y^k,
– all other points of s lie in the interior of Y^k,
– s is not homotopic (with fixed endpoints) to an arc on the boundary of Y^k.

Examples of strands are the arcs u_i^k, f_i^k on Y^k and their inverses $\bar{u}_i^k$, $\bar{f}_i^k$, $i = 1, 2, 3$. These strands will be called the *parameter strands* (Fig. 2.1).

Two strands s, t are called *Y-homotopic* if they lie on the same Y^k and if they are homotopic via a homotopy on Y^k that allows the endpoints to glide on the boundary of Y^k. The following is well known (e.g. [Bu, prop. A.18]).

Remark 3.1 *Any strand on Y^k is Y-homotopic to exactly one of the parameter strands.*

A simple closed curve α on S will be called *ordinary* (with respect to the pants decomposition) if it is neither homotopic to a point nor homotopic to one of the parameter curves $\gamma^1, \ldots, \gamma^{3g-3}$ or their inverses.

An ordinary simple closed curve α on S is called *reduced* if among all curves in its free homotopy class it has the minimal number of intersections with $\gamma^1 \cup \cdots \cup \gamma^{3g-3}$. Note that on a pair of pants any simple closed curve is either homotopic to a point or homotopic to a boundary curve. Hence, we have

Remark 3.2 *Any reduced simple closed curve is a product of strands.*

The basic theorem for the coding is the following.

Theorem 3.3 *Let α, β be reduced simple closed curves on S. If they are homotopic, then they are isotopic by an isotopy that fixes $\gamma^1, \ldots, \gamma^{3g-3}$ as sets.*

More precisely, the statement is that there exists a continuous mapping $H : S \times [0,1] \to S$ with the following properties: 1) $H(x,0) = x$ for all $x \in S$; 2) for all $\tau \in [0,1]$ the mapping $x \mapsto H_\tau(x) = H(x,\tau)$, $x \in S$, is a homeomorphism from S to S that fixes $\gamma^1, \ldots, \gamma^{3g-3}$ as sets; 3) $H_1(\beta) = \alpha$.

The main conclusion of the theorem is that if α and β are homotopic, then the two can be written as products of strands $\alpha = s_1 \cdots s_n$, $\beta = t_1 \cdots t_n$ and at the same time the homotopy can be carried out in such a way that for each $\nu = 1, \ldots, n$, the homotopy restricted to s_ν is a Y-homotopy between s_ν and t_ν.

Proof For convenience, we endow S with a hyperbolic metric in which the curves $\gamma^1, \ldots, \gamma^{3g-3}$ are closed geodesics. We also use the unit disk model $\mathbb{D} = \{(x,y) \in \mathbb{R}^2 \mid x^2 + y^2 < 1\}$ of the hyperbolic plane as universal covering with covering map $\pi : \mathbb{D} \to S$. The following facts from covering theory are needed (see e.g. [Bd2], [Mb]).

Any closed curve a in S, not homotopic to a point, lifts to a covering curve $\tilde{a} : \mathbb{R} \to \mathbb{D}$ with two *endpoints at infinity*, i.e. two limit points $p^- = \lim_{t\to-\infty} \tilde{a}(t), p^+ = \lim_{t\to+\infty} \tilde{a}(t)$ on the boundary $\partial\mathbb{D}$ of $\mathbb{D}$. If $\breve{a}$ with endpoints at infinity q^-, q^+ is another lift of a in $\mathbb{D}$, then we either have $\tilde{a} = \breve{a}$ (up to parametrization) or the two pairs of endpoints at infinity $\{p^-, p^+\}$ and $\{q^-, q^+\}$ are disjoint. Finally, we shall use the *unique lifting of homotopies lemma*, stating that if b is a closed curve homotopic to a, then there exists a unique lift $\tilde{b}$ of b in $\mathbb{D}$ homotopic to $\tilde{a}$ and having the same endpoints at infinity as $\tilde{a}$.

Now we lift α and β to homotopic covering curves $\tilde{\alpha}, \tilde{\beta}$ in $\mathbb{D}$ with the same endpoints at infinity, say p^- and p^+. To find strands $s_1, \ldots, s_n$ and $t_1, \ldots, t_n$ as in the conclusion of the theorem, we walk once around α, starting on one of the parameter curves, say γ^{l_0}, then successively follow a strand s_1, cross a parameter curve γ^{l_1}, follow a strand s_2, cross

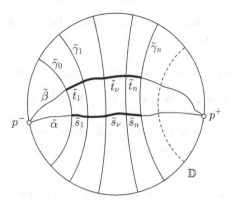

Fig. 3.1. Homotopic lifts in the disk model of the hyperbolic plane.

a parameter curve γ^{l_2}, and so on, until we return to the initial point on $\gamma^{l_n} = \gamma^{l_0}$. We then lift the sequence $s_1, \ldots, s_n$ to a sequence of consecutive arcs $\tilde{s}_1, \ldots, \tilde{s}_n$ on $\tilde{\alpha}$ (so that the product $\tilde{s}_1 \cdots \tilde{s}_n$ is defined), where for $\nu = 1, \ldots, n$, $\tilde{s}_\nu$ connects lifts $\tilde{\gamma}_{\nu-1}, \tilde{\gamma}_\nu$ of $\gamma^{l_{\nu}-1}, \gamma^{l_\nu}$ with each other without crossing any other lifts of $\gamma^1, \ldots, \gamma^{3g-3}$, as shown in Fig. 3.1. For practical reasons we consider one more lift $\tilde{s}_{n+1}$ of s_1 on $\tilde{\alpha}$ going from $\tilde{\gamma}_n$ to the next lift $\tilde{\gamma}_{n+1}$ of γ^{l_1}.

Each $\tilde{\gamma}_\nu$ separates $\mathbb{D}$ into two topological half-planes with p^- and $\tilde{\gamma}_0, \ldots, \tilde{\gamma}_{\nu-1}$ on one side and $\tilde{\gamma}_{\nu+1}, \ldots, \tilde{\gamma}_{n+1}, p^+$ on the other. It follows that $\tilde{\beta}$ also intersects $\tilde{\gamma}_0, \ldots, \tilde{\gamma}_{n+1}$. Let $\bar{\beta}$ be the arc on $\tilde{\beta}$ going from the first intersection of $\tilde{\beta}$ with $\tilde{\gamma}_0$ to the first intersection with $\tilde{\gamma}_n$. The covering transformation

$$T_{\tilde{\alpha}} : \mathbb{D} \to \mathbb{D}$$

with fixed points p^-, p^+ sending $\tilde{\gamma}_0$ to $T_{\tilde{\alpha}}(\tilde{\gamma}_0) = \tilde{\gamma}_n$ and keeping $\tilde{\alpha}$ fixed identifies the two endpoints of $\bar{\beta}$ with each other and so $\pi(\bar{\beta})$ is a closed curve on S. We now easily find a homotopy of $\bar{\beta}$ with endpoints gliding on $\tilde{\gamma}_0, \tilde{\gamma}_n$ sending $\bar{\beta}$ to the arc $\tilde{s}_1 \cdots \tilde{s}_n$ on $\tilde{\alpha}$. Moreover, this can be done such that $T_{\tilde{\alpha}}$ maps the gliding initial point to the gliding endpoint. It follows that $\pi(\bar{\beta})$ is homotopic to α, and so $\pi(\bar{\beta}) = \beta$. Since β is reduced its number of intersections with $\gamma^1, \ldots, \gamma^{3g-3}$ is the same as for α. This implies that $\tilde{\beta}$ intersects each $\tilde{\gamma}_\nu$ in exactly one point and also that $\bar{\beta}$ intersects no lifts of $\gamma^1, \ldots, \gamma^{3g-3}$ other than $\tilde{\gamma}_0, \ldots, \tilde{\gamma}_n$. Now let $\tilde{t}_\nu$ be the arc on $\bar{\beta}$ from $\tilde{\gamma}_{\nu-1}$ to $\tilde{\gamma}_\nu$ and set $t_\nu = \pi(\tilde{t}_\nu)$, $\nu = 1, \ldots, n$. Then $\beta = t_1 \cdots t_n$, where each t_ν is a strand, Y-homotopic to s_ν.

It remains to construct the isotopy of β into α that performs a Y-homotopy from t_ν to s_ν for each ν. For this part of the proof we restrict ourselves to outlining the main steps. For more details we refer the reader e.g. to [Bi] or [St], where similar homotopies are carried out.

In the first step an isotopy fixing $\gamma^1, \ldots, \gamma^{3g-3}$ pointwise is applied to $t_1, \ldots, t_n$ so that the number of intersections of α and β becomes finite. In the next step we remove all possible intersections of α and β. For this we shall work in the universal covering $\mathbb{D}$. Suppose that α intersects β. Then there exists a lift $\breve{\beta}$ of β forming a *D-domain* with $\tilde{\alpha}$, i.e. a simply connected domain $D \subset \mathbb{D}$ whose boundary consists of an arc on $\tilde{\alpha}$ and an arc on $\breve{\beta}$ as shown in Fig. 3.2 (D may or may not intersect lifts of $\gamma^1, \ldots, \gamma^{3g-3}$). To see that such a domain exists we must show that if $\breve{\beta}$ is a lift of β that intersects $\tilde{\alpha}$ in some point p, then $\breve{\beta}$ intersects $\tilde{\alpha}$ a second time. Two cases are possible. Either $\breve{\beta}$ has the same endpoints at infinity as $\tilde{\alpha}$, or the two endpoints at infinity of $\breve{\beta}$ are disjoint from them. In the first case, the above $T_{\tilde{\alpha}}$ satisfies $T_{\tilde{\alpha}}(\breve{\beta}) = \breve{\beta}$, and so $T_{\tilde{\alpha}}(p)$ is another intersection point. In the second case we argue as follows. There exists a lift $\breve{\alpha}$ of α with the same endpoints at infinity as $\breve{\beta}$. Since α is simple, $\breve{\alpha}$ does not intersect $\tilde{\alpha}$. Hence, the two endpoints at infinity of $\breve{\beta}$ lie on the same side of $\tilde{\alpha}$ and so $\breve{\beta}$ intersects $\tilde{\alpha}$ at least twice. This completes the existence proof of a D-domain.

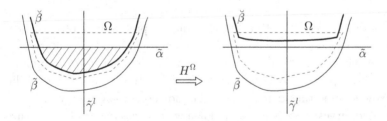

Fig. 3.2. An isotopy removing a D-shaped domain.

It is now easy to see that there exists a D-domain D which furthermore does not contain any parts of lifts of α and β in its interior. Let us thus take such a domain. Then there exists an isotopy H^Ω in $\mathbb{D}$ as shown in Fig. 3.2 that operates as the identity mapping outside an open neighborhood Ω of D and, when applied to $\breve{\beta}$, removes the two intersections of $\breve{\beta}$ and $\tilde{\alpha}$ in Ω. We may take Ω so small that the universal covering map $\pi : \mathbb{D} \to S$ is one-to-one on Ω. This isotopy projects to an isotopy of S that operates as the identity mapping outside $\pi(\Omega)$ and, when applied

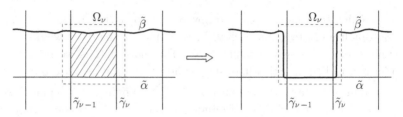

Fig. 3.3. Part of a homotopy in the universal covering.

to β, reduces the number of intersections of α and β by two. It is easy to see that this can be done such that moreover all γ^l remain invariant as sets. After finitely many such isotopies the intersections of α and β are removed.

The lift $\tilde{\beta}$ with the same endpoints at infinity as $\tilde{\alpha}$ now looks as in Fig. 3.1: for $\nu = 1, \ldots, n$, there is a simply connected domain D_ν whose boundary is formed by the arcs $\tilde{s}_\nu, \tilde{t}_\nu$ on $\tilde{\alpha}, \tilde{\beta}$ going from $\tilde{\gamma}_{\nu-1}$ to $\tilde{\gamma}_\nu$, and the arcs on $\tilde{\gamma}_{\nu-1}$, $\tilde{\gamma}_\nu$ going from $\tilde{\alpha}$ to $\tilde{\beta}$. These domains do not contain any lifts of parts of α and β in their interior, and so each D_ν is contained in some open subset Ω_ν on which the universal covering map π is one-to-one. We now perform isotopies on $\Omega_1, \ldots, \Omega_n$ as indicated in Fig. 3.3. This concludes the proof of the theorem. $\square$

Proof of Theorem 2.4 We continue using the preceding notations. Let α be a simple closed curve on S. If α is homotopic—up to orientation—to one of the $\gamma^1, \ldots, \gamma^{3g-3}$, say to γ^l, then the code polygon is $c_1^l c_2^l c_3^l c_4^l$ or $\bar{c}_4^l \bar{c}_3^l \bar{c}_2^l \bar{c}_1^l$. Up to cyclic permutation of the arcs this is the only possibility.

Now assume that α is ordinary, i.e. not homotopic to any of the parameter curves or their inverses. Then we may assume that α is reduced (e.g. by taking the closed geodesic in the free homotopy class). As a reduced curve, α is as a product of Y-strands $\alpha = s_1 s_2 \cdots s_n$, where for each $\nu = 1, \ldots, n$, strands s_ν and $s_{\nu+1}$ arrive at and leave from the corresponding curve γ^{l_ν} on different sides.

In the universal covering $\mathbb{D}$ the lift $\tilde{s}_1 \cdots \tilde{s}_n$ of $s_1 \cdots s_n$ (see the proof of Theorem 3.3) successively connects lifts $\tilde{\gamma}_0, \tilde{\gamma}_1, \ldots, \tilde{\gamma}_n$, with each other as shown in Fig. 3.4. We also look at the lift $\tilde{s}_{n+1}$ of s_1 that comes after $\tilde{s}_n$ going from $\tilde{\gamma}_n$ to the next lift $\tilde{\gamma}_{n+1}$ of γ^1.

Each s_ν is a strand on some Y^{k_ν} and, by Lemma 3.1, s_ν is Y-homotopic to a parameter strand, i.e. one of the arcs $u_i^{k_\nu}, f_i^{k_\nu}$, or their inverses, $i = 1, 2, 3$. Let us call this arc h_ν. The homotopy between s_ν and

h_ν lifts to a homotopy—with endpoints gliding on $\tilde{\gamma}_{\nu-1}$, respectively $\tilde{\gamma}_\nu$—between $\tilde{s}_\nu$ and a lift $\tilde{h}_\nu$ of h_ν that goes from $\tilde{\gamma}_{\nu-1}$ to $\tilde{\gamma}_\nu$.

Let us now consider the curve $\hat{\alpha}$ in $\mathbb{D}$ that goes along $\tilde{h}_1$ from $\tilde{\gamma}_0$ to $\tilde{\gamma}_1$, then along $\tilde{\gamma}_1$ from the endpoint of $\tilde{h}_1$ to the initial point of $\tilde{h}_2$, from there along $\tilde{h}_2$ to $\tilde{\gamma}_2$, and so on, until it arrives at the initial point of $\tilde{h}_{n+1}$ on $\tilde{\gamma}_n$, where $\tilde{h}_{n+1}$ is the lift of h_1 that goes from $\tilde{\gamma}_n$ to $\tilde{\gamma}_{n+1}$ (homotopic to $\tilde{s}_{n+1}$). The image $\varkappa = \pi \circ \hat{\alpha}$ under the universal covering map $\pi : \mathbb{D} \to S$ is a code polygon in the free homotopy class of α.

If $\varkappa'$ is an arbitrary code polygon in the homotopy class of α, then we may apply to it an arbitrarily small homotopy so as to obtain a reduced curve β. This β then is a product of strands $\beta = t_1 \cdots t_m$. By Theorem 3.3, we have $m = n$ and, possibly after cyclic permutation of the listing of the strands, each t_ν is Y-homotopic to s_ν, where the Y-homotopy is the restriction to t_ν of an isotopy between β and α. Lifting this isotopy to the universal covering we obtain homotopic lifts $\tilde{\alpha}, \tilde{\beta}$ together with products of arcs $\tilde{s}_1 \cdots \tilde{s}_n \tilde{s}_{n+1}$ on $\tilde{\alpha}$ and $\tilde{t}_1 \cdots \tilde{t}_n \tilde{t}_{n+1}$ on $\tilde{\beta}$, where for $\nu = 1, \ldots, n+1$, $\tilde{s}_\nu$ and $\tilde{t}_\nu$ are lifts of s_ν, t_ν that connect the *same* lifts $\tilde{\gamma}_{\nu-1}, \tilde{\gamma}_\nu$ of $\gamma^{l_{\nu-1}}$ and γ^{l_ν} with each other (for objects on S the ν-indices are modulo n).

Hence, the above construction applied to β yields exactly the same sequence of $\tilde{h}_\nu$-s and $\tilde{\gamma}_\nu$-s as for α (modulo cyclic permutation). It follows that $\varkappa' = \varkappa$. This completes the proof of the theorem. $\qquad\square$

For any simple closed curve α on S not homotopic to a point we let $\mathcal{C}(\alpha)$ be the code word of the code polygon as given by Theorem 2.4.

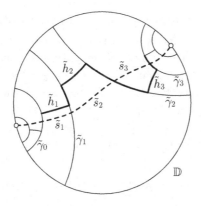

Fig. 3.4. Lift of a code polygon in the unit disk.

The theorem then says that simple curves have the same code if and only if they are homotopic. If α is an ordinary curve, i.e. if it is not homotopic to one of the γ^l or their inverses, then we may abbreviate the code in the form

$$\mathcal{C}(\alpha) = \langle h_1, \overrightarrow{c}^{l_1}, h_2, \overrightarrow{c}^{l_2}, \ldots, h_n, \overrightarrow{c}^{l_n} \rangle, \qquad (3.1)$$

where the h_ν are parameter strands (see Lemma 3.1) and each $\overrightarrow{c}^{l_\nu}$ is a sequence of letters from the list $c_1^{l_\nu}, c_2^{l_\nu}, c_3^{l_\nu}, c_4^{l_\nu}$ or from the list $\overleftarrow{c}_4^{l_\nu}, \overleftarrow{c}_3^{l_\nu}, \overleftarrow{c}_2^{l_\nu}, \overleftarrow{c}_1^{l_\nu}$.

Remark 3.4 *Theorem 2.4 is closely related to the Dehn-Thurston coding theorem, as detailed for example in Penner [Pe, section 1.2], where it is used for the study of train tracks. The Dehn-Thurston code for an ordinary simple closed curve α may be defined in terms of (3.1) as follows.*

Determine, for each $l = 1, \ldots, 3g - 3$, two integer numbers m_l and t_l called "crossing" and "twist" in the following way. The crossing m_l is the number of times $\overrightarrow{c}^l$ is listed in (3.1); this is the same as the number of times a reduced member in the homotopy class of α intersects γ^l. For the twist we let t_l' be the number of times letter c_4^l occurs in (3.1), t_l'' the number of times $\overleftarrow{c}_4^l$ occurs and set $t_l = t_l' - t_l''$. (A closer look shows that we always either have $t_l' = 0$ or $t_l'' = 0$.) The Dehn-Thurston code of α is the sequence $(m_1, \ldots, m_{3g-3}) \times (t_1, \ldots, t_{3g-3})$.

With Theorem 2.4 it becomes not too difficult an exercise to show that the homotopy class of α is uniquely determined by this sequence.

4 Codes and canonical generators

In this section we outline a way to adapt canonical generators of the fundamental group $\pi_1(S)$ of S to a given pasting scheme and to translate the code of a simple closed curve on S into a word in the generators.

We use the notations of Section 2. Consider again the surface $S^{2g-2} =: S'$ obtained after the first step of the pasting. The boundary curves of S' are grouped into pairs $\alpha^1, \tilde{\alpha}^1, \ldots, \alpha^g, \tilde{\alpha}^g$. The surface S is obtained by pasting together the two curves of each pair $\alpha^j, \tilde{\alpha}^j$ according to (2.3) which now reads

$$\gamma^{2g-3+j}(t) = \alpha^j(t) = \tilde{\alpha}^j(\tfrac{1}{2} - t), \quad t \in \mathbb{R}, \quad j = 1, \ldots, g. \qquad (4.1)$$

On S' we choose for $j = 1, \ldots, g$ a simple arc β^j from $\alpha^j(0)$ to $\tilde{\alpha}^j(\tfrac{1}{2})$. This can be done in such a way that the β^j are pairwise disjoint and such that the endpoints are the only points of β^j on the boundary of S'. Each β^j becomes a simple closed curve on S that intersects α^j in exactly one point. Moreover, the curves $\alpha^1, \beta^1, \ldots, \alpha^g, \beta^g$ form a canonical homology basis for S.

Now we choose a point p_0 in the interior of S' that shall serve as base point for $\pi_1(S)$ and draw connecting arcs r_j from p_0 to $\alpha^j(0)$. These must be simple, pairwise disjoint except for the common point p_0, and meet the boundary of S' only at their endpoints $\alpha^j(0)$. On S we then get the following loops at p_0, where the bar is again the notation for the inverse direction.

$$\alpha_j = r_j\,\alpha^j\,\bar{r}_j, \quad \beta_j = r_j\,\beta^j\,\bar{r}_j, \quad j = 1, \ldots, g. \tag{4.2}$$

As is well known the loops $\alpha_1, \ldots, \beta_g$ (more precisely their homotopy classes with base point p_0) form a system of canonical generators of $\pi_1(S)$.

In order to translate code words into words in these generators we consider the collection $\mathcal{P}$ of all initial and endpoints of the arcs u_i^k, f_i^k, c_j^l, used for the coding. For each $p \in \mathcal{P}$ we draw a connecting arc t_p from p_0 to p. For the t_p there are no topological restrictions. If now w is any of the coding arcs u_i^k, f_i^k, c_j^l, say with initial point p and endpoint q, then the following is a loop at p_0,

$$\tau(w) = t_p w \bar{t}_q. \tag{4.3}$$

We denote by $\tau(w)$ the corresponding member of $\pi_1(S)$. For any code polygon $\varkappa = w_1, \ldots, w_N$ and its corresponding code $\langle w_1, \ldots, w_N \rangle$ the product $\tau(w_1) \cdots \tau(w_N)$ lies in the free homotopy class of $\varkappa$. The corresponding product $\tau(w_1) \cdots \tau(w_N)$ is determined by $\varkappa$ only up to cyclic permutation, but words in $\pi_1(S)$ that differ by cyclic permutation are conjugate and so we have a well defined mapping $\mathcal{F}$ which associates to any code $\langle w_1, \ldots, w_N \rangle$ the following conjugacy class in $\pi_1(S)$

$$\mathcal{F}(\langle w_1, \ldots, w_N \rangle) = [\tau(w_1) \cdots \tau(w_N)]_{\pi_1(S)}. \tag{4.4}$$

A "worked out" example of such a translation in genus 2 is given in Section 9.

5 An algorithm for simple curves

In the next two sections we describe an algorithm that enumerates simple curves. This algorithm is adapted to a pasting scheme of S with parameter curves $\gamma^1, \ldots, \gamma^{3g-3}$ and has the following features:

(i) the construction of the curves is such that they are automatically pairwise non-homotopic;

(ii) all curves that differ from a given one by Dehn twists along $\gamma^1, \ldots, \gamma^{3g-3}$ are produced in a single step.

This overview shall become more detailed during this and the next section. The algorithm will, of course, only produce codes, but we shall describe it as an instruction for drawing curves.

Strands and connectors. If a reduced simple curve ω on S is clipped at all its intersection points with $\gamma^1 \cup \cdots \cup \gamma^{3g-3}$ so that it decays into Y-strands, then we may simultaneously apply Y-homotopies to the individual strands without allowing them to cross, so that in the end on each Y^k we have one of the four possible configurations shown in Fig. 5.1, where some of the "roads" may be empty.

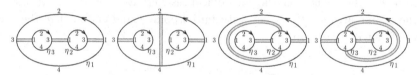

Fig. 5.1. Configurations of strands.

(Cases 2,3,4 are topologically equivalent but the algorithm will distinguish them.) All endpoints of the strands now lie in small neighborhoods of the points p_{ij}^k, called *zones*, and the strands do not form a connected curve anymore. The zone around p_{ij}^k will be called *zone j* and the strands in this new form will be said to be in *zone position*.

The construction of a curve unfolds in three stages. The first stage—carried out by the *core algorithm* of the next section—consists of drawing an ordered sequence of strands:

$$\sigma = \sigma_1, \ldots, \sigma_n, \tag{5.1}$$

where each σ_ν is a strand on some Y^{k_ν} in zone position. The strands are pairwise disjoint and the sequence is such that for $\nu = 1, \ldots, n$ the endpoint of strand σ_ν and the initial point of strand $\sigma_{\nu+1}$ lie on the same parameter curve $\gamma^{l\nu}$ (ν-indices mod n). For the algorithmic syntax of these strands we refer to the next section.

In the second stage a connecting arc along $\gamma^{l\nu}$ —called a *connector*— from the endpoint of σ_ν to the initial point of $\sigma_{\nu+1}$ is inserted in a prescribed way. This will produce a closed curve c homotopic to a simple curve. In the third stage the infinite family of all curves c' is printed out with a printout that contains variables that differ from c by Dehn twists (see below) along $\gamma^1, \ldots, \gamma^{3g-3}$.

Collars. To describe how the connectors are inserted it is useful to look at collars around the parameter curves γ^l. To this end we consider the Cartesian product

$$\Gamma = \mathbb{S}^1 \times [0,1] = \mathbb{R}/[\tau \mapsto \tau + 1] \times [0,1], \tag{5.2}$$

with the two boundary curves, say δ and δ', parametrized periodically with period 1 in the form

$$\delta(t) = (t, 0), \quad \delta'(t) = (\tfrac{1}{2} - t, 0), \quad t \in \mathbb{R}.$$

Now we take copies Γ^l of Γ with corresponding boundary curves δ^l, δ'^l parametrized in the same way, $l = 1, \ldots, 3g - 3$, and replace the pasting condition (2.3) by

$$\delta^l(t) = \eta_{i_l}^{k_l}(t), \quad \delta'^l(t) = \eta_{i'_l}^{k'_l}(t), \quad t \in \mathbb{R}, \quad l = 1, \ldots, 3g - 3. \tag{5.3}$$

This does not change the topology of S but gives us more room to draw the connectors.

Example 5.1 Fig. 5.2 shows the case of a surface of genus 2 with the following pasting scheme

$$\eta_3^1; \eta_2^1, \quad \eta_3^2; \eta_2^2, \quad \eta_1^1; \eta_1^2. \tag{5.4}$$

In the first row the figure shows the two pairs of pants Y^1, Y^2, in the second row we have the collars $\Gamma^1, \Gamma^2, \Gamma^3$.

The arrows of the collars are in the direction of the parametrizations of δ^l and δ'^l. The pasting of the collars to the pairs of pants is with *matching* arrows and such that the zones marked $1, 2, 3, 4$ on the boundary of the collars are pasted to the zones with the same labels on the boundary of the pairs of pants. For better readability we use the following notation in Fig. 5.2

$$\delta_1 = \tilde{\eta}_3^1, \quad \delta'_1 = \tilde{\eta}_2^1, \quad \delta_2 = \tilde{\eta}_3^2, \quad \delta'_2 = \tilde{\eta}_2^2, \quad \delta_3 = \tilde{\eta}_1^1, \quad \delta'_3 = \tilde{\eta}_1^2.$$

Thus, any boundary η_i^k of Y^k is identified with the boundary $\tilde{\eta}_i^k$ of one of the collars.

The connectors of the strands shown in the figure are drawn as pairwise disjoint simple arcs traversing the collars from one zone to the other.

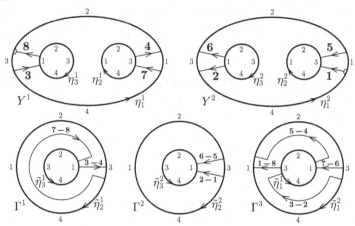

Fig. 5.2. Simple closed curve on a surface of genus 2.

Winding and block words. For any connector λ in Γ^l we define the *quarter winding number* or simply *winding* $\vartheta = \vartheta(\lambda)$ as follows. The absolute value $|\vartheta|$ is the number of quarters of the boundary of Γ^l that an observer will pass if he walks along λ from beginning to end. If λ follows the orientation of the initial boundary (the boundary containing the initial point) we define the sign of ϑ to be positive; if the connector follows the opposite direction the sign is negative. Note that if we would follow λ in its opposite direction, then the role of initial and endpoints would be reversed and the quarter winding number be the same. In Fig. 5.2, the illustrated connectors $2-3$ and $4-5$ have $\vartheta = 2$, for connector $7-8$ we have $\vartheta = 4$ and all remaining quarter winding numbers are zero.

To any curve $\alpha = \sigma_1 \lambda_1 \cdots \sigma_n \lambda_n$ obtained from the sequence (5.1) by inserting connectors λ_ν in Γ^{l_ν}, $\nu = 1, \ldots, n$, we associate a *block word* (cyclic)

$$\mathcal{B}(\alpha) = \langle (h_1, \vartheta_1), \ldots, (h_n, \vartheta_n) \rangle, \qquad (5.5)$$

where $\vartheta_\nu = \vartheta(\lambda_\nu)$ and h_ν is the parameter strand $h_\nu \in \{u_1^{k_\nu}, \ldots, \bar{f}_3^{k_\nu}\}$ in the Y-homotopy class of σ_ν on Y^{k_ν}, $\nu = 1, \ldots, n$.

The block word $\mathcal{B}(\alpha)$ can be translated into the code word $\mathcal{C}(\alpha)$ as follows (where we replace the collars Γ^l by the curves γ^l again): if $\vartheta_\nu > 0$, there is a unique sequence $\overrightarrow{c}^{\,l_\nu} = c^{l_\nu}_{i_\nu}, c^{l_\nu}_{i_\nu+1}, \ldots, c^{l_\nu}_{i_\nu+\vartheta_\nu-1}$ (i-indices modulo 4) of arcs on γ^{l_ν} as in (2.5) such that the corresponding product $c^{l_\nu}_{i_\nu} c^{l_\nu}_{i_\nu+1} \cdots c^{l_\nu}_{i_\nu+\vartheta_\nu-1}$ on γ^{l_ν} connects the endpoints of h_ν and $h_{\nu+1}$ with each other and has quarter winding ϑ_ν. If $\vartheta_\nu < 0$, there is a similar such sequence $\overleftarrow{c}^{\,l_\nu}, \overleftarrow{c}^{\,l_\nu}_{i_\nu-1}, \ldots, \overleftarrow{c}^{\,l_\nu}_{i_\nu-\vartheta_\nu+1}$ in the inverse direction. If $\vartheta_\nu = 0$ we let $\overrightarrow{c}^{\,l_\nu}$ be the empty sequence. With these sequences we define a mapping T in the following way.

$$B = \langle (h_1, \vartheta_1), \ldots (h_n, \vartheta_n) \rangle \longmapsto T(B) = \langle h_1, \overrightarrow{c}^{\,l_1}, \ldots, h_n, \overrightarrow{c}^{\,l_n} \rangle. \quad (5.6)$$

For $\alpha = \sigma_1 \lambda_1 \cdots \sigma_n \lambda_n$ this mapping satisfies

$$T(\mathcal{B}(\alpha)) = \mathcal{C}(\alpha). \quad (5.7)$$

Inserting the connectors. The strands in each output (5.1) of the core algorithm are such that there *exist* connectors $\lambda_1, \ldots, \lambda_n$ as above with the property that $\alpha = \sigma_1 \lambda_1 \cdots \sigma_n \lambda_n$ becomes a simple closed curve with code (5.7). We now describe a procedure that *selects* a sequence of inserted connectors in an automatic way. These will play the role of "default connectors" in the enumeration algorithm.

Let Γ^l be any of the collars that has endpoints of strands from the sequence (5.1) on its boundary and denote by $p_1, \ldots, p_m$, $m = m_l$, the initial and/or endpoints of the strands leaving from or arriving at δ^l. These points lie on the four zones on δ^l and we enumerate them in consecutive order following the orientation of δ. This enumeration is only determined up to cyclic permutation. To make it unique we adopt the following.

Convention 5.2 Point p_1 lies on the first zone of δ^l on which there are points of the sequence $p_1, \ldots, p_m$. Furthermore, p_1 is the first point on this zone (with respect to the orientation of δ^l).

On $\delta^{\prime l}$ there is a similar such sequence $q_1, \ldots, q_m$, but here we enumerate the points such that for each $\mu = 1, \ldots, m$, p_μ and q_μ belong to a pair of consecutive strands, i.e. p_μ (respectively, q_μ) is the endpoint of some strand σ_ν and q_μ (respectively, p_μ) is the initial point of the next strand $\sigma_{\nu+1}$. Since the pairs p_μ, q_μ can be joined by pairwise disjoint connectors, it follows that the enumeration of the q_μ thus defined is also in consecutive order, but *against* the orientation of $\delta^{\prime l}$.

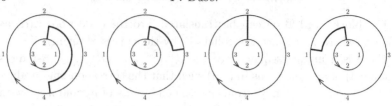

Fig. 5.3. Conventions for the first connector in a collar.

In order to define a selection of connectors in a standardized way we adopt the following.

Convention 5.3 The default connector λ_1 from p_1 to q_1 has quarter winding number $\vartheta \in \{-2, -1, 0, 1\}$.

Fig. 5.3 shows the four possibilities for the case where p_1 lies in zone 2. The thick line is λ_1. The winding $\vartheta_\mu = \vartheta^l_\mu$ of any other connector $c_\mu = c^l_\mu$ in collar Γ^l is now determined by a formula: let i, i' be the zones on $\delta^l, \delta^{\prime l}$ (in this order) that contain the endpoints of λ_1 and j, j' the zones on $\delta^l, \delta^{\prime l}$ (also in this order) that contain the endpoints of c_μ. Furthermore, set $s = 1$ in the special case where q_1 and q_μ lie in the same zone (i.e. $i' = j'$) and q_μ comes later than q_1 with respect to the orientation of $\delta^{\prime l}$; set $s = 0$ in all other cases. Then the formula is as follows, where $\mathrm{mod}_4(k)$ means k modulo 4 with values in $\{1, 2, 3, 4\}$,

$$\vartheta^l_\mu = 2 + \mathrm{mod}_4(3 - i' - i) - \mathrm{mod}_4(j' - i') - \mathrm{mod}_4(j - i + 1) + 4s. \quad (5.8)$$

The formula may be checked by drawing the various cases into Fig. 5.3.

Conventions 5.2, 5.3, together with formula (5.8) define, for any sequence of strands $\sigma = \sigma_1, \ldots, \sigma_n$ as in (5.1) a sequence of numbers $\vartheta_1, \ldots, \vartheta_n$ in the following way: if p_{μ_ν}, q_{μ_ν} are the endpoints of $\sigma_\nu, \sigma_{\nu+1}$ on the boundary of Γ^{l_ν}, then $\vartheta_\nu = \vartheta^{l_\nu}_{\mu_\nu}$. We denote this sequence as follows

$$(\vartheta_1, \ldots, \vartheta_n) = \vartheta(\sigma_1, \ldots, \sigma_n). \quad (5.9)$$

Inserting the connectors $\lambda^{l_\nu}_{\mu_\nu}$ with these quarter winding numbers yields a simple closed curve on S. We denote this curve by $\alpha(\sigma_1, \ldots, \sigma_n)$.

Dehn-twists. For any $\kappa \in \mathbb{Z}$ there is a homeomorphism $D^\kappa : \Gamma \to \Gamma$ (see (5.2)) defined as follows,

$$D^\kappa(t, s) = (t + \kappa s), \quad t \in \mathbb{S}^1 = \mathbb{R}/[\tau \mapsto \tau + 1], \quad s \in [0, 1]. \quad (5.10)$$

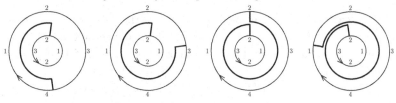

Fig. 5.4. Dehn-twist D applied to the connectors of Fig. 5.3.

This homeomorphism fixes all boundary points of Γ. Fig. 5.4 shows how it affects the default connectors of Fig. 5.3. The quarter winding numbers are now respectively, $\vartheta = 2, 3, 4, 5$.

Taking, for any $l = 1, \ldots, 3g - 3$, the natural homeomorphism $\phi_l : \Gamma \to \Gamma^l$ that identifies Γ with its copy Γ^l in S we get the corresponding homeomorphism $D_l^\kappa = \phi_l \circ D^\kappa \circ \phi_l^{-1} : \Gamma^l \to \Gamma^l$. Since D_l^κ fixes the boundary points of Γ^l we may extend it to a homeomorphism $S \to S$ defining it to be the identity mapping outside Γ^l. The mapping D_l^κ, more precisely any homeomorphism isotopic to it, is called a *Dehn-twist* along γ^l (for properties of Dehn-twists we refer the reader e.g. to [Bi, sect. 2] or [St, sect. 6.3.2]).

Applying D_l^κ to the above curve $\alpha = \boldsymbol{\alpha}(\sigma_1, \ldots, \sigma_n)$ we get a new simple closed curve α' on S in which all of the above connectors λ_μ^l in Γ^l are replaced by connectors with the same endpoints whose quarter winding numbers are $\vartheta_\mu^l + 4\kappa$.

More generally, we may take any sequence of integers $(\kappa_1, \ldots, \kappa_{3g-3}) \in \mathbb{Z}^{3g-3}$ and apply the product $D_1^{\kappa_1} \circ \cdots \circ D_{3g-3}^{\kappa_{3g-3}}$ to α. For $\nu = 1, \ldots, n$, the connector on Γ^{l_ν} between strands σ_ν and $\sigma_{\nu+1}$ then has quarter winding $\vartheta_\nu + 4\kappa_{l_\nu}$. For the curve thus obtained the block word becomes

$$\mathcal{B}(D_1^{\kappa_1} \circ \cdots \circ D_{3g-3}^{\kappa_{3g-3}}(\boldsymbol{\alpha}(\sigma_1, \ldots, \sigma_n)))$$
$$= \langle (h_1, \boldsymbol{\vartheta}_1 + 4\kappa_{l_1}), \ldots, (h_n, \boldsymbol{\vartheta}_n + 4\kappa_{l_n}) \rangle. \quad (5.11)$$

The following ensures feature (ii) of the enumeration algorithm.

Proposition 5.4 *Let $\sigma_1, \ldots, \sigma_n$ be a sequence of strands as in (5.1) and $\alpha = \sigma_1 \lambda_1 \cdots \sigma_n \lambda_n$ with block word $\mathcal{B}(\alpha) = \langle (h_1, \vartheta_1), \ldots, (h_n, \vartheta_n) \rangle$ a simple closed curve obtained from it by inserting connectors. Then there exists $(\kappa_1, \ldots, \kappa_{3g-3}) \in \mathbb{Z}^{3g-3}$ such that, up to free homotopy,*

$$\alpha = D_1^{\kappa_1} \circ \cdots \circ D_{3g-3}^{\kappa_{3g-3}}(\boldsymbol{\alpha}(\sigma_1, \ldots, \sigma_n))$$

and $\vartheta_\nu = \boldsymbol{\vartheta}_\nu + 4\kappa_{l_\nu}$, $\nu = 1, \ldots, n$.

Proof In each collar Γ^l traversed by α there is the default connector λ_1 from p_1 to q_1 defined by Convention 5.3. The connector from p_1 to q_1 belonging to α differs from it by a quarter twist amount of $4\kappa_l$ for some integer κ_l. If Γ^l is not traversed by α we set $\kappa_l = 0$. With $\kappa_1, \ldots, \kappa_{3g-3}$ thus defined we set $D = D_1^{\kappa_1} \circ \cdots \circ D_{3g-3}^{\kappa_{3g-3}}$. In each Γ^l traversed by α the connector of $D^{-1}(\alpha)$ from p_1 to q_1 is homotopic with fixed endpoints to λ_1, i.e. it satisfies Convention 5.3. It follows that the quarter winding numbers of all connectors of $D^{-1}(\alpha)$ in Γ^l are given by (5.8). Therefore, $\alpha(\sigma_1, \ldots, \sigma_n)$ and $D^{-1}(\alpha)$ have the same block word and, hence, the same code word. By Theorem 2.4 they are homotopic and so α and $D(\alpha(\sigma_1, \ldots, \sigma_n))$ are homotopic. This proves the first statement as in Proposition 5.4; the second statement is obvious. $\qquad\square$

Extended homotopy classes and the families $\mathcal{F}_{ij}^k$. Two closed curves α, β are called *extended homotopic* if α is either homotopic to β or to $\bar{\beta}$. The extended homotopy class of a curve α is denoted by $[\alpha]$. For the enumeration algorithm we group these classes into the following families.

Family $\mathcal{F}_0$ consists of the extended homotopy classes of the parameter curves $\gamma^1, \ldots, \gamma^{3g-3}$. To define the remaining families we introduce a lexicographic order on the zones: let z_{ij}^k denote zone j on boundary η_i^k of pair of pants Y^k and write $(k', i', j') \prec (k, i, j)$ if either $k' < k$, or $k' = k$ and $i' < i$, or $k' = k$ and $i' = i$ and $j' < j$. Family $\mathcal{F}_{ij}^k$ then consists of the classes $[\alpha]$ of curves whose code polygon $\mathcal{C}(\alpha)$ has at least one strand in Y^k with endpoints on zone z_{ij}^k, but no strands with endpoints on zone $z_{i'j'}^{k'}$ on $Y^{k'}$ for any $(k', i', j') \prec (k, i, j)$. Many of these families are, of course, empty, but for ease of description we leave it this way.

The algorithm. The following algorithm takes as input a positive integer M. It uses the core algorithm described in the next section. The output is a finite sequence of infinite families of block words.

Algorithm 5.5 For $k = 1, \ldots, 2g - 2$, $i = 1, 2, 3$ and $j = 1, 2, 3, 4$, do the following:

- close the gates on all $z_{i'j'}^{k'}$ with $(k', i', j') \prec (k, i, j)$;
- run $\mathrm{core}(k, i, j)[M]$ to get list $\boldsymbol{\sigma}^1, \ldots, \boldsymbol{\sigma}^R$;
- for $r = 1, \ldots, R$:

  - set $\boldsymbol{\sigma} = \boldsymbol{\sigma}^r$ to get list $\sigma_1, \ldots, \sigma_n$;
  - evaluate $\boldsymbol{\vartheta}(\sigma_1, \ldots, \sigma_n)$ to get the list $\boldsymbol{\vartheta}_1, \ldots, \boldsymbol{\vartheta}_n$;

- print "$\langle (h_1, \boldsymbol{\vartheta}_1 + 4x_{l_1}), \dots, (h_n, \boldsymbol{\vartheta}_n + 4x_{l_n}) \rangle$".

In this algorithm the instruction "close the gates" attributes the value 'closed' to all zones $z_{i'j'}^{k'}$ on the boundaries of $Y^1, \dots, Y^{2g-2}$ for which $(k', i', j') \prec (k, i, j)$; all other zones get the attribute 'open'. The core algorithm will not set any strands with endpoints on closed zones. Each σ^r in line 4 is as in (5.1). The function $\boldsymbol{\vartheta}$ on line 5 is as in (5.9). The h_ν on line 6 are the parameter strands Y-homotopic to σ_ν; the sequence $l_1, \dots, l_n$ is provided by the core algorithm; the symbols $x_1, \dots, x_{3g-3}$ are integer variables.

Let us now state the main properties. By Theorem 2.4, any ordinary simple closed curve α on S is homotopic to a uniquely determined code polygon $\mathcal{P}(\alpha)$ with code word

$$\mathcal{C}(\alpha) = \langle w_1, w_2, \dots, w_N \rangle = \langle h_1, \overrightarrow{c}^{\,l_1}, h_2, \overrightarrow{c}^{\,l_2}, \dots, h_n, \overrightarrow{c}^{\,l_n} \rangle,$$

or, equivalently, with block word

$$\mathcal{B}(\alpha) = T^{-1}(\mathcal{C}(\alpha)) = \langle (h_1, \vartheta_1), \dots, (h_n, \vartheta_n) \rangle,$$

where, for $\nu = 1, \dots, n$, ϑ_ν is the quarter winding number of the connector λ_ν on γ^ν defined by the sequence $\overrightarrow{c}^{\,l_\nu}$ (see (3.1), (5.5)–(5.7)). We call N the *code length* of α and n the *strand length* of α. The code length is denoted by $L_{\mathcal{C}}(\alpha)$.

For each $\mathcal{F}_{ij}^k$ and for any positive integer M, the core algorithm in the next section will enumerate the sequences $\sigma_1, \dots, \sigma_n$ with $n \le M$, for which the corresponding sequence of parameter strands $h_1, \dots, h_n$ occurs in the block word $\mathcal{B}(\alpha)$ of some α with $[\alpha] \in \mathcal{F}_{ij}^k$, $[\alpha]$ being the extended homotopy class of α. Moreover, the enumeration will be such that any of the above cyclic words $\langle (h_1, \boldsymbol{\vartheta}_1 + 4x_{l_1}), \dots, (h_n, \boldsymbol{\vartheta}_n + 4x_{l_n}) \rangle$ is printed out only once.

We now make the following definition. We say that two ordinary simple closed curves α, β on S are *twist-equivalent* (with respect to the given pasting scheme) if there exists a product of Dehn-twists along $\gamma^1, \dots, \gamma^{3g-3}$, $D = D_1^{k_1} \circ \cdots \circ D_{3g-3}^{k_{3g-3}}$, such that β and $D(\alpha)$ are extended-homotopy equivalent. By Proposition 5.4, α and β are twist-equivalent if and only if they have the same sequence of parameter strands $h_1, \dots, h_n$. Together with the above we have therefore the following result.

Proposition 5.6 *Given any positive integer M as input, Algorithm 5.5 prints out exactly once each twist-equivalence class of ordinary simple closed curves with strand length $\le M$.* $\square$

In this proposition, the printout of the equivalence class is given as a word in the variables $x_1, \ldots, x_{3g-3}$. To obtain explicit examples from it one has to substitute each x_l by some integer κ_l so as to obtain a block word $B = \langle (h_1, \vartheta_1 + 4\kappa_{l_1}), \ldots, (h_n, \vartheta_n + 4\kappa_{l_n}) \rangle$ and then build the code polygon $\varkappa = T(B)$ as in (5.6). For the code length of the latter we note the following.

Proposition 5.7 *Let* $B = \langle (h_1, \vartheta_1 + 4\kappa_{l_1}), \ldots, (h_n, \vartheta_n + 4\kappa_{l_n}) \rangle$ *and let* α *be an ordinary simple closed curve in the homotopy class of* $T(B)$. *For* $l = 1, \ldots, 3g - 3$, *denote by* m_l *the number of intersections of* α *with* γ^l. *Then the code length of* α *satisfies*

$$n + \sum_{l=1}^{3g-3} m_l \cdot \max\{(4|\kappa_l| - 5), 0\} \leq L_{\mathbb{C}}(\alpha) \leq n + \sum_{l=1}^{3g-3} m_l(4|\kappa_l| + 5).$$

Proof By Theorem 3.3, m_l is the number of times index l occurs in the sequence $l_1, \ldots, l_n$ in B. By Convention 5.3, the so-called first connector in each collar Γ^l defined by Algorithm 5.5 has quarter winding number in $\{-2, -1, 0, 1\}$. The possible shapes are shown in Fig. 5.3. A glance at this figure and recalling Convention 5.2 yields the following bounds for the quarter winding number of the default connector λ_ν,

$$-5 \leq \vartheta_\nu \leq 5.$$

The proposition follows.   □

Using Proposition 5.7 one may use the algorithm in an obvious way to enumerate all code words of a given code length N.

6 The core algorithm

The core algorithm takes as input a triple (k, i, j), with $k = 1, \ldots, 2g-2$, $i = 1, 2, 3$, $j = 1, 2, 3, 4$, and a positive integer M. The output is the list Σ of all sequences $\sigma = \sigma_1, \ldots, \sigma_n$ of strands in zone position as in (5.1), with $n \leq M$, where at least one σ_ν has an endpoint in zone z_{ij}^k of Y^k, while no strand has endpoints in the zones $z_{i'j'}^{k'}$, with $(k', i', j') \prec (k, i, j)$.

Syntax for strands. In the language of the algorithm a strand σ_ν contained in Y^{k_ν} going from boundary curve $\eta_{i_\nu}^{k_\nu}$ to boundary curve $\eta_{i_\nu^*}^{k_\nu}$ is written as a sequence

$$\sigma_\nu = (h_\nu, s_\leftarrow(\nu, \varepsilon), z_\leftarrow(\nu, \varepsilon), s_\rightarrow(\nu, \varepsilon), z_\rightarrow(\nu, \varepsilon)), \tag{6.1}$$

where h_ν is the name of the parameter strand in the Y-homotopy class of σ_ν and the four functions are as follows, where the variable ε varies through the two values 'ini' $=$ initial point of σ_ν and 'end' $=$ endpoint of σ_ν. Function $s_\leftarrow(\nu, \varepsilon)$ is the label of the "next strand against the arrow", i.e. the first strand an observer on the boundary of Y^{k_ν} will meet when, starting at ε, he walks *against* the orientation of the boundary. Function $z_\leftarrow(\nu, \varepsilon)$ is the zone where this meeting occurs. Similarly, $s_\rightarrow(\nu, \varepsilon)$ is the "next strand in the direction of the arrow" and $z_\rightarrow(\nu, \varepsilon)$ is the corresponding zone.

In the example in the center of Fig. 6.1, e.g. we have $s_\leftarrow(n, \text{ini}) = \nu$, $z_\rightarrow(n, \text{ini}) = 3$, $z_\rightarrow(n, \text{end}) = 2$, etc.

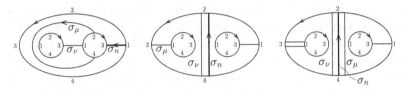

Fig. 6.1. Relative positions of strands.

Convention for the first strand. In order to avoid that different outputs of the core algorithm lead to the same extended homotopy class of closed curves, we adopt the convention that σ_1 lies in Y^k with initial point in zone z_{ij}^k (k, i, j as in the input) and that σ_1 is the first strand in zone z_{ij}^k, i.e. the first strand met by an observer walking along the boundary curve η_i^k in the direction of the arrow and arriving at zone z_{ij}^k.

Level functions. While the core algorithm is running there will be preliminary strands and data at all levels. The following functions will provide the data at level n: for $\nu = 1, \ldots, n$, the ν-th strand at level n is the sequence

$$\sigma_\nu[n] = (h_\nu, s_\leftarrow(\nu, \varepsilon)[n], z_\leftarrow(\nu, \varepsilon)[n], s_\rightarrow(\nu, \varepsilon)[n], z_\rightarrow(\nu, \varepsilon)[n]);$$

– $\boldsymbol{\sigma}[n]$ is the sequence $\sigma_1[n], \ldots, \sigma_n[n]$;
– first$(k, i, j)[n]$ is the label of the first strand at zone z_{ij}^k, i.e. the first endpoint of a strand in zone z_{ij}^k met by an observer walking along η_i^k in the direction of the arrow; if no endpoints are in this zone the value is defined to be 0;

– last$(k, i, j)[n]$ is the label of the last strand at zone z_{ij}^k, respectively 0 if at level n there are no strands at this zone;

– gate$(k, i, j)[n]$ is one of the two values 'open' or 'closed'.

In addition to these, there are three data functions for the bifurcations of the algorithm:

– zones(b) is the list of allowable zones for the initial point of the next strand to be chosen at bifurcation step b;

– currentzone(b) is the currently chosen zone from zones(b);

– $n_c(b)$ is the label of the currently chosen strand at bifurcation step b.

Finally, there are the following variables whose values depend on the momentary status of the running algorithm: 'bif' assumes integer values and is used to indicate the momentary bifurcation level; 'entries' is used to memorize a list; 'returnmodus', 'bifurcate' and 'append' are Boolean variables with the possible values 'true' and 'false'.

The core algorithm is as follows, where the individual instructions will be explained below.

Algorithm 6.1 (Core algorithm)

- Initialize;
- while bif > 0, do the following:
  - if returnmodus = 'true', then choose next zone;
  - set returnmodus = 'false';
  - upgrade strand data;
  - upgrade gates;
  - evaluate next entry;
  - if entry includes the starting zone, then append $\sigma[n]$ to Σ;
  - set $n = n + 1$;
  - if $n < M + 1$ and bifurcate = 'true', then install new bifurcation;
  - if $n = M + 1$, then set returnmodus = 'true' and return to previous open bifurcation;
  - end while;
  - end.

We now detail these instructions. For simplicity the strands are denoted again by σ_ν rather than $\sigma_\nu[n]$.

Initialize. In this step, the various functions are set to the following initial values, where $k = 1, \ldots, 2g - 2$, $i = 1, 2, 3$, $j = 1, 2, 3, 4$:

$\Sigma = \{\,\}$, the empty set;

$\text{gate}(k, i, j)[0] =$ 'closed' if $(k, i, j) \prec (\boldsymbol{k}, \boldsymbol{i}, \boldsymbol{j})$, 'open' otherwise;

$\text{first}(k, i, j)[0] = \text{last}(k, i, j)[0] = 0$;

$n = 1$;

$\text{bif} = 1$; $n_c(1) = 1$;

$\text{zones}(1) = \{\boldsymbol{j}\}$;

$\text{currentzone}(1) = \boldsymbol{j}$;

$\text{returnmodus} =$ 'false'.

When the initialization is accomplished the status of the return modus is returnmodus = 'false' and the algorithm advances to the instruction 'upgrade strand data'.

Upgrade strand data. When this instruction is called, the running variable, n, is interpreted as the label of the *new* strand (currently being "drawn") and all functions[n] have to be defined for this new value of n. The definitions are straightforward; we restrict ourselves to sketching the main points.

The zone $z_{i_n j_n}^{k_n}$ of the initial point of σ_n is available from the preceding steps, and h_n is now set to be the parameter strand on Y^{k_n} whose initial point is $p_{i_n j_n}^{k_n}$. For the data functions we distinguish two cases.

Case 1: there are no endpoints of $\sigma_1, \ldots, \sigma_{n-1}$ in zone $z_{i_n j_n}^{k_n}$. Then there are also no endpoints in the zone $z_{i_n^* j_n^*}^{k_n}$ of the endpoint of σ_n and the algorithm sets $\text{first}(k_n, i_n, j_n)[n] = \text{last}(k_n, i_n, j_n)[n] = n$ and $\text{first}(k_n, i_n^*, j_n^*)[n] = \text{last}(k_n, i_n^*, j_n^*)[n] = n$. For all other (k, i, j) the functions 'first' and 'last' have the same values as for $n - 1$.

For the values of $f(\nu, \varepsilon)[n]$, for $f = s_{\leftarrow}, z_{\leftarrow}, s_{\rightarrow}, z_{\rightarrow}$, the algorithm first makes the preliminary definition $f(\nu, \varepsilon)[n] = f(\nu, \varepsilon)[n - 1]$, $\nu = 1, \ldots, n - 1$, and then scans the zones $z_{i_n j}^{k_n}$ and $z_{i_n^* j}^{k_n}$, for $j = 1, 2, 3, 4$, to add the necessary corrections and to determine the values of $f(n, \varepsilon)[n]$. By means of example, in the case of the first diagram in Fig. 6.1 this is as follows. On the boundary labelled i_n^*, one has $\text{last}(k_n, i_n^*, 2) = \mu$, $\text{first}(k_n, i_n^*, 4) = \mu$, and the corrections are

$$s_{\leftarrow}(\mu, \text{end})[n] = n, \quad z_{\leftarrow}(\mu, \text{end})[n] = z_{i_n^* 3}^{k_n},$$
$$s_{\rightarrow}(\mu, \text{ini})[n] = n, \quad z_{\rightarrow}(\mu, \text{ini})[n] = z_{i_n^* 3}^{k_n}.$$

The new values are

$$s_{\leftarrow}(n, \text{ini})[n] = n, \qquad z_{\leftarrow}(n, \text{ini})[n] = z_{i_n 1}^{k_n},$$
$$s_{\rightarrow}(n, \text{ini})[n] = n, \qquad z_{\rightarrow}(n, \text{ini})[n] = z_{i_n 1}^{k_n}.$$
$$s_{\leftarrow}(n, \text{end})[n] = \mu, \qquad z_{\leftarrow}(n, \text{end})[n] = z_{i_n^* 2}^{k_n},$$
$$s_{\rightarrow}(n, \text{end})[n] = \mu, \qquad z_{\rightarrow}(n, \text{end})[n] = z_{i_n^* 4}^{k_n}.$$

Case 2: there are endpoints of $\sigma_1, \ldots, \sigma_{n-1}$ in zone $z_{i_n j_n}^{k_n}$. In this case, the values of $s_{\leftarrow}(n, \text{ini})[n]$, $z_{\leftarrow}(n, \text{ini})[n]$, $s_{\rightarrow}(n, \text{ini})[n]$, $z_{\rightarrow}(n, \text{ini})[n]$ are already available from previous steps (see the description of 'evaluate next entry'). The remaining data functions are determined in a similar way as in Case 1. By means of example, if $z_{\leftarrow}(n, \text{ini})[n] = z_{i_n j_n}^{k_n}$ and $s_{\leftarrow}(n, \text{ini})[n] = \nu$ (as in the second and the third diagram in Fig. 6.1), then $s_{\rightarrow}(n, \text{end})[n] = \nu$, $z_{\rightarrow}(n, \text{end})[n] = z_{i_n^* j^*}^{k_n}$, $\text{first}(k_n, i_n, j_n)[n] = \text{first}(k_n, i_n, j_n)[n-1]$, $\text{last}(k_n, i_n^*, j_n^*)[n] = \text{last}(k_n, i_n^*, j_n^*)[n-1]$, etc.

Upgrade gates. When the preceding instruction is accomplished the algorithm performs a preliminary upgrade of the gates setting

$$\text{gate}(k, i, j)[n] = \text{gate}(k, i, j)[n-1],$$

for $k = 1, \ldots, 2g - 2$, $i = 1, 2, 3$, $j = 1, 2, 3, 4$. Now assume that the case for the preceding instruction was Case 1. Then σ_n must be prevented from being crossed at later steps and so the following adjustments are made, where k_n, i_n, j_n are the labels of the zone $z_{i_n j_n}^{k_n}$ of the initial point of σ_n:

if $j_n = 1, 3$, then $\text{gate}(k_n, \text{mod}_3(i_n + j_n - 2), 2)[n] =$
$$\text{gate}(k_n, \text{mod}_3(i_n + j_n - 2), 4)[n] = \text{'closed'},$$

if $j_n = 2, 4$, then $\text{gate}(k_n, \text{mod}_3(i_n + 1), j')[n] =$
$$\text{gate}(k_n, \text{mod}_3(i_n + 2), j'')[n] = \text{'closed'},$$

for $j' = 1, 2, 4$, $j'' = 2, 3, 4$.

Evaluate next entry. When the upgrades for σ_n are accomplished the algorithm advances to the determination of the allowable zones for the initial point of σ_{n+1}.

Recall the boundary curve $\eta_{i_n^*}^{k_n}$ of Y^{k_n} containing the endpoint of σ_n and let $Y^{k_{n+1}}$ with boundary curve $\eta_{i_{n+1}}^{k_{n+1}}$ be the adjacent pair of pants, i.e. such that Y^{k_n} and $Y^{k_{n+1}}$ are pasted together along $\eta_{i_n^*}^{k_n}$ and $\eta_{i_{n+1}}^{k_{n+1}}$.

Then σ_{n+1} has to be placed on $Y^{k_{n+1}}$ with initial point on $\eta_{i_{n+1}}^{k_{n+1}}$. There are two cases.

Case 1: except possibly for the initial point of σ_1, there are no endpoints of $\sigma_1, \ldots, \sigma_n$ on $\eta_{i_{n+1}}$.

In this case the algorithm sets bifurcate = 'true'; it sets append = 'true' if $(k_{n+1}, i_{n+1}) = (\boldsymbol{k}, \boldsymbol{i})$ (i.e. if σ_1 lies on $Y^{k_{n+1}}$ with initial point on $\eta_{i_{n+1}}$) and append = 'false' otherwise. The variable 'entries' is set to entries = $\{1, 2, 3, 4\}$. The interpretation of the latter is that in subsequent steps any of the non-closed zones in the list will eventually serve as the initial point of σ_{n+1} (unless $n+1$ exceeds the maximal value M).

Case 2: the complement of Case 1.

In order to explain how the list of entries is determined in this case we first look at an example.

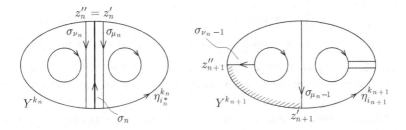

Fig. 6.2. The initial point of σ_{n+1} is confined to the shaded area.

Fig. 6.2 shows the two pairs of pants $Y^{k_n}, Y^{k_{n+1}}$ together with the boundary curves $\eta_{i_n}^{k_n}, \eta_{i_{n+1}}^{k_{n+1}}$ along which they are pasted together in S. At its exit on Y^{k_n} strand σ_n is "bracketed" by the strands with labels

$$\mu_n = s_{\leftarrow}(n, \text{end})[n] \quad \text{and} \quad \nu_n = s_{\rightarrow}(n, \text{end})[n].$$

These strands are followed by $\sigma_{\mu_n-1}, \sigma_{\nu_n-1}$ on $Y^{k_{n+1}}$. If we wish the strands to be connectable into a simple curve on S, then σ_{n+1} must be bracketed by $\sigma_{\nu_n-1}, \sigma_{\mu_n-1}$ on its entry to $Y^{k_{n+1}}$; more precisely, we must have

$$\mu_n - 1 = s_{\rightarrow}(n+1, \text{ini})[n+1] \quad \text{and} \quad \nu_n - 1 = s_{\leftarrow}(n+1, \text{ini})[n+1].$$

This confines the possible entry positions to the shaded area shown in Fig. 6.2.

Let us return to the general case. In order to deal with the fact that σ_1 has no predecessor strand we make the following definition. If $s_{\leftarrow}(n, \text{end})[n] = 1$ and $z_{\leftarrow}(n, \text{end})[n] = \boldsymbol{j}$ (the zone of the initial point of σ_1), then we set $s_\Leftarrow(n,\text{end})[n] = s_\leftarrow(1,\text{ini})[n]$ and $z_\Leftarrow(n,\text{end})[n] = z_\leftarrow(1,\text{ini})[n]$; in all other cases we set $s_\Leftarrow(n,\text{end})[n] = s_\leftarrow(n,\text{end})[n]$ and $z_\Leftarrow(n,\text{end})[n] = z_\leftarrow(n,\text{end})[n]$. We make the same definitions with $\leftarrow, \Leftarrow$ replaced by $\rightarrow, \Rightarrow$. With these notations we set

$$\mu_n = s_\Leftarrow(n,\text{end})[n], \qquad\qquad \nu_n = s_\Rightarrow(n,\text{end})[n],$$
$$z'_n = z^{k_n}_{i_n^* j'_n} = z_\Leftarrow(n,\text{end})[n], \quad z''_n = z^{k_n}_{i_n^* j''_n} = z_\Rightarrow(n,\text{end})[n].$$

Strand σ_{μ_n} has a follower (successor or predecessor) $\sigma_{\mu_{n+1}}$ on $Y^{k_{n+1}}$ at some zone $z'_{n+1} = z^{k_{n+1}}_{i_{n+1} j'_{n+1}}$ given as follows: if z'_n is the zone of the endpoint of σ_{μ_n}, then $\mu_{n+1} = \mu_n + 1$ and z'_{n+1} is the zone of the initial point of $\sigma_{\mu_{n+1}}$; otherwise $\mu_{n+1} = \mu_n - 1$ and z'_{n+1} is the zone of the endpoint of $\sigma_{\mu_{n+1}}$. In the same way σ_{ν_n} has a follower $\sigma_{\nu_{n+1}}$ at some zone $z''_{n+1} = z^{k_{n+1}}_{i_{n+1} j''_{n+1}}$. We write

$$\mu_{n+1} = s_\Rightarrow(n+1,\text{ini})[n+1], \quad \nu_{n+1} = s_\Leftarrow(n+1,\text{ini})[n+1],$$
$$z'_{n+1} = z_\Rightarrow(n+1,\text{ini})[n+1], \quad z''_{n+1} = z_\Leftarrow(n+1,\text{ini})[n+1].$$

(The different directions of the arrows for n and $n+1$ are in accordance with the pasting rule (2.3), where pairs of pants are pasted together with opposite arrows.)

There are two possible cases for σ_{n+1}, shown schematically in Fig. 6.3.

a) $\text{last}(k_{n+1}, i_{n+1}, j''_{n+1})[n] = \nu_{n+1}$;
b) $\text{last}(k_{n+1}, i_{n+1}, j''_{n+1})[n] \neq \nu_{n+1}$ and $z''_{n+1} = z'_{n+1}$.

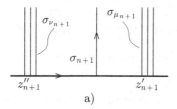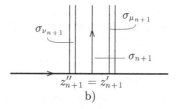

Fig. 6.3. In a) the algorithm bifurcates, in b) it does not.

The algorithm sets bifurcate $=$ 'true' in a), and 'false' in b). If $(k_{n+1}, i_{n+1}) = (\boldsymbol{k}, \boldsymbol{i})$ and if, furthermore, $s_\leftarrow(1,\text{ini})[n] = \nu_{n+1}$ and $s_\rightarrow(1,\text{ini})[n] = \mu_{n+1}$, then strands $\sigma_1, \ldots, \sigma_n$ are connectable into a simple closed curve and the algorithm sets append $=$ 'true', otherwise it sets append $=$ 'false'. In case a) the variable 'entries' is set to

$$\text{entries} = \{j^0, \ldots, j^m\},$$

where $j^l = \mathrm{mod}_4(j''_{n+1} + l)$, and m is such that for $l = 1, \ldots, m-1$ one has $j^l \neq j''_{n+1}, j'_{n+1}$. In case b), the variable 'entries' keeps its current value.

Finally, in Case 1 as well as Case 2, for $b = \mathrm{bif} + 1$ the sequence zones(b) is defined to be the sequence 'entries' from which all j^l with gate(k_{n+1}, i_{n+1}, j^l) = 'closed' have been removed.

Append $\boldsymbol{\sigma}[n]$ to Σ. The current sequence $\boldsymbol{\sigma}[n] = \{\sigma_1[n], \ldots, \sigma_n[n]\}$ is copied to the list Σ.

Install next bifurcation. The variable 'bif' is set to bif = bif + 1. Then, for $b = \mathrm{bif}$, the algorithm sets

$$n_c(b) = n, \quad \mathrm{currentzone}(b) = \text{first member of zones}(b).$$

Return to previous open bifurcation. This is a loop. Beginning with $b = \mathrm{bif}$ it is tested whether or not currentzone(b) is the last member of zones(b). If the result is positive then the variable b is set to $b = b-1$ and the test is repeated. This is continued until either the test is negative for some $b > 0$, or $b = 0$. When the test is negative for some $b > 0$ the procedure stops, the value of 'returnmodus' is returnmodus = 'true' and the algorithm proceeds to 'choose next zone'.

When $b = 0$ is reached the core algorithm exits the while loop and stops altogether. At this moment the writing of the sequence Σ is accomplished.

Choose next zone. This instruction is called for when the test in the above instruction is negative for some $b = b_{\mathrm{open}} > 0$. At this moment the algorithm has returned to the previous still open bifurcation, i.e. to the currently largest value of $b \leq \mathrm{bif}$ for which currentzone(b) is not the last member of zones(b). The following is now carried out. The value of 'bif' is reset to bif = b_{open}; the value of n is reset to $n = n_c(\mathrm{bif})$ (with the reset bif); the value of currentzone(bif) is replaced by the next member of zones(bif).

When the instruction 'choose next zone' is completed the algorithm advances to the next step which is to set returnmodus = 'true' and then continues with the upgrade of the strand data for the currently drawn strand σ_n, with initial point in zone currentzone(bif), where n is again as it was when bifurcation level bif was installed.

This concludes the description of the core algorithm. By construction it prints out, for any given $(\boldsymbol{k}, \boldsymbol{i}, \boldsymbol{j})$ with $\boldsymbol{k} = 1, \ldots, 2g - 2$, $\boldsymbol{i} = 1, 2, 3$, $\boldsymbol{j} = 1, 2, 3, 4$, and any positive integer M, all sequences $\sigma_1, \ldots, \sigma_n$ with $n \leq M$ for which the corresponding sequence of parameter strands $h_1, \ldots, h_n$, occurs in the block word $\mathcal{B}(\alpha)$ of some ordinary simple closed curve α on S whose extended homotopy class $[\alpha]$ belongs to family $\mathcal{F}_{ij}^k$. Since the core algorithm respects Convention 5.2, we have, moreover, that each cyclic word $\langle (h_1, \boldsymbol{\vartheta}_1 + 4x_{l_1}), \ldots, (h_n, \boldsymbol{\vartheta}_n + 4x_{l_n}) \rangle$ obtained from it by Algorithm 5.5 occurs only once.

7 Code length and geodesic length

When S is endowed with a hyperbolic metric we shall represent the homotopy classes of closed curves by geodesics. A closed geodesic α then has two lengths: the arc length—denoted by $\ell(\alpha)$—with respect to the hyperbolic metric, and the code length, $L_{\mathcal{C}}(\alpha)$, defined in terms of the given pasting scheme, and the choice of the parameter strands. In this section we show that the two lengths may be compared with each other by means of inequalities. To save space we restrict ourselves to a rather simple example of such a comparison.

Let us first briefly recall the *Fenchel-Nielsen parameters* (see, for example, [Bu, sects. 1.7, 3.6] for an introduction, [K] for a construction of the canonical generators of the corresponding Fuchsian group, and [Ma] for an algorithmic way to obtain canonical generators out of the Fenchel-Nielsen parameters).

For any triple of positive real numbers x_1, x_2, x_3, there exists a pair of pants Y endowed with a hyperbolic metric such that the boundary curves η_i are geodesics of lengths $\ell(\eta_i) = x_i$, $i = 1, 2, 3$. Letting the parameter strands u_1, u_2, u_3 move in their homotopy classes with endpoints gliding on the boundary so as to obtain positions of minimal lengths, we obtain the *common perpendiculars* $\boldsymbol{u}_1, \boldsymbol{u}_2, \boldsymbol{u}_3$ between the boundary geodesics, i.e., geodesic arcs that meet the boundary perpendicularly at their endpoints. The $\boldsymbol{u}_i$ dissect Y into two isometric right-angled geodesic hexagons. As a consequence, their endpoints separate each boundary geodesic into two arcs of the same length. We also let $\boldsymbol{f}_i$ be the geodesic arc in the homotopy class of f_i that meets η_i orthogonally at its endpoints, $i = 1, 2, 3$.

We now may choose any sequence of positive real numbers $\lambda_1, \ldots, \lambda_{3g-3}$, and take hyperbolic pairs of pants $Y^1, \ldots, Y^{2g-2}$, where the lengths of the boundary geodesics are such that whenever the pasting scheme requests $\eta_{i_l}^{k_l}$ to be pasted together with $\eta_{i_l'}^{k_l'}$, then

$$\ell(\eta_{i_l}^{k_l}) = \ell(\eta_{i_l'}^{k_l'}) = \lambda_l,$$

$l = 1, \ldots, 3g - 3$. If the two boundary geodesics are parametrized on $\mathbb{R}$ with period 1 as earlier, but now in addition with constant speed, then the hyperbolic metrics of $Y^1, \ldots, Y^{2g-2}$ project to a hyperbolic metric on the surface S obtained by this pasting. The parameter curves γ^l are then closed geodesics of lengths λ_l, $l = 1, \ldots, 3g - 3$.

The construction bears an additional degree of freedom: the sequence of *twist parameters*. Let $\tau_1, \ldots, \tau_{3g-3}$ be a sequence of real numbers. Then $\eta_{i_l}^{k_l}$ and $\eta_{i_l'}^{k_l'}$ may be parametrized as requested and such that moreover $\eta_{i_l}^{k_l}(-\frac{1}{2}\tau_l)$ is the endpoint of the perpendicular $\boldsymbol{u}_{i_l}^{k_l}$ on $\eta_{i_l}^{k_l}$ and $\eta_{i_l'}^{k_l'}(-\frac{1}{2}\tau_l)$ is the endpoint of the perpendicular $\boldsymbol{u}_{i_l'}^{k_l'}$ on $\eta_{i_l'}^{k_l'}$. Since the common perpendiculars always bisect the boundary geodesics into two arcs of the same length, we then also have the following, where the lower indices are modulo 3: $\eta_{i_l}^{k_l}(-\frac{1}{2}\tau_l + \frac{1}{2})$ is the endpoint of the perpendicular $\boldsymbol{u}_{i_l+2}^{k_l}$ on $\eta_{i_l}^{k_l}$, and $\eta_{i_l'}^{k_l'}(-\frac{1}{2}\tau_l + \frac{1}{2})$ is the endpoint of the perpendicular $\boldsymbol{u}_{i_l'+2}^{k_l'}$ on $\eta_{i_l'}^{k_l'}$.

With the parametrizations defined in this way we let S be the surface obtained by pasting together $Y^1, \ldots, Y^{2g-2}$ with respect to pasting condition (2.3). The parameters $\lambda_1, \ldots, \lambda_{3g-3}$; $\tau_1, \ldots, \tau_{3g-3}$, are called the Fenchel-Nielsen length respectively twist parameters of S with respect to the given construction. Any compact orientable hyperbolic surface of genus $g \geq 2$ can be obtained in this way, based on a given pasting scheme. Moreover, one may always restrict oneself to taking $|\tau_l| \leq 1/2$, $l = 1, \ldots, 3g - 3$ (see, for example [Bu, theorem 3.6.4]).

It is possible to standardize the choice of the parameter strands, for instance, by suitably adapting them to the twist parameters, but to save space we shall omit this. Here is the comparison which serves as an example.

Theorem 7.1 *Let S be a compact hyperbolic surface, given as above, and assume that parameter strands have been chosen in a certain way. Then there exists a constant $q > 0$ such that any simple closed geodesic α on S satisfies*

$$\frac{1}{q}\,\ell(\alpha) \leq L_{\mathbb{C}}(\alpha) \leq q\,\ell(\alpha).$$

Proof Let ℓ_0 be the smallest and ℓ_1 the largest Fenchel-Nielsen length parameter used for S. If α is one of the parameter geodesics $\gamma^1, \ldots, \gamma^{3g-3}$, or their inverses, then the code length equals 4 and the geodesic length is bounded from below and from above by ℓ_0 and ℓ_1. In this case, the above inequalities are thus established.

Assume, from now on, that α is an ordinary simple closed geodesic. Then α is a product of strands, i.e., starting on some γ^{l_0} it successively intersects transversely a sequence of parameter geodesics $\gamma^{l_0}, \gamma^{l_1}, \ldots$, until it comes back to the starting point on $\gamma^{l_n} = \gamma^{l_0}$. For each $\nu = 1, \ldots, n$, the arc on α from $\gamma^{l_{\nu-1}}$ to γ^{l_ν} is a strand on some Y^{k_ν}. On Y^{k_ν} this strand is homotopic, with endpoints gliding on the boundary, to the common perpendicular $\boldsymbol{w}_\nu$, say, between the boundary geodesics of Y^{k_ν} that are labelled $\gamma^{l_{\nu-1}}$ and γ^{l_ν} on S (thus, $\boldsymbol{w}_\nu$ is one of the $\boldsymbol{u}_i^k, \boldsymbol{f}_i^k$ or their inverses). Lifting α to a covering geodesic $\tilde{\alpha}$ in the universal covering $\mathbb{D}$ of S we have therefore the configuration as in Fig. 3.4, but now slightly modified as shown in Fig. 7.1.

Namely, there is an arc $\bar{\alpha}$ of length $\ell(\bar{\alpha}) = \ell(\alpha)$ on $\tilde{\alpha}$ that, starting on some lift $\tilde{\gamma}_0$ of γ^{l_0} successively crosses lifts $\tilde{\gamma}_1, \tilde{\gamma}_2, \ldots$, of $\gamma^{l_1}, \gamma^{l_2}, \ldots$, ending up on a lift $\tilde{\gamma}_n$ of γ^{l_0} again. We also consider the next lift $\tilde{\gamma}_{n+1}$ of γ^{l_1} met by $\tilde{\alpha}$ after having left $\tilde{\gamma}_n$.

The configuration has the property that, for $\nu = 1, \ldots, n$, the geodesics $\tilde{\gamma}_{\nu-1}$ and $\tilde{\gamma}_{\nu+1}$ always lie on different sides of $\tilde{\gamma}_\nu$. Moreover, for each $\nu = 1, \ldots, n$, the perpendicular $\boldsymbol{w}_\nu$ has a lift $\tilde{\boldsymbol{w}}_\nu$ going from $\tilde{\gamma}_{\nu-1}$ to $\tilde{\gamma}_\nu$, and there is a lift $\tilde{\boldsymbol{w}}_{n+1}$ of $\boldsymbol{w}_1$ from $\tilde{\gamma}_n$ to $\tilde{\gamma}_{n+1}$. Note that each $\tilde{\boldsymbol{w}}_\nu$ is the unique common perpendicular between $\tilde{\gamma}_{\nu-1}$ and $\tilde{\gamma}_\nu$ in $\mathbb{D}$. By hyperbolic trigonometry applied to pairs of pants ([Bu, pg. 454]), there exist constants $0 < \omega_0 < \omega_1$ depending on ℓ_0, ℓ_1 such that

$$\omega_0 \leq \ell(\boldsymbol{w}_\nu) = \ell(\tilde{\boldsymbol{w}}_\nu) \leq \omega_1, \quad \nu = 1, \ldots, n. \tag{7.1}$$

The code word $\mathcal{C}(\alpha)$ is of the form (see (3.1))

$$\mathcal{C}(\alpha) = \langle h_1, \overrightarrow{c_1}, h_2, \overrightarrow{c_2}, \ldots, h_n, \overrightarrow{c_n} \rangle,$$

where for $\nu = 1, \ldots, n$, h_ν is the parameter strand in the homotopy class of $\boldsymbol{w}_\nu$ (endpoints gliding on the boundary) and $\overrightarrow{c_\nu}$ is a sequence of letters from the list $c_1^{l_\nu}, c_2^{l_\nu}, c_3^{l_\nu}, c_4^{l_\nu}$, or from the list $\bar{c}_4^{l_\nu}, \bar{c}_3^{l_\nu}, \bar{c}_2^{l_\nu}, \bar{c}_1^{l_\nu}$.

Let us now apply a homotopy to the corresponding code polygon such that its lift in $\mathbb{D}$ is the right-angled geodesic polygon $\hat{\alpha}$ with sides $\tilde{\boldsymbol{w}}_1, \tilde{\zeta}_1, \tilde{\boldsymbol{w}}_2, \tilde{\zeta}_2, \ldots, \tilde{\boldsymbol{w}}_n, \tilde{\zeta}_n$, as shown in Fig. 7.1, where for $\nu = 1, \ldots, n$, side $\tilde{\zeta}_\nu$ is the arc on $\tilde{\gamma}_\nu$ going from $\tilde{\boldsymbol{w}}_\nu$ to $\tilde{\boldsymbol{w}}_{\nu+1}$.

It is easy to see that there exists a constant $\omega_2 \geq 0$, depending on the

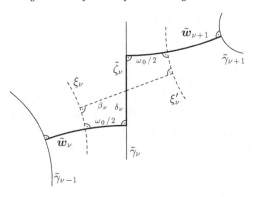

Fig. 7.1. Right-angled geodesic polygon homotopic to $\tilde{\alpha}$.

twist parameters and the choice of the parameter strands, such that we have the following, where $\#(\overrightarrow{c_\nu})$ denotes the number of letters in $\overrightarrow{c_\nu}$:

$$\max\{\frac{1}{4}\ell(\gamma^{l_\nu})(\#(\overrightarrow{c_\nu}) - \omega_2), 0\} \le \ell(\tilde{\zeta}_\nu) \le \frac{1}{4}\ell(\gamma^{l_\nu})(\#(\overrightarrow{c_\nu}) + \omega_2). \quad (7.2)$$

This leads to the following upper bound, where ω_3 is a constant that can be written in terms of the preceding ones,

$$\ell(\alpha) \le \ell(\hat{\alpha}) \le \omega_3 L_{\mathbb{C}}(\alpha). \quad (7.3)$$

For the lower bound we make some use of hyperbolic trigonometry. Perpendicularly to $\tilde{w}_\nu$ and $\tilde{w}_{\nu+1}$ and at distance $\omega_0/2$ from $\tilde{\gamma}_\nu$ we draw the perpendicular geodesics ξ_ν, ξ'_ν, as shown in Fig. 7.1, so as to obtain a crossed right-angled geodesic hexagon consisting of two isometric quadrilaterals with three right angles. For one of these quadrilaterals, let us denote by δ_ν the length of the side on $\tilde{\gamma}_\nu$ and by β_ν the length of the side (dotted) on the perpendicular from ξ_ν to ξ'_ν. Then $2\delta_\nu = \ell(\tilde{\zeta}_\nu)$, and $2\beta_\nu$ is the distance between ξ_ν and ξ'_ν. The hyperbolic trigonometric formula to make use of is

$$\sinh(\beta_\nu) = \sinh(\frac{1}{2}\omega_0)\cosh(\delta_\nu). \quad (7.4)$$

For $\nu = 1, \ldots, n$, the regions around the geodesics $\tilde{\gamma}_\nu$ in $\mathbb{D}$ enclosed by the pairs ξ_ν, ξ'_ν are pairwise disjoint, and therefore

$$\ell(\alpha) \ge \sum_{\nu=1}^{n} 2\beta_\nu. \quad (7.5)$$

If $1 + \#(\overrightarrow{c_\nu}) \le 2\omega_2$, then we content ourselves with the fact that $2\beta_\nu \ge \omega_0$

which leads to the lower bound

$$2\beta_\nu \geq \frac{\omega_0}{2\omega_2}(1 + \#(\vec{c_\nu})). \tag{7.6}$$

If $1 + \#(\vec{c_\nu}) > 2\omega_2$, then by (7.2), $\delta_\nu \geq \#(\vec{c_\nu})\ell_0/16$, and we extract a lower bound for $2\beta_\nu$ from (7.4). For the latter, we first observe that for any constants $a > 0$, $m \geq 1$, the function $f(t) = \operatorname{arcsinh}(a \cdot \cosh(t))$, $t \in \mathbb{R}$, is convex and satisfies the rule $f(mt) - f(0) \geq m(f(t) - f(0))$. Applying this to $a = \ell_0/16$ and $m = \#(\vec{c_\nu})$ we get

$$2\beta_\nu \geq \omega_4(1 + \#(\vec{c_\nu})), \tag{7.7}$$

where ω_4 is again a constant that one can write in terms of the preceding ones. Inequalities (7.5), (7.6), (7.7) together complete the proof of the theorem.   □

Remark. *A similar proof, albeit not exactly the same, is given in [Mz2, proposition 3.5] where the analogue of Theorem 7.1 for the Dehn-Thurston coding is shown.*

8 Drawing the geodesics

In this section we indicate a way to represent the closed geodesics graphically on a fundamental polygon for S, where S is endowed with a hyperbolic metric. For fundamental polygons of Fuchsian groups we refer the reader e.g. to [Bd2]. Much of the notation used here is independent of the remaining sections.

In what follows, $\mathbb{H}$ may be any model of the hyperbolic plane. For the graphical representations in Section 9 we shall, however, use the disk model $\mathbb{D}$.

Assume that S is represented as a quotient $S = \mathbb{H}/\Gamma$, where Γ is a Fuchsian group acting on the hyperbolic plane $\mathbb{H}$. Let $F \subset \mathbb{H}$ be a *convex* geodesic fundamental polygon for Γ and let $T_1, \ldots, T_n$ be a set of side pairing generators of Γ, i.e. the sides of F are grouped into pairs s_i, s_i', $i = 1, \ldots, n$, and for each i the generator T_i sends s_i to s_i' in such a way that $T_i(F)$ is adjacent to F along s_i'. Fig. 8.1 shows an example of genus 2.

Each $g \in \Gamma$ is an isometry, $g : \mathbb{H} \to \mathbb{H}$, with a uniquely determined invariant geodesic a_g, the *axis* of g. Under the canonical projection $\pi : \mathbb{H} \to S = \mathbb{H}/\Gamma$, this axis is mapped to a closed geodesic c_g on S. For any $h \in \Gamma$, the image $h(a_g)$ is the axis of the conjugate hgh^{-1} of g, and we have $\pi(h(a_g)) = \pi(a_g) = c_g$. Hence, c_g is attributed to the conjugacy

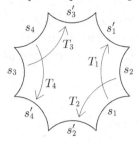

Fig. 8.1. Fundamental polygon and side pairing generators.

class $[g] = \{hgh^{-1} \mid h \in \Gamma\}$ and we may write $c_g = c_{[g]}$. Furthermore, the mapping $[g] \mapsto c_{[g]}$ from the conjugacy classes of Γ to the closed geodesics in S is one-to-one and onto. The representation of $c_{[g]}$ in F is defined to be the set

$$\hat{a}_{[g]} = F \cap \bigcup_{h \in \Gamma} h(a_g).$$

This set may be computed as follows. First, one checks whether a_g meets F. If this is not the case, then one applies the procedure illustrated in Fig. 8.2.

This procedure is the following. In the first step the perpendicular geodesic arc ρ is dropped from some interior point p_0 in F to a_g. This arc intersects one of the sides s_i or s_i', say in p_1. Geodesic a_g is now replaced by $g_1(a_g)$, where $g_1 = T_i^{-1}$ (respectively, T_i). If $g_1(a_g)$ does not meet F, then $g_1(\rho)$ meets the boundary of F in two points, $g_1(p_1)$ and a point p_2 on some other side s_k (respectively, s_k'). One now sets $g_2 = T_k^{-1}$ (respectively, $g_2 = T_k$) and replaces $g_1(a_g)$ by $g_2 g_1(a_g)$, and

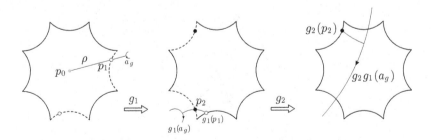

Fig. 8.2. An axis is brought to the fundamental polygon.

so on. Eventually, this leads to a conjugate $g' = hgh^{-1}$ of g whose axis $a_{g'} = h(a_g)$ meets F in two distinct boundary points.

Let q_0, q_1 be these points. We then proceed in a similar way, as illustrated in Fig. 8.3:

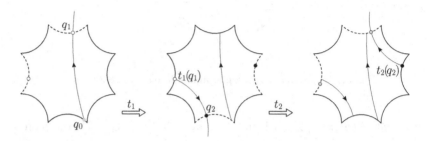

Fig. 8.3. An axis is "chopped" into the fundamental polygon.

Denote by $\tilde{a}_{g'}$ the geodesic segment on $a_{g'}$ with initial point q_0 passing through q_1 and being of length $\ell(c_g)$, the length of c_g. The projection $\pi(\tilde{a}_{g'})$ goes exactly once around c_g.

Now let $t_1 \in \{T_1^{\pm 1}, \ldots, T_n^{\pm 1}\}$ be the generator, or its inverse, that maps the side of F carrying q_1 to the corresponding pairing side. Chop off the geodesic segment $q_0 q_1$ from $\tilde{a}_{g'}$ to obtain the part $\tilde{a}_{g'}^1$ of $\tilde{a}_{g'}$ that lies outside F, and map the latter to the segment $t_1(\tilde{a}_{g'}^1)$ that enters into F at $t_1(q_1)$ and leaves F at some subsequent boundary point q_2. Then chop off segment $t_1(q_1)q_2$ and continue, until the sum of the lengths of the arcs in F obtained in this way equals $\ell(c_g)$.

In the practical implementation of the above procedure a difficulty, due to rounding errors, may arise when an arc comes close to a vertex of F (e.g. if the geodesic should pass through the vertex but does not do so numerically). This is bypassed by the following modification.

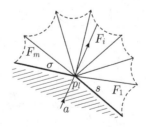

Fig. 8.4. A numerically critical vertex.

Let p_l be a vertex of F that is critical for arc a in the sense that it is numerically uncertain which of the two adjacent sides, say s and σ, the procedure should use. We then look at all the images $g(F)$, $g \in \Gamma$ that have vertex p_l in common. These images, together with F, tile a neighborhood of p_l as shown in Fig. 8.4 without overlapping, and we enumerate them counterclockwise: $F, F_1, F_2, \ldots, F_m$. As a leaves F near p_l it may cross several F_i in rapid succession. But there is a first $F_i = g_i(F)$ into which a enters and then leaves on some side that does not meet vertex p_l (the shaded area in Fig. 8.4). We then chop off the outside part a' of a as described above but now apply g_i^{-1} to it rather than the generator, or its inverse, that correspond to s, respectively σ.

9 Examples of genus 2

In this section we work out an example of genus 2. We begin with the pasting scheme, then define canonical generators of the fundamental group and introduce the access arcs that will allow us to express the arcs for the coding in terms of the fundamental group and vice versa. We then assume that the surface—S—is endowed with a hyperbolic metric in which the parameter curves, $\gamma^1, \gamma^2, \gamma^3$, are closed geodesics. In this case, S is the quotient of the hyperbolic plane by the action of a Fuchsian group Γ and we provide the formulae that compute canonical generators of Γ in terms of the Fenchel-Nielsen coordinates of S. The section ends with some pictures of the Birman-Series set that have been computed based on these formulae and the algorithms of the preceding sections.

The pasting scheme is

$$\eta_3^1; \eta_2^1, \quad \eta_3^2; \eta_2^2, \quad \eta_1^1; \eta_1^2. \tag{9.1}$$

Fig. 9.1a) shows the pairs of pants Y^1, Y^2. The pasting is such that the zones $1, 2, 3, 4$, of one boundary curve always go, in this order, to the zones $3, 2, 1, 4$ of the other. The curves $\eta_3^1, \eta_3^2, \eta_1^1$, yield, respectively, the oriented curves $\gamma^1, \gamma^2, \gamma^3$. In this section the latter are renamed c, d, e, for simplicity. The curves $\alpha_1, \beta_1, \alpha_2, \beta_2$, drawn on Y^1, Y^2 form so-called *canonical generators* of the fundamental group of S, i.e. generators with the relation $\alpha_1 \beta_1 \alpha_1^{-1} \beta_1^{-1} \alpha_2 \beta_2 \alpha_2^{-1} \beta_2^{-1} = 1$, where 1 is the neutral element of the fundamental group.

Fig. 9.1b) shows the arcs u_i^k, f_i^k, c_j^l, for the coding, but now renamed $u_i, v_i, f_i, g_i, c_j, d_j, e_j$, for better readability, $i = 1, 2, 3$; $j = 1, 2, 3, 4$. The arrows on c_j, d_j, e_j correspond to the orientations of c, d, e on S. Hence, for the oriented arcs drawn in Fig. 9.1b) the pasting is with matching arrows.

Fig. 9.1c) shows the access arcs that connect the base point q_0 for the fundamental group with the endpoints of the coding arcs, r_j, s_j, t_j, $j = 1, 2, 3, 4$. The access arcs allow us to assign to each coding arc a well determined element of the fundamental group. For instance, r_2 leads from q_0 to the initial point of c_2 and $\bar{r}_3$ leads from the endpoint of c_2 back to q_0. The corresponding element in the fundamental group is β_1^{-1}.

Fig. 9.1d), finally shows a sequence of closed curves $z_1, \ldots, z_4$ based at some point p "opposite" to q_0. Cutting S open along these we get a canonical fundamental domain.

The following list yields the members of the fundamental group associated to the coding arcs via the access arcs given in Fig. 9.1c). In this list the symbol $\mathbf{1}$ denotes again the neutral element of the fundamental group.

$$
\begin{aligned}
& u_1 \to \alpha_1, \ u_2 \to \beta_1 \alpha_1^{-1} \beta_1^{-1}, \ u_3 \to \mathbf{1}, \\
& v_1 \to \mathbf{1}, \ v_2 \to \beta_2 \alpha_2 \beta_2^{-1}, \ v_3 \to \beta_2 \alpha_2^{-1} \beta_2^{-1}, \\
& f_1 \to \beta_1, \ f_2 \to \beta_1 \alpha_1^{-1} \beta_1^{-1} \alpha_1, \ f_3 \to \beta_1 \alpha_1 \beta_1^{-1} \alpha_1^{-1}, \\
& g_1 \to \beta_2^{-1}, \ g_2 \to \alpha_2 \beta_2 \alpha_2^{-1} \beta_2^{-1}, \ g_3 \to \alpha_2^{-1} \beta_2 \alpha_2 \beta_2^{-1}. \\
& c_2 \to \beta_1^{-1}, \ d_3 \to \beta_2, \ e_4 \to \alpha_2 \beta_2 \alpha_2^{-1} \beta_2^{-1}, \\
& c_j \to \mathbf{1}, \ d_j \to \mathbf{1}, \ e_j \to \mathbf{1}, \text{ in the remaining cases.}
\end{aligned}
\tag{9.2}
$$

Conversely, we may represent the generators in the following way, where the symbol $\sim$ denotes homotopy with fixed base point q_0 and the bar denotes inverse direction.

$$
\begin{aligned}
\alpha_1 &\sim \bar{e}_2 \bar{e}_1 u_1 u_3, \\
\beta_1 &\sim \bar{u}_3 \bar{c}_4 \bar{c}_3 \bar{c}_2 \bar{c}_1 u_3, \\
\alpha_2 &\sim e_3 e_4 \bar{v}_3 \bar{v}_1, \\
\beta_2 &\sim v_1 \bar{d}_1 \bar{d}_2 \bar{d}_3 \bar{d}_4 \bar{v}_1.
\end{aligned}
\tag{9.3}
$$

This list allows us to compute the corresponding generators of the Fuchsian group as a function of the Fenchel-Nielsen parameters $\lambda_1, \lambda_2, \lambda_3$, τ_1, τ_2, τ_3. When these values are given we think of Y^1, Y^2 as being endowed with a hyperbolic metric such that the boundary curves are geodesics of lengths $\ell(c) = \lambda_1$, $\ell(d) = \lambda_2$, $\ell(e) = \lambda_3$. We denote by $\boldsymbol{u}_i, \boldsymbol{v}_i$ the common perpendiculars (see Section 7) in the homotopy classes, with endpoints gliding on the boundary of the u_i, v_i, respectively. Their lengths are given by the following formulae, where by abuse of nota-

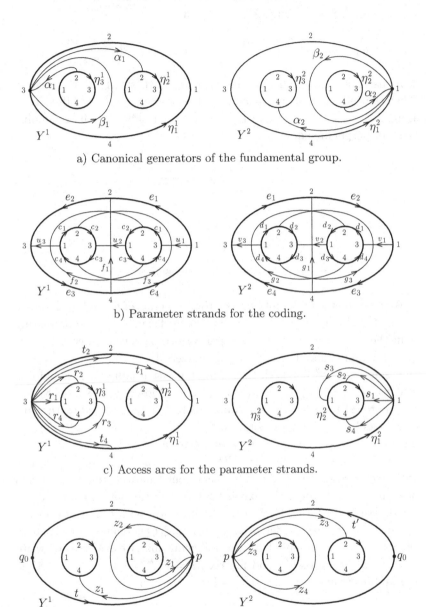

a) Canonical generators of the fundamental group.

b) Parameter strands for the coding.

c) Access arcs for the parameter strands.

d) Cutting along $z_1, \ldots, z_4$ yields a canonical fundamental domain.

Fig. 9.1. Fundamental group and coding in genus 2.

tion u_i, v_i also denote the lengths (absolute values) of the arcs and
$c_i = \cosh(\frac{1}{2}\lambda_i)$, $s_i = \sinh(\frac{1}{2}\lambda_i)$, $i = 1, 2, 3$:

$$u_1 = u_3 = \text{arccosh}\frac{c_1 + c_1 c_3}{s_1 s_3}, \quad v_1 = v_3 = \text{arccosh}\frac{c_2 + c_2 c_3}{s_2 s_3}. \quad (9.4)$$

Now assume, for simplicity, that τ_1, τ_2, τ_3 have small positive values.
Then the endpoints of arcs u_3 and v_1 on geodesic $e = \gamma^3$ are shifted away
from each other by $\tau_3 \lambda_3$ and there are similar shifts for the remaining
pairs of perpendiculars.

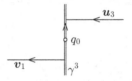

Fig. 9.2. A twist along γ^3.

We let q_0 be the midpoint between the endpoints of u_3 and v_1 as shown
in Fig. 9.2. In order to obtain an explicit Fuchsian group Γ representing
S in the form $S = \mathbb{H}/\Gamma$, we fix a pair of orthogonal oriented geodesics
ξ, ζ in $\mathbb{H}$ and think of π as being the universal covering map $\pi : \mathbb{H} \to S$
that sends ζ to $\gamma^3 = e$ such that the intersection point z_0 of ξ and ζ is
sent to $\pi(z_0) = q_0$ ($\pi(\xi)$ is in general not a closed geodesic on S). In
the figures at the end of this section, where $\mathbb{H}$ is represented by the unit
disk model $\mathbb{D}$, in complex notation $\mathbb{D} = \{z = x + iy \in \mathbb{C} \mid |z| < 1\}$, we
have $z_0 = 0$, ξ goes from -1 to 1, and ζ from $-i$ to i.

For $t \in \mathbb{R}$ we denote by $R(t), U(t) : \mathbb{H} \to \mathbb{H}$ the hyperbolic isometries
with axes ξ respectively, ζ and displacement length $|t|$ that shift points
of $\mathbb{H}$ in the direction of the axis if $t > 0$ and in the opposite direction if
$t < 0$. We also let $H_0 : \mathbb{H} \to \mathbb{H}$ be the half turn with center of rotation
z_0. In the unit disk model $\mathbb{D}$, and with z_0, ξ, ζ as defined above, these
isometries are represented by the following matrices, where we write
$s_t = \sinh[t]$, $c_t = \cosh[t]$ and matrix $M = \left[\begin{smallmatrix} a & b \\ c & d \end{smallmatrix}\right]$ operates on $z \in \mathbb{C}$ as
the Möbius transformation $M[z] = (az + b)/(cz + d)$.

$$H_0 = \begin{bmatrix} i & 0 \\ 0 & -i \end{bmatrix},$$

$$R(t) = \begin{bmatrix} 1 + c_t & s_t \\ s_t & 1 + c_t \end{bmatrix}, \quad U(t) = \begin{bmatrix} 1 + c_t & is_t \\ -is_t & 1 + c_t \end{bmatrix}. \quad (9.5)$$

It is a routine exercise to show that with the above settings the covering transformations A_1, B_1, A_2, B_2 corresponding to $\alpha_1, \beta_1, \alpha_2, \beta_2$—and hence generators of Γ—are as follows, where we abbreviate $\vartheta_j = \tau_j \lambda_j$, $j = 1, 2, 3$.

$$
\begin{aligned}
A_1 &= U(\tfrac{1}{2}\lambda_3 + \tfrac{1}{2}\vartheta_3)R(\boldsymbol{u}_1)U(\vartheta_1)R(\boldsymbol{u}_3)U(\tfrac{1}{2}\vartheta_3)H_0, \\
B_1 &= U(\tfrac{1}{2}\vartheta_3)R(\boldsymbol{u}_3)U(-\lambda_1)R(-\boldsymbol{u}_3)U(-\tfrac{1}{2}\vartheta_3), \\
A_2 &= U(-\tfrac{1}{2}\lambda_3 - \tfrac{1}{2}\vartheta_3)R(-\boldsymbol{v}_3)U(-\vartheta_2)R(-\boldsymbol{v}_1)U(-\tfrac{1}{2}\vartheta_3)H_0, \\
B_2 &= U(-\tfrac{1}{2}\vartheta_3)R(-\boldsymbol{v}_1)U(\lambda_2)R(\boldsymbol{v}_1)U(\tfrac{1}{2}\vartheta_3).
\end{aligned}
\tag{9.6}
$$

In order to construct a canonical fundamental domain we consider the curves z_1, z_2, z_3, z_4 as shown in Fig. 9.1d), whose common base point, p, is the endpoint of arc $t = e_3 e_4$ or, equivalently, the endpoint of $t' = \bar{e}_2 \bar{e}_1$. It is now easy to check that

$$
t z_1 \bar{t} \sim \beta_1 \alpha_1^{-1} \beta_1^{-1}, \quad t z_1 z_2 \bar{t} \sim \alpha_1^{-1} \beta_1^{-1}, \quad t z_1 z_2 \bar{z}_1 \bar{t} = \beta_1^{-1},
$$

and similarly

$$
t' z_3 \bar{t}' \sim \beta_2 \alpha_2^{-1} \beta_2^{-1}, \quad t' z_3 z_4 \bar{t}' \sim \alpha_2^{-1} \beta_2^{-1}, \quad t' z_3 z_4 \bar{z}_3 \bar{t}' \sim \beta_2^{-1}.
$$

From this we get the following vertices of a canonical fundamental polygon in $\mathbb{H}$.

$$
p_1 = U(-\tfrac{1}{2}\lambda_3)[q_0], \qquad p_5 = U(\tfrac{1}{2}\lambda_3)[q_0],
$$

$$
p_2 = B_1 A_1^{-1} B_1^{-1}[p_1], \quad p_3 = A_1^{-1} B_1^{-1}[p_1], \quad p_4 = B_1^{-1}[p_1],
$$
$$
p_6 = B_2 A_2^{-1} B_2^{-1}[p_5], \quad p_7 = A_2^{-1} B_2^{-1}[p_5], \quad p_8 = B_2^{-1}[p_5].
\tag{9.7}
$$

In the implementation the twist parameters were restricted to the interval $[-\tfrac{1}{2}, \tfrac{1}{2}]$ and the selection of the parameter strands was made such that the u_i and v_i have minimal possible lengths.

Such a polygon is used in Fig. 9.3 showing the first 255 simple closed geodesics on a compact Riemann surface of genus 2 drawn on a canonical fundamental domain in $\mathbb{D}$. Here "first" is meant with respect to the order given by the enumeration algorithm. The Fenchel-Nielsen length parameters of this example are $\ell(\gamma^1) = 0.3$, $\ell(\gamma^2) = 0.3$, $\ell(\gamma^3) = 1.3$; the twist parameters are equal to 0. Sides $p_2 p_3$ and $p_6 p_7$ are near the boundary of $\mathbb{D}$ and practically not visible. The pasting scheme for the sides is $p_1 p_2 \leftrightarrow p_3 p_4$, $p_2 p_3 \leftrightarrow p_4 p_5$, etc.

The figure shows four large (white) areas into which no simple geodesics enter. The occurrence of such areas is caused by the presence of small closed geodesics, and was first described qualitatively by

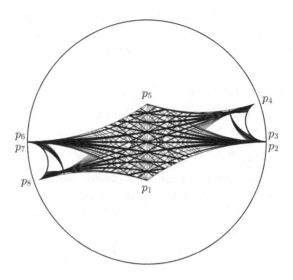

Fig. 9.3. A family of simple closed geodesics of a Riemann surface of genus 2.

Jorgensen [J], whose paper is at the origin of the study of the Birman-Series set (see the corresponding reference in [BS3]).

The examples shown on the next two pages grew out of an attempt to visualize the Birman-Series set of the surface $F48$ obtained from the (2,3,8)-triangle group. Birman and Series [BS2] have shown that the union of the complete simple geodesics—the Birman-Series set—is nowhere dense. In [BuP] it was shown that there exists a constant c_g depending only on the genus, such that on any Riemann surface of genus g the complement of this set contains a disk of radius c_g. An estimate of c_g is not known, but it is certainly a small number. In fact, a printout of the first 1500 simple closed geodesics (with the ordering given by the geodesic length) on $F48$ showed a fundamental domain that was completely black! For a better visualization of the geodesics—rather than printing them out individually—we re-arranged them into families with the same length. Some of these families are shown in Fig. 9.4 drawn on a canonical octagon and in Fig. 9.5 on a symmetric octagon (with opposite sides identification). The function tr is defined as $\mathrm{tr}(\gamma) = \cosh \frac{1}{2}\ell(\gamma)$. This is half the trace of the corresponding element of the Fuchsian group.

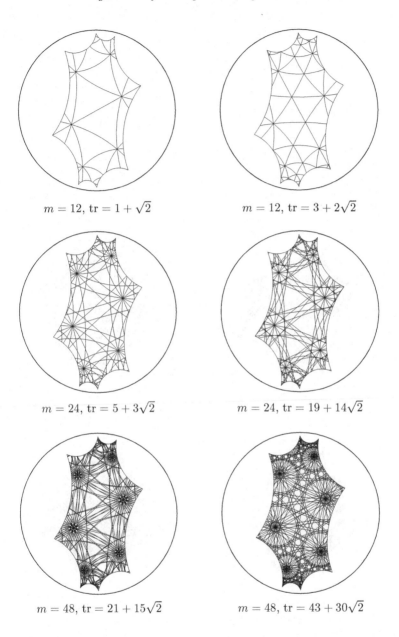

$m = 12, \text{tr} = 1 + \sqrt{2}$

$m = 12, \text{tr} = 3 + 2\sqrt{2}$

$m = 24, \text{tr} = 5 + 3\sqrt{2}$

$m = 24, \text{tr} = 19 + 14\sqrt{2}$

$m = 48, \text{tr} = 21 + 15\sqrt{2}$

$m = 48, \text{tr} = 43 + 30\sqrt{2}$

Fig. 9.4. Families of simple closed geodesics of the surface $F2$ drawn on a canonical fundamental domain; m is the multiplicity, tr is the half trace.

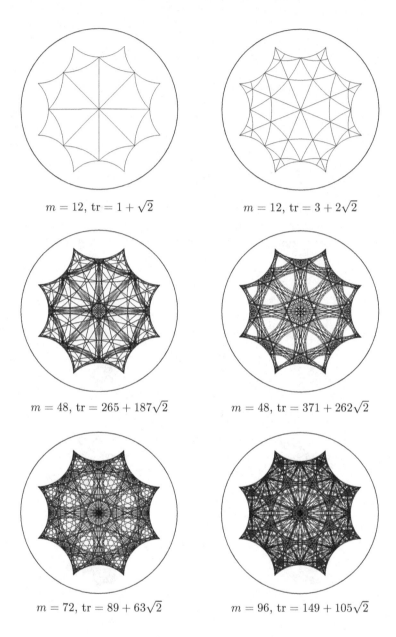

$m = 12$, tr $= 1 + \sqrt{2}$

$m = 12$, tr $= 3 + 2\sqrt{2}$

$m = 48$, tr $= 265 + 187\sqrt{2}$

$m = 48$, tr $= 371 + 262\sqrt{2}$

$m = 72$, tr $= 89 + 63\sqrt{2}$

$m = 96$, tr $= 149 + 105\sqrt{2}$

Fig. 9.5. Families of simple closed geodesics of the surface $F2$ drawn on a symmetric fundamental domain; m is the multiplicity, tr is the half trace.

10 Concluding remarks

We conclude with some remarks and additional literature. The statements in this section are given without proofs.

Price-performance ratio. The strategy of Algorithm 5.5 is to find simple closed curves by trying out all sequences of parameter strands $h = (h_1, \ldots, h_M)$ having the property that by inserting suitable connectors between the pairs $h_\nu, h_{\nu+1}$ one obtains an arc $h_1 \lambda_1 \cdots h_{M-1} \lambda_{M-1} h_M$ that has no self-intersections (after a slight homotopy, if necessary). Let us call these sequences *h-sequences*. While looking at all of them, the algorithm sorts out the subsequences $(h_1, \ldots, h_n)$ that allow a connector from h_n to h_1 to produce a simple closed curve.

The number of elementary steps needed for each $(h_1, \ldots, h_M)$ is of order M^2. How many h-sequences are there in comparison with the number of simple closed curves that are obtained?

Let us consider the quotient $\wp(M) = a(M)/b(M)$, where $a(M)$ is the number of h-sequences of length M that are used, and $b(M)$ is the number of base words of strand length $\leq M$ that are obtained. For the implementation used for the figures in Section 9 we got, for example,

$$\wp(8) = \frac{1278}{262}, \quad \wp(16) = \frac{44348}{5900}, \quad \wp(24) = \frac{447906}{49566}.$$

This is only slowly growing. It would be interesting to find the growth rate of $\wp(M)$ in general. It would also be interesting to know whether the function $\wp$ may be used to define a concept of density for the simple closed geodesics on a hyperbolic surface within the set of all complete simple geodesics.

Computation time. The computation time of Algorithm 5.5 may be estimated by estimating the number of h-sequences. For this it is useful to define, for any h-sequence $h = (h_1, \ldots, h_M)$, the class $\beta(h)$ of all curves homotopic to $h_1 \lambda_1 \cdots h_{M-1} \lambda_{M-1} h_M$ (as above) with endpoints gliding on the respective parameter curves. The algorithm is such that the homotopy classes $\beta(h)$ are pairwise distinct. It suffices therefore to estimate the number of homotopy classes. This may be done as follows.

Select for each $Y^1, \ldots, Y^{2g-2}$ one of the configurations of streets as shown in Fig. 5.1 and draw a number of strands on each street, using altogether M strands. For each parameter curve γ^l the number of strands arriving on one side of γ^l must be equal to the number of strands arriving on the other. To fulfill this condition we partition M into a sum

$M = M_1 + \cdots + M_{3g-3}$ and require, for $l = 1, \ldots, 3g - 3$, that M_l strands arrive at γ^l on either side. There are at most M^{3g-3} ways to do this, and for any partition and any Y^k there is at most one distribution of strands into the given streets on Y^k that fulfills this requirement. Furthermore, for any l, the ends of the strands on one side of γ^l are paired with the ends of the strands on the other by cyclic permutation, and there are only M_l such permutations. This altogether shows that the number of h-sequences of length M is at most $\sigma_g M^{6g-6}$, where σ_g is a constant depending only on the genus.

Taking into account that the number of elementary steps needed to produce an h-sequence $(h_1, \ldots, h_M)$ is of order M^2, we conclude that Algorithm 5.5 runs in polynomial time with a polynomial of degree $\leq 6g - 4$.

Historical remarks and additional literature. The coding of simple closed curves and families thereof, based on a pairs of pants decomposition, was proposed by Dehn [De1] (English translation by J. Stillwell in [De2]) who saw in it advantages over the coding via fundamental domains. A thorough study of Dehn's method was later given by Göritz [Goe].

Algorithms for simple curves were given by Reinhart [Re1, Re2], Zieschang [Zi1, Zi2], Chillingworth [Ch1, Ch2] and Birman-Series [BS1, BS3]. All of them use again a coding based on fundamental polygons and they are decision algorithms determining whether or not a given word corresponds to a simple curve. Among the criteria for simplicity, probably the easiest is the one in [BS1, theorem B], where the surface has non-empty boundary and the fundamental group is a free group. A recent interesting problem is to *recognize patterns* in the words to say whether or not a closed curve is simple. Patterns in two-generator groups that are related to Farey numbers are described in [GiK1]. Patterns that completely characterize the simple curves on the once punctured torus are given in Series [Ser] in terms of cutting sequences and in Buser-Semmler [BSe] in terms of the fundamental group. For twice punctured tori, patterns are described in terms of a coding based on train tracks in Keen-Parker-Series [KPS, theorem 5.3]. For closed surfaces such patterns do not seem to exist in the literature as yet.

Cohen and Lustig [CL, Lu] and Tan [Ta] use modifications of the Birman-Series algorithm to determine the number of self-intersections of non-simple words.

Combining an enumeration of the simple curves with Theorem 7.1

(where the constants have to be made explicit) one may compute the *systole* of a hyperbolic surface, i.e. the length of the shortest non-trivial closed geodesic. However, a much more direct algorithm for computing systoles is given in Akrout [Ak].

Finally, we point out a coding of closed curves by cutting sequences with an application to palindromes in the paper by Jane Gilman and Linda Keen [GiK2] in this book.

References

[Ak] Akrout, H. Un processus effectif de détermination des systoles pour les surfaces hyperboliques. *Geom. Dedicata* **121** (2006), 1–8.

[Bd1] Beardon, A.F. The geometry of discrete groups, in *Discrete Groups and Automorphic Functions* (Proc. Conf., Cambridge, 1975), Academic Press, London, 1977, pp. 163–198.

[Bd2] Beardon, A. F. *The Geometry of Discrete Groups*, Graduate Texts in Mathematics **91** Springer-Verlag, New York, 1995.

[Bi] Birman, J. S. The algebraic structure of surface mapping class groups, in *Discrete Groups and Automorphic Functions* (Proc. Conf., Cambridge, 1975), Academic Press, London, 1977, pp. 163–198.

[BS1] Birman, J. S. and Series, C. An algorithm for simple curves on surfaces. *J. London Math. Soc. (2)* **29** (1984), 331–342.

[BS2] Birman, J. S. and Series, C. Geodesics with bounded intersection number on surfaces are sparsely distributed. *Topology* **24**(2) (1985), 217–225.

[BS3] Birman, J. S. and Series, C. Dehn's algorithm revisited, with applications to simple curves on surfaces, in *Combinatorial Group Theory and Topology* (Alta, Utah, 1984), Ann. of Math. Stud. **111**, Princeton Univ. Press, Princeton, NJ, 1987, pp. 451–478

[Bu] Buser, P. *Geometry and Spectra of compact Riemann Surfaces*, Progress in Mathematics **106**, Birkhäuser Boston Inc., Boston, MA, 1992.

[BuP] Buser, P. and Parlier, H. The distribution of simple closed geodesics on a Riemann surface, in *Complex Analysis and its Applications*, OCAMI Stud. **2**, Osaka Munic. Univ. Press, Osaka, 2007, pp. 3–10.

[BSe] Buser, P. and Semmler, K.-D. The geometry and spectrum of the one-holed torus. *Comment. Math. Helv.* **63**(2) (1988), 259–274,

[Ch1] Chillingworth, D. R. J. Winding numbers on surfaces. I. *Math. Ann.* **196** (1972), 218–249.

[Ch2] Chillingworth, D. R. J. Winding numbers on surfaces. II. *Math. Ann.* **199** (1972), 131–153.

[CL] Cohen, M. and Lustig, M. Paths of geodesics and geometric intersection numbers. I, in *Combinatorial Group Theory and Topology* (Alta, Utah, 1984), Ann. of Math. Stud. **111**, Princeton Univ. Press, Princeton, NJ, 1987, pp. 479–500.

[De1] Dehn, M. Über Kurvensysteme auf zweiseitigen Flächen mit Anwendung auf das Abbildungsproblem. Autogr. Vortrag im Math. Kolloquium, Breslau, 11. Februar 1922.

[De2] Dehn, M. *Papers on Group Theory and Topology*, translated from the German and with introductions and an appendix by John Stillwell, with an appendix by Otto Schreier, Springer-Verlag, New York, 1987.

[GiK1] Gilman, J. and Keen, L. Word sequences and intersection numbers, in *Complex Manifolds and Hyperbolic Geometry* (Guanajuato, 2001), Contemp. Math. **311**, Amer. Math. Soc., Providence, RI, 2002, pp. 231–249.

[GiK2] Gilman, J. and Keen, L. Cutting sequences and palindromes, in *Geometry of Riemann Surfaces* (Proc. Conf., Anogia, 2007), London Mathematical Society Lecture Note Series **368**, Cambridge University Press, Cambridge, 2010, pp. 205–227.

[Goe] Goeritz, L. Normalformen der Systeme einfacher Kurven auf orientierbaren Flächen. *Abh. Math. Sem. Univ. Hamburg* **9** (1933), 223–243.

[Ha] Harvey, W. J. (ed.). Discrete Groups and Automorphic Functions (Proc. Conf., Cambridge, 1975), Academic Press, London, 1977.

[H] Huber, H. Zur analytischen Theorie hyperbolischer Raumformen und Bewegungsgruppen. *Math. Ann.* **138** (1959), 1–26.

[J] Jorgensen, T. Simple geodesics on Riemann surfaces. *Proc. Amer. Math. Soc.* **86** (1982), no. 1, 120–122.

[K] Keen, L. Intrinsic moduli on Riemann surfaces. *Ann. of Math. (2)* **84** (1966), 404–420.

[KPS] Keen, L., Parker, J. R. and Series, C. Combinatorics of simple closed curves on the twice punctured torus. *Israel J. Math.* **112** (1999), 29–60.

[Lu] Lustig, M. Paths of geodesics and geometric intersection numbers. II, in *Combinatorial Group Theory and Topology* (Alta, Utah, 1984), Ann. of Math. Stud. **111**, Princeton Univ. Press, Princeton, NJ, 1987, pp. 501–543.

[Mb] Macbeath, A. M. Topological background, in *Discrete Groups and Automorphic Functions* (Proc. Conf., Cambridge, 1975), Academic Press, London, 1977, pp. 1–45.

[Ma] Maskit, B. Matrices for Fenchel-Nielsen coordinates. *Ann. Acad. Sci. Fenn. Math.* **26**(2) (2001), 267–304.

[MR1] McShane, G. and Rivin, I. Simple curves on hyperbolic tori. *C. R. Acad. Sci. Paris Sér. I Math.* **320** (1995), 1523–1528.

[MR2] McShane, G. and Rivin, I. A norm on homology of surfaces and counting simple geodesics. *Internat. Math. Res. Notices* **2** (1995), 61–69 (electronic).

[Mz1] Mirzakhani, M. Simple geodesics and Weil-Petersson volumes of moduli spaces of bordered Riemann surfaces. *Invent. Math.* **167**(1) (2007), 179–222.

[Mz2] Mirzakhani, M. Growth of the number of simple closed geodesics on hyperbolic surfaces. *Ann. of Math. (2)* **168**(1) (2008), 97–125.

[Pe] Penner, R.C. with Harer, J.L. *Combinatorics of Train Tracks*, Annals of Mathematics Studies **125**, Princeton University Press, Princeton, 1992.

[Re1] Reinhart, B.L. Simple curves on compact surfaces. *Proc. Nat. Acad. Sci. U.S.A.* **46** (1960), 1242–1243.

[Re2] Reinhart, B.L. Algorithms for Jordan curves on compact surfaces. *Ann. of Math. (2)* **75** (1962), 209–222.

[Ri] Rivin, I. Simple curves on surfaces. *Geom. Dedicata* **87** (2001), 345–360.

[Sel] Selberg, A. Harmonic analysis and discontinuous groups in weakly symmetric Riemannian spaces with applications to Dirichlet series. *J. Indian Math. Soc. (N.S.)* **20** (1956), 47–87.

[Ser] Series, C. The geometry of Markoff numbers. *Math. Intelligencer* **7**(3) (1985), 20–29.

[St] Stillwell, J. *Classical Topology and Combinatorial Group Theory*, second edition, Graduate Texts in Mathematics **72**, Springer-Verlag, New York, 1993.

[Ta] Tan, S. P. Self-intersections of curves on surfaces. *Geom. Dedicata* **62**(2) (1996), 209–225.

[Zi1] Zieschang, H. Algorithmen für einfache Kurven auf Flächen. *Math. Scand.* **17** (1965), 17–40.

[Zi2] Zieschang, H. Algorithmen für einfache Kurven auf Flächen, II. *Math. Scand.* **25** (1969), 49–58.

Matings in holomorphic dynamics

Shaun Bullett [1]

School of Mathematical Sciences, Queen Mary University of London
s.r.bullett@qmul.ac.uk

Abstract

Matings between pairs of holomorphic dynamical systems occur in several different contexts. Polynomial maps can be mated with polynomial maps to yield rational maps, Fuchsian groups can be mated with Fuchsian groups to yield quasifuchsian Kleinian groups, and certain polynomial maps can be mated with certain Fuchsian groups to yield holomorphic correspondences. We consider some of the examples, methods of construction, results and open questions in all three areas. In a final section we present yet another class of examples, matings between Fuchsian groups and holomorphic correspondences. We show how the modular group, which is 'rigid' as a group, has a deformation space in this context, and we describe some of the features of this space.

1 Introduction

We shall be considering *holomorphic dynamical systems* of several kinds, all acting on the Riemann sphere $\hat{\mathbb{C}}$ or some subset of it:

• Fuchsian (respectively Kleinian) groups: discrete subgroups of $PSL_2(\mathbb{R})$ (respectively $PSL_2(\mathbb{C})$) acting by Möbius transformations.
• Iterated rational maps: maps of the form $z \to p(z)/q(z)$ where $p(z)$ and $q(z)$ are polynomials.
• Iterated holomorphic correspondences: multivalued 'maps' of the form $z \to w$ where z and w satisfy some polynomial relation $P(z,w)=0$.

The basic idea of a *mating* is as follows. Suppose we are given two holomorphic dynamical systems F_i $(i = 1,2)$, each acting on a closed

Note: The author thanks the Heilbronn Institute, University of Bristol, for its support during the preparation of the final version of this article.

simply-connected subset A_i of $\hat{\mathbb{C}}$ in such a way that the boundary ∂A_i is invariant under F_i. Moreover suppose that there is a homeomorphism $h : \partial A_1 \to \partial A_2$ which *topologically conjugates* $F_1|_{\partial A_1}$ to $F_2|_{\partial A_2}$. We can then glue the two systems together to form a system on $X = A_1 \cup_h A_2$. If X is homeomorphic to a sphere, we call the new system a *topological mating* between F_1 and F_2, and we can ask whether there is a holomorphic dynamical system F on $\hat{\mathbb{C}}$ which is topologically conjugate to this glued system via a homeomorphism which is holomorphic on the interiors of A_1 and A_2. If such an F exists we say that $F_1|_{A_1}$ and $F_2|_{A_2}$ are *matable* and that F is a *mating* between them.

For the different types of holomorphic dynamical system we take appropriate variations on this basic idea. For example, suppose we wish to mate a polynomial map p_1 with a polynomial map p_2 of the same degree n. In general the filled Julia sets A_1 and A_2 of p_1 and p_2 will not have homeomorphic boundaries, but if A_1 and A_2 are both locally connected, then rather than using a homeomorphism to glue them together we may use the smallest closed equivalence relation generated by regarding each $p_i|_{\partial A_i}$ as a quotient of the map $(z \to z^n)|_{S^1}$ (where S^1 denote the unit circle). And when mating a quadratic polynomial map p with the modular group $PSL_2(\mathbb{Z})$ we shall see that we have to take a wedge of *two* copies of the filled Julia of p for our set A_1 in order to make the dynamics on the boundary match that of the modular group on its limit set $\hat{\mathbb{R}}$. But, once we have made these minor adjustments to definitions, the surprise is how often pairs of holomorphic dynamical systems turn out to be matable, and also the types of the holomorphic dynamical systems that realise these matings. When they exist:

• Matings between Fuchsian groups acting on the complex upper half-plane with limit set $\hat{\mathbb{R}}$ are realised by *quasifuchsian* groups (discrete subgroups of $PSL_2(\mathbb{C})$ which have limit sets which are *quasicircles*).

• Matings between polynomial maps acting on their filled Julia sets are realised by rational maps.

• Matings between Fuchsian groups acting on the complex upper half-plane with limit set $\hat{\mathbb{R}}$ and polynomial maps acting on their filled Julia sets are realised by holomorphic correspondences.

The word 'mating' was introduced by Douady and Hubbard, who in the early 1980s observed the phenomenon of mating between polynomials, in computer pictures of Julia sets of rational maps, and set out to prove its existence. This observation led to a great deal of subsequent work, by Douady and Hubbard themselves and by others. The same term is applied below in the context of a much older topic in Kleinian group theory, that of Bers' Simultaneous Uniformization Theorem [Ber], and to a rather newer topic, that of iterated holomorphic correspondences which fit together the dynamics of polynomial groups and Fuchsian groups, the first examples of which were exhibited in [BP1]. I make no attempt to present a comprehensive survey of all these areas, but rather try to introduce the main ideas, indicating the parallels between the examples of matings in the different areas, and mentioning some of the major results, methods and outstanding questions. More detailed information concerning combinations of holomorphic dynamical systems, and the contents of 'Sullivan's Dictionary' between the theories of iterated rational maps and Kleinian groups, can be found in [Pil] and [MNTU].

2 The Measurable Riemann Mapping Theorem

We first introduce a technical tool that will be a common theme in our construction of matings in different contexts. Recall that the *Riemann Mapping Theorem* asserts that if U is a bounded open simply-connected subset of $\mathbb{C}$, then there exists a conformal homeomorphism $\phi : U \to \mathbb{D}$, where $\mathbb{D}$ denotes the open unit disc in $\mathbb{C}$. The Riemann mapping ϕ is unique up to post-composition by conformal homeomorphisms of $\mathbb{D}$, that is to say fractional linear maps which send the unit disc $\mathbb{D}$ to itself.

The *Measurable Riemann Mapping Theorem* asserts the analogous result when the initial data includes a measurable function $\mu : U \to \mathbb{C}$, and rather than a conformal homeomorphism ϕ from U to $\mathbb{D}$, what we seek is a *quasiconformal* homeomorphism $f : U \to \mathbb{D}$ which at almost every point $z \in U$ has *complex dilatation* $\mu(z)$.

Before we state this theorem formally, we first recall that a homeomorphism f between open sets in $\mathbb{C}$ is said to be K-*quasiconformal* if it sends infinitesimally small round circles to infinitesimally small ellipses which have ratio of major axis length to minor axis length less than or equal to K. The Beltrami differential

$$\mu(f) = \frac{\partial f}{\partial \bar{z}} \Big/ \frac{\partial f}{\partial z}$$

is known as the *complex dilatation* of f. It is a standard result that f is K-quasiconformal, with $K = (1 + k)/(1 - k)$, if and only if $\mu(f)$ is defined almost everywhere and has essential supremum $||\mu||_\infty = k < 1$.

The Measurable Riemann Mapping Theorem is due to Morrey, Bojarski, Ahlfors and Bers [Mor, AB]. The statement below is taken from [Dou] (Théorème 5): it is expressed in terms of functions defined on the whole of $\mathbb{C}$, but can be adapted to suit other situations, for example when the domain of μ is a bounded simply-connected open subset U of $\mathbb{C}$ and we seek a quasiconformal homeomorphism $f : U \to \mathbb{D}$ or indeed when the domain of μ is the Riemann sphere $\hat{\mathbb{C}} = \mathbb{C} \cup \{\infty\}$ and we seek a quasiconformal homeomorphism $f : \hat{\mathbb{C}} \to \hat{\mathbb{C}}$.

The Measurable Riemann Mapping Theorem

Let μ be any L^∞ function $\mathbb{C} \to \mathbb{C}$ with $||\mu||_\infty = k < 1$. Then there exists an orientation-preserving quasiconformal homeomorphism $f : \mathbb{C} \to \mathbb{C}$ which has complex dilatation $\mu(f)$ equal to μ almost everywhere on $\mathbb{C}$. This homeomorphism is unique if we require that $f(0) = 0$ and $f(1) = 1$. Furthermore if μ depends analytically (respectively continuously) on a parameter λ then the homeomorphism f also depends analytically (respectively continuously) on λ.

3 Matings between Fuchsian groups

Let G_1 be a *geometrically finite Fuchsian group*, that is to say a discrete subgroup of $PSL_2(\mathbb{R})$, acting by fractional linear maps on the upper half $\mathcal{U}$ of the complex plane, with a fundamental domain having a finite number of sides. Suppose the limit set of this action is $\hat{\mathbb{R}} = \mathbb{R} \cup \{\infty\}$. The group G_1 also acts on the lower half-plane $\mathcal{L}$ and the limit set of this action is also $\hat{\mathbb{R}}$. Let G_2 be another geometrically finite discrete subgroup of $PSL_2(\mathbb{R})$, isomorphic to G_1 as an abstract group, and such that the action of G_2 on $\mathcal{U}$ is topologically conjugate to that of G_1.

Bers' Simultaneous Uniformisation Theorem

Given subgroups G_1 and G_2 of $PSL_2(\mathbb{R})$ with the properties described above, there exists a discrete subgroup G of $PSL_2(\mathbb{C})$ the action of which on the Riemann sphere $\hat{\mathbb{C}}$ has the following properties:

(i) The limit set of the action is a quasicircle $\Lambda \subset \hat{\mathbb{C}}$.

(ii) On one component, Ω_1, of $\hat{\mathbb{C}} \setminus \Lambda$ the action of G is conformally conjugate to the action of G_1 on $\mathcal{U}$.

(iii) On the other component, Ω_2, of $\hat{\mathbb{C}} \setminus \Lambda$ the action of G is conformally conjugate to the action of G_2 on $\mathcal{L}$.

In the situation described by (i), (ii) and (iii) we may call the *Kleinian* group G (discrete subgroup of $PSL_2(\mathbb{C})$) acting on $\hat{C}$ a *mating* between the Fuchsian group G_1 acting on $\mathcal{U}$ and the Fuchsian group G_2 acting on $\mathcal{L}$. The action of G is a *holomorphic realisation* of the dynamical system obtained by gluing together the actions of G_1 on $\mathcal{U}$ and G_2 on $\mathcal{L}$ by means of a (topological) homeomorphism from the boundary $\partial \mathcal{U}$ of $\mathcal{U}$ to the boundary $\partial \mathcal{L}$ of $\mathcal{L}$ which conjugates the action of G_1 on $\partial \mathcal{U}$ to that of G_2 on $\partial \mathcal{L}$.

Bers' Simultaneous Uniformisation Theorem is a consequence of the Measurable Riemann Mapping Theorem, as we shall now see. Since G_1 is geometrically finite, the orbit space $\mathcal{L}/G_1$ is a Riemann surface with a finite number of marked cone points and puncture points. The orbit space $\mathcal{L}/G_2$ is a Riemann surface with a combinatorially identical set of data. It follows by standard Riemann surface theory that there exists a quasiconformal diffeomorphism $h : \mathcal{L}/G_1 \to \mathcal{L}/G_2$, sending marked points to marked points. The complex dilatation μ of h, when composed with the orbit projection, yields a G_1-equivariant L^∞ function $\mu : \mathcal{L} \to \mathbb{C}$, which we may extend to the whole of $\hat{\mathbb{C}}$ by defining $\mu(z)$ to be zero on $\hat{\mathbb{C}} \setminus \mathcal{L} = \overline{\mathcal{U}}$. An equivalent description, in terms of measurable fields of infinitesimal ellipses, is that the field of ellipses defined by μ on $\mathcal{L}/G_1$ is pulled back to a G_1-equivariant ellipse field on $\mathcal{L}$ and extended to the rest of $\hat{\mathbb{C}}$ by the standard (round) circle field on $\overline{\mathcal{U}}$. By the measurable Riemann Mapping Theorem there now exists a quasiconformal diffeomorphism $\phi : \hat{\mathbb{C}} \to \hat{\mathbb{C}}$ having complex dilatation μ. But, as is easily verified by the chain rule, each element of $G = \phi G_1 \phi^{-1}$ has complex dilatation zero, and so maps infinitesimal round circles to infinitesimal round circles. Thus G is a group of conformal automorphisms of $\hat{\mathbb{C}}$, that is to say a subgroup of $PSL_2(\mathbb{C})$. The limit set of G is the set $\Lambda = \phi(\hat{\mathbb{R}})$, which is a quasicircle by definition, since it is the image of a round circle under ϕ. Moreover ϕ provides a conformal conjugacy between the actions of G_1 on $\mathcal{U}$ and G on $\Omega_1 = \phi(\mathcal{U})$, and $\phi \circ h^{-1} : \mathcal{L} \to \Omega_2 = \phi(\mathcal{L})$ provides a conformal conjugacy between the actions of G_2 on $\mathcal{L}$ and G on $\Omega_2 = \phi(\mathcal{L})$, where here $h : \mathcal{L} \to \mathcal{L}$ denotes the lift of our quasiconformal diffeomorphism $h : \mathcal{L}/G_1 \to \mathcal{L}/G_2$.

3.1 A family of examples: once-punctured torus groups

Consider discrete representations in $PSL_2(\mathbb{R})$ of the free group F_2 on generators X and Y. Let A and B be elements of $PSL_2(\mathbb{R})$ representing X and Y, and restrict attention to the case that A and B are hyperbolic and their commutator $ABA^{-1}B^{-1}$ is parabolic. A generic representation of this kind has fundamental domain a quadrilateral in the upper half-

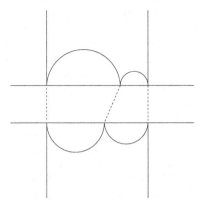

Fig. 3.1. Fundamental domains for different actions of a punctured torus group on the upper and lower half-planes: one can construct a conjugacy between the two actions on the real axis by sending the vertices of one fundamental domain to the vertices of the other, as shown, and extending equivariantly to all the vertices of the respective tilings.

plane, with all four vertices on the (completed) real line and all four sides geodesics in the hyperbolic metric, that is to say arcs of semicircles orthogonal to the real line: the group elements A and B identify pairs of opposite sides of this quadrilateral, and the orbit space is a punctured torus. The cross-ratio of the four vertices of a fundamental domain is a conjugacy invariant of the group (as a subgroup of $PSL_2(\mathbb{R})$). Any two representations G_1 and G_2 of this kind (which in general will have different cross-ratios) provide examples of groups to which we may apply Bers' Theorem. We consider G_1 acting on the upper half-plane and G_2 acting on the lower. Their actions on their common limit set, the completed real axis, will be topologically conjugate, but in general this conjugacy will not be smooth (figure 3.1). Bers' Theorem tells us that we can realise the topological mating obtained by gluing these two actions together as a *holomorphic* dynamical system, a Kleinian group (discrete subgroup of $PSL_2(\mathbb{C})$) which has as its limit set a quasicircle.

On the boundary of this family of examples we find some phenomena that we shall also observe in matings between polynomials and in matings between groups and polynomials. Start with a mating of the kind described above, and choose a simple closed geodesic l on the punctured torus Ω_1/G_1: such geodesics are in one-to-one correspondence with $\mathbb{Q} \cup \{\infty\}$, the p/q geodesic corresponding to a straight line of slope p/q on a square lattice (figure 3.2). Following Maskit [Mas], we may deform G_1,

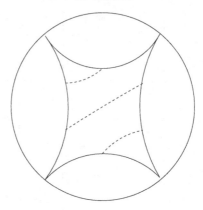

Fig. 3.2. Fundamental domain for a punctured torus group, drawn on the Poincaré disc, with the geodesic for $p/q = 1/2$ marked.

keeping G_2 fixed, by an isotopy contracting the length of this geodesic to zero. The effect on the upper half-plane is to contract to a point each component of the union of the lifts of l to $\mathcal{U}$. We obtain a Kleinian group $G \subset PSL_2(\mathbb{C})$ which has limit set Λ a proper quotient of $\hat{\mathbb{R}}$. On the component Ω_2 of its ordinary set, the action of G remains conformally conjugate to that of G_2 on $\mathcal{L}$. However Ω_1 has now been pinched into (infinitely many) components, each topologically a disc, and the limit set Λ of G is now the closure of a countable infinity of topological circles. A group G of this kind is called a *singly cusped b-group*.

One can obtain a *doubly cusped* group G by contracting a geodesic of slope p'/q' on $\mathcal{L}/G_2$ at the same time as contracting a geodesic of slope p/q on $\mathcal{U}/G_1$. Such a group G can be considered as a *mating* between two singly cusped groups. Note that the cases $p'/q' = -p/q$ are excluded as in these the ends of each curve to be contracted in $\mathcal{U}$ are also the ends of a curve to be contracted in $\mathcal{L}$, so when we perform the contraction certain simple closed curves on $\hat{\mathbb{C}}$ are pinched to points and we no longer have a topological sphere.

Corresponding to a lamination of *irrational* slope ν on the torus $\mathcal{U}/G_1$ is what is known as a *singly degenerate* representation $G \subset PSL_2(\mathbb{C})$, which has ordinary set with just one component (the image of $\mathcal{L}$) and limit set a dendrite. We remark that whereas a singly cusped G is geometrically finite, this is no longer the case for a singly degenerate G. One can further modify a singly degenerate G by contracting a geodesic of rational slope on $\mathcal{L}/G_2$, in which case the dendrite becomes a circle-packing. Finally, corresponding to a lamination of irrational slope ν on $\mathcal{U}/G_1$ and a lamination of irrational slope ν' on $\mathcal{L}/G_2$ (with $\nu' \neq -\nu$) there is a *doubly degenerate* representation, with limit set the entire Riemann sphere $\hat{\mathbb{C}}$. We can think of such a G as a *mating* between two singly degenerate (dendrite) representations. In the next section we shall see analogous behaviour in examples of matings between polynomial maps.

The first examples of doubly degenerate Kleinian groups were due to Jørgensen [Jør1] (see also his 1975 preprint, published as [Jør2]). A well-illustrated account of this phenomenon can be found in the book *Indra's Pearls* [MSW]. The *Classification Theorem for Kleinian Once-punctured Torus Groups* [Min] states that every once-punctured torus group is uniquely expressible as a mating of Fuchsian groups or as one of the pinched types listed above. The proof of this classification was a major achievement of 3-dimensional hyperbolic manifold theory, completed by Minsky's proof of Thurston's *Ending Lamination Conjecture* in this case [Min]. McMullen [McM2] proved that the limit set Λ of a Kleinian once-punctured torus group, degenerate or otherwise, is always a quotient of the circle $\hat{\mathbb{R}}$, with the action of G on Λ a quotient of the unperturbed action of G_1 on $\hat{\mathbb{R}}$, and that as a corollary Λ is always locally connected (this is definitely not the case for Julia sets of polynomials [Mil3]). McMullen proved earlier, [McM1], that cusped groups occur at a dense set of points on the boundary of moduli space.

4 Matings between polynomial maps

We restrict attention to quadratic polynomials to simplify descriptions of matings and also because a more extensive set of results is known for these than for polynomials of higher degree.

Recall that every quadratic polynomial with coefficients in $\mathbb{C}$ is conjugate to a unique polynomial of the form $q_c : z \to z^2 + c$, and that the *Mandelbrot set M* is defined to be the set of all values of $c \in \mathbb{C}$ for which the Julia set of q_c is connected. Equivalently M is the set of all values of c such that the orbit $(q_c^n(0))_{n \geq 0}$ of the critical point 0 is bounded. The *filled Julia set $K(q_c)$* of q_c is the set of all $z \in \mathbb{C}$ which have bounded orbits under q_c. The *Julia set* of q_c is the boundary of $K(q_c)$. More generally, for a rational map $z \to f(z) = p(z)/q(z)$, the *Julia set* of f is the set of points $z \in \hat{\mathbb{C}}$ at which the set of forward iterates $(f^n)_{n \geq 0}$ of f fails to be a normal family [Mil3].

When $c \in M$, there is a conformal bijection ϕ_c between the basin of attraction of the fixed point at infinity, $\hat{\mathbb{C}} \setminus K(q_c)$, and the open unit disc $\mathbb{D}$, conjugating the map q_c on $\hat{\mathbb{C}} \setminus K(q_c)$ to the map $z \to z^2$ on $\mathbb{D}$. The map ϕ_c is unique, once we require it to be tangent to the map $z \to 1/z$ at ∞. Its inverse ϕ_c^{-1} is known as the *Böttcher coordinate* for $\hat{\mathbb{C}} \setminus K(q_c)$, and the image under ϕ_c^{-1} of a radial line $\{re^{2\pi i\theta} : 0 \le r < 1\}$ is called the *external ray* of argument θ for q_c.

By the criterion of Carathéodory, the Böttcher coordinate extends to a continuous map from the unit circle S^1 onto $J(q_c)$ if and only if $J(q_c)$ is locally connected. It is well known [Mil3] that a sufficient condition for $J(q_c)$ to be locally connected is that q_c is *hyperbolic*, that is to say every critical point of q_c lies in the basin of attraction of some periodic orbit. Given two hyperbolic quadratic polynomials q_c and $q_{c'}$ with locally connected Julia sets, we glue a copy K of $K(q_c)$ to a copy K' of $K(q_{c'})$ along their boundaries, using the smallest closed equivalence relation for which $\phi_c^{-1}(e^{2\pi it}) \sim \phi_{c'}^{-1}(e^{-2\pi it})$ for each $t \in [0,1)$. If $(K \cup K')/\sim$ is a topological sphere, the map induced on this sphere by gluing q_c on K to $q_{c'}$ on K' is called a *topological mating* of q_c with $q_{c'}$. If this map is topologically conjugate to a *rational map* f on $\hat{\mathbb{C}}$, via a conjugacy which is conformal on the interiors of K and K', we say the q_c and $q_{c'}$ are *matable*, and that f is a *mating*.

We shall not concern ourselves in detail with the case that $J(q_c)$ is not locally connected, but remark that whether or not $J(q_c)$ is locally connected it is possible to define a *formal mating* by gluing together copies of q_c on $\mathbb{C}$ and $q_{c'}$ on $\mathbb{C}$ along a *circle at infinity*, using the identification $\infty.e^{2\pi it} \to \infty.e^{-2\pi it}$. One can then quotient by the smallest closed equivalence relation for which any two points on the same ray are equivalent, and ask whether the resulting space is a topological sphere.

We shall consider several methods which have been used to show that polynomials are matable.

4.1 Thurston's Criterion and the Quadratic Mating Theorem

A (topological) branched covering map $f : S^2 \to S^2$ is said to be *post-critically finite* if the forward orbits of all branch points are finite. For example a topological mating of two polynomial maps for which all critical points are periodic is such a map. Thurston's Theorem [DH2] provides a criterion for a post-critically finite branched covering map to be *equivalent* to a rational map, and a proof of uniqueness of this rational map (up to Möbius conjugation) when it does exist. For a statement of the criterion, and the definition of 'equivalent' in this context see [DH2]: what concerns us here is that given a topological mating between two postcritically finite hyperbolic polynomials, once we have decided whether they are *topologically matable* (that is whether $(K \cup K')/_\sim$ is a topological sphere) Thurston's Theorem provides a method to decide whether they are *matable*, that is to say whether there is a *rational* map realising the mating. The question of whether $(K \cup K')/_\sim$ is a topological sphere turns out to be purely combinatorial: in the case of hyperbolic quadratic polynomials q_c and $q_{c'}$, it can be shown that the space $(K \cup K')/_\sim$ is a sphere if and only if c and c' are not in conjugate limbs of the Mandelbrot set. If c and c' are in conjugate limbs, then certain pairs of external rays in the *formal* mating have both end points in common, so collapsing them pinches certain simple closed curves on S^2 to points and yields a space that cannot be a topological sphere. For quadratic polynomials, it can be proved by an application of Thurston's Theorem that there is no obstruction to realising a topological mating by a rational map. Using this technique Tan Lei and Mary Rees proved results which together gave a complete solution to the question of matability for quadratic polynomials with periodic critical orbits.

The Quadratic Mating Theorem (Mary Rees, Tan Lei)

Two quadratic polynomials q_c and $q_{c'}$ with periodic critical orbits are matable if and only if c and c' do not belong to conjugate limbs of the Mandelbrot set. [Ree1, Tan]

Using *quasiconformal surgery*, this theorem can be extended to all *hyperbolic* quadratic polynomials. Shishikura [Shi] further extended it to the case where the critical points of q_c (or of both q_c and $q_{c'}$) are *preperiodic*, and so one (or both) of $K(q_c)$ and $K(q_{c'})$ is a dendrite. An explicit case where both are dendrites is examined in detail by Milnor in [Mil2], which is also an excellent source for further information on matings between quadratic polynomials.

4.2 Quadratic-like maps and quasiconformal surgery

To construct matings between polynomials which are not both postcritically finite, we can no longer use Thurston's Theorem. However there are cases in which we can either construct a mating directly, using conformal surgery, or we can deform an existing mating to obtain the one we seek.

Before describing an example of the use of surgery, we first observe that there is a sense in which any quadratic polynomial q_c with connected Julia set can be regarded as a mating between itself and $q_0 : z \to z^2$, since the basin of ∞ of q_c is a copy of the filled Julia set of q_0 (the unit disc $\mathbb{D}$) and on this basin the map q_c is conjugate to the map q_0 on $\mathbb{D}$. A more general notion than that of a polynomial map is that of a *polynomial-like map* [DH1]. This is a proper holomorphic surjection $p : V \to U$ where U and V are simply-connected open sets in $\mathbb{C}$ with $U \supset \overline{V}$. Such a map p has a well-defined *filled Julia set* $K(p) = \bigcap_{n \geq 0} p^{-n}(\overline{V})$ and a well-defined *degree* d. The *Straightening Theorem* of Douady and Hubbard [DH1] asserts that given any polynomial-like map p there is a genuine polynomial map P which is *hybrid equivalent* to p in the sense that there is a quasiconformal conjugacy h between p and P on neighbourhoods of $K(p)$ and $K(P)$ such that h is conformal on the interior of $K(p)$. In the case that $K(p)$ is connected, we can think of P as a *mating* between the polynomial-like map p and the polynomial $z \to z^d$ on their respective filled Julia sets.

Let $int(M_0)$ denote the interior of the *main cardioid* in the Mandelbrot set, that is to say $int(M_0)$ is the set of values c for which q_c has an attracting fixed point other than ∞. These are also the values of c for which the Julia set $J(q_c)$ is a quasicircle, and for $c \in int(M_0)$ the map q_c is conjugate on a neighbourhood of $J(q_c)$ to q_0 on a neighbourhood of the unit circle. Because it is a nice illustration of how one can use the Measurable Riemann Mapping Theorem to construct matings of polynomials in a very similar manner to the way we constructed matings of Fuchsian groups, we adapt Douady and Hubbard's method to show that for any $c \in M$ and $c' \in int(M_0)$ the map q_c can be mated with $q_{c'}$.

Theorem 4.1 *For every $c \in M$ and $c' \in int(M_0)$ the quadratic maps q_c and $q_{c'}$ are matable.*

Proof We first associate an annulus A to q_c. An *equipotential* for q_c is the image of a circle $\{Re^{2\pi it} : 0 \leq t < 1\}$ under ϕ_c^{-1} (where ϕ_c is the Böttcher coordinate conjugating q_c on $\hat{\mathbb{C}} \setminus K(q_c)$ to q_0 on $\mathbb{D}$). It is a smooth Jordan curve parameterised by *external angle* t. The region bounded by such an equipotential is a simply-connected domain V, mapped two-to-one by q_c onto a larger domain $U \supset \overline{V}$ which also has boundary an equipotential parametrised by external angle. Let A denote the annulus $U \setminus \overline{V}$, and denote its inner and outer boundaries by

$\partial_1 A$ and $\partial_2 A$ respectively. The map q_c sends $\partial_1 A$ two-to-one onto $\partial_2 A$. Next, to $q_{c'}$ we associate an annulus B which lies *inside* $K(q_{c'})$ (whereas A lies *outside* $K(q_c)$). This is where we use the hypothesis that $c' \in M_0$. We take a small round circle $\mathcal{C}$ around the attracting fixed point of $q_{c'}$ and let $\partial_2 B = q_{c'}^{-n}(\mathcal{C})$ and $\partial_1 B = q_{c'}^{-(n+1)}(\mathcal{C})$, for a sufficiently large value of n. Since $J(q_{c'})$ is a topological circle, on which the action of $q_{c'}$ is the doubling map, $\partial_2 B$ and $\partial_1 B$ are simple closed curves bounding an annulus B, and $q_{c'}$ maps $\partial_1 B$ two-to-one onto $\partial_2 B$.

By the (classical) Riemann Mapping Theorem there exists a conformal homeomorphism h from the outer component of $\hat{\mathbb{C}} \setminus A$ (the component containing ∞) to the inner component of $\hat{\mathbb{C}} \setminus B$. Since the boundaries $\partial_2 A$ and $\partial_2 B$ of the two regions are smooth curves, h extends to a diffeomorphism $\partial_2 A \to \partial_2 B$.

Lemma 4.2 *The diffeomorphism* $h : \partial_2 A \to \partial_2 B$ *extends to a quasiconformal homeomorphism* h *from* A *to* B *which on boundaries conjugates* $q_c : \partial_1 A \to \partial_2 A$ *to* $q_{c'} : \partial_1 B \to \partial_2 B$.

This Lemma can be proved as follows. Cut the annulus A open along a smooth path linking a point $z_0 \in \partial_2 A$ to one of the two points $q_c^{-1}(z_0) \in \partial_1 A$ (for example cut A along the ray of argument zero), and similarly cut the annulus B along a smooth curve linking $h(z_0) \in \partial_2$ to one of the two points $q_{c'}^{-1} h(z_0) \in \partial_1 B$. By arbitrarily choosing a smooth diffeomorphism between these two curves, we reduce the problem of finding a quasiconformal homeomorphism $h : A \to B$, with the prescribed properties, to the problem of extending a specified smooth diffeomorphism between the boundaries of two rectangles to a quasiconformal homeomorphism between their interiors. But a diffeomorphism between smooth curves is *quasisymmetric* and it is well-known that every quasisymmetric homeomorphism between the boundaries of two rectangles extends to a quasiconformal homeomorphism of interiors - indeed there is a *conformally natural* extension [DE].

Given the quasiconformal homeomorphism h provided by the Lemma, define a branched covering map $f : \hat{\mathbb{C}} \to \hat{\mathbb{C}}$ by setting:

- $f(z) = q_c(z)$ for z in the inner component of $\hat{\mathbb{C}} \setminus A$;
- $f(z) = h^{-1} q_{c'} h(z)$ for $z \in A$ and for z in the outer component of $\hat{\mathbb{C}} \setminus A$.

By construction f is continuous. The dilatation of h defines an ellipse field on A, which we can then pull back inductively under q_c to define an ellipse field on $\bigcup_{n \geq 0} q_c^{-n}(A)$. Extend this to an ellipse field μ on the whole of $\hat{\mathbb{C}}$ by setting it to be the standard field of infinitesimal round circles elsewhere. By construction this ellipse field is invariant under pull-back by f. The Measurable Riemann Mapping Theorem now yields a diffeomorphism $g : \hat{\mathbb{C}} \to \hat{\mathbb{C}}$ which has dilatation μ, and conjugating f by g yields a *conformal* branched covering map $\hat{\mathbb{C}} \to \hat{\mathbb{C}}$. $\qquad\square$

4.3 Pinching, and other techniques for realising matings

The interior, M_0, of the main cardioid of the Mandelbrot set may be parametrised by the multiplier $re^{2\pi i\theta}$ ($r < 1$) of the attracting fixed point. Following along a rational ray $\theta = p/q$ to the boundary of M_0 corresponds in the dynamical plane to pinching to a single point a 'q-pronged star' centred on the attracting fixed point, and also pinching each of its iterated inverse images. For example Douady's rabbit may be obtained from a quadratic map with an attracting (but not super-attracting) fixed point by contracting 3-pronged stars. This process is analogous to the pinching of geodesics described earlier in the case of Kleinian groups in that one is pinching a simple closed curve on the space of grand orbits, but the proofs are technically more difficult in the absence of the 3-dimensional hyperbolic geometry available in that case: one can contract a star to a smaller version by defining an appropriate ellipse field on a suitable neighbourhood and then applying the Measurable Riemann Mapping Theorem to 'straighten' the ellipses into circles, but the final step of contracting to a point is a technically difficult exercise because it involves taking a limit of K-quasiconformal homeomorphisms with K tending to infinity. Cui [Cui] and Haïssinsky [Hai1, Hai2] developed powerful methods of proof, and Haïssinsky and Tan Lei [HT] have applied these to show that in many cases matings between hyperbolic polynomials may be deformed into matings between parabolic polynomials.

Another technique for proving the existence of matings is to identify a candidate rational map and show that it is indeed a mating. By applying Yoccoz puzzle-piece methods to identify candidates, Yampolsky and Zakeri [YZ] were able to demonstrate the existence of matings between Siegel disc quadratic maps. There are many other topics concerning matings between polynomial maps that we do not have space to go into here, for example questions concerning *shared matings* (rational maps that can be expressed as matings in more than one way), questions of *continuity* and *non-continuity* of matings, and questions about the local and global structure of moduli spaces of rational maps. See, for example, [Ree2, Ree3, Eps, Mil1].

5 Matings between polynomial maps and Fuchsian groups

The first hint that it might be possible to mate polynomials of degree 2 with the modular group is the existence of a homeomorphism

$$h : \hat{\mathbb{R}}_{\geq 0} = [0, \infty] \to [0, 1]$$

conjugating the pair of maps on $[0, \infty]$:

$$\{x \to x + 1, \ x \to x/(x+1)\}$$

(which generate $PSL_2(\mathbb{Z})$) to the pair of maps on $[0, 1]$:

$$\{t \to t/2, \ t \to (t+1)/2\}$$

(the inverse binary shift). This conjugacy h (known as the 'Minkowski Question Mark Function') sends $x \in \mathbb{R}$ represented by the continued fraction $[x_0; x_1, x_2, \ldots]$ to $t \in I$ represented by the binary expression consisting of x_0 copies of 1, followed by x_1 copies of 0, followed by x_2 copies of 1, and so on.

If the Julia set $J(q_c)$ of $q_c : z \to z^2 + c$ is connected and locally connected then the Böttcher coordinate $\phi_c^{-1} : \mathbb{D} \to \hat{\mathbb{C}} \setminus K(q_c)$ extends to a continuous surjection $S^1 \to J(q_c)$, which semi-conjugates the map $z \to z^2$ on S^1 (the binary shift) to the map q_c on $J(q_c)$. We deduce that we may use the homeomorphism h described above to glue the action of q_c^{-1} on $J(q_c)$ to that of $\{z \to z + 1, \ z \to z/z + 1\}$ on $[0, \infty]/_{0 \sim \infty}$.

Now consider how the pair of maps $\tau_1 : x \to x + 1$ and $\tau_2 : x \to x/(x+1)$ behave on the rest of the real axis. Restricted to $[-\infty, 0]$ as domain and codomain they define a two-to-one map, conjugate to the binary shift on $[0, 1]$, and restricted to $[-\infty, 0]$ as domain and $[0, \infty]$ as codomain they define a bijection.

Thus if we identify together the points $-\infty, 0$ and ∞ of $[-\infty, \infty]$, to give a 'figure of eight', the pair of maps τ_1, τ_2 induce a $(2 : 2)$ correspondence made up of three parts: on the circle $[-\infty, 0]/_\sim$ a $(2 : 1)$ correspondence conjugate to the binary shift, on the circle $[0, \infty]/_\sim$ a $(1 : 2)$ correspondence conjugate to the inverse shift, and from the circle $[-\infty, 0]/_\sim$ to the circle $[0, \infty]/_\sim$ a $(1 : 1)$ correspondence.

We now take two copies K_1 and K_2 of $K(q_c)$ and glue them together at the boundary point of external angle 0 on each, to form a space $K_1 \vee K_2$. Each point $z \in K(q_c)$ has a 'conjugate' point z' defined by $q_c(z') = q_c(z)$, where $z' \neq z$ unless z is the critical point 0 of q_c. Consider the $(2:2)$ correspondence defined on $K_1 \vee K_2$ by sending

- $z \in K_1$ to $q_c(z) \in K_1$ and to $z' \in K_2$;
- $z \in K_2$ to the pair of points $q_c^{-1}(z) \in K_2$.

It is an elementary exercise to check that this correspondence on $K_1 \vee K_2$ can be glued to that defined by $z \to z+1$ and $z \to z/(z+1)$ on the complex upper half-plane using the surjections $[-\infty, 0]/_{-\infty \sim 0} \to \partial K_1$ and $[0, \infty]/_{0 \sim \infty} \to \partial K_2$ defined above. Thus we have a topological mating between the action of the modular group on the upper half-plane and our $(2:2)$ correspondence on $K_1 \vee K_2$. The question is whether this topological mating can be realised by a *holomorphic correspondence* on $\hat{\mathbb{C}}$, that is to say a multivalued map $f : z \to w$ defined by a polynomial equation $P(z, w) = 0$, where in this case P is of bidegree $(2:2)$ (that is to say of degree 2 in z and also in w).

Definition *Let $q_c : z \to z^2 + c$ be a quadratic map with connected filled Julia set $K(q_c)$. A holomorphic correspondence $f : z \to w$ of bidegree $(2:2)$ is called a mating between q_c and the modular group $PSL_2(\mathbb{Z})$ if:*

(a) there exists a completely invariant open simply-connected region $\Omega \subset \hat{\mathbb{C}}$ and a conformal bijection h from Ω to the upper half-plane conjugating the two branches of $f|_\Omega$ to the pair of generators $z \to z+1$, $z \to z/(z+1)$ of $PSL_2(\mathbb{Z})$;

(b) the complement of Ω is the union of two closed sets Λ_- and Λ_+, which intersect in a single point and are equipped with homeomorphisms $h_\pm : \Lambda_\pm \to K(q_c)$, conformal on interiors, respectively conjugating f restricted to Λ_- as domain and codomain to q_c on $K(q_c)$, and conjugating f restricted to Λ_+ as domain and codomain to q_c^{-1} on $K(q_c)$.

In [BP1] the following one (complex) parameter family of correspondences $z \to w$ was introduced:

$$(1) \qquad \left(\frac{az+1}{z+1} \right)^2 + \left(\frac{az+1}{z+1} \right) \left(\frac{aw-1}{w-1} \right) + \left(\frac{aw-1}{w-1} \right)^2 = 3.$$

It was proved in [BP1] that the parameter value $a = 4$ gives a correspondence which is a mating between $PSL_2(\mathbb{Z})$ and the quadratic polynomial $q_{-2} : z \to z^2 - 2$, and the following conjecture was made.

Conjecture 5.1 *The family (1) of $(2:2)$ correspondences contains*

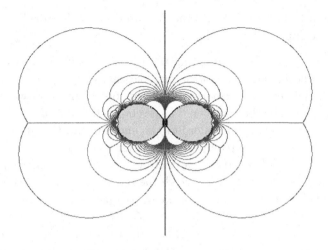

Fig. 5.1. The mating between $q_{1/4} : z \to z^2 + 1/4$ and $PSL_2(\mathbb{Z})$.

matings between $PSL_2(\mathbb{Z})$ and every quadratic polynomial having a connected Julia set, that is to say every $z \to z^2 + c$ with $c \in M$, the Mandelbrot set.

The correspondence in the family (1) given by the parameter value $a = 7$ is plotted in figure 5.1. In [BP1] this correspondence was conjectured to be a mating between $q_{1/4}$ and $PSL_2(\mathbb{Z})$: it follows from recent results of Bullett and Haïsinsky [BHai] (see below) that this is indeed the case. In the figure the 'filled Julia sets' Λ_- and Λ_+ are the two regions shown in grey: on these the correspondence is conjugate to $z \to \sqrt{z - 1/4}$ and $z \to z^2 + 1/4$ respectively. On their complement, $\Omega = \hat{\mathbb{C}} \setminus (\Lambda_- \cup \Lambda_+)$, the action of the correspondence is conjugate to the standard action of $PSL_2(\mathbb{Z})$ on the open upper half-plane $\mathcal{U}$: the tiling of Ω illustrated in figure 5.1 corresponds to the tiling of $\mathcal{U}$ by copies of the fundamental domain $\{z : 0 \le Re(z) \le 1/2, Im(z) > 0, |z - 1| \ge 1\}$. As one varies the parameter a, the sets Λ_- and Λ_+ change shape, appearing to run through all the topological types of filled Julia sets $K(q_c)$ for $c \in M$, the Mandelbrot set. While Λ_- and Λ_+ remain connected, the action of $PSL_2(\mathbb{Z})$ on their complement Ω appears to remain conformally equivalent to the standard action of $PSL_2(\mathbb{Z})$ on the upper half-plane (in [BP1] this is proved to be the case when a is *real*). These observations, and computer plots of parameter space suggesting a resemblance between the Mandelbrot set and the set of correspondences in the family (1) for which Λ_- and Λ_+ are connected, provided strong evidence to support Conjecture 1.

However, difficulties in adapting the theory of *polynomial-like maps* [DH1] to one of 'pinched polynomial-like maps' prevented a general proof. A different question turned out to be easier to answer. The modular group may be considered as a representation of the free product $C_2 * C_3$ of cyclic groups, of orders two and three, in $PSL_2(\mathbb{C})$. Up to conjugacy there is a one (complex) parameter family of such representations, and in the parameter space there is a set $\mathcal{D}$, homeomorphic to a once-punctured closed disc, for which the representation is discrete and faithful. The modular group corresponds to a particular *boundary* point of $\mathcal{D}$. Let r be any representation of $C_2 * C_3$ corresponding to a parameter value in the *interior* $\mathcal{D}^\circ$ of $\mathcal{D}$. The ordinary set $\Omega(r)$ of the Kleinian group defined by such a representation r is connected and the limit set $\Lambda(r)$ is a Cantor set. In [BHar] the notion of a mating between such a representation r of $C_2 * C_3$ and a quadratic polynomial $q_c : z \to z^2 + c$ was introduced: Λ_- and Λ_+ are now disjoint, and their complement Ω is associated to $\Omega(r)$ by the property that the grand orbit space of Ω under the correspondence is conformally isomorphic to the orbit space of $\Omega(r)$ under the Kleinian group generated by $C_2 * C_3$ and the unique involution χ which fixes C_2 and sends each element of C_3 to its inverse (see [BHar]: in the case of the representation $PSL_2(\mathbb{Z})$ of $C_2 * C_3$ the involution χ is $z \to 1/z$). The correspondence plotted in figure 5.2 is an example of such a mating.

Theorem 5.2 *For every quadratic map $q_c : z \to z^2 + c$ with $c \in M$, the Mandelbrot set, and every faithful discrete representation r of $C_2 * C_3$ in $PSL(2, \mathbf{C})$ having connected ordinary set, there exists a polynomial relation $p(z, w) = 0$ defining a $(2 : 2)$ correspondence which is a mating between q_c and r.*

The proof [BHar] of this theorem is similar in spirit to the proof of Theorem 4.1 outlined in the previous section: we associate an annulus A and boundary identification data to each $q_c : z \to z^2 + c$, and we associate another annulus B with similar boundary identification data to each representation r of $C_2 * C_3$ in $PSL_2(\mathbb{C})$ which has a connected ordinary set. We then show that there is a quasiconformal homeomorphism carrying one annulus and data to the other and use this to paste together the different parts of a topological mating. Finally we apply

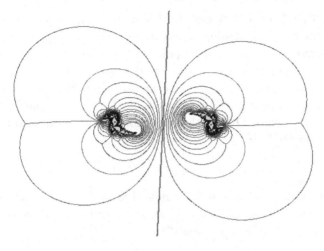

Fig. 5.2. A mating between a generic faithful discrete representation of $C_2 * C_3$ in $PSL_2(\mathbb{C})$ and a quadratic map with an attracting orbit of period 6. This example is a correspondence in the family (2) referred to in the text.

the Measurable Riemann Mapping Theorem to guarantee the existence of an invariant complex structure (see [BHar] for details).

It was also shown in [BHar] that the matings f constructed in the proof of Theorem 5.2 are conjugate to members of the following two-parameter generalisation of the family (1):

$$(2) \qquad \left(\frac{az+1}{z+1}\right)^2 + \left(\frac{az+1}{z+1}\right)\left(\frac{aw-1}{w-1}\right) + \left(\frac{aw-1}{w-1}\right)^2 = 3k.$$

The reason for this is that the method of construction produces not just a correspondence $f : z \to w$, but also an extension of the involution σ on Ω to an involution J on $\hat{\mathbb{C}}$, exchanging Λ_+ with Λ_-, indeed conjugating f to f^{-1} on the whole of $\hat{\mathbb{C}}$, and with the crucial property that $z \to \{z, J \circ f(z)\}$ is an equivalence relation (where 'o' denotes composition). It follows that $J \circ f$ is the *deleted covering correspondence* $Cov_0^Q : z \to w$ of some rational map Q, that is to say $J \circ f$ is defined by

$$\frac{Q(w) - Q(z)}{w - z} = 0.$$

Thus

$$f = J \circ Cov_0^Q$$

for some involution J and rational map Q. We can choose coordinates so that either Q or J (but not both) is in a standard form. Counting singular points of the mating f tells us that Q has one double critical point, corresponding to the fixed point of the rotation ρ on Ω, and two single critical points, corresponding to the critical points of q_c on Λ_+ and Λ_-. Thus, up to pre- and post-composition by Möbius transformations, Q is equivalent to the polynomial $Q(z) = z^3 - 3z$, and by taking coordinates in which Q has this expression we may write the relation defining f in the form

$$(Jw)^2 + (Jw)z + z^2 = 3.$$

Finally if we now put J into the standard form $J(z) = -z$ (necessarily pre-composing our normalised expression for Q by a Möbius transformation) we obtain the expression (2).

As we remarked earlier in this Section, the moduli space $\mathcal{D}$ of faithful discrete representations r of $C_2 * C_3$ is homeomorphic to a once-punctured disc. It is closely related to the moduli space of once-punctured torus groups considered in Section 3. Indeed the orbifold of the ordinary set of any representation r corresponding to an interior point of $\mathcal{D}$ is commensurable with a hyperbolic torus having a single cone point of angle $2\pi/3$ rather than a puncture point, and to each $p/q \in \mathbb{Q} \cup \{\infty\}$ one can associate a simple closed geodesic on this orbifold [BHai], just as in the case of a punctured torus. Contracting the p/q geodesic corresponds to deforming r to a representation corresponding to a boundary cusp of $\mathcal{D}$. For example contracting the geodesic associated to 0 takes one to the representation of $C_2 * C_3$ in $PSL_2(\mathbb{C})$ defined by $PSL_2(\mathbb{Z})$, the modular group, which has limit set $\hat{\mathbb{R}}$. Contracting the geodesic associated to p/q takes one to a representation $r_{p/q}$ of $C_2 * C_3$ in $PSL_2(\mathbb{C})$ which has limit set a *circle-packing*.

In [BHai], Haïssinsky's geodesic-pinching technology is applied to the matings constructed in Theorem 5.2. Applied to the geodesic of rotation number 0 in a mating between a generic faithful representation r of $C_2 * C_3$ and q_c, it yields a proof of Conjecture 5.1 for a large class of quadratic polynomials q_c, including all hyperbolic quadratic polynomials and all quadratic polynomials with parabolic points, for example $q_{1/4}$, as illustrated in figure 5.1 (see [BHai] for the precise definition of this class).

The same method can be applied to pinch a geodesic of any rational rotation number p/q in a mating between r and q_c, at least for certain values of c, but the technicalities become more complicated and a complete proof is only given in [BHai] for the following case:

Theorem 5.3 *For all $p/q \in \mathbb{Q}$ the family (2) contains a mating between the circle-packing representation $r_{p/q}$ of $C_2 * C_3$ in $PSL_2(\mathbb{C})$ and q_0 : $z \to z^2$.*

We remark that there is a topological obstruction to the existence of a mating between a circle-packing representation $r_{p/q}$ and any q_c for which c is in the $(1 - p/q)$-limb of the Mandelbrot set. This is for precisely the same reason that matings between quadratic polynomials q_c and $q_{c'}$ with c and c' in conjugate limbs of M are excluded, namely that the pinching process that would realise such a mating would pinch certain simple closed curves on the sphere to points.

5.1 *Open questions*

1. Is q_0 matable with a *totally degenerate* representation of $C_2 * C_3$ (a representation r_ν with ν irrational)? If such a mating exists and has locally connected limit set, then the combinatorial structure of the mating would be that of two copies of the unit disc carrying the action of $z \to z^2$, glued together along Cantor sets corresponding to *Sturmian orbits* [BS] of rotation number ν in opposite directions on their boundaries, with the remaining parts of the boundaries of these discs 'folded in' to form dendrites.

2. Can one mate a preperiodic q_c (which has limit set a dendrite) with a *totally degenerate* representation of $C_2 * C_3$? This would be a correspondence analogue of a doubly degenerate Kleinian group or a rational map which is a mating of two preperiodic quadratic polynomials.

3. More generally is it true that every discrete faithful representation r of $C_2 * C_3$ in $PSL_2(\mathbb{C})$ is matable with every q_c having $c \in M$, provided only that if c lies in the ρ-limb of M, or (if ρ is irrational) that c lies on the boundary of the cardioid at internal angle ρ, then r is not the circle-packing (or totally degenerate) representation of $C_2 * C_3$ corresponding to the boundary point of moduli space indexed by $1 - \rho$?

Examples of matings between higher degree polynomials and Hecke groups (representations of $C_2 * C_n$ analogous to the modular group) are considered in [BF]. For the general theory of iterated holomorphic correspondences see [BP2].

6 Deforming the modular group: quasifuchsian correspondences

A finitely generated Fuchsian group of G, generated by (say)

$$z \to \frac{a_j z + b_j}{c_j z + d_j} \quad (1 \le j \le n)$$

has the same grand orbits as those of the $(n : n)$ holomorphic correspondence defined by the polynomial relation $P(z, w) = 0$ where

$$P(z, w) = \prod_{j=1}^{n} (w(c_j z + d) - (a_j z + b)).$$

Any deformation of G as a group necessarily retains the reducibility of $P(z, w) = 0$ into factors which are linear in z and w, but we have more freedom when we deform G as a correspondence. As an example, we shall examine a family of correspondence deformations of the modular group $PSL_2(\mathbb{Z})$, with chosen generators $z \to z + 1$ and $z \to z/(z + 1)$. The deformations will be such that the limit set of the correspondence remains a topological circle (indeed a *quasicircle*), on which the action of the correspondence continues to be topologically conjugate to the action of $PSL_2(\mathbb{Z})$ on $\hat{\mathbb{R}}$ throughout. Note that as a *quasifuchsian group* (a Kleinian group with limit set a topological circle) $PSL_2(\mathbb{Z})$ is *rigid*, and no such deformations are possible. A detailed discussion of this family of examples will be presented in [Bul2].

As already observed, the modular group is a representation of the free product $C_2 * C_3$ of cyclic groups of orders 2 and 3. Writing σ for the involution $z \to -1/z$ and ρ for the order three element $z \to -1/(z + 1)$, our chosen generators $z \to z + 1$ and $z \to z/(z + 1)$ for $PSL_2(\mathbb{Z})$ are $\sigma\rho$ and $\sigma\rho^{-1}$. By a suitable change of coordinates, we may view the triple $1, \rho, \rho^{-1}$ as the *covering correspondence* Cov^{Q_0} of the function $Q_0(z) = z^3$, that is to say the correspondence $z \to w$ defined by

$$Q_0(w) - Q_0(z) = 0$$

and so we may view the pair ρ, ρ^{-1} as the *deleted covering correspondence* $Cov_0^{Q_0}$ of $Q_0(z) = z^3$, that is to say the correspondence defined by

$$\frac{Q_0(w) - Q_0(z)}{w - z} = 0.$$

Thus, up to Möbius conjugacy, we may view $PSL_2(\mathbb{Z})$ as a $(2 : 2)$ holomorphic correspondence f_0 defined by a composition

$$f_0 = J \circ Cov_0^{Q_0}$$

where $Q_0(z) = z^3$ and $J(= \sigma)$ is an involution which has the property that the fixed points of f_0 are parabolic, with the derivative of f equal to $+1$ at each of these fixed points. While there are many choices for J, all give the same correspondence f_0 up to Möbius conjugacy. This is just a translation into the language of correspondences of the fact that, up to conjugacy, $PSL_2(\mathbb{Z})$ is the unique representation of $C_2 * C_3$ for which the fixed points of $\sigma\rho$ and $\sigma\rho^{-1}$ are parabolic, each with derivative equal to $+1$.

We deform the correspondence $f_0 = J \circ Cov_0^{Q_0}$ by perturbing $Q_0(z) = z^3$ to $Q_\epsilon(z) = z^3 - 3\epsilon z$, at the same time perturbing the involution J in such a way that the fixed points of $f_\epsilon = J \circ Cov_0^{Q_\epsilon}$ remain parabolic. We now look for a normal form for this family of deformations. Note that provided $\epsilon \neq 0$ the rational function Q_ϵ is determined up to equivalence (that is, up to pre- and post-multiplication by Möbius transformations) by the property that it has one double and two single critical points. Therefore Q_ϵ is equivalent to $Q_1(z) = z^3 - 3z$ (provided $\epsilon \neq 0$), and thus every correspondence of the form $J \circ Cov_0^{Q_\epsilon}$ (where $\epsilon \neq 0$) is conjugate to one of the form $J \circ Cov_0^{Q_1}$. But in the previous Section, in a different context, we have already seen that correspondences $z \to w$ of the form $J \circ Cov_0^{Q_1}$ are conjugate to members of the two-parameter family:

$$(2) \qquad \left(\frac{az+1}{z+1}\right)^2 + \left(\frac{az+1}{z+1}\right)\left(\frac{aw-1}{w-1}\right) + \left(\frac{aw-1}{w-1}\right)^2 = 3k.$$

It remains to identify the restrictions imposed on a and k by the requirement that the fixed points of the correspondence (2) be parabolic. But the fixed points are parabolic if and only if they are double (as fixed points) and a straightforward calculation [Bul2] shows that the correspondence (2) has two double fixed points (rather than the generic four single ones) if and only if:

$$(3) \qquad\qquad k = 1 - \frac{(a-7)^2}{48}.$$

In this normalisation, the group $PSL_2(\mathbb{Z})$ corresponds to the parameter values $k = 0$, $a = 7 + 4\sqrt{3}$. It can be proved (see [Sam]) that for sufficiently small perturbations of a from this value, with k required to continue to satisfy (3), the limit set Λ of the correspondence (2) remains a quasicircle, the action of the correspondence on Λ remains topologically conjugate to that of the modular group on $\hat{\mathbb{R}}$, and the action of the correspondence on one component, Ω_1, of $\hat{\mathbb{C}} \setminus \Lambda$, remains conformally conjugate to that of the modular group on the complex

upper half-plane. However while the action of the correspondence on
the other component, Ω_2, of $\hat{\mathbb{C}} \setminus \Lambda$, remains 'discrete' in the sense that
the grand orbit space is Hausdorff, this action is no longer conjugate to
that of a group.

To see why one might expect such behaviour, we first consider the
geometry of the standard action of the modular group interpreted as
the action of the unperturbed correspondence $J \circ Cov_0^{Q_0}$. As $Q_0(z) = z^3$
has fixed points at 0 and ∞, the subset of $\hat{\mathbb{C}}$ defined by $\{z : -\pi/3 \leq Arg(z) \leq \pi/3\}$ is a transversal D_{Q_0} for Q_0, that is to say a fundamental
domain for ρ, and the parabolic condition on the fixed points of $J \circ Cov_0^{Q_0}$
may be achieved as follows. Draw any circle which has centre on the
positive real axis and which touches the two lines bounding D_{Q_0}. Take
J to be the involution which has its fixed points α and β the points
where this circle cuts the real axis. Writing D_J for the region exterior
to this circle, D_J is a fundamental domain for J, and the pair $\{D_{Q_0}, D_J\}$
satisfy the conditions of the Klein Combination Theorem (their union
covers $\hat{\mathbb{C}}$) so their intersection is a fundamental domain for a faithful
discrete action of $C_2 * C_3$ on $\hat{\mathbb{C}}$. The limit set of this action is the closure
of the orbit of the points P_1 and P_2 where the circle ∂D_J touches the
boundary ∂D_{Q_0}. This limit set is a round circle, and what we see is an
action of $PSL_2(\mathbb{Z})$ which is Möbius conjugate to the standard action on
the upper half-plane (completed at ∞). Note that although we have a
choice of transversal D_{Q_0}, and also a choice of fundamental domain D_J,
all choices give the same action of $PSL_2(\mathbb{Z})$ up to Möbius conjugacy.

Now replace $Q_0(z) = z^3$ by $Q_1(z) = z^3 - 3z$. The double critical point
$z = 0$ of Q_0 is split into two single critical points $z = \pm 1$ of Q_1, but
the double critical point $z = \infty$ of Q_0 remains unchanged. The subset
$[-\infty, -2]$ of the negative real axis is mapped by $Cov_0^{Q_1}$ to a smooth curve
asymptotic to the lines $Arg(z) = \pm\pi/3$ and crossing the positive real
axis orthogonally at $z = 1$. The region to the right of this curve is easily
seen to be a transversal D_{Q_1} for the function Q_1 (a maximal set on which
Q_1 is injective) and thus the analogue of a fundamental domain for the
correspondence $Cov_0^{Q_1}$. We choose an involution J in the same way as
before, that is to say we choose a circle which has centre on the real axis
and which is tangent to ∂D_{Q_1}, and we let J be the involution which has
as its fixed points the two points where this circle intersects the real axis.
As before we let D_J denote the region exterior to the circle. Now D_J
and D_{Q_1} again satisfy the conditions of the Klein Combination Theorem
(which works just as well for correspondences as for groups [Bul1]), so our

correspondence has a 'faithful discrete action' on $\hat{\mathbb{C}}$, with fundamental domain $D = D_J \cap D_{Q_1}$, and limit set the closure of the grand orbit of the pair of points P_1 and P_2 where ∂D_J touches ∂D_{Q_1}. It can be shown [Sam] that in these circumstances the limit set Λ is a quasicircle, and that the action of the correspondence on the component Ω_1 of $\hat{\mathbb{C}} \setminus \Lambda$ containing $z = \infty$ is conformally conjugate to that of the modular group on the upper half-plane. This is because in a neighbourhood of $z = \infty$ the covering correspondence of Q_1 is conjugate to a pair of rotations $\{\rho, \rho^{-1}\}$ through $\pm 2\pi/3$ since Q_1, being a polynomial of degree 3, has a double critical point at $z = \infty$.

Although the construction described in the previous paragraph works for any involution J which has fixed points on the real axis and fundamental domain a round disc whose boundary touches ∂D_{Q_1}, this is only part of the story, as it only gives those members of the family (2) which correspond to *real* values of a. We call these *real* correspondence perturbations of the modular group. To get the full one complex parameter family of deformations we must consider transversals D_{Q_1} for Q_1 which are not symmetric with respect to the real axis, and involutions J whose fixed points are no longer on the real axis, but which still have a fundamental domain D_J with boundary ∂D_J (no longer necessarily a round circle), touching ∂D_{Q_1} at a pair of parabolic fixed points of $J \circ Cov_0^{Q_1}$. The full case is treated in [Bul2], where it is shown that there is an open region $\mathcal{Q}$ in moduli space where Λ remains a quasicircle and the correspondence remains a mating between the modular group on one component of $\hat{\mathbb{C}} \setminus \Lambda$ and a 'quasifuchsian correspondence' on the other. In [Bul2] we call this region $\mathcal{Q}$, which we conjecture to be simply-connected, a *quasifuchsian slice* of moduli space.

To understand the action of the correspondence $f = J \circ Cov_0^{Q_1}$ on the component Ω_2 of $\hat{\mathbb{C}} \setminus \Lambda$, we must consider the behaviour of the forward and backward *singular* points of f: these are the points z which have a single image under f or f^{-1} rather than two images (the case for generic z). The forward singular points of f are the *co-critical* points p_1 and p_2 of Q_1, and the backward singular points are $q_1 = J(p_1)$ and $q_2 = J(p_2)$. In the normalisation where Q_1 is the function $Q_1(z) = z^3 - 3z$ the point p_1 is $z = 2$ and p_2 is $z = -2$.

Note that J is a branch of $f \circ f^{-1} \circ f$, and also a branch of $f^{-1} \circ f \circ f^{-1}$, so for each of $i = 1, 2$ the singular points p_i and q_i are always on the same grand orbit of f. We say that f satisfies a *critical relation* if there is any other grand orbit relationship between p_1, p_2, q_1 and q_2. To be more precise, we say that f satisfies a *critical relation* if either p_1 and p_2 lie on the same grand orbit of f (in which case q_1 and q_2 will also lie on this grand orbit) or for at least one of $i = 1$ and $i = 2$ the points p_i and q_i are related by some branch of mixed iteration of f and f^{-1} other than J. It transpires that a necessary and sufficient condition for the second possibility is that the grand orbit of p_i contains a fixed point of J.

Theorem 6.1 *[Bul2] There exists a countable set $\mathcal{C}$ of parameter values in $\mathcal{Q}$ at which the correspondence f satisfies a critical relation. In the normalisation (2) these form a subset $\{a_n : n \geq 0\}$, of the positive real a-axis. At the parameter value a_n the backward singular point q_1 is mapped by a branch of f^n to the forward singular point p_1. The correspondences f given by all parameter values in $\mathcal{Q}$ other than these a_n are quasiconformally conjugate to one another.*

The points a_n accumulate on the boundary of $\mathcal{Q}$ at $a = 7$, where the correspondence f is the mating of $q_{1/4} : z \to z^2 + 1/4$ with $PSL_2(\mathbb{Z})$ described in [BP1] and pictured in figure 5.1 in Section 5 above.

We outline some of the ingredients of the proof of this theorem. First we examine how the correspondence f acts on Ω_2. In the normalisation in which f is written $J \circ Cov_0^{Q_1}$ (with Q_1 the function $Q_1(z) = z^3 - 3z$), consider a triangular 'tile' Δ which has one side the segment P_1P_2 of ∂D_J running between the fixed points P_1 and P_2 of f, and as its other two sides the images of P_1P_2 under $Cov_0^{Q_1}$, which are a pair of curves P_1P_3 and P_2P_3. See figure 6.1: if the fixed points of J are on the real axis then ∂D_J is a round circle, as illustrated, and the edges of Δ are the arc P_1P_2 of that circle, and the smooth curves shown as dashed lines P_1P_3 and P_2P_3 in the figure, but in general the shape of ∂D_J, and hence of the 'triangle' Δ, may be much distorted. The $(2 : 2)$ correspondence $Cov_0^{Q_1}$ maps Δ to itself. On the boundary of Δ its action is topologically that of a rotation ρ of order 3 and its inverse ρ^{-1}. But on the interior of Δ the action of the $(2 : 2)$ correspondence $Cov_0^{Q_1}$ is best described as opening up a slit from $p_1(= -2)$ to $p_2(= +2)$ and then closing down the resulting circle to the same line segment in a different way. The correspondence $f = J \circ Cov_0^{Q_1}$ acts on Δ by this action followed by a rotation J through an angle π around the mid point of one edge of Δ to send it to the neighbouring tile $J(\Delta)$. The inverse correspondence f^{-1} sends $J(\Delta)$ back to Δ, also opening up a slit and closing it down again. Each of the other tiles in the global orbit making up Ω_2 is an image of $J(\Delta)$ under a unique branch of f^n for some $n > 0$ or an image of Δ under

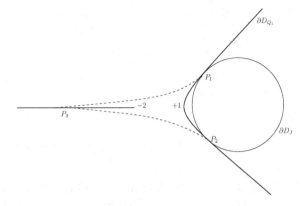

Fig. 6.1. The tile Δ with vertices P_1, P_2, P_3.

a unique branch of f^{-n} for some $n > 0$: the action of the correspondence f permuting the tiles on Ω_2 is combinatorially identical to the action of the generators $\sigma\rho$ and $\sigma\rho^{-1}$ of the modular group $PSL_2(\mathbb{Z})$, permuting copies of the ideal triangle $\Delta = \{z : |Re(z)+1/2| \leq 1/2, |z+1/2| \geq 1/2\}$ on the upper half-plane.

Choose a (generic) base point $\underline{a}$ in $\mathcal{Q}$ and denote by $f_{\underline{a}}$ the correspondence at this parameter value. For convenience in drawing pictures we suppose $\underline{a}$ to be close to (but not equal to) the parameter value $a = a_0$ at which our correspondence f_a is the modular group. The orbifold $\mathcal{O}$ of grand orbits of $f_{\underline{a}}$ on Ω_2 is a punctured sphere with three cone points, each of cone angle π, corresponding to the fixed point of J and the singular points p_1 and p_2. If we mark $\mathcal{O}$ with a path from the cusp to the cone point corresponding to the fixed point of J, and make sure that this path avoids the other two cone points, then we can recover a triangular tile Δ (as in figure 6.1) as a triple branched cover of $\mathcal{O}$, with each edge of Δ a double cover of the marked path in $\mathcal{O}$. Once we have Δ we can recover the whole of Ω_2, and the action of $f_{\underline{a}}$ on it, by analytic continuation (laying out adjacent tiles inductively). Thus $f_{\underline{a}}$ is determined by the complex structure of $\mathcal{O}$, together with the marking. The parameter values in $\mathcal{Q}$ at which there are critical relations form a set $\mathcal{C}$ of isolated points, so given any value $a \in \mathcal{C}$ we can find a path from $\underline{a}$ to a in $\mathcal{Q} \setminus \mathcal{C}$ along which the complex structure on $\mathcal{O}$ is continuously deformed. But if we mark one more curve on $\mathcal{O}$, a path from the cusp to one of the cone points corresponding to p_1 and p_2, and we double cover the marked orbifold, taking as ramification points the cone points and the puncture

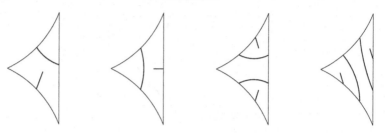

Fig. 6.2. Patterns $P_{1,1}, \rho(P_{1,1}), P_{1,2}$ and $\rho(P_{1,2})$: the vertical edge is E_1.

point, we obtain the orbifold of a (marked) punctured torus group, as considered in Subsection 3.1. The deformation theory of these groups is well-understood, and we can deduce that every quasiconformal deformation of $f_{\underline{a}}$ can be achieved in such a way as to preserve a simple closed geodesic $\gamma_{p/q}$ of rational slope p/q, or a geodesic lamination γ_ν of slope ν, on the punctured torus.

Under the procedure of quotienting by the elliptic involution on the punctured torus to obtain $\mathcal{O}$, and then passing to a triple cover and cutting along a marked curve to obtain Δ, a simple closed geodesic on the punctured torus becomes a *pattern* $\mathcal{P}$ on Δ with the following properties:

(a) $\mathcal{P}$ is invariant under Cov^{Q_1};

(b) every arc in $\mathcal{P}$ meets the interval $[p_1, p_2]$ exactly once;

(c) every arc in $\mathcal{P}$ which ends in the interior of Δ, does so at either p_1 or p_2;

(d) the end points of arcs on the edge of Δ containing the fixed point of the involution J form a set invariant under J.

We call such a $\mathcal{P}$ an *invariant pattern*. Write E_1 for the edge of Δ containing the fixed point of J, and write E_2, E_3 for the other two edges (in anticlockwise order from E_1). In [Bul2] a standard set of invariant patterns $\{P_{m,n}\}$ is defined, for pairs (m, n) with $0 \leq m \leq n$. These patterns have the following properties, which characterise $P_{m,n}$ up to isotopies of Δ which keep its vertices fixed, but which are allowed to move the singular points p_1 and p_2:

(i) Each edge of Δ meets n arcs of $P_{m,n}$;

(ii) m of the arcs starting on E_1 continue to E_2, and the remaining $n - m$ continue to E_3.

See figure 6.2 for examples. It is shown in [Bul2] that every $\gamma_{p/q}$

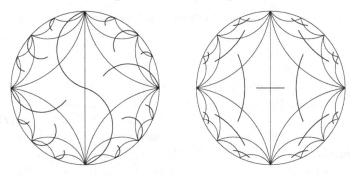

Fig. 6.3. The global laminations corresponding to $P_{1,1}$ and $\rho(P_{1,1})$.

gives rise on the tile Δ to one of these patterns $P_{m,n}$ or to a rotated version $\rho^{\pm 1}(P_{m,n})$ (where ρ is a rotation through $2\pi/3$). If we transfer the pattern to each of the other tiles making up Ω_2 by applying the appropriate branch of f^n to $J(\Delta)$ or of f^{-n} to Δ, we obtain a *global lamination* on Ω_2. Such laminations are of two types: those where each leaf meets only a finite number of tiles, and those where each leaf meets infinitely many tiles (see figure 6.3 for an example of each type). It is a straightforward exercise to deform the complex structure of $f_{\underline{a}}$ to pinch each component of a global lamination of the first kind to a point, and obtain a critical relation corresponding to a point of $\mathcal{C}$, that is to say in the interior of $\mathcal{Q}$. Moreover our knowledge of the deformation space of punctured tori can be applied to show that every critical relation corresponding to a point of $\mathcal{C}$ can be obtained in this way. Conjecturally, laminations of the second kind correspond to a dense set of points on the boundary of $\mathcal{Q}$, but it seems quite a difficult technical problem to apply Haïssinsky's technology to pinch leaves which meet the boundary of Ω_2.

To complete the classification of the points of $\mathcal{C}$, we must examine which patterns give rise to the same critical relation. This is a matter of keeping track of markings: as we move once around any loop in $\mathcal{Q}\setminus\mathcal{C}$ the global lamination is isotoped *globally*, the boundary of Δ is isotoped to a new position (but with the vertices unchanged), and the intersection of the new lamination with the original Δ will in general be a new pattern.

Proposition 6.2 *Under the action of $\pi_1(\mathcal{Q}\setminus\mathcal{C})$ every pattern is equivalent to one of the following:*

(i) $P_{0,0}$, $P_{0,1}$, $P_{1,1} = \rho(P_{0,1})$, $\rho^{-1}(P_{0,1}) = \rho(P_{1,1})$;

(ii) $P_{m,n}$ for some coprime pair of non-negative integers (m,n) with $1 \leq m < n$;

(iii) $\rho(P_{1,n})$ or $\rho^{-1}(P_{n-1,n})$ for some $n > 1$.

For a proof of this proposition see [Bul2]. It is also shown there that the global laminations associated to those patterns which have leaves which each meet only a *finite* number of copies of Δ are $P_{0,0}$, $\rho^{-1}(P_{0,1})$ and those of type (iii), so these are the only patterns associated to critical relations in the interior of $\mathcal{Q}$. Moreover it is not hard to see that the patterns $\rho(P_{1,n})$ and $\rho^{-1}(P_{n-1,n})$ give rise to the same critical relation. The proof of Theorem 6.1 is completed by showing that these give the *real* critical relations described in the statement. The pattern $P_{0,0}$ corresponds to an arc from p_1 to p_2 inside Δ and gives the group case, corresponding to the parameter value a_0. Figure 6.4 illustrates the dynamics of the correspondence at the parameter value a_1, where the next critical relation occurs. Here the normalisation is chosen to put the fixed points of J at the origin and infinity. We see the action of $PSL_2(\mathbb{Z})$ on the outer component, Ω_1, of $\hat{\mathbb{C}} \setminus \Lambda$: the action of the correspondence on the inner component, Ω_2, is that obtained from the $PSL_2(\mathbb{Z})$ action by pinching to points the connected components of the global lamination corresponding to $\rho(P_{1,1})$ (illustrated on the right in figure 6.3). After pinching, the tile boundaries in the right-hand picture in figure 6.3 become the vertical diameter and its images under the correspondence in figure 6.4: these are alternate lines in the figure. We remark that the limit set in figure 6.4 is not a *round* circle - indeed by the Schwarz reflection principle it cannot be a smooth curve.

The global laminations corresponding to the (non-rotated) $P_{m,n}$ with $(m,n) \neq (0,0)$ have leaves which all meet infinitely many tiles and therefore all have at least one end on the limit set Λ. We remark that pinching either of the laminations $P_{0,1}$ or $P_{1,1}$ (the latter is illustrated on the left in figure 6.3) deforms the generic correspondence to the mating of the quadratic map $q_{1/4}$ with $PSL_2(\mathbb{Z})$ illustrated in figure 5.1.

Finally we remark that if instead of splitting just one of the double critical points of $Q_0(z) = z^3$, we split both, and replace Q_0 by a generic rational map Q of degree 3, the construction we have outlined in this section yields quasifuchsian correspondences that do not act as a group on either of the components of $\hat{\mathbb{C}} \setminus \Lambda$. A generic such Q has four single critical points, and by choosing Q appropriately we can create a real one-parameter family of Fuchsian correspondence deformations of the

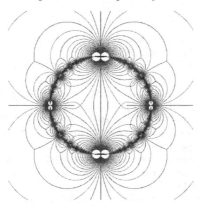

Fig. 6.4. Dynamics of the correspondence at the parameter value a_1.

modular group which all have limit set Λ a *round* circle and where the actions on the two components of its complement are reflections of one another in this round circle.

6.1 Open questions

1. Is it possible to realise holomorphic examples of all the pinched correspondences given by pinching the laminations $P_{m,n}$ described above? Is the only restriction to realising two such pinched correspondences simultaneously on the two sides of the limit set the condition that patterns are not of the form $P_{m,n}$ on one side and $P_{n-m,n}$ on the other? Are the conjectured degenerate representations corresponding to irrational laminations all realisable, either on just one component of $\hat{\mathbb{C}} \setminus \Lambda$ or on both? The latter would be a type of *doubly degenerate* correspondence.

2. Given a mating of $PSL_2(\mathbb{Z})$ with a quadratic polynomial q_c, as in Section 5, we can deform the representation of $PSL_2(\mathbb{Z})$ as a correspondence. Can we deform it to a degenerate correspondence? When $K(q_c)$ is a dendrite, this would yield yet another type of *doubly degenerate* correspondence.

References

[AB] Ahlfors, L. and Bers, L. Riemann's mapping theorem for variable metrics. *Annals of Math.* **72** (1960), 385–404.

[Ber] Bers, L. Simultaneous uniformization. *Bull. Amer. Math. Soc.* **66** (1960), 94–97.

[Bul1] Bullett, S. A combination theorem for covering correspondences and an application to mating polynomial maps with Kleinian groups. *AMS Journal of Conformal Geometry and Dynamics* **4** (2000), 75–96.

[Bul2] Bullett, S. Deformations of the modular group as a quasifuchsian correspondence. In preparation, 2008.

[BF] Bullett, S. and Freiberger, M. Holomorphic correspondences mating Chebyshev-like maps and Hecke groups. *Ergodic Theory Dynam. Systems* **25** (2005), 1057–1090.

[BHai] Bullett, S. and Haïssinsky, P. Pinching holomorphic correspondences. *AMS Journal of Conformal Geometry and Dynamics* **11** (2007), 65–89.

[BHar] Bullett, S. and Harvey, W. Mating quadratic maps with Kleinian groups via quasiconformal surgery. *Electronic Research Announcements of the AMS* **6** (2000), 21–30.

[BP1] Bullett, S. and Penrose, C. Mating quadratic maps with the modular group. *Invent. Math.* **115** (1994), 483–511.

[BP2] Bullett, S. and Penrose, C. Regular and limit sets for holomorphic correspondences. *Fundamenta Math.* **167** (2001), 111–171.

[BS] Bullett, S. and Sentenac, P. Ordered orbits of the shift, square roots and the devil's staircase. *Math. Proc. Cam. Phil. Soc.* **115** (1994), 451–481.

[Cui] Cui, G.-Z. Dynamics of rational maps, topology, deformation and bifurcation. Preprint 2002.

[Dou] Douady, A. Le théorème d'intégrabilité des structures presque complexes, in *The Mandelbrot Set, Theme and Variations*, ed. Tan Lei, London Mathematical Society Lecture Notes **274**, Cambridge University Press, 2000, pp. 307–324.

[DE] Douady, A. and Earle, C. J. Conformally natural extensions of homeomorphisms of the circle. *Acta Math.* **157** (1986), 23–48.

[DH1] Douady, A. and Hubbard, J. H. On the dynamics of polynomial-like mappings. *An. de l'École Norm. Sup.* **18** (1985), 287–343.

[DH2] Douady, A. and Hubbard, J. H. A proof of Thurston's topological characterization of rational functions. *Acta Math.* **168** (1992), 11–47.

[Eps] Epstein, A. L. Bounded hyperbolic components of quadratic rational maps. *Ergodic Theory Dynam. Systems* **20** (2000), 727–748.

[Hai1] Haïssinsky, P. Chirurgie parabolique. *C.R. Acad. Sci. Paris* **327** (1998), 195–198.

[Hai2] Haïssinsky, P. Pincement de polynômes. *Comment. Math. Helv.* **77** (2002), 1–23.

[HT] Haïssinsky, P. and Tan Lei, Matings of geometrically finite polynomials. *Fund. Math.* **181** (2004), 143–188.

[Jør1] Jørgensen, T. Compact 3-manifolds of constant negative curvature fibering over the circle. *Ann. Math.* **106** (1977), 51–72.

[Jør2] Jørgensen, T. On pairs of once-punctured tori, in *Kleinian Groups and Hyperbolic 3-Manifolds*, ed. Y. Komori, V. Markovic and C. Series, Lond. Math. Soc. Lecture Notes **299**, Cambridge University Press, 2003, pp. 183–207.

[Mas] Maskit, B. Parabolic elements in Kleinian groups. *Ann. Math.* **117** (1983), 659–668.

[McM1] McMullen, C. Cusps are dense. *Ann. Math.* **133** (1991), 217–247.

[McM2] McMullen, C. Local connectivity, Kleinian groups and geodesics on the blowup of the torus. *Invent. Math* **146** (2001), 35–91.

[Mil1] Milnor, J. Rational maps with two critical points. *Experimental Math.* **9** (2000), 481–522.

[Mil2] Milnor, J. Pasting together Julia sets: a worked out example of mating. *Experimental Math.* **13** (2004), 55–92.

[Mil3] Milnor, J. *Dynamics in One Complex Variable*, third edition, Annals of Mathematics Studies No. 160, Princeton University Press, 2006.

[Min] Minsky, Y. The classification of punctured torus groups. *Ann. Math.* **149** (1999), 559–626.

[Mor] Morrey, C. B. On the solutions of quasi-linear elliptic partial differential equations. *Trans. Amer. Math. Soc.* **43** (1938), 126–166.

[MNTU] Morosowa, S., Nishimura, Y., Taniguchi, M. and Ueda, T. *Holomorphic Dynamics*, Cambridge Studies in Advanced Mathematics **66**, Cambridge University Press, 2000.

[MSW] Mumford, D., Series, C. and Wright, D. *Indra's Pearls: the Vision of Felix Klein*, Cambridge University Press, 2002.

[Pil] Pilgrim, K. *Combinations of Complex Dynamical Systems*, Lecture Notes in Mathematics No. 1827, Springer-Verlag, Berlin and Heidelberg, 2003.

[Ree1] Rees, M. Realization of matings of polynomials as rational maps of degree two. Manuscript, 1986.

[Ree2] Rees, M. Components of degree two hyperbolic rational maps. *Invent. math.* **100** (1990), 357–382.

[Ree3] Rees, M. *Views of Parameter Space: Topographer and Resident*, Astérisque **28** (2003).

[Sam] Samarasinghe, M. PhD Thesis, Queen Mary, University of London, 2008.

[Shi] Shishikura, M. On a theorem of M. Rees for matings of polynomials, in *The Mandelbrot Set, Theme and Variations*, ed. Tan Lei, Lond. Math. Soc. Lecture Notes **274**, Cambridge University Press, 2000, pp. 289–305

[Tan] Tan Lei, Matings of quadratic polynomials. *Ergodic Theory Dynam. Systems* **12** (1992), 589–620.

[YZ] Yampolsky, M. and Zakeri, S. Mating Siegel quadratic polynomials. *J. Amer. Math. Soc.* **14** (2001), 25–78.

Equisymmetric strata of the singular locus of the moduli space of Riemann surfaces of genus 4

Antonio F. Costa [1]

Departamento de Matemáticas Fundamentales, UNED
acosta@mat.uned.es

Milagros Izquierdo [2]

Matematiska institutionen, Linköpings universitet
miizq@mai.liu.se

Abstract

The moduli space $\mathcal{M}_g$ is the space of analytic equivalence classes of Riemann surfaces of a fixed genus g. The space $\mathcal{M}_g$ has a natural equisymmetric stratification, each stratum consists of the Riemann surfaces with an automorphisms group acting in a given topological way. Using Fuchsian groups we describe the equisymmetrical stratification for the moduli space of surfaces of genus 4.

1 Introduction

Harvey [Har] alluded to the existence of the *equisymmetric stratification* of $\mathcal{M}_g$. Broughton [Bro1] showed that the equisymmetric stratification is indeed a stratification of $\mathcal{M}_g$ by irreducible algebraic subvarieties whose interior, if it is non-empty, is a smooth, connected, locally closed algebraic subvariety of $\mathcal{M}_g$, Zariski dense in the stratum. The equisymmetric stratification gives a contractible simplicial complex $\mathcal{X}$ such that the modular group acts on $\mathcal{X}$. In this paper we find the stratification of the moduli space of Riemann surfaces of genus four into smooth locally closed subvarieties, the *equisymmetric strata* [Bro1] such that each stratum consists of *equisymmetric* surfaces. Two closed Riemann surfaces $X, \overline{X}$ of genus g are called *equisymmetric* if their automorphism groups determine conjugate finite subgroups of the modular group of genus g.

The equisymmetric strata are in correspondence with topological equivalence classes of orientation preserving actions of a finite group G on a surface X. In this way one can represent the modular group as a "cell complex" of groups. An analysis of the stratification may give results in the homology and cohomology of the modular group.

Each stratum corresponds with a finite subgroup of the modular group represented as the full group of automorphisms of some compact Riemann surface. To find such full automorphisms groups we need to use the list of finite maximal signatures for Fuchsian groups in [Sin2].

We obtain that the equisymmetric stratification of $\mathcal{M}_4$ has 41 equisymmetric types of strata described in the last Theorem of this paper. Recently, Kimura [Kim], and earlier Bogopolski [Bo], found the topological classification of groups of automorphisms of Riemann surfaces of genus 4. Bogopolski [Bo] found as well the maximal isotropy groups of strata in $\mathcal{M}_4$ but it is important to remark that such information does not provide the equisymmetric stratification that we are presenting now. The results of [Bo] and [Kim] agree with our results and could shorten drastically our exposition but we present here the calculations for sake of completeness. Finally and related with this line of research, we point out that Breuer [Bre] has found the automorphism groups for Riemann surfaces of genera up to 48.

2   Riemann surfaces and Fuchsian groups

A Fuchsian group Γ is a discrete subgroup of the group $\mathcal{G} = \mathrm{Aut}(\mathcal{D})$ of conformal automorphisms of $\mathcal{D}$. The algebraic structure of a Fuchsian group and the geometric structure of its quotient orbifold are given by the signature of Γ:

$$s(\Gamma) = (g;\ m_1, ..., m_r). \qquad (2.1)$$

The orbit space $\mathcal{D}/\Gamma$ is an orbifold with underlying surface of genus g, having r cone points. The integers m_i are called the periods of Γ and they are the orders of the cone points of $\mathcal{D}/\Gamma$. The group Γ is called the *fundamental group* of the orbifold $\mathcal{D}/\Gamma$.

A group Γ with signature (2.1) has a *canonical presentation*:

$$\langle x_1, ..., x_r, a_1, b_1, ..., a_g, b_g |$$

$$x_i^{m_i}, i = 1, ..., r, x_1 ... \, x_r a_1 b_1 a_1^{-1} b_1^{-1} ... \, a_g b_g a_g^{-1} b_g^{-1} \rangle. \qquad (2.2)$$

The last relation in the above presentation is called the long relation.

The generators $x_1, ..., x_r$, are called the *elliptic generators*. Any elliptic element in Γ is conjugated to a power of some of the elliptic generators.

The hyperbolic area of the orbifold $\mathcal{D}/\Gamma$ coincides with the hyperbolic area of an arbitrary fundamental region of Γ and equals:

$$\mu(\Gamma) = 2\pi(2g - 2 + \sum_{i=1}^{r}(1 - \frac{1}{m_i})), \tag{2.3}$$

Given a subgroup Γ' of index N in a Fuchsian group Γ, one can calculate the structure of Γ' in terms of the structure of Γ and the action of Γ on the Γ'-cosets:

Theorem 2.1 *([Sin1]) Let Γ be a Fuchsian group with signature (2.1) and canonical presentation (2). Then Γ contains a subgroup Γ' of index N with signature*

$$s(\Gamma') = (h; m'_{11}, m'_{12}, ..., m'_{1s_1}, ..., m'_{r1}, ..., m'_{rs_r}) \tag{2.4}$$

if and only if there exists a transitive permutation representation $\theta : \Gamma \rightarrow \Sigma_N$ satisfying the following conditions:

   (i) *The permutation $\theta(x_i)$ has precisely s_i cycles of lengths less than m_i, the lengths of these cycles being $m_i/m'_{i1}, ..., m_i/m'_{is_i}$.*
   (ii) *The Riemann-Hurwitz formula*

$$\mu(\Gamma')/\mu(\Gamma) = N. \tag{2.5}$$

The map $\theta : \Gamma \rightarrow \Sigma_N$ is the *monodromy* of the covering $f : \mathcal{D}/\Gamma' \rightarrow \mathcal{D}/\Gamma$. Moreover $\Gamma'^{-1}(Stb(1))$, where $1, 2, ..., N$ are the labels of the sheets of the covering f.

A Fuchsian group Γ without elliptic elements is called a *surface group* and it has signature $(h; -)$. Given a Riemann surface represented as the orbit space $X = \mathcal{D}/\Gamma$, with Γ a surface Fuchsian group, a finite group G is a group of automorphisms of X if and only if there exists a Fuchsian group Δ and an epimorphism $\theta : \Delta \rightarrow G$ with $ker(\theta) = \Gamma$. The Fuchsian group Δ is the lifting of G to the universal covering $\pi : \mathcal{D} \rightarrow \mathcal{D}/\Gamma$ and is called the *universal covering transformations group* of (X, G).

Let Γ be a Fuchsian group with signature (2.1). Then the Teichmüller space $T(\Gamma)$ of Γ is homeomorphic to a complex ball of dimension $d(\Gamma) = 3g - 3 + r$ (see [Nag]). Let $\Gamma' \leq \Gamma$ be Fuchsian groups, the inclusion mapping $\alpha : \Gamma' \rightarrow \Gamma$ induces an embedding $T(\alpha) : T(\Gamma) \rightarrow T(\Gamma')$ defined by $[r] \mapsto [r\alpha]$. See [Nag] and [Sin2].

The modular group of Γ is the quotient group $\mathrm{Mod}(\Gamma) =$

Aut(Γ)/Inn(Γ), where Inn(Γ) is the normal subgroup of Aut(Γ) consisting of all inner automorphisms of Γ. The *moduli space* of Γ is the quotient $\mathcal{M}(\Gamma) = T(\Gamma)/\text{Mod}(\Gamma)$ endowed with the quotient topology.

A signature s of Fuchsian groups is called *finitely maximal* if for any Fuchsian group Γ with signature s and a group Γ' containing it, d(Γ') < d(Γ). By Theorem 5.1.2 of [BEGG] given a finitely maximal signature there is a Fuchsian group with such signature which is not contained in any other Fuchsian group. Then to decide whether a given finite group can be the full group of automorphism of some compact Riemann surface we use the list of finite maximal signatures. To decide whether a signature is finite maximal we will need all pairs of signatures $s(\Gamma)$ and $s(\Gamma')$ for some Fuchsian groups Γ and Γ' such that $\Gamma' \leq \Gamma$ and d(Γ) = d(Γ'). The full list of such pairs of groups was obtained by Singerman in [Sin2].

3 Strata of Riemann surfaces of genus 4

Two closed Riemann surfaces $X, \overline{X}$ of genus 4 are called *equisymmetric* if their automorphisms groups determine conjugate subgroups of the modular group of genus 4.

An *(effective and orientable) action* of a finite group G on a Riemann surface X is a representation $\epsilon : G \to \text{Aut}(X)$. Two actions ϵ, ϵ' of G on a Riemann surface X are *(weakly) topologically equivalent* if there is a $w \in \text{Aut}(G)$ and an $h \in Hom^+(X)$ such that $\epsilon'(g) = h\epsilon w(g)h^{-1}$. The equisymmetric strata are in correspondence with topological equivalence classes of orientation preserving actions of a finite group G on a surface X. See [Bro1] pages 103–104.

Each stratum corresponds with a finite subgroup of the modular group represented as the full group of automorphisms of some compact Riemann surface. To find such full automorphisms groups we need to use the list of finite maximal signatures for Fuchsian groups in [Sin2].

Remark. In order to shorten the calculations we will use the well-known fact that a Riemann surface of genus four does not admit an automorphism of prime order p for $p \neq 2, 3, 5$.

Using the Riemann-Hurwitz and the above Remark we obtain the following possible signatures for the Fuchsian groups Δ.

Lemma 3.1 *Possible signatures for the Fuchsian groups Δ that can admit a surface subgroup of genus four, where the ones marked by asterisks correspond to non-maximal signatures.*

$|G|$	$s(\Delta)$	$|G|$	$s(\Delta)$	$|G|$	$s(\Delta)$
2	$(2;2,2)$,	2	$(1;2,\overset{6}{\ldots},2)$,	2	$(0;2,\overset{10}{\ldots},2)$,
3	$(2;-)$,	3	$(1;3,3,3)$,	3	$(0;3,\overset{6}{\ldots},3)$,
4	$(0;2,2,2,2,4,4)$,	4	$(0;2,4,4,4,4)$,	4	$(1;4,4)^{*}$,
4	$(1;2,2,2)$,	4	$(0;2,\overset{7}{\ldots},2)$,	5	$(0;5,5,5,5)$,
6	$(1;2,2)^{*}$,	6	$(0;2,\overset{6}{\ldots},2)$,	6	$(0;2,6,6,6)$,
6	$(0;2,2,2,3,6)$,	6	$(0;2,2,3,3,3)$,	6	$(0;3,3,6,6)^{*}$,
8	$(0;2,2,2,2,4)$,	8	$(0;2,2,8,8)^{*}$,	8	$(0;2,4,4,4)$,
9	$(0;9,9,9)^{*}$,	9	$(0;3,3,3,3)^{*}$,	10	$(0;5,10,10)^{*}$,
10	$(0;2,2,5,5)^{*}$,	12	$(1;2)$	12	$(0;6,6,6)^{*}$,
12	$(0;3,12,12)^{*}$,	12	$(0;2,2,4,4)^{*}$,	12	$(0;4,6,12)$,
12	$(0;2,2,3,6)$,	12	$(0;2,2,2,2,2)$,	12	$(0;2,3,3,3)$,
15	$(0;5,5,5)^{*}$,	15	$(0;3,5,15)$,	18	$(0;2,9,18)$,
18	$(0;2,2,3,3)^{*}$,	18	$(0;2,2,2,6)$,	20	$(0;2,2,2,5)$,
20	$(0;2,10,10)^{*}$,	20	$(0;4,4,5)^{*}$,	24	$(0;3,4,6)$,
24	$(0;2,2,2,4)$,	24	$(0;4,4,4)^{*}$,	24	$(0;3,3,12)^{*}$,
24	$(0;2,8,8)^{*}$,	24	$(0;2,6,12)^{*}$,	27	$(0;3,3,9)^{*}$,
27	$(0;2,6,9)$,	30	$(0;2,5,10)^{*}$,	32	$(0;2,4,16)$,
36	$(0;2,2,2,3)$,	36	$(0;3,3,6)^{*}$,	36	$(0;2,6,6)^{*}$,
36	$(0;3,4,4)^{*}$,	40	$(0;2,4,10)$,	45	$(0;3,3,5)^{*}$,
48	$(0;2,3,24)$,	48	$(0;2,4,8)^{*}$,	54	$(0;2,3,18)$,
60	$(0;2,3,15)$,	60	$(0;2,5,5)^{*}$,	72	$(0;2,4,6)$,
72	$(0;2,3,12)$,	72	$(0;3,3,4)^{*}$,	90	$(0;2,3,10)^{*}$,
108	$(0;2,3,9)$,	120	$(0;2,4,5)$,	144	$(0;2,3,8)$

non-maximal signature

Each action of a finite group G on a surface X_4 is determined by an epimorphism $\theta : \Delta \to G$ from a Fuchsian group Δ such that $ker(\theta) = \Gamma$, where $X_4 = \mathcal{D}/\Gamma$ and Γ is a **surface** Fuchsian group. The condition Γ to be a surface Fuchsian group imposes that the order of the image under θ of an elliptic generator x_i of Δ is the same as the order of x_i and $\theta(x_1)\theta(x_2)\ldots\theta(x_{r-1}) = \theta(x_r)^{-1}$. Two epimorphisms $\theta_1, \theta_2 : \Delta \to G$ define two topologically equivalent actions of G on X if and only if there exist automorphisms $\phi : \Delta \to \Delta$, $w : G \to G$ such that $\theta_2 = w \circ \theta_1 \circ \phi^{-1}$. See Proposition 2.2 in [Bro2], [Smi] and [Zie].

Let $\mathcal{B}$ be the subgroup of $\text{Aut}(\Delta)$ induced by orientation preserving homeomorphisms of the orbifold $\mathcal{D}/\Delta$. Then two different epimorphisms $\theta_1, \theta_2 : \Delta \to G$ define the same class of G-actions if and only if they lie in the same $\mathcal{B} \times \text{Aut}(G)$-class.

We are interested in finding elements of $\mathcal{B} \times \text{Aut}(G)$ that make our epimorphisms $\theta_1, \theta_2 : \Delta \to G$ equivalent. We can produce the automorphism $\phi \in \mathcal{B}$ ad hoc. We do not need the full knowledge of the action of $\mathcal{B}$ since in the cases where we need the action of $\mathcal{B}$ to prove that two epimorphisms are equivalent, this action will be very simple. In most of the cases the only elements $\mathcal{B}$ we need are compositions of the automorphisms ϕ_j, $j = 1, ..., r - 1$, defined by $\phi_j(x_j) = x_{j+1}$, $\phi_j(x_{j+1}) = x_{j+1}^{-1} x_j x_{j+1}$, where x_j and x_{j+1} are the generators moved by the automorphism.

In the following we calculate all possible classes of epimorphisms $\theta : \Delta \to G$. Observe that, if Δ' is a non-maximal Fuchsian subgroup of Δ with no possible epimorphisms $\theta' : \Delta' \to G'$, then there are no epimorphisms $\theta : \Delta \to G$. In all the calculations $[\theta(x_1)\theta(x_2)\ldots\theta(x_{r-1})]^{-1} = \theta(x_r)$.

We separate the cases according to the order of the group G. Of course there is the obvious case $G = \{1\}$ that corresponds to all surfaces of genus 4.

1. $|G| = 2$. There are three classes of epimorphisms $\theta : \Delta \to \mathbb{Z}_2$. One class for each one of the signatures $s(\Delta_1) = (2; 2, 2)$, $s(\Delta_2) = (1; 2, 2, 2, 2, 2, 2)$ and $s(\Delta_3) = (0; 2, 2, 2, 2, 2, 2, 2, 2, 2, 2)$. Observe that the Fuchsian groups Δ_3 provide the *hyperelliptic locus*. See [BSS] for the connectedness of the hyperelliptic locus.

2. $|G| = 3$. There are four classes of epimorphisms $\theta : \Delta \to \mathbb{Z}_3 = \langle a/a^3 = 1 \rangle$: one class for the signature $s(\Delta_1) = (2; -)$ (non-maximal), one class for the signature $s(\Delta_2) = (1; 3, 3, 3)$ and two classes for $s(\Delta_3) = (0; 3, 3, 3, 3, 3, 3)$. The last two epimorphisms $\theta : \Delta_3 \to \mathbb{Z}_3$ are defined by

$\theta_1(x_{2i}) = a$ and $\theta_1(x_{2i-1}) = a^{-1}$, $1 \le i \le 3$, and $\theta_2(x_i) = a$, $1 \le i \le 6$. The Fuchsian groups Δ_3 yield the *cyclic trigonal locus*. Thus, the cyclic trigonal locus consists of two disconnected components $\mathcal{C}_1^3$, $\mathcal{C}_2^3$, see [Gon]. Note that from [Gon] we have that the two actions of Z_3 cannot occur on a Riemann surface then $\mathcal{C}_1^3$ and $\mathcal{C}_2^3$ are disjoint.

3. $|G| = 4$. *[i]* The Fuchsian groups admitting an epimorphism have signatures: $s(\Delta_1) = (1;4,4)$, $s(\Delta_2) = (1;2,2,2)$, $s(\Delta_3) = (0;2,2,2,2,4,4)$, $s(\Delta_4) = (0;2,4,4,4,4)$ and $s(\Delta_5) = (0;2,2,2,2,2,2,2)$. The groups $s(\Delta_1)$ are non-maximal. The epimorphisms $\overline{\theta} : \Delta_1 \to Z_4$ lift to epimorphisms $\theta : \Delta \to D_4$, where $s(\Delta) = (0;2,2,2,2,4)$, which are studied in case 6*[ii]*.

[ii] There is one class of epimorphisms $\theta : \Delta_2 \to Z_2 \times Z_2 = \langle a,b/a^2 = b^2 = (ab)^2 = 1\rangle$, with representative $\theta(a_1) = \theta(b_1) = a$, $\theta(x_1) = a$, $\theta(x_2) = b$ and $\theta(x_3) = ab$.

[iii] There is one class of epimorphisms $\theta : \Delta_3 \to Z_4$ defined by $\theta(x_1) = \theta(x_2) = \theta(x_3) = \theta(x_4) = a^2$, $\theta(x_5) = a$, $\theta(x_6) = a^3$. The surfaces uniformized by $\mathrm{Ker}\theta$ belong to the hyperelliptic locus.

[iv] There is one class of epimorphisms $\theta : \Delta_4 \to Z_4$ defined by $\theta(x_1) = a^2$, $\theta(x_2) = \theta(x_3) = \theta(x_4) = a$, $\theta(x_5) = a^3$.

[v] There are two classes of epimorphisms $\theta : \Delta_5 \to Z_2 \times Z_2$. One class provided by epimorphsims $\theta_1(x_i) = a$, $1 \le i \le 5$, $\theta_1(x_6) = b$. The second class given by epimorphisms $\theta_1(x_{2i}) = a$ and $\theta_1(x_{2i-1}) = b$. Observe that the surfaces given by the epimorphisms of the first type are hyperelliptic but not the surfaces given by the second epimorphism.

4. $|G| = 5$. There are three classes of epimorphisms $\theta : \Delta \to Z_5$, with $s(\Delta) = (0;5,5,5,5)$. One class is provided by epimorphisms $\theta_1(x_1) = a$, $\theta_1(x_2) = a^2$, $\theta_1(x_3) = a^3$, $\theta_1(x_4) = a^4$. The second class is given by epimorphisms $\theta_2(x_1) = \theta_2(x_2) = \theta_2(x_3) = a$, $\theta_2(x_4) = a^2$. The third class has representative $\theta_3(x_1) = \theta_3(x_3) = a$, $\theta_3(x_2) = \theta_3(x_4) = a^4$. Observe that the *cyclic pentagonal locus* consists of three connected components $\mathcal{C}_1^5$, $\mathcal{C}_2^5$, $\mathcal{C}_3^5$ (see [Gon]). The groups Δ here inducing the strata $\mathcal{C}_1^5$ and $\mathcal{C}_3^5$ are non-maximal. The groups associated to the stratum $\mathcal{C}_1^5$ are subgroups of groups Λ with signature $s(\Lambda) = (0;2,2,5,5)$. The groups associated to $\mathcal{C}_3^5$ are subgroups of groups Λ with signature $s(\Lambda) = (0;2,2,2,5)$.

5. $|G| = 6$. *[i]* Signature $s(\Delta_1) = (0;2,6,6,6)$. There is essentially one epimorphism $\theta : \Delta_1 \to Z_6$ defined by $\theta(x_1) = a^3$, $\theta(x_2) = \theta(x_3) = a$. The surfaces uniformized by $\mathrm{Ker}\theta$ belong to the hyperelliptic but not the trigonal locus. See also [IY].

[ii] Signature $s(\Delta_2) = (0; 2, 2, 3, 3, 3)$. Now, there is one epimorphism $\theta : \Delta_2 \to Z_6$ up to renaming the generator: $\theta(x_1) = \theta(x_2) = a^3$, $\theta(x_3) = \theta(x_4) = a^2$. The surfaces uniformized by $\mathrm{Ker}\theta$ belong to the cyclic trigonal locus, furthermore they belong to $\mathcal{C}_2^3$. See also [IY].

The epimorphisms $\theta : \Delta_2 \to D_3 = \langle a, s | a^3 = s^2 = (sa)^2 = 1 \rangle$ are defined by $\theta(x_1) = sa^i$, $\theta(x_2) = sa^j$, $\theta(x_3) = a^\varepsilon$, $\theta(x_4) = a^\delta$, where $i, j \in \{0, 1, 2\}$, $\varepsilon, \delta \in \{1, 2\}$ and $j - i + \varepsilon + \delta \neq 0 (\mathrm{mod}\,3)$. Now, applying a suitable conjugation in D_3 all the epimorphisms are conjugated to one of the following three: $\theta_1(x_1) = s$, $\theta_1(x_2) = s$, $\theta_1(x_3) = a$, $\theta_1(x_4) = a$; $\theta_2(x_1) = s$, $\theta_2(x_2) = sa$, $\theta_2(x_3) = a$, $\theta_2(x_4) = a^2$; and $\theta_3(x_1) = s$, $\theta_3(x_2) = sa$, $\theta_3(x_3) = a^2$, $\theta_3(x_4) = a^3$. The element $\phi_2^2 \times c_s \in \mathcal{B} \times \mathrm{Aut}(D_3)$, where $\phi_2^2(x_2) = x_3^{-1} x_2 x_3$, $\phi_2^2(x_3) = x_3^{-1} x_2^{-1} x_3 x_2 x_3$ and $c_s(s) = s, c_s(a) = a^2$, takes the epimorphism θ_2 to θ_1. Finally $\phi_2^2 \phi_3 \phi_2^2 \times c_s \in \mathcal{B} \times \mathrm{Aut}(D_3)$, where $\phi_3 \phi_2^2(x_2) = x_3^{-1} x_2 x_3$, $\phi_3 \phi_2^2(x_3) = x_4$ and $\phi_3 \phi_2^2(x_4) = x_4^{-1} x_3^{-1} x_2^{-1} x_3 x_2 x_3 x_4$, takes the epimorphism θ_3 to θ_1.

These epimorphisms define one stratum in the component $\mathcal{C}_1^3$ of the cyclic trigonal locus. See also [IY].

[iii] Consider the signature $s(\Delta_3) = (0; 2, 2, 2, 3, 6)$. There is essentially one epimorphism $\theta : \Delta_3 \to Z_6$ defined by $\theta(x_1) = \theta(x_2) = \theta(x_3) = a^3$, $\theta(x_4) = a^2$. The surfaces uniformized by $\mathrm{Ker}\theta$ belong to the hyperelliptic but not the trigonal locus. See also [IY].

[iv] The signature $s(\Delta_4) = (0; 2, 2, 2, 2, 2, 2)$. There are epimorphisms $\theta : \Delta_4 \to D_3$: $\theta(x_1) = s$, $\theta(x_2) = sa^i$, $\theta(x_3) = sa^j$, $\theta(x_4) = sa^h$, $\theta(x_5) = sa^k$, with $i, j, h, k \in \{0, 1, 2\}, i+j+h+k > 0$. These epimorphisms induce one stratum in $\mathcal{M}_4$. First of all, after applying suitable conjugations by elements in D_3 and the combination of braids ϕ_i^2 all the epimorphisms can be classified in the following table:

1) $\theta_1(x_1) = s$, $\theta_1(x_2) = s$, $\theta_1(x_3) = s$, $\theta_1(x_4) = s$, $\theta_1(x_5) = sa$
2) $\theta_2(x_1) = s$, $\theta_2(x_2) = s$, $\theta_2(x_3) = s$, $\theta_2(x_4) = sa$, $\theta_2(x_5) = sa$
3) $\theta_3(x_1) = s$, $\theta_3(x_2) = s$, $\theta_3(x_3) = sa$, $\theta_3(x_4) = s$, $\theta_3(x_5) = sa$
4) $\theta_4(x_1) = s$, $\theta_4(x_2) = sa$, $\theta_4(x_3) = s$, $\theta_4(x_4) = s$, $\theta_4(x_5) = sa$
5) $\theta_5(x_1) = sa$, $\theta_5(x_2) = s$, $\theta_5(x_3) = s$, $\theta_5(x_4) = s$, $\theta_5(x_5) = sa$
6) $\theta_6(x_1) = s$, $\theta_6(x_2) = s$, $\theta_6(x_3) = sa$, $\theta_6(x_4) = sa$, $\theta_6(x_5) = s$
7) $\theta_7(x_1) = s$, $\theta_7(x_2) = sa$, $\theta_7(x_3) = s$, $\theta_7(x_4) = sa$, $\theta_7(x_5) = s$
8) $\theta_8(x_1) = sa$, $\theta_8(x_2) = s$, $\theta_8(x_3) = s$, $\theta_8(x_4) = sa$, $\theta_8(x_5) = s$
9) $\theta_9(x_1) = s$, $\theta_9(x_2) = sa$, $\theta_9(x_3) = sa$, $\theta_9(x_4) = s$, $\theta_9(x_5) = s$
10) $\theta_0(x_1) = sa$, $\theta_0(x_2) = sa$, $\theta_0(x_3) = s$, $\theta_0(x_4) = s$, $\theta_0(x_5) = s$

All these epimorphisms are equivalent: for instance $\phi_5 \cdot \phi_4$ commutes θ_1 with θ_2, $c_s \cdot \phi_3^2 \cdot \phi_4$ commutes θ_1 with θ_3. Therefore the surfaces $\mathcal{D}/\mathrm{Ker}(\theta)$ form one stratum in $\mathcal{M}_4$.

[v] Fuchsian groups with signature $s(\Delta_5) = (0; 3, 3, 6, 6)$ are non-maximal. The epimorphisms $\bar{\theta} : \Delta_5 \to Z_6$ lift to epimorphisms $\theta : \Delta \to Z_6 \times Z_2$ and $\theta : \Delta \to D_6$, where $s(\Delta) = (0; 2, 2, 3, 6)$, which are studied in case 9*[iii]*.

[vi] The groups $s(\Delta_6)$ with signature $s(\Delta_6) = (1; 2, 2)$ are non-maximal. The epimorphisms $\bar{\theta} : \Delta_6 \to D_3$ lift to epimorphisms $\theta : \Delta \to D_6$, where $s(\Delta) = (0; 2, 2, 2, 2, 2, 2)$, which are studied in case 9*[iv]*.

6. $|G| = 8$. *[i]* Groups with signature $s(\Delta_1) = (0; 2, 2, 8, 8)$ are non-maximal. The epimorphisms $\bar{\theta} : \Delta_1 \to Z_8$ lift to epimorphisms $\theta : \Delta \to D_8$, where $s(\Delta) = (0; 2, 2, 2, 8)$. These epimorphisms are studied in case 11*[i]*.

[ii] Signature $s(\Delta_2) = (0; 2, 2, 2, 2, 4)$. There are two classes of epimorphisms $\theta : \Delta_2 \to D_4 = \langle a, s | a^4 = s^2 = (sa)^2 = 1 \rangle$. One class is formed by the epimorphisms equivalent to $\theta_1 : \Delta_2 \to D_4$ defined as $\theta_1(x_1) = \theta_1(x_2) = a^2$, $\theta_1(x_3) = s$, $\theta_1(x_4) = sa$. The second class consists of epimorphisms equivalent to $\theta_2 : \Delta_2 \to D_4$ defined by $\theta_2(x_1) = \theta_2(x_2) = s$, $\theta_2(x_3) = sa$, $\theta_2(x_4) = sa^2$. By Theorem 7 the stratum associated to epimorphisms θ_1 belongs to the hyperelliptic locus but not the stratum given by the epimorphisms θ_2.

[iii] The signature $s(\Delta_3) = (0; 2, 4, 4, 4)$. There is a unique epimorphism $\theta : \Delta_3 \to Q = \langle t, s | t^2 = s^2 = (st)^2 \rangle$. The epimorphism is defined, up to renaming the generators of Q, by $\theta(x_2) = s$, $\theta(x_3) = t$, $\theta(x_4) = (st)$. By Theorem 7 the Riemann surfaces uniformized by $\mathrm{Ker}(\theta)$ belong to the hyperelliptic locus.

7. $|G| = 9$. *[i]* Fuchsian groups with signature $s(\Delta_1) = (0; 9, 9, 9)$ are non-maximal. The epimorphisms $\bar{\theta} : \Delta_1 \to Z_9$ lift to epimorphisms $\theta : \Delta \to Z_{18}$, where $s(\Delta) = (0; 2, 9, 18)$.

[ii] Groups with signature $s(\Delta_2) = (0; 3, 3, 3, 3)$ are non-maximal. The epimorphisms $\bar{\theta} : \Delta_2 \to Z_3 \times Z_3$ lift to epimorphisms $\theta : \Delta \to G_{18}$, where $s(\Delta) = (0; 2, 2, 3, 3)$, studied in cases 12*[iv]* and 18*[ii]*.

8. $|G| = 10$. *[i]* Fuchsian groups with signature $s(\Delta_1) = (0; 2, 2, 5, 5)$. First of all, epimorphisms $\bar{\theta}_1 : \Delta_1 \to D_5$, $\bar{\theta}_1(x_1) = s$, $\bar{\theta}_1(x_2) = s$, $\bar{\theta}_1(x_3) = a$, extend to epimorphisms $\theta_1 : \Delta \to D_{10}$, where $s(\Delta) = (0; 2, 2, 2, 5)$. These epimorphisms are studied in case 13*[i]*. Epimorphisms $\bar{\theta}_3 : \Delta_1 \to Z_{10} = \langle a | a^{10} = 1 \rangle$ extend to epimorphisms $\theta_1 : \Delta \to D_{10}$, where $s(\Delta) = (0; 2, 2, 2, 5)$, studied in case 13*[i]*. These epimorphisms define the stratum C_3^5 which belongs to the hyperelliptic locus by Theorem 7. Now, epimorphisms $\bar{\theta}_3 : \Delta_1 \to D_5$, $\bar{\theta}_3(x_1) = s$,

$\bar{\theta}_3(x_2) = sa$, $\bar{\theta}_1(x_3) = a$ provide maximal Fuchsian groups. These epimorphisms determine the stratum $\mathcal{C}_1^5$.

[ii] Signature $s(\Delta_2) = (0; 5, 10, 10)$. Epimorphisms $\bar{\bar{\theta}} : \Delta_2 \to Z_{10}$, $\bar{\bar{\theta}}(x_1) = a^2$, $\bar{\bar{\theta}}(x^2) = a^9$, extend to epimorphisms $\bar{\theta} : \overline{\Delta} \to Z_{10} \rtimes_2 Z_4$, where $s(\overline{\Delta}) = (0; 2, 10, 10)$, studied in case 13*[ii]* and 19. Epimorphisms $\theta : \Delta_2 \to Z_{10}$, $\theta(x_1) = a^2$, $\theta(x^2) = a$ induce one maximal action of the cyclic group, thus one stratum consisting of one surface in $\mathcal{C}_2^5$.

9. $|G| = 12$. *[i]* First, consider the signature $s(\Delta_1) = (0; 4, 6, 12)$. The epimorphisms $\theta : \Delta \to Z_{12} = \langle a | a^{12} = 1 \rangle$ are defined by $\theta(x_1) = a^{\pm 3}$, $\theta(x_2) = a^{\pm 2}$. All these epimorphisms are conjugated by automorphisms of Z_{12}. By Theorem 7 we get one surface R_4 belonging to the hyperelliptic locus.

[ii] Consider the signature $s(\Delta_2) = (0; 2, 3, 3, 3)$. The only group of order 12 generated by three elements of order 3 is $A_4 = \langle a, s/a^3 = s^2 = (as)^3 = 1 \rangle$ (see [CM]). There are epimorphisms $\theta : \Delta_2 \to A_4$, for instance $\theta(x_1) = s$, $\theta(x_2) = a$, $\theta(x_3) = as$ and $\theta(x_4) = sas$. Again, all these epimorphisms are conjugated by automorphisms of A_4.

[iii] Now, consider the signature $s(\Delta_3) = (0; 2, 2, 3, 6)$. There are epimorphisms from Δ_3 onto both $D_6 = \langle a, s | a^6 = s^2 = (sa)^2 = 1 \rangle$ and $Z_6 \times Z_2 = \langle a, s | a^6 = s^2 = [a, s] = 1 \rangle$. There is one epimorphism $\theta : \Delta_3 \to Z_6 \times Z_2$ up to renaming the generators. θ is defined by $\theta(x_1) = s$, $\theta(x_2) = sa^3$, $\theta(x_3) = a^{\pm 2}$. The epimorphisms $\theta : \Delta_3 \to D_6$ are defined by $\theta(x_1) = sa^i$, $\theta(x_2) = sa^{i+3}$, $\theta(x_3) = a^{\pm 2}$, $i \in \{0, \ldots, 5\}$. Now, applying a suitable conjugation in D_6 all the epimorphisms are conjugated to $\theta_1(x_1) = s$, $\theta_1(x_2) = sa^3$, $\theta_1(x_3) = a^2$. Therefore there are two strata of cyclic trigonal surfaces, one determined by the unique class of actions of D_6, the second determined by the unique action of $Z_6 \times Z_2$. Moreover, the surfaces with automorphisms group D_6 lie in $\mathcal{C}_1^3$ while the surfaces with $\mathrm{Aut}(X_4) = Z_6 \times Z_2$ lie in $\mathcal{C}_2^3$. See [IY].

[iv] The epimorphisms $\theta : \Delta_5 \to D_6$, with $s(\Delta_4) = (0; 2, 2, 2, 2, 2)$, are defined by $\theta(x_1) = sa^i$, $\theta(x_2) = sa^j$, $\theta(x_3) = sa^h$ and $\theta(x_4) = sa^k$, where $j - i + k - h \equiv 3 \bmod 6$. All these epimorphisms are conjugated by elements of D_6. They define one stratum in $\mathcal{M}_4$.

[v] Fuchsian groups with signature $s(\Delta_5) = (0; 6, 6, 6)$ are non-maximal. The epimorphisms $\bar{\theta} : \Delta_5 \to Z_6 \times Z_2$ lift to epimorphisms $\theta : \Delta \to Z_3 \times D_4$, where $s(\Delta) = (0; 2, 6, 12)$, and to epimorphisms $\phi_1 : \Lambda_1 \to Z_3 \times A_4$, with $s(\Lambda_1) = (0; 3, 3, 6)$. This epimorphism is studied in case 24*[i]*.

[vi] Groups with signature $s(\Delta_6) = (0; 3, 12, 12)$ are non-maximal. The

epimorphisms $\bar{\theta} : \Delta_6 \to Z_{12}$ lift to epimorphisms $\theta : \Delta \to Z_3 \times D_4$, where $s(\Delta) = (0; 2, 6, 12)$, studied in case 14/$vi$/.

/vii/ Groups with signature $s(\Delta_7) = (0; 2, 2, 4, 4)$. There could be one epimorphism $\bar{\theta} : \Delta_7 \to T = Z_3 \rtimes Z_4 = \langle a, t | a^3 = t^4 = 1, t^3 at = a^2 \rangle$ up to renaming the generators of the group T. The action of $\theta(\Delta_7)$ on the $\langle t^2 \rangle$-cosets leaves 12 fixed cosets which is geometrically impossible. There is no such epimorphism.

/$viii$/ Groups with signature $s(\Delta_8) = (1; 2)$ are non-maximal. The epimorphisms $\bar{\theta} : \Delta_8 \to A_4$ lift to epimorphisms $\theta : \Delta \to \Sigma_4$, where $s(\Delta) = (0; 2, 2, 2, 4)$, studied in case 14/$i$/.

10. $|G| = 15$. /i/ First, consider the signature $s(\Delta_1) = (0; 5, 5, 5)$. There is no such epimorphism, since the only possible epimorphism from a Fuchsian group Δ_1 should be onto Z_{15}, which is not generated by elements of order 5.

/ii/ Now, consider the signature $s(\Delta_2) = (0; 3, 5, 15)$. There is one class of epimorphisms $\theta : \Delta_2 \to Z_{15}$ defined by $\theta(x_1) = a^{\pm 5}$, $\theta(x_2) = a^{\pm 2i}$, where $i \in \{1, 2\}$. The automorphisms w_2, w_4, w_8 of Z_{15}, with $w_j(a) = a^j$, commute the above epimorphisms θ. Now, $\theta(x_1)$ leaves five $\langle a^5 \rangle$-cosets fixed and $\theta(x_3)$ fixes one $\langle a^5 \rangle$-coset. In the same way $\theta(x_2)$ and $\theta(x_3)$ leave four $\langle a^3 \rangle$-cosets fixed together. Thus, there is a unique cyclic trigonal and pentagonal surface T_4 of genus 4 with $\text{Aut}(T_4) = Z_{15}$. See also [IY].

11. $|G| = 16$. /i/ Consider the signature $s(\Delta_1) = (0; 2, 2, 2, 8)$. All the epimorphisms $\theta : \Delta_1 \to D_8 = \langle a, s | a^8 = s^2 = (sa)^2 = 1 \rangle$ are equivalent to the epimorphism: $\theta(x_1) = a^4$, $\theta(x_2) = sa^4$, $\theta(x_3) = sa$ by applying $\phi_1 \times w_j$, $\phi_2 \times w_j$ or $\phi_2 \cdot \phi_1 \times w_j$, $j \in \{1, 2, 3\}$, $w_j(a) = a^{2j+1}, w_j(s) = s$ in $\mathcal{B} \times \text{Aut}(D_8)$. Using Theorem 7 to study the action of $\theta(\Delta_1)$ on the $\langle a^4 \rangle$-cosets, we obtain that the surfaces uniformized by $\text{Ker}(\theta)$ form one stratum of hyperelliptic surfaces in $\mathcal{M}_4$.

/ii/ Signature $s(\Delta_2) = (0; 2, 16, 16)$. The group Δ_2 is non-maximal: there is essentially one epimorphism $\bar{\theta}$ from Δ_2 to a group of order 16: $\bar{\theta} : \Delta_2 \to Z_{16} = \langle a | a^{16} = 1 \rangle$. This epimorphism lifts to the epimorphism $\theta : \Delta \to QD_{16} = \langle a, s | a^16 = s^2 = sasa^9 = 1 \rangle$ studied in case 17.

/iii/ Signature $s(\Delta_3) = (0; 4, 4, 8)$. The group Δ_3 is non-maximal: there is essentially one epimorphism $\bar{\theta}$ from Δ_3 to a group of order 16: $\bar{\theta} : \Delta_2 \to Q_{16} = \langle \bar{a}, \bar{t} | \bar{a}^8 = \bar{t}^2 = 1, \bar{t}^3 \bar{a} \bar{t} = \bar{a}^{-1}, \bar{a}^4 = \bar{t}^2 \rangle$. This epimorphism lifts to the epimorphism $\theta : \Delta \to QD_{16} = \langle a, s | a^{16} = s^2 = sasa^9 = 1 \rangle$ studied in case 17, where $\bar{a} = a^2$ and $\bar{t} = sa$.

12. $|G| = 18$. /i/ Consider the signature $s(\Delta_1) = (0; 2, 2, 2, 6)$. There is no group of order 18 generated by three involutions and containing elements of order 6 [CM]. Hence there are no epimorphisms of this type.

[ii] Now, consider the signature $s(\Delta_2) = (0; 2, 9, 18)$. The epimorphisms $\theta : \Delta_2 \to \mathbb{Z}_{18} = \langle a \mid a^{18} \rangle$ are defined as $\theta(x_1) = a^9$, $\theta(x_2) = a^{\pm 2i}$, where $i \in \{1, 2, 4\}$, all of them are equivalent by automorphisms of $\mathbb{Z}_{18}$. Studying $\theta^{-1}\langle a^9 \rangle$ and $\theta^{-1}\langle a^6 \rangle$ we see that there is a unique hyperelliptic non trigonal surface S_4 with $\mathrm{Aut}(S_4) = \mathbb{Z}_{18}$.

[iii] Groups with signature $s(\Delta_3) = (0; 3, 6, 6)$ are non-maximal. The epimorphisms $\bar{\theta}_1 : \Delta_3 \to \mathbb{Z}_6 \times \mathbb{Z}_3 = \langle a, b, t | a^3 = b^3 = t^2 = [a, b] = [a, t] = [t, b] = 1 \rangle$, defined as $\bar{\theta}_1(x_1) = b$, $\bar{\theta}_1(x_2) = tab$, and $\bar{\theta}_2 : \Delta_3 \to \mathbb{Z}_3 \times \mathrm{D}_3 = \langle a, b, s | a^3 = b^3 = s^2 = [a, b] = (sa)^2 = [s, b] = 1 \rangle$, defined as $\bar{\theta}_2(x_1) = b$, $\bar{\theta}_2(x_2) = sab$, lift to an epimorphism $\theta : \Delta \to \mathbb{Z}_6 \times \mathrm{D}_3$, where $s(\Delta) = (0; 2, 6, 6)$. This epimorphism is studied in case 18*[v]*.

There is an epimorphism $\bar{\theta}_3 : \Delta_3 \to \mathbb{Z}_3 \times \mathrm{D}_3$ defined as $\bar{\theta}_3(x_1) = a^2 b$, $\bar{\theta}_3(x_2) = sab$, that lifts to an epimorphism $\bar{\phi}_2 : \Delta \to \mathrm{D}_3 \times \mathrm{D}_3$, with $s(\Delta) = (0; 2, 6, 6)$. The last epimorphism lifts to the one studied in case 18*[ii]*.

[iv] Groups with signature $s(\Delta_4) = (0; 2, 2, 3, 3)$ are extensions of groups with signature $= (0; 3, 3, 3, 3)$. We consider the epimorphisms $\theta_{1,2} : \Delta_4 \to \mathbb{Z}_3 \times \mathrm{D}_3$, $\theta_1(x_1) = sa^i$, $\theta_1(x_2) = sa^j$, $i \neq j \in \{0, 1, 2\}$, $\theta_1(x_3) = a^{i-j}b$, $\theta_1(x_4) = b^2$; $\theta_2(x_1) = s$, $\theta_2(x_2) = sa$, $\theta_2(x_3) = ab$, $\theta_2(x_4) = ab^2$; and $\theta_3 : \Delta_4 \to (\mathbb{Z}_3 \times \mathbb{Z}_3) \rtimes \mathbb{Z}_2 = \langle a, b, s | a^3 = b^3 = s^2 = (sa)^2 = (sb^2) = [a, b] = 1 \rangle$ defined as $\theta_3(x_1) = sa$, $\theta_3(x_2) = sb$, $\theta_3(x_3) = ab$, $\theta_3(x_4) = b^2$. The epimorphisms θ_2 and θ_3 lift to epimorphisms $\theta : \Delta \to \mathrm{D}_3 \times \mathrm{D}_3$, with $s(\Delta) = (0; 2, 2, 2, 3)$, studied in case 18*[ii]*.

The epimorphisms $\theta_1 : \Delta_4 \to \mathbb{Z}_3 \times \mathrm{D}_3$ yield maximal actions of $\mathbb{Z}_3 \times \mathrm{D}_3$. First of all, all these epimorphisms are conjugated by elements a^h, s or sa^h in $\mathbb{Z}_3 \times \mathrm{D}_3$, so the actions of $\mathbb{Z}_3 \times \mathrm{D}_3$ are topologically equivalent and the surfaces determined by Δ_4 are equisymmetric. Now (see [IY]), by Theorem 7, $s(\theta_1^{-1}(\langle b \rangle)) = (0; 3, 3, 3, 3, 3, 3)$. Thus, the surfaces $\mathcal{D}/\mathrm{Ker}(\theta_1)$ form one equisymmetric stratum of cyclic trigonal Riemann surfaces in $\mathcal{C}_2^3$.

13. $|G| = 20$. *[i]* Signature $s(\Delta_1) = (0; 2, 2, 2, 5)$. Consider epimorphisms $\theta : \Delta_1 \to \mathrm{D}_{10} = \langle a, s | a^{10} = s^2 = sa^2 = 1 \rangle$. Applying $\phi_1 \times w_j$, $\phi_2 \times w_j$ or $\phi_2 \cdot \phi_1 \times w_j$, $j \in \{3, 7, 9\}$, $w_j(a) = a^j, w_j(s) = s$ in $\mathcal{B} \times \mathrm{Aut}(\mathrm{D}_{10})$ we obtain that any such epimorphism is equivalent to $\theta(x_1) = a^5$, $\theta(x_2) = sa^5$, $\theta(x_3) = sa^2$. The surfaces uniformized by $\mathrm{Ker}(\theta)$ form the stratum $\mathcal{C}_2^5$ of pentagonal surfaces in the hyperelliptic locus.

[ii] Groups with signature $s(\Delta_2) = (0; 2, 10, 10)$ are non-maximal. There is one epimorphism $\overline{\theta}_1 : \Delta_2 \to Z_{10} \times Z_2 = \langle a, s | a^{10} = s^2 = [s, a] = 1 \rangle$ up to renaming the generators of $Z_{10} \times Z_2$: $\overline{\theta}_1(x_1) = s$, $\overline{\theta}_1(x_2) = sa$. This epimorphism lifts to an epimorphism $\theta : \Delta \to Z_5 \rtimes_2 D_4 = \langle a, s, t | a^5 = t^4 = s^2 = (st)^2 = [s, a] = t^3 ata = 1 \rangle$, studied in case 19.

[iii] Groups with signature $s(\Delta_3) = (0; 4, 4, 5)$. There is one epimorphism $\overline{\theta}_2 : \Delta_3 \to Z_5 \rtimes_2 Z_4 = \langle a, t | a^5 = t^4 = t^3 ata = 1 \rangle$ up to renaming the generators of $Z_5 \rtimes_2 Z_4$: $\overline{\theta}_2(x_1) = t$, $\overline{\theta}_2(x_2) = t^3 a$. This epimorphism lifts to the epimorphism $\theta : \Delta \to Z_5 \times D_4$ above.

The essentially unique epimorphism $\overline{\theta}_3 : \Delta_3 \to Z_5 \rtimes_4 Z_4$ lifts to an epimorphism $\theta : \Delta \to \Sigma_5$, where $s(\Delta) = (0; 2, 4, 5)$, studied in case 27.

14. $|G| = 24$. *[i]* Groups with signature $s(\Delta_1) = (0; 2, 2, 2, 4)$. There is one class of epimorphisms $\theta : \Delta_1 \to \Sigma_4 = \langle s, t / s^2 = t^4 = (ts)^3 = 1 \rangle$ after conjugation by elements in Σ_4 for instance $\theta(x_1) = s$, $\theta(x_2) = t^2 st^2$, $\theta(x_3) = st^2 st$. These epimorphisms are elevations of the epimorphism in case 9*[viii]*.

[ii] Signature $s(\Delta_2) = (0; 3, 4, 6)$ (see [CIY1]). There is only one such epimorphism $\theta : \Delta_2 \to \langle 2, 3, 3 \rangle = Q \rtimes Z_3 = \langle a, s, t / a^3 = t^4 = s^4 = (st)^4 = 1, s^2 = t^2, a^2 sa = t, a^2 ta = st \rangle$, see [CM]. After renaming the generators of $\langle 2, 3, 3 \rangle$ this epimorphism is given by $\theta(x_1) = sta$, $\theta(x_2) = s$. By Theorem 1 this surface is hyperelliptic but not trigonal. See [CIY1] and [IY].

[iii] There are no epimorphisms from a Fuchsian group with signature $s(\Delta_3) = (0; 4, 4, 4)$ onto a group of order 24 because there is no group of order 24 generated by two elements of order four whose product has order four. See [CM].

[iv] There are no epimorphisms from a Fuchsian group with signature $s(\Delta_4) = (0; 3, 3, 12)$ onto a group of order 24 generated by two elements of order 3. The only such group is the binary tetrahedral group, which does not have elements of order 12. See [CM].

[v] There are no epimorphisms from a Fuchsian group with signature $s(\Delta_5) = (0; 2, 8, 8)$ onto a group of order 24 since there is no group of order 24 generated by two elements of order 8. See [CM].

[vi] Groups with signature $s(\Delta_6) = (0; 2, 6, 12)$ are non-maximal. Groups Δ_6 admit epimorphisms only onto $D_4 \times Z_3$. The epimorphisms $\overline{\theta} : \Delta_6 \to D_4 \times Z_3 = \langle \overline{a}, s | \overline{a}^{12} = s^2 = s\overline{a}s\overline{a}^5 = 1 \rangle$ with $\overline{\theta}(x_1) = s$ and $\overline{\theta}(x_2) = s\overline{a}$ lift to epimorphisms $\theta : \Delta \to Z_3 \times \Sigma_4$, where $s(\Delta) = (0; 2, 3, 12)$, that are studied in case 17.

15. $|G| = 27$. There are no Riemann surfaces with 27 automorphisms since the group $Z_9 \rtimes Z_3$ is not generated by order three elements.

16. $|G| = 30$. As case 10*[i]*.

17. $|G| = 32$. Consider the signature $s(\Delta) = (0; 2, 4, 16)$. There is only one group of order 32 generated by one element of order 2 and one element of order 4 with product of order 16: $QD_{16} = \langle a, s | a^{16} = s^2 = sasa^9 = 1 \rangle$. The epimorphisms $\theta : \Delta \to QD_{16}$ are defined by: $\theta(x_1) = s$ (or $\theta(x_1) = sa^8$), $\theta(x_2) = sa^i, i \equiv 1 \bmod 2$. Any such epimorphism is conjugated to $\theta_1(x_1) = s$, $\theta_1(x_2) = sa$, by applying a combination of the following automorphisms of QD_{16}: $\overline{w}(a) = a, \overline{w}(s) = sa^8$, $w_j(s) = s, w_j(a) = a^j, j \equiv 1/i \bmod 16, i \equiv 1 \bmod 2$. Applying Theorem 1 to $\theta_1^{-1}(\langle a^8 \rangle)$ we get one equisymmetric stratum consisting of the single hyperelliptic surface P_4.

18. $|G| = 36$. *[i]* First, consider Fuchsian groups Δ_1 with signature $(0; 2, 4, 12)$. There is no epimorphism from Δ_1 onto a group of order 36. See [CIY1] and [IY].

[ii] Secondly, consider Fuchsian groups Δ_2 with signature $(0; 2, 2, 2, 3)$. The only group of order 36 generated by 3 involutions is $D_3 \times D_3 = \langle a, b, s, t / a^3 = b^3 = s^2 = t^2 = [a, b] = [s, b] = [t, a] = (sa)^2 = (tb)^2 = 1 \rangle$. Consider epimorphism $\theta : \Delta_2 \to D_3 \times D_3$ like $\theta(x_1) = s$, $\theta(x_2) = tb$, $\theta(x_3) = sta$ and $\theta(x_4) = a^2 b$. The surfaces $\mathcal{D}/\mathrm{Ker}(\theta)$ are equisymmetric. First, observe that there is an epimorphism $\theta : \Delta \to D_3 \times D_3$ as above if and only if $\theta(x_4) = a^{\varepsilon} b^{\delta}$, where $\varepsilon, \delta \in \{1, 2\}$. Otherwise the action of $\theta(x_4)$ on the $\langle a \rangle$- and $\langle b \rangle$-cosets is geometrically impossible.

Conjugating with a suitable element of $D_3 \times D_3$ or by the automorphism $w(s) = t$, $w(t) = s$, $w(a) = b$ and $w(b) = a$ of $D_3 \times D_3$ any epimorphism $\theta : \Delta \to D_3 \times D_3$ is taken to any of the following three

1. $\theta_0(x_1) = s, \theta_0(x_2) = t, \theta_0(x_3) = stab$
2. $\theta_1(x_1) = t, \theta_1(x_2) = stab, \theta_1(x_3) = sa^2$
3. $\theta_2(x_1) = stab, \theta_2(x_2) = sa^2, \theta_2(x_3) = tb^2$

Now, $\phi_2 \cdot \phi_1$ takes the epimorphism θ_0 to the epimorphism θ_1 and $\phi_1 \cdot \phi_2$ takes the epimorphism θ_0 to the epimorphism θ_2.

The surfaces $\mathcal{D}/\mathrm{Ker}(\theta)$ form one stratum in $\mathcal{M}_4$ lying in $\mathcal{C}_1^3$. They admit two trigonal morphisms. See [CIY2] and [IY].

[iii] Signature $s(\Delta_3) = (0; 3, 3, 6)$ is studied in case 24*[i]*.

[iv] Signature $s(\Delta_4) = (0; 3, 4, 4)$ is studied in case 24*[ii]*.

[v] Groups with signature $s(\Delta_5) = (0; 2, 6, 6)$ are extensions of groups with signature $s(\Lambda) = (0; 3, 6, 6)$. First of all, the non-maximal groups with signature $(0; 2, 6, 6)$ are subgroups of groups Δ with signature

$s(\Delta) = (0; 2, 4, 6)$. They are given by the epimorphism $\overline{\phi}_2 : \Delta_5 \rightarrow$ $D_3 \times D_3$ defined as $\overline{\phi}_2(x_1) = stab^2$, $\overline{\phi}_2(x_2) = sb$. Its elevation to an epimorphism $\theta : \Delta \rightarrow (Z_3 \times Z_3) \rtimes D_4$ is studied in case 24*[ii]*.

The maximal action induced by the group Δ_5 with signature $(0; 2, 6, 6)$ produces a surface in the family given in case 12*[iv]*. Δ_5 is defined by an elevation of both $\overline{\theta}_1 : \Delta_3 \rightarrow Z_6 \times Z_3$ and $\overline{\theta}_2 : \Delta_3 \rightarrow Z_3 \times D_3$ in case 12*[iii]* to $\theta : \Delta_5 \rightarrow Z_6 \times D_3 = \langle a, b, s, t/a^3 = b^3 = s^2 = t^2 = (st)^2 = [a, b] = [s, b] = [t, a] = (sa)^2 = [t, b] = 1 \rangle$, defined by $\theta(x_1) = sa^2$, $\theta(x_2) = tab^2$ and $\theta(x_3) = stab$. Note that $\phi_2 \in \mathcal{B}$ interchanges the conjugacy classes of elements of order 6 in $Z_6 \times D_3$. By Theorem 1 the surface $Z_4 = \mathcal{D}/Ker(\theta)$ is a trigonal, non-hyperelliptic Riemann surface lying in $\mathcal{C}_2^3$. See [IY].

19. $|G| = 40$. Signature $s(\Delta) = (0; 2, 4, 10)$. There is one class of epimorphisms $\theta : \Delta \rightarrow Z_5 \rtimes_2 D_4 = \langle a, s, t | a^5 = t^4 = s^2 = (st)^2 = [s, a] = t^3 ata = 1 \rangle$ up to automorphisms of $Z_5 \rtimes_2 D_4$: the lifting of $\overline{\theta}_1$ and $\overline{\theta}_2$ in 13*[ii]*–*[iii]*. Such an epimorphism is given by $\theta(x_1) = st$, $\theta(x_2) = t^3 a$. These epimorphisms yield one stratum consisting of one surface, $V_4 = \mathcal{D}/ker(\theta)$. By applying Theorem 1 to the $\langle t^2 a^5 \rangle$-cosets we get that this surface is hyperelliptic (in $\mathcal{C}_5^2$).

20. $|G| = 45$. As case 10*[i]*.

21. $|G| = 48$. *[i]* The signature $s(\Delta_1) = (0; 2, 3, 24)$ is as case 14*[iv]*.
[ii] The signature $s(\Delta_2) = (0; 2, 4, 8)$ is as cases 14*[iii]* and 14*[v]*.

22. $|G| = 54$. As case 15. See [Sin2].

23. $|G| = 60$. *[i]* Clearly there is no epimorphism from a Fuchsian group Δ_1 with signature $(0; 2, 3, 15)$ onto a group of order 60, since the only group of order 60 generated by elements of order 2 and 3 is A_5.
[ii] Signature $s(\Delta_2) = (0; 2, 5, 5)$. There is a unique epimorphism $\overline{\theta}$: $\Delta_2 \rightarrow A_5$ up to conjugation by an element in Σ_5: $\overline{\theta}(x_1) = (1, 3)(4, 2)$, $\overline{\theta}(x_2) = (1, 2, 3, 4, 5)$. This epimorphism lifts to an epimorphism θ : $\Delta_2 \rightarrow \Sigma_5$, commented in case 27.

24. $|G| = 72$. This case was studied in [CIY1] and [IY].
[i] Consider Fuchsian groups Δ_1 with signature $(0; 2, 3, 12)$. The group Δ_1 contains the group Λ_1 with signature $s(\Lambda_1) = (0; 3, 3, 6)$. There is one class of epimorphisms $\theta_1 : \Delta_1 \rightarrow \Sigma_4 \times Z_3 = \langle a, b, \overline{s} | a^3 = b^3 = (\overline{s})^2 = [a, b] = [\overline{s}, b] = (a\overline{s})^4 = 1 \rangle$ with representative $\theta_1(x_1) = \overline{s}$, $\theta_1(x_2) = ab$, $\theta_1(x_3) = a^2 \overline{s} b$. Therefore θ_1 yields a stratum consisting of one trigonal Riemann surface $X_4 = \mathcal{D}/Ker(\theta_1)$, $X_4 \in \mathcal{C}_2^3$.
[ii] Consider Fuchsian groups Δ_2 with signature $(0; 2, 4, 6)$. The group Δ_2 contains the group Λ_2 with signature $s(\Lambda_2) = (0; 3, 4, 4)$ and one

group $\overline{\Lambda}_2$ with signature $s(\overline{\Lambda}_2) = (0; 2, 6, 6)$ as subgroups of index 2. There is one class of epimorphisms $\theta_2 : \Delta_2 \to (\mathbb{Z}_3 \times \mathbb{Z}_3) \rtimes D_4 = \langle a, b, s, t | a^3 = b^3 = s^2 = t^4 = [a, b] = (sa)^2 = (sb)^2 = 1, t^3 a t = a^{-1}, t^3 b t = b^{-1} \rangle$ with representative $\theta_2(x_1) = s$, $\theta_2(x_2) = ta$, $\theta_2(x_3) = stb$. Therefore θ_2 yields a stratum consisting of one cyclic trigonal Riemann surface $Y_4 = \mathcal{D}/\mathrm{Ker}(\theta_2)$, $Y_4 \in \mathcal{C}_1^3$.

[iii] The signature $(0; 3, 3, 4)$ is as case 14*[iii]*.

25. $|G| = 90$. As case 10*[i]* (see [Sin2]).

26. $|G| = 108$. As case 15 (see [Sin2]).

27. $|G| = 120$. As case 23*[ii]* by [Sin2]. The required epimorphism $\theta : \Delta_2 \to \Sigma_5$ is the lifting of the epimorphism $\overline{\theta} : \Delta_2 \to A_5$ given by $\overline{\theta}(x_1) = (1, 3)(4, 2)$, $\overline{\theta}(x_2) = (1, 2, 3, 4, 5)$. Thus θ yields an equisymmetric stratum in $\mathcal{C}_1^5 \subset \mathcal{M}_4$ formed by a single pentagonal Riemann surface $U_4 = \mathcal{D}/\mathrm{Ker}(\theta)$. The surface U_4 is known as *Bring's curve*, the only cyclic pentagonal surface in $\mathcal{M}_4$ admitting several, indeed 6, pentagonal morphisms.

28. $|G| = 144$. As a consequence of cases 14*[iii]* and 14*[v]*, there is no epimorphism Δ with signature $(0; 2, 3, 8)$ onto a group of order 144 (see [Sin2]).

Summarizing the above discussion, we have found the following equisymmetric stratification of $\mathcal{M}_4$:

Theorem 3.2 *In the following list we shall describe the strata of the equisymmetic stratification of the moduli space $\mathcal{M}_4$ of Riemann surfaces of genus 4. We shall say that a stratum is given by Fuchsian groups with signature s if the surfaces in such strata are uniformized by surface Fuchsian groups that are finite index normal subgroups of Fuchsian groups with signature s. The signatures describe some topological properties of the action of the automorphism group of the surfaces in a given stratum. There is of course the stratum containing all genus four surfaces that corresponds to the Fuchsian group of signature $(4; -)$, this stratum is the only one containing the surfaces with only the identity automorphism.*

1) One stratum induced by Fuchsian groups with signature $(2; 2, 2)$.

2) One stratum induced by Fuchsian groups with signature $(1; 2, \overset{6}{...}, 2)$.

3) The stratum of hyperelliptic surfaces induced by Fuchsian groups with signature $(0; 2, \overset{10}{...}, 2)$.

4) One stratum induced by Fuchsian groups with signature $(1; 3, 3, 3)$.

5) Two strata $\mathcal{C}_1^3$, $\mathcal{C}_2^3$ of cyclic trigonal Riemann surfaces induced by Fuchsian groups with signature $(0; 3, 3, 3, 3, 3, 3)$.

6) One stratum induced by Fuchsian groups with signature $(1; 2, 2, 2)$.

7) One stratum of hyperelliptic surfaces induced by Fuchsian groups with signature $(0; 2, 2, 2, 2, 4, 4)$.

8) One stratum induced by Fuchsian groups with signature $(0; 2, 4, 4, 4, 4)$.

9) Two strata induced by Fuchsian groups with signature $(0; 2, .\overset{7}{.}., 2)$, one of them in the hyperelliptic locus.

10) One stratum $\mathcal{C}_2^5$ of cyclic pentagonal surfaces induced by Fuchsian groups with signature $(0; 5, 5, 5, 5)$.

11) One stratum of hyperelliptic surfaces induced by Fuchsian groups with signature $(0; 2, 6, 6, 6)$.

12) Two strata of cyclic trigonal surfaces induced by Fuchsian groups with signature $(0; 2, 2, 3, 3, 3)$.

13) One stratum of hyperelliptic surfaces induced by Fuchsian groups with signature $(0; 2, 2, 2, 3, 6)$.

14) One stratum induced by Fuchsian groups with signature $(0; 2, .\overset{6}{.}., 2)$.

15) Two strata induced by Fuchsian groups with signature $(0; 2, 2, 2, 2, 4)$, one of them in the hyperelliptic locus.

16) One stratum of hyperelliptic surfaces induced by Fuchsian groups with signature $(0; 2, 4, 4, 4)$.

17) One stratum $\mathcal{C}_1^5$ of cyclic pentagonal surfaces induced by Fuchsian groups with signature $(0; 2, 2, 5, 5)$.

18) One cyclic pentagonal surface Q_4 induced by a Fuchsian group with signature $(0; 5, 10, 10)$ in the stratum $\mathcal{C}_2^5$.

19) One hyperelliptic surface W_4 induced by a Fuchsian group with signature $(0; 4, 6, 12)$.

20) One stratum induced by Fuchsian groups with signature $(0; 2, 3, 3, 3)$.

21) Two strata of cyclic trigonal surfaces induced by Fuchsian groups with signature $(0; 2, 2, 3, 6)$.

22) One stratum induced by Fuchsian groups with signature $(0; 2, 2, 2, 2, 2)$.

23) One cyclic pentagonal surface T_4 induced by a Fuchsian group with signature $(0; 3, 5, 15)$ in the family $\mathcal{C}_2^5$.

24) One stratum of hyperelliptic surfaces induced by Fuchsian groups with signature $(0; 2, 2, 2, 8)$.

25) One hyperelliptic surface S_4 induced by a Fuchsian group with signature $(0; 2, 9, 18)$.

26) One stratum of cyclic trigonal surfaces induced by Fuchsian groups with signature $(0; 2, 2, 3, 3)$.

27) One stratum C_2^5 *of cyclic pentagonal, hyperelliptic surfaces induced by Fuchsian groups with signature* $(0; 2, 2, 2, 5)$.

28) One stratum of surfaces induced by Fuchsian groups with signature $(0; 2, 2, 2, 4)$.

29) One hyperelliptic surface R_4 *induced by a Fuchsian group with signature* $(0; 3, 4, 6)$.

30) One hyperelliptic surface P_4 *induced by a Fuchsian group with signature* $(0; 2, 4, 16)$.

31) One uniparametric family of cyclic trigonal surfaces induced by Fuchsian groups with signature $(0; 2, 2, 2, 3)$.

32) One cyclic trigonal surface Z_4 *induced by a Fuchsian group with signature* $(0; 2, 6, 6)$.

33) One cyclic pentagonal, hyperelliptic surface V_4 *induced by a Fuchsian group with signature* $(0; 2, 4, 10)$.

34) One cyclic trigonal surface X_4 *induced by a Fuchsian group with signature* $(0; 2, 3, 12)$.

35) One cyclic trigonal surface Y_4 *induced by a Fuchsian group with signature* $(0; 2, 4, 6)$.

36) One cyclic pentagonal surface U_4 *induced by a Fuchsian group with signature* $(0; 2, 4, 5)$ *in the family* C_1^5.

Acknowledgement. *The authors wish to express their gratitude to the referee and editors for their many helpful suggestions.*

Notes

1 Partially supported by MTM2008-00250.
2 Partially supported by the Swedish Research Council (VR)

Bibliography

[Bo] Bogopolski, O. V. Classification of actions of finite groups on orientable surface of genus four. *Siberian Advances in Mathematics* **7**(4) (1997), 9–38.

[Bre] Breuer, T. *Characters and Automorphism Groups of Compact Riemann Surfaces*, London Mathematical Society Lecture Note Series **280**, Cambridge University Press, Cambridge, 2000.

[Bro1] Broughton, A. The equisymmetric stratification of the moduli space and the Krull dimension of mapping class groups. *Topology Appl.* **37** (1990), 101–113.

[Bro2] Broughton, A. Classifying finite group actions on surfaces of low genus. *J. Pure Appl. Algebra* **69** (1990), 233–270.

[BEGG] Bujalance, E., Etayo, J. J., Gamboa, J. M. and Gromadzki, G. *Automorphisms Groups of Compact Bordered Klein Surfaces, A Combinatorial Approach*, Lecture Notes in Math. **1439**, Springer Verlag, 1990.

[BSS] Buser, P., Seppälä, M. and Silhol, R. Triangulations and moduli spaces of Riemann surfaces with group actions. *Manuscripta Math.* **88** (1995), 209–224.

[CIY1] Costa, A. F., Izquierdo, M. and Ying, D. On Riemann with non-unique cyclic trigonal morphism. *Manuscripta Math.* **118** (2005), 443–453.

[CIY2] Costa, A. F., Izquierdo, M. and Ying D. On the family of cyclic trigonal Riemann surfaces of genus 4 with several trigonal morphisms. *RACSAM* **101** (2007), 81–86.

[CM] Coxeter, H. S. M. and Moser, W. O. J. *Generators and Relations for Discrete Groups*, Springer-Verlag, Berlin, 1957.

[Gon] González-Díez, G. On prime Galois covering of the Riemann sphere. *Ann. Mat. Pure Appl.* **168** (1995), 1–15.

[Har] Harvey, W. On branch loci in Teichmüller space. *Trans. Amer. Math. Soc.* **153** (1971), 387–399.

[IY] Izquierdo, M. and Ying, D. Equisymmetric strata of the moduli space of cyclic trigonal Riemann surfaces of genus 4. *Glasgow Math, J.* **51**(1) (2009), 19–29.

[Kim] Kimura, H. Classification of automorphisms groups, up to topological equivalencce, of compact Riemann surfaces of genus 4. *J. Algebra* **264** (2003), 26–54.

[MSSV] Magaard, K., Shaska, T., Shpectorov, S. and Völklein, H. The locus of curves with prescribed automorphism group. Communications in arithmetic fundamental groups (Kyoto, 1999/2001). *Sūrikaisekikenkyūsho Kōkyūroku* No. 1267 (2002), 112–141.

[MSV] Magaard, K., Shpectorov, S. and Völklein, H. A. GAP package for braid orbit computation and applications. *Experiment. Math.* **12** (2003), 385–393.

[Nag] Nag, S. *The Complex Theory of Teichmüller Spaces*, Wiley-Interscience, 1988.

[Sin1] Singerman, D. Subgroups of Fuchsian groups and finite permutation groups. *Bull. London Math. Soc.* **2** (1970), 319–323.

[Sin2] Singerman, D. Finitely maximal Fuchsian groups. *J. London Math. Soc.* **6** (1972), 29–38.

[Smi] Smith, P. A. Abelian actions on 2-manifolds. *Michigan Math. J.* **14** (1967), 257–275.

[Wol] Wolfart, J. Regular dessins, endomorphisms of Jacobians, and transcendence, in *A Panorama of Number Theory or The View from Baker's Garden*, ed. G. Wüstholz, Cambridge University Press, 2002, pp. 107–120.

[Zie] Zieschang, H. (1981) *Finite Groups of Mapping Classes of Surfaces*, Lectures Notes in Mathematics **875**, Springer-Verlag, Berlin, 1981.

Diffeomorphisms and automorphisms of compact hyperbolic 2-orbifolds

Clifford J. Earle

Department of Mathematics, Cornell University
cliff@math.cornell.edu

Introductory remarks

This paper, a sequel to [E], owes its existence to the 2007 conference for Bill Harvey. In his lecture on the opening day, Gabino González-Diez mentioned an example from a remark in [E]. Later he asked me about its proof, which is not given in [E]. Theorem 1.1 restates the example, and Appendix I provides the missing proof.

Most of the paper explores connections between [E] and the interesting papers [MH] by Maclachlan and Harvey and [BH1] and [BH2] by Birman and Hilden. Some of their results about homeomorphisms of Riemann surfaces have analogues for diffeomorphisms of two-dimensional compact hyperbolic orbifolds. These are stated below as corollaries of Theorems 1.2 and 1.3. It seems appropriate to include them here because both Harvey and Maclachlan were present at the conference.

The main result, Theorem 1.2, is proved here twice. One proof uses Theorem 1 of [E], which depends on ideas from [EE] and [ES]. The other, in Appendix III, relies on [DE] and [EM]. The papers [DE], [EE], [EM], and [ES] are collaborations with four different coauthors. I am indebted to them all. Two of them, Jim Eells and Adrien Douady, are no longer with us. I dedicate this paper to their memories.

I also thank the organizers of the Harvey conference for the opportunity to take part in that stimulating event.

1 Statement of results

1.1 The example

Our first theorem is a slightly altered version of the aforementioned example. (For the original, see Remark (c) in §3.1 of [E].) Our alteration is a change of variables. The variables x and y in [E] are replaced by $z := \omega x$ and $w := \omega y$, where $\omega := \exp 2\pi i/3$.

Theorem 1.1 *For any real number $t > 0$, let X_t be the compact Riemann surface of genus two determined by the equation*

$$w^2 = z(z^2 - \omega)(z^2 + tz - 1). \tag{1.1}$$

The Riemann surface X_t has an anti-holomorphic automorphism of order 4, but it has no anti-holomorphic involution unless $t = 1$.

Theorem 1.1 provides examples (all X_t with $t \neq 1$) of Riemann surfaces that are isomorphic to their conjugate surfaces but cannot be defined by polynomial equations with real coefficients. For a discussion of this issue, see Silhol [Si]. Theorem 2 of [E] proves the existence of some examples of higher genus. For more recent related work, see [FG] and the literature quoted there.

Many explicit hyperelliptic examples of any even genus were given independently by Shimura in [Sh]. Shimura's examples are generic among hyperelliptic surfaces with a certain anti-holomorphic automorphism of order 4. The surfaces X_t in Theorem 1.1 are not. Their set of Weierstrass points has special properties. These will be used when we prove Theorem 1.1 in Appendix I.

1.2 Hyperbolic 2-orbifolds

To state the other theorems, we need some definitions. We use the open unit disk $D := \{z \in \mathcal{C} : |z| < 1\}$ with its Poincaré metric as a model for the hyperbolic plane. Let G be the group of all hyperbolic isometries of D. By definition, a *hyperbolic 2-orbifold* is the quotient of D by a subgroup Γ of G that acts properly discontinuously on D. We say that two such orbifolds are *isomorphic* if they are the quotients of D by conjugate subgroups of G.

As G contains orientation reversing maps, a hyperbolic Riemann surface and its conjugate surface have isomorphic hyperbolic orbifold structures. As Γ may contain orientation reversing maps, hyperbolic 2-orbifolds need not be orientable.

Let X and Y be hyperbolic 2-orbifolds, determined by discrete groups Γ_X and Γ_Y, and let $p \colon X \to Y$ be continuous. By definition, p is a *smooth orbifold covering* if and only if there is a commutative diagram

$$
\begin{array}{ccc}
D & \xrightarrow{\;\widetilde{p}\;} & D \\
{\scriptstyle \pi_X}\downarrow & & {\scriptstyle \pi_Y}\downarrow \\
X & \xrightarrow{\;p\;} & Y
\end{array}
\tag{1.2}
$$

in which the vertical arrows are the quotient maps by Γ_X and Γ_Y, and $\widetilde{p}$ is a (C^∞) diffeomorphism of D onto itself. We call $\widetilde{p}$ a *lift* of f.

By definition, a *diffeomorphism of X onto Y* is an injective smooth orbifold covering $f\colon X \to Y$. The set $\mathrm{Diff}(X)$ of diffeomorphisms of X onto itself is a group under composition. The subgroup $\mathrm{Diff}_0(X)$ consists, by definition, of the f in $\mathrm{Diff}(X)$ having a lift $\widetilde{f}$ that commutes with every element of Γ_X. It inherits a topology from the C^∞ topology on the group $\mathrm{Diff}(D)$ of diffeomorphisms of D onto itself (see Corollary 2.5 in §2).

Let $p\colon X \to Y$ be a smooth orbifold covering. Suppose $f \in \mathrm{Diff}(X)$ and $g \in \mathrm{Diff}(Y)$. We say that f is a *lift* of g if the diagram

$$
\begin{array}{ccc}
X & \xrightarrow{\;f\;} & X \\
{\scriptstyle p}\downarrow & & {\scriptstyle p}\downarrow \\
Y & \xrightarrow{\;g\;} & Y
\end{array}
\tag{1.3}
$$

commutes. We say that f in $\mathrm{Diff}(X)$ is *p-equivariant* if it is a lift of some g in $\mathrm{Diff}(Y)$.

With these definitions, the final sentence of Theorem 1 in [E] can be restated in the following more geometric way.

Theorem 1.2 *If X is a compact hyperbolic 2-orbifold, then $\mathrm{Diff}_0(X)$ is contractible. In particular, every f in $\mathrm{Diff}_0(X)$ is isotopic to the identity through diffeomorphisms in $\mathrm{Diff}_0(X)$.*

The proof in §3 uses some basic properties of discontinuous groups of hyperbolic isometries that are proved in §2. These properties also have the following geometric consequence, which is proved in §4. Analogous results, for orbifolds whose defining groups contain no orientation reversing maps, were proved by Maclachlan and Harvey in [MH].

Theorem 1.3 *Let X and Y be compact hyperbolic 2-orbifolds. If $p\colon X \to Y$ is a smooth orbifold covering, then each g in $\mathrm{Diff}_0(Y)$ has a unique lift $\widehat{g}$ in $\mathrm{Diff}_0(X)$. The map $g \mapsto \widehat{g}$ is an isomorphism of $\mathrm{Diff}_0(Y)$ onto the group of p-equivariant maps in $\mathrm{Diff}_0(X)$.*

Because of Theorem 1.3, Theorem 1.2 has p-equivariant versions that we state as corollaries. They are inspired by Theorems 1 and 2 of Birman-Hilden [BH2] and Corollary 12 in Maclachlan-Harvey [MH]. The proofs are in §5.

Corollary 1.4 *Let X and Y be compact hyperbolic 2-orbifolds, and let $p: X \to Y$ be a smooth orbifold covering. The set of maps in $\mathrm{Diff}(Y)$ that have lifts in $\mathrm{Diff}_0(X)$ is contractible. In particular, every such map is isotopic to the identity through diffeomorphisms with lifts in $\mathrm{Diff}_0(X)$.*

Corollary 1.5 *Let X, Y, and p be as above. The group of p-equivariant maps in $\mathrm{Diff}_0(X)$ is contractible. In particular, every such map is isotopic to the identity through p-equivariant diffeomorphisms in $\mathrm{Diff}_0(X)$.*

The proof of Theorem 1 in [E] uses the Teichmüller theoretic methods first developed in [EE] and [ES], so the proof of Theorem 1.2 in §3 depends indirectly on Teichmüller theory. Appendix III provides a direct proof that depends instead on the methods of [DE] and [EM]. For the reader's convenience, Appendix II proves some well-known facts about the group $\mathrm{Diff}(D)$.

2 Some facts about compact hyperbolic 2-orbifolds and their diffeomorphisms

The results in this section are well known for classical Fuchsian groups. We give their proofs because our groups may contain reflections. This influences the statement and proof of Lemma 2.1.

As in §1.2, we consider groups Γ of hyperbolic isometries acting properly discontinuously on the Poincaré disk D, producing a hyperbolic orbifold D/Γ.

Lemma 2.1 *If $\pi: D \to D/\Gamma$ is the quotient map and $f \in \mathrm{Diff}(D)$, then $\pi f = \pi$ if and only if $f \in \Gamma$.*

Proof If $\gamma \in \Gamma$, then $\pi\gamma = \pi$ by the definition of π. Conversely, if $f \in \mathrm{Diff}(D)$ and $\pi f = \pi$, then for each z in D there is at least one γ in Γ such that $f(z) = \gamma(z)$. We must see how γ depends on z.

Let E be the discrete set of elliptic fixed points of Γ in D, and let $\Omega := D \setminus E$. As Ω is connected and dense in D, we need only show that for each z in Ω there is γ in Γ such that $f = \gamma$ in a neighborhood of z.

We first consider the set Ω_0 of z in Ω such that $\gamma(z) \neq z$ for all γ in Γ except the identity. Choose z in Ω_0. Because Γ acts properly discontinuously, z has a neighborhood U such that $\gamma(U) \cap U$ is empty for all nontrivial γ in Γ. (In particular, the set Ω_0 is open.)

Choose γ in Γ so that $\gamma(z) = f(z)$, and set $V := U \cap f^{-1}(\gamma(U))$. Then V is a neighborhood of z. If $\tilde{z} \in V$, $\tilde{\gamma} \in \Gamma$, and $f(\tilde{z}) = \tilde{\gamma}(\tilde{z})$, then $\gamma^{-1}(\tilde{\gamma}(\tilde{z})) \in U$. Therefore $\tilde{\gamma} = \gamma$, and $f = \gamma$ in V.

It follows readily that for each connected component W of Ω_0 there is γ_W in Γ such that $f = \gamma_W$ in W.

Finally, suppose $z \in \Omega \setminus \Omega_0$. Then Γ contains the reflection σ in some geodesic passing through z, and σ is the only nontrivial element of Γ that fixes z. Choose an open z-centered Poincaré disk U such that $\gamma(U) \cap U$ is empty for all nontrivial γ in Γ except σ. The set of fixed points of σ in U is a geodesic segment F that separates U into two open semi-disks U_+ and $U_- := \sigma(U_+)$.

As U_+ and U_- are connected subsets of Ω_0, there are γ_+ and γ_- in Γ such that f equals γ_+ in U_+ and γ_- in U_-. Clearly, $\gamma_+ = \gamma_-$ on F, so either $\gamma_- = \gamma_+$ in U or $\gamma_- = \gamma_+\sigma$ in U. In the latter case, $f = \gamma_+$ in U_+ and $f = \gamma_+\sigma$ in U_-. As f is injective, that is impossible. Therefore $f = \gamma_+$ in U. $\qquad\square$

Corollary 2.2 *Let* $X := D/\Gamma_X$ *and* $Y := D/\Gamma_Y$. *Then* f *in* $\mathrm{Diff}(D)$ *is a lift of some smooth orbifold covering* $p\colon X \to Y$ *if and only if* $f\gamma f^{-1} \in \Gamma_Y$ *for all* γ *in* Γ_X. *The covering* p *is a diffeomorphism if and only if* Γ_Y *and the group* $f\Gamma_X f^{-1}(:= \{f\gamma f^{-1} : \gamma \in \Gamma_X\})$ *are equal.*

Proof Suppose $f\gamma f^{-1} \in \Gamma_Y$ for all γ in Γ_X. Then $f\Gamma_X f^{-1}$ is a subgroup of Γ_Y, and f is obviously a lift of a diffeomorphism p_f of X onto the orbifold $D/f\Gamma_X f^{-1}$. It is also a lift of an orbifold covering $p\colon X \to Y$, and p is injective (hence a diffeomorphism) if and only if $f\Gamma_X f^{-1} = \Gamma_Y$.

Conversely, let $\tilde{p}$ be a lift of an orbifold covering $p\colon X \to Y$. Choose γ in Γ_X. As the diagram (1.2) commutes, $\pi_Y \tilde{p}\gamma\tilde{p}^{-1} = \pi_Y$. Therefore, by Lemma 2.1, $\tilde{p}\gamma\tilde{p}^{-1} \in \Gamma_Y$ for all γ in Γ_X. $\qquad\square$

If D/Γ is a hyperbolic orbifold, we shall denote by $\mathrm{Diff}_0(\Gamma)$ the centralizer of Γ in $\mathrm{Diff}(D)$. That is,

$$\mathrm{Diff}_0(\Gamma) := \{f \in \mathrm{Diff}(D) : f\gamma = \gamma f \text{ for all } \gamma \text{ in } \Gamma\}.$$

The following corollaries explain both the role of $\mathrm{Diff}_0(\Gamma)$ and our choice of notation.

Corollary 2.3 *If $X := D/\Gamma$, each f in $\mathrm{Diff}_0(\Gamma)$ is the lift of exactly one diffeomorphism of X. That diffeomorphism belongs to $\mathrm{Diff}_0(X)$.*

Proof By Corollary 2.2, each f in $\mathrm{Diff}_0(\Gamma)$ is the lift of some g in $\mathrm{Diff}(X)$. Let $\pi\colon D \to X$ be the quotient map. The equation $g\pi = \pi f$ determines g uniquely, and $g \in \mathrm{Diff}_0(X)$ because its lift f belongs to $\mathrm{Diff}_0(\Gamma)$. $\square$

By definition, the *obvious map* from $\mathrm{Diff}_0(\Gamma)$ to $\mathrm{Diff}_0(X)$ sends f in $\mathrm{Diff}_0(\Gamma)$ to the unique g in $\mathrm{Diff}_0(X)$ such that f is a lift of g.

As a subgroup of $\mathrm{Diff}(D)$, $\mathrm{Diff}_0(\Gamma)$ inherits the C^∞ topology. As we shall see, the obvious map is a surjective isomorphism from $\mathrm{Diff}_0(\Gamma)$ to $\mathrm{Diff}_0(X)$, so it induces a topology on $\mathrm{Diff}_0(X)$. The proof requires another lemma. We shall denote the closed unit disk and its boundary by $\overline{D}$ and S^1 respectively.

Lemma 2.4 *If the hyperbolic 2-orbifold D/Γ is compact, then every f in $\mathrm{Diff}_0(\Gamma)$ has a continuous extension to $\overline{D}$ that fixes S^1 pointwise.*

Proof Let ρ be the Poincaré distance in D.

By hypothesis, $\rho(f(\gamma(z)), \gamma(z)) = \rho(\gamma(f(z)), \gamma(z)) = \rho(f(z), z)$ for all z in D and γ in Γ. Hence $z \mapsto \rho(f(z), z)$ is an automorphic function with respect to Γ. As D/Γ is compact, there is a number K such that $\rho(f(z), z) < K$ for all z in D.

Let $\widehat{f}\colon \overline{D} \to \overline{D}$ equal f in D and the identity on S^1. We claim that $\widehat{f}$ is continuous. The only issue is at points of S^1. Let the sequence (z_n) converge to ζ in S^1. We must show that $(\widehat{f}(z_n))$ converges to ζ.

We may assume the z_n are in D. As $\rho(\widehat{f}(z_n), z_n) < K$ for all n, the sequences (z_n) and $(\widehat{f}(z_n))$ have the same limit ζ in S^1. $\square$

Corollary 2.5 *If $X := D/\Gamma$ is compact, then the obvious map from $\mathrm{Diff}_0(\Gamma)$ to $\mathrm{Diff}_0(X)$ is a surjective isomorphism. In particular, $\mathrm{Diff}_0(X)$ has a unique topology such that the obvious map is a homeomorphism.*

Proof By definition, each member of $\mathrm{Diff}_0(X)$ has at least one lift in $\mathrm{Diff}_0(\Gamma)$, so the obvious map is surjective. It is clearly a group homeomorphism. It is injective because if f in $\mathrm{Diff}_0(\Gamma)$ is a lift of the identity, then $f \in \Gamma$ by Lemma 2.1 and, by Lemma 2.4, f is the identity on S^1, hence on D. $\square$

3 Proof of Theorem 1.2

First we shall find a compact hyperbolic Riemann surface W such that the given orbifold X is the quotient of W by a finite group of hyperbolic isometries.

Let $X := D/\Gamma_X$, and let X^+ be the quotient of D by the group Γ_{X^+} of orientation preserving maps in Γ_X. As Γ_{X^+} has finite index in Γ_X and X is compact, X^+ is also compact. Therefore, Γ_{X^+} is a finitely generated Fuchsian group, and it has a normal subgroup N of finite index that contains no elements of finite order (see [BN] and [F]).

If $\Gamma_{X^+} = \Gamma_X$, we set $\Gamma_W := N$. If $\Gamma_{X^+} \neq \Gamma_X$, we choose γ in $\Gamma_X \setminus \Gamma_{X^+}$ and we set $\Gamma_W := N \cap \gamma N \gamma^{-1}$. In either case, Γ_W is a normal subgroup of Γ_X, its index in Γ_X is finite, and it contains no orientation reversing elements. The orbifold $W := D/\Gamma_W$ is therefore a compact Riemann surface.

The finite group $H := \Gamma_X/\Gamma_W$ acts on W as a group of hyperbolic isometries (holomorphic or anti-holomorphic automorphisms), and $W/H = X$. The quotient map $p\colon W \to X$ is a smooth orbifold covering.

Since $\Gamma_W \subset \Gamma_X$, $\mathrm{Diff}_0(\Gamma_X) \subset \mathrm{Diff}_0(\Gamma_W)$.

Lemma 3.1 *The obvious map of* $\mathrm{Diff}_0(\Gamma_W)$ *to* $\mathrm{Diff}_0(W)$ *sends* $\mathrm{Diff}_0(\Gamma_X)$ *to the group* $N_0(H) := \{g \in \mathrm{Diff}_0(W) : ghg^{-1} \in H$ *for all h in H}.*

Proof Let f in $\mathrm{Diff}_0(\Gamma_W)$ be the lift of g in $\mathrm{Diff}_0(W)$. By Lemma 2.4, f has a continuous extension to $\overline{D}$ that fixes S^1 pointwise.

If $g \in N_0(H)$, choose any γ in Γ_X. Let h in H be the diffeomorphism of W induced by γ. As $\widehat{h} := ghg^{-1}$ belongs to H, we can choose some $\widehat{\gamma}$ in Γ_X that induces $\widehat{h}$. As $ghg^{-1}\widehat{h}^{-1}$ is the identity on W, $f\gamma f^{-1}\widehat{\gamma}^{-1}$ belongs to Γ_W, by Lemma 2.1. Therefore, $f\gamma f^{-1} \in \Gamma_X$. As $f\gamma f^{-1}$ and γ are equal on S^1, they are equal in D. Therefore, $f \in \mathrm{Diff}_0(\Gamma_X)$.

Conversely, suppose $f \in \mathrm{Diff}_0(\Gamma_X)$. Choose any h in H and any γ in $\mathrm{Diff}_0(\Gamma_X)$ that induces h. Then $f\gamma f^{-1}$ is a lift of ghg^{-1}. As $f\gamma f^{-1} = \gamma$, $ghg^{-1} = h$, and $g \in N_0(H)$. $\qquad\square$

Theorem 1.2 now follows readily. The obvious maps from $\mathrm{Diff}_0(\Gamma_X)$ to $\mathrm{Diff}_0(X)$ and from $\mathrm{Diff}_0(\Gamma_W)$ to $\mathrm{Diff}_0(W)$ are bijective by Corollary 2.5. They are homeomorphisms by decree. By Lemma 3.1, $\mathrm{Diff}_0(X)$, $\mathrm{Diff}_0(\Gamma_X)$, and $N_0(H)$ are homeomorphic. All are contractible because $N_0(H)$ is, by Theorem 1 of [E].

Finally, the continuity of the evaluation map $(f, x) \mapsto f(x)$ from

$\mathrm{Diff}_0(X) \times X$ to X (see Appendix II) makes it easy to produce isotopies in $\mathrm{Diff}_0(X)$.

For any f in $\mathrm{Diff}_0(X)$, let $t \mapsto f_t$ be a continuous map from $[0,1]$ to $\mathrm{Diff}_0(X)$ such that f_0 is f and f_1 is the identity. The map $(t,x) \mapsto f_t(x)$ from $[0,1] \times X$ to X is an isotopy from f_0 to the identity, and each $f_t \in \mathrm{Diff}_0(X)$ by construction. $\square$

4 Proof of Theorem 1.3

Set $X := D/\Gamma_X$ and $Y := D/\Gamma_Y$. Choose some lift $\widetilde{p}$ in $\mathrm{Diff}(D)$ of the given smooth orbifold covering $p\colon X \to Y$.

Lemma 4.1 *The automorphism $\widetilde{g} \mapsto \widetilde{p}^{-1}\widetilde{g}\widetilde{p}$ of $\mathrm{Diff}(D)$ maps $\mathrm{Diff}_0(\Gamma_Y)$ into $\mathrm{Diff}_0(\Gamma_X)$. The image of $\mathrm{Diff}_0(\Gamma_Y)$ is the set of $\widetilde{f}$ in $\mathrm{Diff}_0(\Gamma_X)$ that are lifts of p-equivariant maps in $\mathrm{Diff}_0(X)$.*

Proof Let $\widetilde{g}$ belong to $\mathrm{Diff}_0(\Gamma_Y)$. As $\widetilde{p}$ is a lift of $p\colon X \to Y$, Corollary 2.2 implies that $\widetilde{p}\gamma\widetilde{p}^{-1} \in \Gamma_Y$ for all γ in Γ_X. Therefore $\widetilde{g}$ commutes with $\widetilde{p}\gamma\widetilde{p}^{-1}$ for all γ in Γ_X, so $\widetilde{p}^{-1}\widetilde{g}\widetilde{p} \in \mathrm{Diff}_0(\Gamma_X)$. Set $\widetilde{f} := \widetilde{p}^{-1}\widetilde{g}\widetilde{p}$.

Let f in $\mathrm{Diff}_0(X)$ and g in $\mathrm{Diff}_0(Y)$ be the maps induced by $\widetilde{f}$ and $\widetilde{g}$ respectively. We claim that f is a lift of g. By construction,

$$p f \pi_X = p \pi_X \widetilde{f} = \pi_Y \widetilde{p}\widetilde{f} = \pi_Y \widetilde{g}\widetilde{p} = g \pi_Y \widetilde{p} = g p \pi_X,$$

so $pf = gp$ as required. Therefore, f is p-equivariant.

We have shown that $\widetilde{g} \mapsto \widetilde{p}^{-1}\widetilde{g}\widetilde{p}$ maps $\mathrm{Diff}_0(\Gamma_Y)$ into the set of $\widetilde{f}$ in $\mathrm{Diff}_0(\Gamma_X)$ that are lifts of p-equivariant maps in $\mathrm{Diff}_0(X)$.

Now let $\widetilde{f}$ in $\mathrm{Diff}_0(\Gamma_X)$ be the lift of a p-equivariant map in $\mathrm{Diff}_0(X)$. To complete the proof of the lemma, we must show that $\widetilde{p}\widetilde{f}\widetilde{p}^{-1}$ belongs to $\mathrm{Diff}_0(\Gamma_Y)$.

Let f be the map in $\mathrm{Diff}_0(X)$ induced by $\widetilde{f}$. By assumption, f is p-equivariant, so there is g in $\mathrm{Diff}(Y)$ such that $gp = pf$.

Set $\widetilde{g} := \widetilde{p}\widetilde{f}\widetilde{p}^{-1}$. It is easy to verify that $\widetilde{g}$ is a lift of g. In addition, since $\widetilde{f}$ commutes with Γ_X, $\widetilde{g}$ commutes with the subgroup Γ_W of Γ_Y consisting of the isometries $\widetilde{p}\gamma\widetilde{p}^{-1}$, γ in Γ_X.

The orbifold $W := D/\Gamma_W$ is compact, since $\widetilde{p}$ induces a diffeomorphism of X onto W. Therefore, by Lemma 2.4, $\widetilde{g}$ has a continuous extension to $\overline{D}$ that fixes S^1 pointwise.

Choose any γ in Γ_Y, and consider $\widetilde{g}\gamma\widetilde{g}^{-1}$. It belongs to Γ_Y because $\widetilde{g}$ is a lift of g. It equals γ on S^1, hence in D, because of the boundary behavior of γ and $\widetilde{g}$. Therefore, $\widetilde{g}$ belongs to $\mathrm{Diff}_0(\Gamma_Y)$. $\square$

By Corollary 2.5, we can interpret the map $\widetilde{g} \mapsto \widetilde{p}^{-1} \widetilde{g} \widetilde{p}$ in Lemma 4.1 as an injective map of $\text{Diff}_0(Y)$ into $\text{Diff}_0(X)$. This map carries each g in $\text{Diff}_0(Y)$ to a p-equivariant lift f in $\text{Diff}_0(X)$, and its image is the group of all p-equivariant maps in $\text{Diff}_0(X)$. Theorem 1.3 follows at once.

5 Proof of Corollaries 1.4 and 1.5

Suppose g in $\text{Diff}(Y)$ has a lift f in $\text{Diff}_0(X)$. As f is p-equivariant, it is the lift of some g_0 in $\text{Diff}_0(Y)$, by Theorem 1.3. Clearly $g = g_0$, so the set of g in $\text{Diff}(Y)$ with lifts in $\text{Diff}_0(X)$ is precisely $\text{Diff}_0(Y)$. Corollary 1.4 therefore follows immediately from Theorem 1.2.

To prove Corollary 1.5, we consider the map $\phi: \text{Diff}_0(Y) \to \text{Diff}_0(X)$ induced by the map $\widetilde{g} \mapsto \widetilde{p}^{-1} \widetilde{g} \widetilde{p}$ of $\text{Diff}_0(\Gamma_Y)$ into $\text{Diff}_0(\Gamma_X)$.

It is well known that $\text{Diff}(D)$ is a topological group (see Proposition 5.6 in Appendix II). Therefore $\widetilde{g} \mapsto \widetilde{p}^{-1} \widetilde{g} \widetilde{p}$ is a homeomorphism of $\text{Diff}(D)$ onto itself, and it maps $\text{Diff}_0(\Gamma_Y)$ homeomorphically onto its image in $\text{Diff}_0(\Gamma_X)$. That means ϕ is a homeomorphism of $\text{Diff}_0(Y)$ onto its image in $\text{Diff}_0(X)$. By the proof of Theorem 1.3, that image is precisely the group of p-equivariant maps in $\text{Diff}_0(X)$. Corollary 1.5 now also follows directly from Theorem 1.2.

Appendix I: Proof of Theorem 1.1

For $0 < t < \infty$ we rewrite the defining equation (1.1) of the compact Riemann surface X_t as

$$w^2 = z(z + \omega^2)(z - \omega^2)(z - r)(z + r^{-1}), \quad z \text{ and } w \text{ in } \widehat{\mathcal{C}}, \qquad (5.1)$$

where $\widehat{\mathcal{C}}$ is the Riemann sphere and $r = r(t) := (\sqrt{t^2 + 4} - t)/2$. Thus, $0 < r < 1$, $t = r^{-1} - r$, and $t = 1$ when $r = (\sqrt{5} - 1)/2$.

The form of equation (5.1) shows that X_t is a hyperelliptic Riemann surface of genus two (see Example IV.11.11 in [FK]). Consider the map

$$j(z, w) := (z, -w), \quad z \text{ and } w \text{ in } \widehat{\mathcal{C}}, \qquad (5.2)$$

of X_t onto itself. By Proposition III.7.9 of [FK] and its corollaries, j is the hyperelliptic involution of X_t, every holomorphic automorphism of X_t commutes with j, and the Weierstrass points of X_t are the fixed points of j. These are the points where $w = 0$ or ∞, so $z = 0, \infty, \pm \omega^2, r$, or $-1/r$.

Recall that $\omega = \exp 2\pi i/3$ and $\omega^3 = 1$. For each X_t, the formula

$$f(z, w) := (-\bar{z}^{-1}, i\omega^2 \bar{z}^{-3} \bar{w}), \quad z \text{ and } w \text{ in } \widehat{\mathcal{C}}, \qquad (5.3)$$

defines an anti-holomorphic automorphism f of order 4. This proves the first assertion of Theorem 1.1.

Observe that $f^2 (:= f \circ f)$ is the hyperelliptic involution j. In particular, f commutes with j.

Let J be the group of order two that j generates. The map $(z, w) \mapsto z$ from X_t to the Riemann sphere $\widehat{C}$ is an explicit model for the quotient map from X_t to X_t/J. This gives $\widehat{C}$ a hyperbolic orbifold structure. The quotient map is locally injective except at the Weierstrass points of X_t. The *orbifold points* in $\widehat{C}$ are the images of the Weierstrass points. Explicitly, they are $0, \infty, \pm\omega^2, r$, and $-r^{-1}$.

Equation (5.3) shows that f induces the involution $\widehat{f}(z) = -1/\bar{z}$ of $\widehat{C}$. The set *Orb* of orbifold points is invariant under $\widehat{f}$.

Now suppose X_t has an anti-holomorphic involution g. Consider the holomorphic automorphism $h := fg$ of X_t onto itself. Since h commutes with j, it induces a holomorphic automorphism $\widehat{h}$ of $\widehat{C}$. Since $g = f^{-1}h$ is an involution and $f^2 = j$, h cannot be j or the identity map. Therefore $\widehat{h}$ is not the identity. Using very elementary methods, we shall show that this is impossible unless $t = 1$.

Lemma 5.1 *The sets* $\{0, \infty, \omega^2, -\omega^2\}$ *and* $\{r, -1/r\}$ *are* $\widehat{h}$*-invariant.*

Proof By its construction, $\widehat{h}$ leaves the set *Orb* invariant. As it is a Möbius transformation, $\widehat{h}$ also preserves the cross-ratio of any ordered quadruplet of distinct points of *Orb*.

Direct examination of cases shows that $\{0, \infty, \omega^2, -\omega^2\}$ is the only 4-point subset of *Orb* such that the cross-ratio of its points is -1, $1/2$, or 2, depending on their ordering. That property is $\widehat{h}$-invariant. $\square$

The next lemma describes the group H of all Möbius transformations that map the set $\{0, \infty, \omega^2, -\omega^2\}$ to itself. Although it is well known, we sketch a proof. The idea is that 0 and ∞ are the north and south poles of the sphere, and $\pm\omega^2$ are antipodal points on the equator.

Lemma 5.2 *The group H is the dihedral group of order 8 generated by*

$$A(z) := -z \quad and \quad B(z) := \frac{z + \omega^2}{-\omega z + 1}, \quad z \ in\ \widehat{C}.$$

Proof Since $B(0) = \omega^2$, $B(\omega^2) = \infty$, $B(\infty) = -\omega^2$, and $B(-\omega^2) = 0$, it is easy to see that A and B belong to H and generate a dihedral group of order 8. It is also easy to show that any member of H that fixes 0 is either A or the identity. The lemma follows readily. $\qquad\square$

Lemma 5.3 *The subgroup H_{Orb} of H that leaves the set Orb invariant is trivial if $t \neq 1$. If $t = 1$, H_{Orb} has order two and is generated by AB^3.*

Proof Since $A(r) \notin Orb$, $A \notin H_{Orb}$. Since B^2 is the map $z \mapsto -\omega/z$, $B^2(r) \notin Orb$. Therefore $B \notin H_{Orb}$. Thus, the only possible nontrivial elements of H_{Orb} are AB, AB^2, or AB^3. Consider the numbers

$$AB(r) = \frac{r + \omega^2}{\omega r - 1}, \quad AB^2(r) = \frac{\omega}{r}, \quad \text{and} \quad AB^3(r) = \frac{-r + \omega^2}{\omega r + 1}.$$

The first of them is real if and only if $r^2 - r - 1 = 0$. Since $0 < r < 1$, that is impossible. The second is not real for any real r. The third is real if and only if $r^2 + r - 1 = 0$. Since $0 < r < 1$, that happens if and only if $r = (\sqrt{5} - 1)/2$ (equivalently, $t = 1$). When $t = 1$, AB^3 interchanges r and $-1/r$. The lemma follows. $\qquad\square$

Corollary 5.4 *The Riemann surface X_t has no anti-holomorphic involution unless $t = 1$.*

Proof If X_t has an anti-holomorphic involution, then the group H_{Orb} is not trivial. $\qquad\square$

We shall examine the case $t = 1$ more deeply. By Lemma 5.3, any anti-holomorphic involution of X_1 must induce the involution $\widehat{g} := \widehat{f}^{-1}\widehat{h}$ of $\widehat{\mathcal{C}}$, with $\widehat{h} = AB^3$. Therefore

$$\widehat{g}(z) = \frac{\bar{\omega}\bar{z} + 1}{\bar{z} - \omega}, \qquad z \in \widehat{\mathcal{C}}.$$

The fixed-point set of $\widehat{g}$ is the circle $C := \{z \in \mathcal{C} : |z - \omega^2|^2 = 2\}$, so $\widehat{g}$ is inversion in C. The orbifold points ω^2 and ∞ are symmetric with respect to C, and so are 0 and $-\omega^2$. As $t = 1$, C crosses $\mathbb{R}$ at the orbifold points r and $-r^{-1}$.

Now we change variables. Let T be a Möbius transformation such that $T(C)$ is the extended real axis, $T(r) = \infty$, and $T(-r^{-1}) = 0$. We choose T so that $\alpha := T(\omega^2)$ and $\beta := T(0)$ are in the upper half plane $\mathcal{H}^+$. By symmetry, $T(\infty) = \bar{\alpha}$ and $T(-\omega^2) = \bar{\beta}$.

In terms of the coordinate $\zeta := T(z)$, the orbifold points of X_1/J are at 0, ∞, α, $\bar{\alpha}$, β, and $\bar{\beta}$. Therefore X_1 has a defining equation

$$\sigma^2 = \zeta(\zeta - \alpha)(\zeta - \bar{\alpha})(\zeta - \beta)(\zeta - \bar{\beta})$$

and the anti-holomorphic involution $g(\zeta, \sigma) := (\bar{\zeta}, \bar{\sigma})$.

Appendix II: $C^\infty(D, \mathcal{C})$ and $\mathrm{Diff}(D)$

The facts collected here are well known. We have provided some proofs and references for the reader's convenience.

In the remainder of this paper, fg will denote the product of the functions f and g. The composition of f and g, when it exists, will be denoted by $f \circ g$.

The C^∞ topology on the space $C^\infty(D, \mathcal{C})$

Let $C(D, \mathcal{C})$ be the space of continuous complex valued functions on D. Its C^0 topology of uniform convergence on compact sets is metrizable (see for example [A1], Chapter 5, Section 5.2).

A sequence (f_n) in $C(D, \mathcal{C})$ is said to *converge continuously* to f in $C(D, \mathcal{C})$ if

$$\lim_{n \to \infty} f_n(z_n) = f(z) \text{ whenever } \lim_{n \to \infty} z_n = z \text{ in } D.$$

It is well known that $f_n \to f$ in the C^0 topology if and only if (f_n) converges continuously to f (see [Ar1] or [C], §§174–180). It follows readily that the maps $(f, g) \mapsto fg$ from $C(D, \mathcal{C}) \times C(D, \mathcal{C})$ to $C(D, \mathcal{C})$ and $(f, z) \mapsto f(z)$ from $C(D, \mathcal{C}) \times D$ to $\mathcal{C}$ are continuous.

Let $C^\infty(D, \mathcal{C})$ be the space of C^∞ complex valued functions on D. Its C^∞ topology is defined by imposing the C^0 topology on the function and its partial derivatives of all orders. This topology is obviously metrizable, and $f_n \to f$ if and only if f_n converges continuously to f and δf_n converges continuously to δf for all partial derivatives δ of all orders. The map $f \mapsto \delta f$ is continuous for all δ. The maps $(f, g) \mapsto fg$ and $(f, z) \mapsto f(z)$ remain continuous.

Let $C^\infty(D, D)$ be the set of f in $C^\infty(D, \mathcal{C})$ such that $f(D) \subset D$. It inherits the C^∞ topology from $C^\infty(D, \mathcal{C})$.

Lemma 5.5 *The map $(f, g) \mapsto f \circ g$ from $C^\infty(D, \mathcal{C}) \times C^\infty(D, D)$ to $C^\infty(D, \mathcal{C})$ is continuous.*

Proof Let $(f_n, g_n) \to (f, g)$ and $z_n \to z$ as $n \to \infty$. Then $g_n(z_n) \to g(z)$ and

$$(f_n \circ g_n)(z_n) = f_n(g_n(z_n)) \to f(g(z)) = (f \circ g)(z)$$

as $n \to \infty$. Similarly,

$$(f_n \circ g_n)_z(z_n) = (f_n)_z(g_n(z_n))(g_n)_z(z_n) + (f_n)_{\bar{z}}(g_n(z_n))\overline{(g_n)_{\bar{z}}(z_n)}$$
$$\to f_z(g(z))g_z(z) + f_{\bar{z}}(g(z))\overline{g_{\bar{z}}(z)} = (f \circ g)_z(z)$$

as $n \to \infty$. By similar applications of the chain rule and induction on the order of the derivative, it follows that $\delta(f_n \circ g_n)$ converges continuously to $\delta(f \circ g)$ for all partial derivatives δ of all orders. $\qquad\square$

The diffeomorphism group of D

We now apply the preceding results to the group $\mathrm{Diff}(D)$ of C^∞ diffeomorphisms of D onto itself. This group inherits the C^∞ topology from $C^\infty(D, \mathcal{C})$.

Proposition 5.6 *With its C^∞ topology, $\mathrm{Diff}(D)$ is a topological group.*

Proof Because of Lemma 5.5, we need only prove that $f \mapsto f^{-1}$ is a continuous map of $\mathrm{Diff}(D)$ to itself.

First, we study the complex derivatives of f^{-1}. If the $\mathbb{R}$-linear map $z \mapsto w := az + b\bar{z}$ from $\mathcal{C}$ to $\mathcal{C}$ is invertible, then $|a| \neq |b|$ and the inverse map is $w \mapsto (\bar{a}w - b\bar{w})/(|a|^2 - |b|^2)$. The first order complex derivatives of f^{-1} therefore satisfy

$$(f^{-1})_z = \left(\frac{\overline{f_z}}{|f_z|^2 - |f_{\bar{z}}|^2} \right) \circ f^{-1} \quad \text{and} \quad (f^{-1})_{\bar{z}} = \left(\frac{-f_{\bar{z}}}{|f_z|^2 - |f_{\bar{z}}|^2} \right) \circ f^{-1}.$$

By induction,

$$\frac{\partial^{j+k}}{\partial z^j \partial \bar{z}^k}(f^{-1}) = \varphi_{j,k}(f) \circ (f^{-1}), \tag{5.4}$$

where $f \mapsto \varphi_{j,k}(f)$ is a continuous map from $\mathrm{Diff}(D)$ to $C^\infty(D, \mathcal{C})$.

Now let the sequence (f_n) in $\mathrm{Diff}(D)$ converge to f in $\mathrm{Diff}(D)$. By Theorem 4 of Arens [Ar2], the sequence (f_n^{-1}) converges continuously to f^{-1}. It follows readily from (5.4) that $\delta(f_n^{-1})$ converges continuously to $\delta(f^{-1})$ for all partial derivatives δ of all orders. $\qquad\square$

Appendix III: The barycentric isotopy

We give here a more direct proof of Theorem 1.2, using a tool from the Earle-McMullen paper [EM]. We shall call it the barycentric isotopy. Its basic properties are described in [EM].

In particular, Definition 5.8 and Proposition 5.9 below can be found in the proof of Theorem 1.1 in [EM]. We state them here, with some added details, for the reader's convenience. Proposition 5.11 is based on paragraph (3) on page 147 of [EM], where a special case is considered.

The construction

First we recall two methods for determining homeomorphisms of the closed unit disk $\overline{D}$. Both behave well with respect to the group G of hyperbolic isometries of D. We shall denote the group of orientation preserving maps in G by G_+.

Method one is to solve a Beltrami equation. Let M be the open unit ball in $L^\infty(D, \mathcal{C})$. For each μ in M there is a unique homeomorphism w_μ of $\overline{D}$ onto itself that fixes 1, i, and -1 and is quasiconformal in D with complex dilatation μ. (See [A2], especially Chapter V and §1.4 of the first supplementary chapter.)

For μ in M and g in G, set

$$g^*(\mu) := \begin{cases} (\mu \circ g)\overline{g'}/g' & \text{if } g \in G_+, \\ (\overline{\mu} \circ g)g_z/\overline{g_z} & \text{if } g \in G \setminus G_+. \end{cases} \qquad (5.5)$$

The following statement is a special case of Lemma 4 in [DE]. It follows readily from the chain rule, as a homeomorphism g of D belongs to G if and only if either g or $\overline{g}$ is holomorphic.

Lemma 5.7 *For μ in M and g in G, $w_\mu \circ g \circ w_\mu^{-1} \in G$ if and only if $g^*(\mu) = \mu$.*

Method two is the barycentric extension introduced in [DE]. It extends any homeomorphism $\alpha\colon S^1 \to S^1$ to a homeomorphism $\mathrm{ex}(\alpha)$ of $\overline{D}$ onto itself, and

$$\mathrm{ex}(g \circ \alpha \circ h) = g \circ \mathrm{ex}(\alpha) \circ h \qquad (5.6)$$

for all g and h in G and all homeomorphisms $\alpha\colon S^1 \to S^1$. When α is the identity and h is g^{-1}, (5.6) says that $\mathrm{ex}(id)$ commutes with g. As g in G is arbitrary, it follows readily that $\mathrm{ex}(id)$ is the identity on $\overline{D}$.

Now let $\mathrm{QC}_0(D)$ be the group of quasiconformal homeomorphisms of

D onto itself whose homeomorphic extensions to $\overline{D}$ fix S^1 pointwise. In [EM], the two methods above are combined in the following way to obtain a map from $QC_0(D) \times [0,1]$ to $QC_0(D)$.

Choose f in $QC_0(D)$; let μ be its complex dilatation. For t in $[0,1]$, let $\alpha_{t\mu}$ be the restriction of $w_{t\mu}$ to S^1. As in [EM], let $\beta_{t\mu} = \text{ex}(\alpha_{t\mu}^{-1})$. Since $\alpha_{t\mu}^{-1}$ has a quasiconformal extension to D, $\beta_{t\mu}$ is quasiconformal in D, by Theorem 2 in [DE].

Definition 5.8 The *barycentric isotopy* sends (f,t) in $QC_0(D) \times [0,1]$ to the quasiconformal map f_t obtained by restricting $\beta_{t\mu} \circ w_{t\mu}$ to D.

Proposition 5.9 *If* $f \in QC_0(D)$, *then* f_0 *is the identity,* $f_1 = f$, *and* $f_t \in QC_0(D)$ *for all t in* $[0,1]$. *If* f *in* $QC_0(D)$ *and* g *in* G *commute, then* f_t *and* g *commute for all t in* $[0,1]$.

Proof Let f in $QC_0(D)$ be given. As w_0 is the identity on $\overline{D}$ and $\text{ex}(id)$ is the identity, f_0 is the identity. As w_μ is the continuous extension of f to $\overline{D}$, $\beta_\mu = \text{ex}(id)$ and $f_1 = f$. By construction, $f_t \in QC_0(D)$ for all t in $[0,1]$.

If g in G commutes with f, then $g^*(\mu) = \mu$, by Lemma 5.7. Choose any t in $[0,1]$. By (5.5), $g^*(t\mu) = t\mu$, so $g_t := w_{t\mu} \circ g \circ w_{t\mu}^{-1}$ belongs to G, by Lemma 5.7. On S^1 we have $\alpha_{t\mu}^{-1} \circ g_t = g \circ \alpha_{t\mu}^{-1}$. By (5.6), this implies $\beta_{t\mu} \circ g_t = g \circ \beta_{t\mu}$. Therefore

$$g \circ f_t = g \circ \beta_{t\mu} \circ w_{t\mu} = \beta_{t\mu} \circ g_t \circ w_{t\mu} = \beta_{t\mu} \circ w_{t\mu} \circ g = f_t \circ g. \quad \square$$

A direct proof of Theorem 1.2

The second part of Proposition 5.9 makes the barycentric isotopy applicable to hyperbolic 2-orbifolds. In [EM] it was applied to hyperbolic Riemann surfaces. We apply it here to compact hyperbolic 2-orbifolds.

As in §2, let Γ be a discrete subgroup of G such that D/Γ is compact, and let $\text{Diff}_0(\Gamma)$ be the group of diffeomorphisms of D onto itself that commute with every element of Γ. Because of the following lemma, the barycentric isotopy is defined on $\text{Diff}_0(\Gamma) \times [0,1]$.

Lemma 5.10 *If D/Γ is compact, then* $\text{Diff}_0(\Gamma)$ *is contained in* $QC_0(D)$.

Proof The compactness of D/Γ implies that each f in $\text{Diff}_0(\Gamma)$ is quasiconformal. By Lemma 2.4 in §2, the continuous extension of f to $\overline{D}$ fixes S^1 pointwise. $\square$

Proposition 5.11 *If D/Γ is compact, then the barycentric isotopy maps* $\mathrm{Diff}_0(\Gamma) \times [0,1]$ *into* $\mathrm{Diff}_0(\Gamma)$, *and this mapping is continuous with respect to the C^∞ topology on* $\mathrm{Diff}_0(\Gamma)$. *Therefore,* $\mathrm{Diff}_0(\Gamma)$ *is contractible.*

Proof Choose (f, t) in $\mathrm{Diff}_0(\Gamma) \times [0,1]$. The complex dilatation of f is a C^∞ function μ on D, with $\|\mu\|_\infty < 1$. The same is true of $t\mu$, so $w_{t\mu}$ is a C^∞ diffeomorphism in D (see, for example, Theorem 2 of [D]). By Theorem 1 of [DE], barycentric extensions are always real-analytic diffeomorphisms in D, so $f_t \in \mathrm{Diff}(D)$, by Definition 1. As f commutes with each γ in Γ, f_t has the same property, by Proposition 5.9. Thus, $f_t \in \mathrm{Diff}_0(\Gamma)$, and $(f, t) \mapsto f_t$ maps $\mathrm{Diff}_0(\Gamma) \times [0,1]$ into $\mathrm{Diff}_0(\Gamma)$.

We must prove the continuity. As f will vary, we denote its complex dilatation by μ_f. By (5.5), if $f \in \mathrm{Diff}_0(\Gamma)$, then $|\mu_f|$ attains its maximum inside a compact fundamental domain for Γ. Therefore, $(f, t) \mapsto t\mu_f$ is continuous as a map from $\mathrm{Diff}_0(\Gamma) \times [0,1]$ to $L^\infty(D, \mathcal{C})$. Its image is contained in the open unit ball M.

Set $\mathcal{M}(D) := C^\infty(D, \mathcal{C}) \cap M$, and give it the topology generated by the union of its $C^\infty(D, \mathcal{C})$-open subsets and its M-open subsets. The map $(f, t) \mapsto t\mu_f$ from $\mathrm{Diff}_0(\Gamma) \times [0,1]$ to $\mathcal{M}(D)$ is continuous.

We shall consider the maps to $\beta_{t\mu_f}$ and $w_{t\mu_f}$ separately in that order. By the corollary to Theorem 8 in [AB], $\nu \mapsto w_\nu$ is a continuous map from $\mathcal{M}(D)$ to the homeomorphism group of $\overline{D}$, endowed with the topology of uniform convergence. Restricting to S^1, we find that $(f, t) \mapsto \alpha_{t\mu_f}$ is a continuous map from $\mathrm{Diff}_0(\Gamma) \times [0,1]$ to the homeomorphism group $\mathcal{H}(S^1)$ of S^1, with its uniform topology. As that topology makes $\mathcal{H}(S^1)$ a topological group (see [Ar2]), the map $(f, t) \mapsto \alpha_{t\mu_f}^{-1}$ is continuous.

By Proposition 2 of [DE], the barycentric extension defines a continuous map $\alpha \mapsto \mathrm{ex}(\alpha)|_D$ from $\mathcal{H}(S^1)$ to $\mathrm{Diff}(D)$. Hence $(f, t) \mapsto \beta_{t\mu_f}|_D$ is a continuous map from $\mathrm{Diff}_0(\Gamma) \times [0,1]$ to $\mathrm{Diff}(D)$.

Now we consider $w_{t\mu_f}$. By the theory of uniformly elliptic partial differential equations, the map $\nu \mapsto w_\nu|_D$ from $\mathcal{M}(D)$ to the diffeomorphism group $\mathrm{Diff}(D)$ is continuous. (See, for example, the Continuity theorem in §8 of [ES].) Therefore, the composed map $(f, t) \mapsto w_{t\mu_f}|_D$ from $\mathrm{Diff}_0(\Gamma) \times [0,1]$ to $\mathrm{Diff}(D)$ is continuous.

The continuity of $(f, t) \mapsto f_t = \beta_{t\mu_f}|_D \circ w_{t\mu_f}|_D$ follows at once, as $\mathrm{Diff}(D)$ is a topological group. $\qquad\square$

Bibliography

[A1] Ahlfors, L. V. *Complex Analysis*, third edition, McGraw-Hill, New York, 1979.

[A2] Ahlfors, L. V. *Lectures on Quasiconformal Mappings*, second edition, Amer. Math. Soc., Providence, RI, 2006.

[AB] Ahlfors, L. and Bers, L. Riemann's mapping theorem for variable metrics. *Ann. of Math.* **72** (1960), 385–404.

[Ar1] Arens, R. A topology for spaces of transformations. *Ann. of Math.* **47** (1946), 480–495.

[Ar2] Arens, R. Topologies for homeomorphism groups. *Amer. J. Math.* **68** (1946), 593–610.

[BH1] Birman, J. S. and Hilden, H. M. On the mapping class groups of closed surfaces as covering spaces, in *Advances in the Theory of Riemann Surfaces*, ed. L.V. Ahlfors *et al.* Princeton Univ. Press, Princeton, 1971, pp. 81–115.

[BH2] Birman, J. S. and Hilden, H. M. On isotopies of homeomorphisms of Riemann surfaces, *Ann. of Math.* **97** (1973), 424–439.

[BN] Bundgaard, S. and Nielsen, J. On normal subgroups with finite index in F-groups. *Mat. Tidsskr.* (1951), 56–58.

[C] Carathéodory, C. *Theory of Functions of a Complex Variable*, Vol. I, Second English Edition, Chelsea, New York, 1958.

[D] Douady, A. Le théorème d'intégrabilité des structures presque complexes (d'après des notes de X. Buff), in *The Mandelbrot Set, Theme and Variations*, ed. Tan Lei, Cambridge Univ. Press, Cambridge, 2000, pp. 307–324.

[DE] Douady, A. and Earle, C. J. Conformally natural extension of homeomorphisms of the circle. *Acta Math.* **157** (1986), 23–48.

[E] Earle, C. J. On the moduli of closed Riemann surfaces with symmetries, in *Advances in the Theory of Riemann Surfaces*, ed. L. V. Ahlfors *et al.*, Princeton Univ. Press, Princeton, 1971, pp. 119–130.

[EE] Earle, C. J. and Eells, J. A fibre bundle description of Teichmüller theory. *J. Differential Geometry* **3** (1969), 19–43.

[EM] Earle, C. J. and McMullen, C. Quasiconformal isotopies, in *Holomorphic Functions and Moduli, Vol. I*, ed. D. Drasin *et al.*, Springer–Verlag, New York, 1988, pp. 143–154.

[ES] Earle, C. J. and Schatz, A. Teichmüller theory for surfaces with boundary. *J. Differential Geometry* **4** (1970), 169–186.

[FK] Farkas, H. and Kra, I. *Riemann Surfaces, second edition*, Springer-Verlag, New York, 1992.

[F] Fox, R. H. On Fenchel's conjecture about F-groups. *Mat. Tidsskr.* 1952, 61–65.

[FG] Fuertes, Y. and González-Diez, G. Fields of moduli and definition of hyperelliptic covers. *Arch. Math.* **86** (2006), 398–408.

[MH] Maclachlan, C. and Harvey, W. J. On mapping-class groups and Teichmüller spaces. *Proc. London Math. Soc.* **30** (1975), 496–512.

[Sh] Shimura, G. On the field of rationality for an abelian variety. *Nagoya Math. J.* **45** (1972), 167–178.

[Si] Silhol, R. Moduli problems in real algebraic geometry, in *Real Algebraic Geometry*, ed. M. Coste *et al.*, Springer-Verlag, Berlin, 1972, pp. 110–119.

Holomorphic motions and related topics

Frederick P. Gardiner

Department of Mathematics, Brooklyn College
and Department of Mathematics, CUNY Graduate Center
frederick.gardiner@gmail.com

Yunping Jiang

Department of Mathematics, Queens College
and Department of Mathematics, CUNY Graduate Center
Yunping.Jiang@qc.cuny.edu

Zhe Wang

Department of Mathematics, CUNY Graduate Center
wangzhecuny@gmail.com

1 Introduction

Suppose $\overline{\mathbb{C}}$ is the extended complex plane and $E \subset \overline{\mathbb{C}}$ is a subset. For any real number $r > 0$, we let Δ_r be the disk centered at the origin in $\mathbb{C}$ with radius r and Δ be the disk of unit radius. A map

$$h(c, z) : \Delta \times E \to \overline{\mathbb{C}}$$

is called a holomorphic motion of E parametrized by Δ and with base point 0 if

(i) $h(0, z) = z$ for all $z \in E$,
(ii) for every $c \in \Delta$, $z \mapsto h(c, z)$ is injective on $\overline{\mathbb{C}}$, and
(iii) for every $z \in E$, $c \mapsto h(c, z)$ is holomorphic for c in Δ.

We think of $h(c, z)$ as moving through injective mappings with the parameter c. It starts out at the identity when c is equal to the base point 0 and moves holomorphically as c varies in Δ.

We always assume E contains at least three points, p_1, p_2 and p_3. Then since the points $h(c, p_1), h(c, p_2)$ and $h(c, p_3)$ are distinct for each $c \in \Delta$, there is a unique Möbius transformation B_c that carries these three points to $0, 1$, and ∞. Since B_c depends holomorphically on c,

156

$\tilde{h}(c, z) = h(c, B_c(z))$ is also a holomorphic motion and it fixes the points $0, 1, \infty$. We shall call it a normalized holomorphic motion.

Holomorphic motions were introduced by Màñé, Sad and Sullivan in their study of the structural stability problem for the complex dynamical systems, [MSS]. They proved the first result in the topic which is called the λ-lemma and which says that any holomorphic motion $h(c, z)$ of E parametrized by Δ and with base point 0 can be extended uniquely to a holomorphic motion of the closure $\overline{E}$ of E parametrized by Δ and with the same base point. Moreover, $h(c, z)$ is continuous in (c, z) and for any fixed c, $z \mapsto h(c, z)$ is quasiconformal on the interior of $\overline{E}$. Subsequently, holomorphic motions became an important topic with applications to quasiconformal mapping, Teichmüller theory and complex dynamics. After Màñé, Sad and Sullivan proved the λ-lemma, Sullivan and Thurston [ST] proved an important extension result. Namely, they proved that any holomorphic motion of E parametrized by Δ and with base point 0 can be extended to a holomorphic motion of $\overline{\mathbb{C}}$, but parametrized by a smaller disk, namely, by Δ_r for some universal number $0 < r < 1$. They showed that r is independent of E and independent of the motion. By a different method and published in the same journal with the Sullivan-Thurston paper, Bers and Royden [BR] proved that $r \geq 1/3$ for all motions of all closed sets E parameterized by Δ. They also showed that on $\overline{\mathbb{C}}$ the map $z \mapsto h(c, z)$ is quasiconformal with dilatation no larger than $(1 + |c|)/(1 - |c|)$. All of these authors raised the question as to whether $r = 1$ for any holomorphic motion of any subset of $\overline{\mathbb{C}}$ parametrized by Δ and with base point 0. In [S] Slodkowski gave a positive answer by using results from the theory of polynomial hulls in several complex variables. Other authors [AM] [D] have suggested alternative proofs.

In this article we give an expository account of a recent proof of Slodkowski's theorem presented by Chirka in [C]. (See also Chirka and Rosay [CR].) The method involves an application of Schauder's fixed point theorem [CH] to an appropriate operator acting on holomorphic motions of a point and showing that this operator is compact. The compactness depends on the smoothing property of the Cauchy kernel acting on vector fields tangent to holomorphic motions. The main theorem is the following.

Theorem 1.1 (The Holomorphic Motion Theorem) *Suppose*

$$h(c, z) : \Delta \times E \to \overline{\mathbb{C}}$$

is a holomorphic motion of a closed subset E of $\overline{\mathbb{C}}$ parameterized by the unit disk. Then there is a holomorphic motion

$$H(c, z) : \Delta \times \overline{\mathbb{C}} \to \overline{\mathbb{C}}$$

which extends $h(c, z) : \Delta \times E \to \overline{\mathbb{C}}$. Moreover, for any fixed $c \in \Delta$, $H(c, \cdot) : \overline{\mathbb{C}} \to \overline{\mathbb{C}}$ is a quasiconformal homeomorphism whose quasiconformal dilatation

$$K(H(c, \cdot)) \leq \frac{1 + |c|}{1 - |c|}.$$

The Beltrami coefficient of $H(c, \cdot)$ given by

$$\mu(c, z) = \frac{\partial H(c, z)}{\partial \overline{z}} \Big/ \frac{\partial H(c, z)}{\partial z}$$

is a holomorphic function from Δ into the unit ball of the Banach space $\mathcal{L}^\infty(\mathbb{C})$ of all essentially bounded measurable functions on $\mathbb{C}$.

To prove this result we study the modulus of continuity of functions in the image of the Cauchy kernel operator. Then we follow the proof given by Chirka in [C] who introduces a non-linear operator to which the Schauder fixed point theorem [CH] applies. The existence of a fixed point of this operator implies the existence of a holomorphic extension to any disk of radius $r < 1$ and then a normal families argument allows one to take the limit as r approaches 1.

After proving this theorem, we show that tangent vectors to holomorphic motions have $|\epsilon \log \epsilon|$ moduli of continuity and then show how this type of continuity for tangent vectors can be combined with Schwarz's lemma and integration over the holomorphic variable to produce Hölder continuity on the mappings. At one point the argument requires obtaining a lower bound for the Poincaré metric on the Riemann sphere punctured at $0, 1$ and ∞. The method for obtaining this lower bound is described by Ahlfors in [A1]. A slightly improved version is given by Keen and Lakic in [KL].

We also prove that Kobayashi's and Teichmüller's metrics on the Teichmüller space $T(R)$ of a Riemann surface coincide. The proof we give is very similar to the one given in [GL]. This result was observed by Earle, Kra and Krushkal in [EKK] and had been proved earlier by Royden [R] for Riemann surfaces of finite analytic type and by Gardiner [G1] [G2] for surfaces of infinite type.

Acknowledgement. The authors wish to thank all of the members of our complex analysis seminar at the Graduate Center of CUNY for helpful discussions. In particular, we owe special thanks to Linda Keen, Nikola Lakic, Jun Hu, and Sudeb Mitra. Also the referee has provided many helpful suggestions and corrections.

2 The $\mathcal{P}$-operator and the modulus of continuity

Let $\mathcal{C} = \mathcal{C}(\mathbb{C})$ denote the Banach space of complex valued, bounded, continuous functions ϕ on $\mathbb{C}$ with the supremum norm

$$||\phi|| = \sup_{c \in \mathbb{C}} |\phi(c)|.$$

We use $\mathcal{L}^\infty$ to denote the Banach space of essentially bounded measurable functions ϕ on $\mathbb{C}$ with $\mathcal{L}^\infty$-norm

$$||\phi||_\infty = \operatorname{ess\,sup}_{\mathbb{C}} |\phi(\zeta)|.$$

For the theory of quasiconformal mapping we are more concerned with the action of $\mathcal{P}$ on $\mathcal{L}^\infty$. Here the $\mathcal{P}$-operator is defined by

$$\mathcal{P}f(c) = -\frac{1}{\pi} \int\int_{\mathbb{C}} \frac{f(\zeta)}{\zeta - c}\, d\xi d\eta, \quad \zeta = \xi + i\eta, \tag{2.1}$$

where $f \in \mathcal{L}^\infty$ and has a compact support in $\mathbb{C}$. Then

$$\mathcal{P}f(c) \longrightarrow 0 \quad \text{as} \quad c \longrightarrow \infty.$$

Furthermore, if f is continuous and has compact support, one can show that

$$\frac{\partial(\mathcal{P}f)}{\partial \bar{c}}(c) = f(c), \quad c \in \mathbb{C}, \tag{2.2}$$

and by using the notion of generalized derivative [AB] equation (2.2) is still true Lebesgue almost everywhere if we only know that f has compact support and is in $\mathcal{L}^p, p \geq 1$.

For the benefit of the reader we sketch the proof of (2.2) in the case that f is C^1 with compact support. In that case differentiation under the integral sign in (2.1) is permissible and so

$$\frac{\partial(\mathcal{P}f)}{\partial \bar{c}}(c) = -\frac{1}{\pi}\frac{\partial}{\partial \bar{c}} \int\int_{\mathbb{C}} \frac{f(\zeta)}{\zeta - c}\, d\xi d\eta = -\frac{1}{\pi}\frac{\partial}{\partial \bar{c}} \int\int_{\mathbb{C}} \frac{f(\zeta + c)}{\zeta}\, d\xi d\eta$$

$$= -\frac{1}{\pi} \int\int_{\mathbb{C}} \frac{f_{\bar{c}}(\zeta + c)}{\zeta}\, d\xi d\eta = -\frac{1}{\pi} \int\int_{\mathbb{C}} \frac{f_{\bar{\zeta}}(\zeta)}{\zeta - c}\, d\xi d\eta =$$

$$\frac{1}{2\pi i} \int \int_{\mathbb{C}} \frac{f_{\bar{\zeta}}(\zeta)}{\zeta - c} \, d\zeta d\bar{\zeta}$$

$$= -\frac{1}{2\pi i} \int \int_{\mathbb{C}} \frac{df \, d\zeta}{\zeta - c} = -\lim_{\epsilon \to 0} \frac{1}{2\pi i} \int \int_{|\zeta - c| > \epsilon} \frac{df \, d\zeta}{\zeta - c} =$$

$$-\lim_{\epsilon \to 0} \frac{1}{2\pi i} \int_{|\zeta - c| = \epsilon} \frac{f \, d\zeta}{\zeta - c} = \lim_{\epsilon \to 0} \frac{1}{2\pi} \int_0^{2\pi} f(c + \epsilon e^{i\theta}) d\theta = f(c).$$

We refer to [A2] for the verification that this relation still holds when $f \in L^p$ for $p > 2$.

We now show the classical result that $\mathcal{P}$ transforms $\mathcal{L}^\infty$ functions with compact support in $\mathbb{C}$ to Hölder continuous functions with Hölder exponent $1 - 2/p$ for every $p > 2$, [A2]. We also show that $\mathcal{P}$ carries $\mathcal{L}^\infty$ functions with compact support to functions with an $|\epsilon \log \epsilon|$ modulus of continuity.

Lemma 2.1 *Suppose $p > 2$ and*

$$\frac{1}{p} + \frac{1}{q} = 1,$$

so that $1 < q < 2$. Then for any real number $R > 0$, there is a constant $A_R > 0$ such that for any $f \in \mathcal{L}^\infty$ with a compact support contained in Δ_R

$$\|\mathcal{P}f\| \leq A_R \|f\|_\infty$$

and

$$|\mathcal{P}f(c) - \mathcal{P}f(c')| \leq A_R \|f\|_\infty |c - c'|^{1 - \frac{2}{p}}, \quad \forall c, c' \in \mathbb{C}.$$

Proof The norm

$$\|\mathcal{P}f\| = \sup_{c \in \mathbb{C}} \frac{1}{\pi} \left| \int \int_{\mathbb{C}} \frac{f(\zeta)}{\zeta - c} d\xi d\eta \right| \leq \sup_{c \in \mathbb{C}} \frac{1}{\pi} \int \int_{\Delta_R} \frac{|f(\zeta)|}{|\zeta - c|} d\xi d\eta.$$

So

$$\|\mathcal{P}f\| \leq \|f\|_\infty \sup_{c \in \mathbb{C}} \frac{1}{\pi} \int \int_{\Delta_R} \frac{1}{|\zeta - c|} d\xi d\eta \leq C_1 \|f\|_\infty,$$

where

$$C_1 = \frac{1}{\pi} \int \int_{\Delta_R} \frac{1}{|\zeta|} d\xi d\eta = 2R < \infty.$$

Next

$$|\mathcal{P}f(c) - \mathcal{P}f(c')| = \frac{1}{\pi}\left| \int\int_{\mathbb{C}} f(\zeta)\left(\frac{1}{\zeta - c} - \frac{1}{\zeta - c'}\right) d\xi d\eta\right|$$

$$\leq \frac{|c - c'|}{\pi} \int\int_{\Delta_R} \frac{|f(\zeta)|}{|\zeta - c||\zeta - c'|} d\xi d\eta$$

$$\leq \frac{|c - c'|}{\pi}\left(\int\int_{\Delta_R} |f(\zeta)|^p d\xi d\eta\right)^{\frac{1}{p}}\left(\int\int_{\Delta_R} \left|\frac{1}{(\zeta - c)(\zeta - c')}\right|^q d\xi d\eta\right)^{\frac{1}{q}}$$

$$\leq \pi^{\frac{1}{p}-1} R^{\frac{2}{p}}|c - c'|\,\|f\|_\infty\left(\int\int_{\Delta_R} \left|\frac{1}{(\zeta - c)(\zeta - c')}\right|^q d\xi d\eta\right)^{\frac{1}{q}} \leq$$

$$C_2\|f\|_\infty |c - c'|^{\frac{2}{q}-1},$$

where

$$C_2 = \pi^{\frac{1}{p}-1} R^{\frac{2}{p}}\left(\int\int_{\mathbb{C}} \left(\frac{1}{|z||z-1|}\right)^q dx dy\right)^{\frac{1}{q}} < \infty, \quad z = x + iy.$$

Hence $A_R = \max\{C_1, C_2\}$ satisfies the requirements of the lemma. $\quad\square$

Next we prove a stronger form of continuity.

Lemma 2.2 *Suppose the compact support of $f \in \mathcal{L}^\infty$ is contained in Δ. Then $\mathcal{P}f$ has an $|\epsilon \log \epsilon|$ modulus of continuity. More precisely, there is a constant B depending on R such that*

$$|\mathcal{P}f(c) - \mathcal{P}f(c')| \leq \|f\|_\infty B|c - c'| \log\frac{1}{|c - c'|}, \quad \forall\, c, c' \in \Delta_R, \ |c - c'| < \frac{1}{2}.$$

Proof Since

$$|\mathcal{P}f(c) - \mathcal{P}f(c')| = \frac{1}{\pi}\left| \int\int_{\mathbb{C}} f(\zeta)\left(\frac{1}{\zeta - c} - \frac{1}{\zeta - c'}\right) d\xi d\eta\right|$$

$$\leq \frac{1}{\pi} \int\int_{\mathbb{C}} |f(\zeta)|\left|\frac{1}{\zeta - c} - \frac{1}{\zeta - c'}\right| d\xi d\eta$$

$$\leq \frac{|c - c'|\,\|f\|_\infty}{\pi} \int\int_{\Delta} \frac{1}{|\zeta - c||\zeta - c'|} d\xi d\eta,$$

if we put $\zeta' = \zeta - c = \xi' + i\eta'$, then

$$|\mathcal{P}f(c) - \mathcal{P}f(c')| \leq \frac{|c - c'|\,\|f\|_\infty}{\pi} \int\int_{\Delta_{1+R}} \frac{1}{|\zeta'||\zeta' - (c' - c)|} d\xi' d\eta'.$$

The substitution $\zeta'' = \zeta'/(c'-c) = \xi'' + i\eta''$ yields

$$|\mathcal{P}f(c) - \mathcal{P}f(c')| \le \frac{|c-c'|\|f\|_\infty}{\pi} \int\int_{\Delta_{\frac{1+R}{|c'-c|}}} \frac{1}{|\zeta''||\zeta''-1|} d\xi'' d\eta''.$$

Since $|c-c'| < 1/2$, we have $(1+R)/|c'-c| > 2$. This implies that

$$|\mathcal{P}f(c) - \mathcal{P}f(c')|$$

$$\le \frac{|c-c'|\|f\|_\infty}{\pi} \left(\int\int_{\Delta_2} (same) + \int\int_{\Delta_{\frac{1+R}{|c'-c|}} - \Delta_2} (same) \right).$$

Let

$$C_3 = \int\int_{\Delta_2} \frac{1}{|\zeta''||\zeta''-1|} d\xi'' d\eta''.$$

Then

$$|\mathcal{P}f(c) - \mathcal{P}f(c')| \le$$

$$\frac{|c-c'|C_3\|f\|_\infty}{\pi} + \frac{|c-c'|\|f\|_\infty}{\pi} \int\int_{\Delta_{\frac{1+R}{|c'-c|}} - \Delta_2} \frac{1}{|\zeta''||\zeta''-1|} d\xi'' d\eta''.$$

If $|\zeta''| > 2$ then $|\zeta''-1| > |\zeta''|/2$, and so

$$\frac{1}{\pi} \int\int_{\Delta_{\frac{1+R}{|c'-c|}} - \Delta_2} \frac{1}{|\zeta''||\zeta''-1|} d\xi'' d\eta'' \le \frac{1}{\pi} \int\int_{\Delta_{\frac{1+R}{|c'-c|}} - \Delta_2} \frac{2}{|\zeta''|^2} d\xi'' d\eta''$$

$$\le \frac{1}{\pi} \int_0^{2\pi} \int_2^{\frac{1+R}{|c'-c|}} \frac{2}{r^2} r dr d\theta = 4 \int_2^{\frac{1+R}{|c'-c|}} \frac{1}{r} dr$$

$$= 4\left(\log \frac{1+R}{|c'-c|} - \log 2 \right) = 4(-\log|c-c'| + \log(1+R) - \log 2).$$

Thus,

$$|\mathcal{P}f(c) - \mathcal{P}f(c')| \le$$

$$\frac{|c-c'|C_3\|f\|_\infty}{\pi} + 4|c-c'|\|f\|_\infty(-\log|c-c'| + \log(1+R) - \log 2)$$

$$= -|c-c'|\log|c-c'| \left(\frac{4\pi \log(1+R) + C_3\|f\|_\infty - 4\pi \log 2}{-\pi \log|c-c'|} + 4\|f\|_\infty \right)$$

$$\le B\left(-|c-c'|\log|c-c'| \right),$$

where

$$B = \frac{4\pi \log(1+R) + C_3\|f\|_\infty - 4\pi \log 2}{\pi \log 2} + 4\|f\|_\infty.$$

□

Now we have the following theorem.

Theorem 2.3 *For any $f \in \mathcal{L}^\infty$ with a compact support in $\mathbb{C}$, $\mathcal{P}f$ has an $|\epsilon \log \epsilon|$ modulus of continuity. More precisely, for any $R > 0$, there is a constant $C > 0$ depending on R such that*

$$|\mathcal{P}f(c) - \mathcal{P}f(c')| \leq C\|f\|_\infty |c-c'| \log \frac{1}{|c-c'|}, \quad \forall\, c, c' \in \Delta_R, \ |c-c'| < \frac{1}{2}.$$

Proof Suppose the compact support of f is contained in the disk Δ_{R_0}. Then $g(c) = f(R_0 c)$ has the compact support which is contained in the unit disk Δ. Then

$$\mathcal{P}g(c) = -\frac{1}{\pi}\int\int_\mathbb{C} \frac{g(\zeta)}{\zeta - c} d\xi d\eta = -\frac{1}{\pi}\int\int_\mathbb{C} \frac{f(R_0\zeta)}{\zeta - c} d\xi d\eta = \frac{1}{R_0}\mathcal{P}f(R_0 c),$$

and this implies that

$$\mathcal{P}f(c) = R_0 \mathcal{P}g\left(\frac{c}{R_0}\right).$$

Thus

$$|\mathcal{P}f(c) - \mathcal{P}f(c')| = R_0|\mathcal{P}g(\frac{c}{R_0}) - \mathcal{P}g(\frac{c'}{R_0})|$$

$$\leq R_0 B\|f\|_\infty\left(-\left|\frac{c}{R_0} - \frac{c'}{R_0}\right| \log\left|\frac{c}{R_0} - \frac{c'}{R_0}\right|\right)$$

$$= B\|f\|_\infty\left(-|c-c'|(\log|c-c'| - \log R_0)\right)$$

$$= -|c-c'| \log|c-c'| B\|f\|_\infty\left(1 - \frac{\log R_0}{\log|c-c'|}\right)$$

$$\leq C\|f\|_\infty(-|c-c'| \log|c-c'|),$$

where

$$C = B(1 + \frac{\log R_0}{\log 2}).$$

□

3 Extensions of holomorphic motions for $0 < r < 1$

As an application of the modulus of continuity for the $\mathcal{P}$-operator, we first prove that for any holomorphic motion of a set E parameterized by Δ, and for any r with $0 < r < 1$, there is an extension to $\Delta_r \times \overline{\mathbb{C}}$. We take the idea of the proof from the recent papers of Chirka [C] and Chirka and Rosay [CR].

Theorem 3.1 *Suppose E is a subset of $\overline{\mathbb{C}}$ consisting of a finite number of points. Suppose $h(c,z) : \Delta \times E \to \overline{\mathbb{C}}$ is a holomorphic motion. Then for every $0 < r < 1$, there is a holomorphic motion $H_r(c,z) : \Delta_r \times \overline{\mathbb{C}} \to \overline{\mathbb{C}}$ which extends $h(c,z) : \Delta_r \times E \to \overline{\mathbb{C}}$.*

Without loss of generality, suppose

$$E = \{z_0 = 0, z_1 = 1, z_\infty = \infty, z_2, \cdots, z_n\}$$

is a subset of $n + 2 > 3$ points in the Riemann sphere $\overline{\mathbb{C}}$. Let Δ^c be the complement of the unit disk in the Riemann sphere $\overline{\mathbb{C}}$, U be a neighborhood of Δ^c in $\overline{\mathbb{C}}$ and suppose

$$h(c,z) : U \times E \to \overline{\mathbb{C}}$$

is a holomorphic motion of E parametrized by U and with base point ∞. Define

$$f_i(c) = h(c, z_i) : U \to \overline{\mathbb{C}}$$

for $i = 0, 1, 2, \cdots, n, \infty$. We assume the motion is normalized so

$$f_0(c) = 0, \quad f_1(c) = 1, \quad \text{and} \quad f_\infty(c) = \infty, \quad \forall c \in U.$$

Then we have

a) $f_i(\infty) = z_i, i = 2, \cdots, n$,
b) for any $i = 2, \cdots, n$, $f_i(c)$ is holomorphic on U and
c) for any fixed $c \in U$, $f_i(c) \neq f_j(c)$ and $f_i(c) \neq 0, 1$, and ∞ for $2 \leq i \neq j \leq n$.

Since Δ^c is compact, $f_i(c)$ is a bounded function on Δ^c for every $2 \leq i \leq n$ and so there is a constant $C_4 > 0$ such that

$$|f_i(c)| \leq C_4, \text{ for all } c \in \Delta^c \text{ and all } i \text{ with } 2 \leq i \leq n.$$

Moreover, there is a number $\delta > 0$ such that

$\mid f_i(c) - f_j(c) \mid > \delta$, for all i and j with $2 \leq i \neq j \leq n$, and for all $c \in \Delta^c$.

We extend the functions $f_i(c)$ on Δ^c to continuous functions on the Riemann sphere $\overline{\mathbb{C}}$ by defining

$$f_i(c) = f_i\left(\frac{1}{\overline{c}}\right), \quad \text{for all } c \in \overline{\Delta}.$$

We still have

$\mid f_i(c) - f_j(c) \mid > \delta$, for all i and j with $2 \leq i \neq j \leq n$ and for all $c \in \overline{\mathbb{C}}$

and

$|f_i(c)| \leq C_4$ for all i and j with $2 \leq i \neq j \leq n$ and for all $c \in \overline{\mathbb{C}}$.

Since $f_i(c)$ is holomorphic in U and $f_i(\infty) = z_i$, the series expansion of $f_i(c)$ at ∞ is

$$f_i(c) = z_i + \frac{a_1}{c} + \frac{a_2}{c^2} + \cdots + \frac{a_n}{c^n} + \cdots, \quad \forall c \in \Delta^c.$$

This implies that

$$f_i(c) = f_i\left(\frac{1}{\overline{c}}\right) = z_i + a_1\overline{c} + a_2(\overline{c})^2 + \cdots a_n(\overline{c})^n + \cdots, \quad \forall c \in \overline{\Delta}.$$

We have that

$$\frac{\partial f_i}{\partial \overline{c}}(c) = a_1 + 2a_2\overline{c} + \cdots + na_n(\overline{c})^{n-1} + \cdots$$

exists at $c = 0$ and is a continuous function on $\overline{\Delta}$. Furthermore, $(\partial f_i/\partial \overline{c})(c) = 0$ for $c \in \left(\overline{\Delta}\right)^c$. Since $\overline{\Delta}$ is compact, there is a constant $C_5 > 0$ such that

$$\mid \frac{\partial f_i}{\partial \overline{c}}(c) \mid \leq C_5, \quad \forall c \in \overline{\mathbb{C}}, \quad \forall 2 \leq i \leq n.$$

Pick a C^∞ function $0 \leq \lambda(x) \leq 1$ on $\mathbb{R}^+ = \{x \geq 0\}$ such that $\lambda(0) = 1$ and $\lambda(x) = 0$ for $x \geq \delta/2$. Define

$$\Phi(c, w) = \sum_{i=2}^{n} \lambda(|w - f_i(c)|)\frac{\partial f_i}{\partial \overline{c}}(c), \quad (c, w) \in \overline{\mathbb{C}} \times \mathbb{C}. \qquad (3.1)$$

Lemma 3.2 *The function $\Phi(c, w)$ has the following properties:*

i) only one term in the sum (3.1) defining $\Phi(c, w)$ can be nonzero,
ii) $\Phi(c, w)$ is uniformly bounded by C_5 on $\overline{\mathbb{C}} \times \mathbb{C}$,

  iii) $\Phi(c,w) = 0$ *for* $(c,w) \in \left((\overline{\Delta})^c \times \mathbb{C} \right) \cup \left(\overline{\mathbb{C}} \times (\overline{\Delta}_R)^c \right)$ *where* $R = C_4 + \delta/2$,

  iv) $\Phi(c,w)$ *is a Lipschitz function in* w-*variable with a Lipschitz constant* L *independent of* $c \in \hat{\mathbb{C}}$.

Proof Item i) follows because if a point w is within distance $\delta/2$ of one of the values $f_i(c)$, it must be at distance greater than $\delta/2$ from any of the other values $f_j(c)$. Item ii) follows from item i) because there can be only one term in (3.1) which is nonzero and that term is bounded by the bound on $\frac{\partial f_j(c)}{\partial \overline{c}}$. Item iii) follows because if $c \in (\overline{\Delta})^c$, then $(\partial f_i/\partial \overline{c})(c) = 0$, and if $w \in (\overline{\Delta}_R)^c$, then $\Phi(c,w) = 0$. To prove item iv), we note that there is a constant $C_6 > 0$ such that $|\lambda(x) - \lambda(x')| \leq C_6|x - x'|$. Since $|(\partial f_i/\partial \overline{c})(c)| \leq C_5$,

$$|\Phi(c,w) - \Phi(c,w')| \leq C_6 C_5 \sum_{i=2}^{n} \Big| \, |w - f_i(c)| - |w' - f_i(c)| \, \Big|. \qquad (3.2)$$

Since only one of the terms in the sum (3.1) for $\Phi(c,w)$ is nonzero and possibly a different term is nonzero in the sum for $\Phi(c,w')$, we obtain

$$|\Phi(c,w) - \Phi(c,w')| \leq 2C_6 C_5 |w - w'|.$$

Thus $L = 2C_5 C_6$ is a Lipschitz constant independent of $c \in \hat{\mathbb{C}}$.   □

Since $\Phi(c, f(c))$ is an $\mathcal{L}^\infty$ function with a compact support in $\overline{\Delta}$ for any $f \in \mathcal{C}$, we can define an operator $\mathcal{Q}$ mapping functions in $\mathcal{C}$ to functions in $\mathcal{L}^\infty$ with compact support by

$$\mathcal{Q}f(c) = \Phi(c, f(c)), \quad f(c) \in \mathcal{C}.$$

Since $\Phi(c,w)$ is Lipschitz in the w variable with a Lipschitz constant L independent of $c \in \overline{\mathbb{C}}$, we have

$$|\mathcal{Q}f(c) - \mathcal{Q}g(c)| = |\Phi(c, f(c)) - \Phi(c, g(c))| \leq L|f(c) - g(c)|.$$

Thus

$$||\mathcal{Q}f - \mathcal{Q}g||_\infty \leq L||f - g||,$$

and $\mathcal{Q} : \mathcal{C} \to \mathcal{L}^\infty$ is a continuous operator.

  From Lemma 2.1,

$$||\mathcal{P}f|| \leq A_1 ||f||_\infty$$

for any $f \in \mathcal{L}^\infty$ whose compact support is contained in Δ, and so the composition $\mathcal{K} = \mathcal{P} \circ \mathcal{Q}$, where

$$\mathcal{K}f(c) = -\frac{1}{\pi} \int \int_{\mathbb{C}} \frac{\Phi(\zeta, f(\zeta))}{\zeta - c} d\xi d\eta, \quad \zeta = \xi + i\eta,$$

is a continuous operator from $\mathcal{C}$ into itself.

Lemma 3.3   *There is a constant $D > 0$ such that*

$$\|\mathcal{K}f\| \leq D, \quad \forall f \in \mathcal{C}.$$

Proof   Since $\Phi(c, w) = 0$ for $c \in \Delta^c$ and since $\Phi(c, w)$ is bounded by C_5, we have that

$$|\mathcal{K}f(c)| = \left| \frac{1}{\pi} \int \int_{\overline{\mathbb{C}}} \frac{\Phi(\zeta, f(\zeta))}{\zeta - c} d\xi d\eta \right| = \left| \frac{1}{\pi} \int \int_{\Delta} \frac{\Phi(\zeta, f(\zeta))}{\zeta - c} d\xi d\eta \right|$$

$$\leq \frac{1}{\pi} \int \int_{\Delta} \frac{|\Phi(\zeta, f(\zeta))|}{|\zeta - c|} d\xi d\eta$$

$$\leq \frac{C_5}{\pi} \int \int_{\Delta} \frac{1}{|\zeta - c|} d\xi d\eta \leq 2C_5 = D,$$

where $\zeta = \xi + i\eta$.   □

Lemma 3.4   *Suppose $p > 2$ and q is the dual number between 1 and 2 satisfying*

$$\frac{1}{p} + \frac{1}{q} = 1.$$

Then for any $f \in \mathcal{C}$, $\mathcal{K}f$ is α-Hölder continuous for

$$0 < \alpha = \frac{2}{q} - 1 < 1$$

with a Hölder constant $H = A_1 C_5$ independent of f.

Proof   From Lemma 2.1,

$$|\mathcal{K}f(c) - \mathcal{K}f(c')| = |\mathcal{P}(\mathcal{Q}f)(c) - \mathcal{P}(\mathcal{Q}f)(c')|$$

$$\leq A_1 \|\mathcal{Q}f\|_\infty |c - c'|^\alpha \leq A_1 C_5 |c - c'|^\alpha = H|c - c'|^\alpha.$$

□

The above two lemmas imply that $\mathcal{K} : \mathcal{C} \to \mathcal{C}$ is a continuous compact operator. Now for any $z \in \mathbb{C}$, let

$$\mathcal{B}_z = \{f \in \mathcal{C} \mid \|f\| \le |z| + D\}.$$

It is a bounded convex subset in $\mathcal{C}$. The continuous compact operator $z + \mathcal{K}$ maps $\mathcal{B}_z$ into itself. From the Schauder fixed point theorem [CH], $z + \mathcal{K}$ has a fixed point in $\mathcal{B}_z$. That is, there is a $f_z \in \mathcal{B}_z$ such that

$$f_z(c) = z + \mathcal{K}f_z(c), \quad \forall c \in \mathbb{C}.$$

Since $\mathcal{Q}f(c)$ has a compact support in $\overline{\Delta}$ for any $f \in \mathcal{C}$, $\mathcal{K}f_z(c) \to 0$ as $c \to \infty$. So f_z can be extended continuously to ∞ such that $f_z(\infty) = z$.

Lemma 3.5 *The solution $f_z(c)$ is the unique fixed point of the operator $z + \mathcal{K}$.*

Proof Suppose $f_z(c)$ and $g_z(c)$ are two solutions. Take

$$\phi(c) = f_z(c) - g_z(c) = \mathcal{K}(f_z)(c) - \mathcal{K}(g_z)(c).$$

Then $\phi(c) \to 0$ as $c \to \infty$. Now

$$\frac{\partial \phi}{\partial \bar{c}}(c) = \frac{\partial f_z}{\partial \bar{c}}(c) - \frac{\partial g_z}{\partial \bar{c}}(c) = \Phi(c, f_z(c)) - \Phi(c, g_z(c)).$$

So by Lemma 3.2

$$\frac{\partial \phi}{\partial \bar{c}}(c) = 0, \quad \forall c \in \Delta^c.$$

Since $\Phi(c, w)$ is Lipschitz in w-variable with a Lipschitz constant L,

$$\left|\frac{\partial \phi}{\partial \bar{c}}(c)\right| = |\Phi(c, f_z(c)) - \Phi(c, g_z(c))| \le L|f_z(c) - g_z(c)| = L|\phi(c)|.$$

Assuming that $\phi(c)$ is not equal to zero, define

$$\psi(c) = -\frac{\frac{\partial \phi}{\partial \bar{c}}(c)}{\phi(c)},$$

and otherwise, define $\psi(c)$ to be equal to zero. Then $\psi(c)$ is a function in $\mathcal{L}^\infty$ with a compact support in $\overline{\Delta}$. So we have $\mathcal{P}\psi$ in $\mathcal{C}$ such that

$$\frac{\partial \mathcal{P}\psi}{\partial \bar{c}}(c) = \psi(c).$$

Consider $e^{\mathcal{P}\psi} \cdot \phi$. Then

$$\frac{\partial(e^{\mathcal{P}\psi} \cdot \phi)}{\partial \bar{c}} \equiv 0.$$

This means that $e^{\mathcal{P}\psi} \cdot \phi$ is holomorphic on the complex plane $\mathbb{C}$.

When $c \longrightarrow \infty$, $\mathcal{P}\psi \longrightarrow 0$ and $\phi(c) \longrightarrow 0$. This implies that $e^{\mathcal{P}\psi} \cdot \phi$ is bounded on $\mathbb{C}$. So $e^{\mathcal{P}\psi} \cdot \phi$ is a constant function. But $\phi(\infty) = 0$, so $e^{\mathcal{P}\psi} \cdot \phi \equiv 0$. Thus $\phi(c) \equiv 0$ and $f_z(c) = g_z(c)$ for all $c \in \mathbb{C}$. $\qquad\square$

For $z_i \in E$, $2 \leq i \leq n$, consider

$$\mathcal{K}f_i(c) = -\frac{1}{\pi} \int \int_{\mathbb{C}} \frac{\Phi(\zeta, f_i(\zeta))}{\zeta - c} d\xi d\eta,$$

where $\zeta = \xi + i\eta$. From the definition of $\Phi(c, w)$, we have that

$$\Phi(\zeta, f_i(\zeta)) = \frac{\partial f_i}{\partial \bar{\zeta}}(\zeta).$$

So

$$\mathcal{K}f_i(c) = -\frac{1}{\pi} \int \int_{\mathbb{C}} \frac{\frac{\partial f_i}{\partial \bar{\zeta}}(\zeta)}{\zeta - c} d\xi d\eta.$$

This implies that

$$\frac{\partial \mathcal{K}f_i}{\partial \bar{c}}(c) = \frac{\partial f_i}{\partial \bar{c}}(c)$$

and that

$$\frac{\partial(f_i - \mathcal{K}f_i)}{\partial \bar{c}}(c) \equiv 0.$$

So $f_i(c) - \mathcal{K}f_i(c)$ is holomorphic on $\mathbb{C}$. When $c \longrightarrow \infty$, $f_i(c) \longrightarrow z_i$ and $\mathcal{K}f_i(c) \longrightarrow 0$. So $f_i(c) - \mathcal{K}f_i(c)$ is bounded. Therefore it is a constant function. We get that

$$f_i(c) = z_i + \mathcal{K}f_i(c).$$

Thus from Lemma 3.5, $f_i(c) = f_{z_i}(c)$ for all $c \in \overline{\mathbb{C}}$.

By defining $H(c, z) = f_z(c)$ for $(c, z) \in \overline{\Delta}^c \times \mathbb{C} \setminus \{0, 1\}$ and $H(c, 0) = 0$ and $H(c, 1) = 1$ and $H(c, \infty) = \infty$, we get a map

$$H(c, z) = f_z(c) : \overline{\Delta}^c \times \overline{\mathbb{C}} \to \overline{\mathbb{C}},$$

which is an extension of

$$h(c, z) : \overline{\Delta}^c \times E \to \overline{\mathbb{C}}.$$

Lemma 3.6 *The map*

$$H(c, z) = f_z(c) : \overline{\Delta}^c \times \overline{\mathbb{C}} \to \overline{\mathbb{C}},$$

is a holomorphic motion.

Proof First $H(\infty, z) = f_z(\infty) = z$ for all $z \in \overline{\mathbb{C}}$. From the fixed point equation

$$H(c, z) = z + \mathcal{K}H(c, z),$$

$$\frac{\partial H(c, z)}{\partial \overline{c}} = \Phi(c, H(c, z)).$$

Since $\Phi(c, w) = 0$ for all $c \in \overline{\Delta}^c$,

$$\frac{\partial H(c, z)}{\partial \overline{c}} = 0, \quad \forall c \in \overline{\Delta}^c.$$

Thus, for any fixed $z \in \overline{\mathbb{C}}$, $H(c, z) : \overline{\Delta}^c \to \overline{\mathbb{C}}$ is holomorphic.

For any two points z and $z' \in \overline{\mathbb{C}}$, we claim that $H(c, z) \neq H(c, z')$ for all $c \in \mathbb{C}$. This implies that for any fixed $c \in \overline{\Delta}^c$, $H(c, z)$ is an injective map on $z \in \overline{\mathbb{C}}$ and that $H(c, z)$ is a holomorphic motion. To prove the claim take any two points $z, z' \in \overline{\mathbb{C}}$. Assume there is a point $c_0 \in \overline{\mathbb{C}}$ such that $H(c_0, z) = H(c_0, z')$. If $c_0 = \infty$, then $z = z'$, because by assumption the holomorphic motion starts out at the identity. If $c_0 \neq \infty$, then

$$f_z(c_0) - f_{z'}(c_0) = (z - z') + \mathcal{K}f_z(c_0) - \mathcal{K}f_{z'}(c_0),$$

and we can repeat the same argument given in Lemma 3.5.

Let $\phi(c) = f_z(c) - f_{z'}(c)$. Then $\phi(c_0) = 0$. However,

$$\frac{\partial \phi}{\partial \overline{c}}(c) = \frac{\partial f_z}{\partial \overline{c}}(c) - \frac{\partial f_{z'}}{\partial \overline{c}}(c) = \Phi(c, f_z(c)) - \Phi(c, f_{z'}(c)).$$

This implies that

$$\frac{\partial \phi}{\partial \overline{c}}(c) = 0$$

for $c \in \overline{\Delta}^c$. Since $\Phi(c, w)$ is Lipschitz in w-variable with a Lipschitz constant L,

$$\left| \frac{\partial \phi}{\partial \overline{c}}(c) \right| = |\Phi(c, f_z(c)) - \Phi(c, f_{z'}(c))| \leq L|f_z(c) - f_{z'}(c)| = L|\phi(c)|.$$

If $\phi(c) \neq 0$, define

$$\psi(c) = -\frac{\frac{\partial \phi}{\partial \overline{c}}(c)}{\phi(c)},$$

otherwise, define $\psi(c) = 0$. Then

$$\frac{\partial e^{\mathcal{P}\psi} \cdot \phi}{\partial \overline{c}}(c) \equiv 0.$$

So $e^{\mathcal{P}\psi} \cdot \phi$ is holomorphic on $\mathbb{C}$. When $c \longrightarrow \infty$, $\mathcal{P}\psi(c) \longrightarrow 0$ and $\phi(c) \longrightarrow z - z'$. So $e^{\mathcal{P}\psi(c)} \cdot \phi(c)$ is bounded on $\mathbb{C}$. This implies that $e^{\mathcal{P}\psi(c)} \cdot \phi(c)$ is a constant function. Since $\phi(c_0) = 0$, $e^{\mathcal{P}\psi(c)} \cdot \phi(c) \equiv 0$. So $z = z'$. $\qquad\square$

Proof [Proof of Theorem 3.1] Suppose

$$h(c, z) : \Delta \times E \to \overline{\mathbb{C}}$$

is a holomorphic motion. For every $0 < r < 1$, consider $\alpha_r(c) = r/c$. Let $U_r = \alpha_r(\Delta_r) \supset \overline{\Delta}^c$. Then

$$h_r(\alpha_r^{-1}(c), z) : U_r \times E \to \overline{\mathbb{C}}$$

is a holomorphic motion. From Lemmas 3.5 and 3.6, it can be extended to a holomorphic motion

$$\tilde{H}_r(c, z) : \overline{\Delta}^c \times \overline{\mathbb{C}} \to \overline{\mathbb{C}}.$$

Then

$$H_r(c, z) = \tilde{H}(\alpha_r(c), z) : \Delta_r \times \overline{\mathbb{C}} \to \overline{\mathbb{C}}$$

is a holomorphic motion which is an extension of $h(c, z)$ on $\Delta_r \times E$. $\qquad\square$

4 Controlling quasiconformal dilatation

To control the quasiconformal dilatation of a holomorphic motion there are two methods available. One is given by the Bers-Royden paper [BR] and the other is obtained by combining methods given in the Bers-Royden paper and in the Sullivan-Thurston paper [ST]. We discuss the latter method first.

Consider a set of four points $S = \{z_1, z_2, z_3, z_4\}$ in $\overline{\mathbb{C}}$. These points are distinct if and only if the cross ratio

$$Cr(S) = \frac{z_1 - z_3}{z_1 - z_4} : \frac{z_2 - z_3}{z_2 - z_4} = \frac{z_1 - z_3}{z_1 - z_4} \frac{z_2 - z_4}{z_2 - z_3}$$

is not equal to $0, 1$, or ∞. If one of these points is equal to ∞, say z_4, then this cross ratio becomes a ratio

$$Cr(S) = \frac{z_1 - z_3}{z_2 - z_3}.$$

Suppose $H : \overline{\mathbb{C}} \mapsto \overline{\mathbb{C}}$ is an orientation-preserving homeomorphism such that $H(\infty) = \infty$. Then one of the definitions of quasiconformality [LV] of H is that

$$\limsup_{r \to 0} \frac{\sup_{|z-a|=r} |H(z) - H(a)|}{\inf_{a \in \mathbb{C}} |z-a|=r} |H(z) - H(a)|} < \infty.$$

In [ST] Sullivan and Thurston used this definition to prove the following theorem.

Theorem 4.1 *Suppose $H(c, z) : \Delta \times \overline{\mathbb{C}} \to \overline{\mathbb{C}}$ is a normalized holomorphic motion of $\overline{\mathbb{C}}$ parametrized by Δ and with base point 0. Then for each $c_0 \in \Delta$, the map $H(c_0, \cdot) : \overline{\mathbb{C}} \mapsto \overline{\mathbb{C}}$ is quasiconformal.*

Proof Let $a \in \mathbb{C}$ be any point. Let $z_3 = a$. Let z_1 and z_2 be two distinct points in $\mathbb{C}$ not equal to a and $z_4 = \infty$. Then the cross ratio $Cr(S) = (z_1 - z_3)/(z_2 - z_3)$.

Now consider $z_1(c) = H(c, z_1)$, $z_2(c) = H(c, z_2)$, $z_3(c) = H(c, z_3)$, and $z_4(c) = H(c, z_4) = \infty$ and $S(c) = \{z_1(c), z_2(c), z_3(c), z_4(c)\}$. The cross ratio

$$Cr(S(c)) = \frac{z_1(c) - z_3(c)}{z_2(c) - z_3(c)}.$$

Since $H(c, z)$ is a holomorphic motion, $Cr(S(c)) : \Delta \mapsto \mathbb{C} \setminus \{0, 1\}$ is a holomorphic function. Then it decreases the hyperbolic distances from ρ_Δ to $\rho_{0,1}$. So

$$\rho_{0,1}(Cr(S(c_0)), Cr(S)) \le \rho_\Delta(0, c_0) = \log \frac{1 + |c_0|}{1 - |c_0|}.$$

This implies that there is a constant $K = K(c_0) > 0$ such that for any $|Cr(S)| = 1$,

$$|Cr(S(c_0))| \le K.$$

So we have that

$$\limsup_{r \to 0} \frac{\sup_{|z-a|=r} |H(c_0, z) - H(c_0, a)|}{\inf_{a \in \mathbb{C}} |z-a|=r} |H(c_0, z) - H(c_0, a)|} < \infty,$$

that is, $H(c_0, z)$ is quasiconformal. □

Suppose $\mathcal{L}^\infty(W)$ is the Banach space of all essentially bounded measurable functions on W equipped with $\|\cdot\|_\infty$-norm. Bers and Royden [BR] proved the following theorem.

Theorem 4.2 *Suppose $h(c, z) : \Delta \times E \to \hat{\mathbb{C}}$ is a holomorphic motion of E parametrized by Δ and with base point 0 and E has nonempty interior W, then the Beltrami coefficient of $h(c, \cdot)|_W$ given by*

$$\mu(c, z) = \frac{\partial h(c, z)|_W}{\partial \bar{z}} \bigg/ \frac{\partial h(c, z)|_W}{\partial z}$$

is a holomorphic function mapping $c \in \Delta$ into the unit ball of the Banach space $\mathcal{L}^\infty(W)$.

Proof Since the dual of the Banach space $\mathcal{L}^1(W)$ of integrable functions on W is $\mathcal{L}^\infty(W)$, to prove $\mu(c, \cdot)$ is a holomorphic map, it suffices to show that the function

$$c \mapsto \Psi(c) = \int \int_W \alpha(z) \mu_c(z) dx dy$$

is holomorphic in Δ for every $\alpha(z) \in \mathcal{L}^1(W)$. Furthermore, it suffices to check this for every $\alpha(z) \in \mathcal{L}^1(W)$ with a compact support in W.

Suppose $\alpha(z) \in \mathcal{L}^1(W)$ has a compact support $\operatorname{supp}(\alpha)$ in W. There is an $\epsilon > 0$ such that the ϵ-neighborhood $U_\epsilon(\operatorname{supp}(\alpha)) \subset W$. From Theorem 4.1, $h(c, \cdot)$ is quasiconformal, and it is differentiable a.e. in W. Thus

$$\Psi(c) = \int \int_{\operatorname{supp}(\alpha)} \alpha(z) \frac{h_x(c, z) + i h_y(c, z)}{h_x(c, z) - i h_y(c, z)} dx dy$$

$$= \int \int_{\operatorname{supp}(\alpha)} \alpha(z) \frac{1 + i \frac{h_y(c, z)}{h_x(c, z)}}{1 - i \frac{h_y(c, z)}{h_x(c, z)}} dx dy$$

$$= \int \int_{\operatorname{supp}(\alpha)} \alpha(z) \lim_{\lambda \to 0} \frac{1 + i \sigma_c(z, \lambda)}{1 - i \sigma_c(z, \lambda)} dx dy,$$

where

$$\sigma_c(z, \lambda) = \frac{h(c, z + i\lambda) - h(c, z)}{h(c, z + \lambda) - h(c, z)}.$$

For any fixed $z \neq 0, 1, \infty$ and λ small,

$$\varrho(c) = \sigma_c(z, \lambda) : \Delta \mapsto \overline{\mathbb{C}} \setminus \{0, 1, \infty\}$$

is a holomorphic function of $c \in \Delta$. So it decreases the hyperbolic distances on Δ and on $\overline{\mathbb{C}} \setminus \{0, 1, \infty\}$. Since $\varrho(0) = i$, there is a number $0 < r < 1$ such that for

$$|\sigma_c(z, \lambda) - i| \leq \frac{1}{2}, \quad |c| < r.$$

Therefore

$$\left|\frac{1+i\sigma_c(z,\lambda)}{1-i\sigma_c(z,\lambda)}\right| = \left|\frac{-i+\sigma_c(z,\lambda)}{i+\sigma_c(z,\lambda)}\right| \le \frac{\frac{1}{2}}{\frac{3}{2}} = \frac{1}{3}.$$

By the Lebesgue dominated convergence theorem, for $|c| < r$, the sequence of holomorphic functions

$$\Psi_n(c) = \int\int_{\text{supp}(\alpha)} \alpha(z)\frac{1+i\sigma_c(z,\frac{1}{n})}{1-i\sigma_c(z,\frac{1}{n})}dxdy$$

converges uniformly to $\Psi(c)$ as $n \to \infty$. Thus $\Psi(c)$ is holomorphic for $|c| < r$ and this implies that

$$\mu(c,\cdot) : \{c \mid |c| < r\} \to \mathcal{L}^\infty(W)$$

is holomorphic.

Now consider arbitrary $c_0 \in \Delta$. Let $s = 1 - |c_0|$ and let

$$E_0 = h(c_0, E), \ W_0 = h(c_0, W)$$

and

$$g(\tau,\zeta) = h(c_0 + s\tau, z), \quad \zeta = h(c_0, z).$$

Then W_0 is the interior of E_0 since $h(c, z)$ is a quasiconformal homeomorphism. Also

$$g : \Delta \times E_0 \to \overline{\mathbb{C}}$$

is a holomorphic motion. So the Beltrami coefficient of g is a holomorphic function on $\{\tau \mid |\tau| < r\}$. Hence the Beltrami coefficient of h is a holomorphic function on $\{c \mid |c - c_0| < sr\}$. This concludes the proof. ◻

Theorem 4.3 *Suppose $h(c,z) : \Delta \times E \to \overline{\mathbb{C}}$ is a holomorphic motion of E parametrized by Δ and with base point 0 and suppose E has nonempty interior W. Then for each $c \in \Delta$, the map $h(c,z)|_W$ is a K-quasiconformal homeomorphism of W into $\overline{\mathbb{C}}$ with*

$$K \le \frac{1+|c|}{1-|c|}.$$

Proof Since $c \mapsto \mu(c,\cdot)$ mapping from Δ to the unit ball of $\mathcal{L}^\infty(W)$ is a holomorphic map and since $\mu(0,\cdot) = 0$, from the Schwarz's lemma, $\|\mu\|_\infty \le |c|$. This implies that the quasiconformal dilatation of $h(c,\cdot)$ is less than or equal to $K = \frac{1+|c|}{1-|c|}$. ◻

5 Extension of holomorphic motions for $r = 1$

Theorem 5.1 (Slodkowski's Theorem) *Suppose* $h(c, z) : \Delta \times E \to \overline{\mathbb{C}}$ *is a holomorphic motion. Then there is a holomorphic motion*

$$H(c, z) : \Delta \times \overline{\mathbb{C}} \to \overline{\mathbb{C}}$$

which extends $h(c, z) : \Delta \times E \to \overline{\mathbb{C}}$.

Proof Suppose E is a subset of $\overline{\mathbb{C}}$. Suppose $h(c, z) : \Delta \times E \to \overline{\mathbb{C}}$ is a holomorphic motion. Let E_1, E_2... be a sequence of nested subsets consisting of finite number of points in E. Suppose

$$\{0, 1, \infty\} \subset E_1 \subset E_2 \subset \cdots \subset E$$

and suppose $\cup_{i=1}^{\infty} E_i$ is dense in E. Then $h(c, z) : \Delta \times E_i \to \overline{\mathbb{C}}$ is a holomorphic motion for every $i = 1, 2, \ldots$.

From Theorem 3.1, for any $0 < r < 1$ and $i \geq 1$, there is a holomorphic motion $H_i(c, z) : \Delta_r \times \overline{\mathbb{C}} \mapsto \overline{\mathbb{C}}$ such that $H_i | \Delta_r \times E_i = h | \Delta_r \times E_i$. From Theorem 4.3, $z \mapsto H_i(c, z)$ is $(1 + |c|/r)/(1 - |c|/r)$-quasiconformal and fixes $0, 1, \infty$ for all $i > 0$. So for any $|c| \leq r$, the functions $z \mapsto H_i(c, z)$ form a normal family and there is a subsequence $H_{i_k}(c, \cdot)$ converging uniformly (in the spherical metric) to a $(1 + |c|/r)/(1 - |c|/r)$-quasiconformal homeomorphism $H_r(c, \cdot) : \overline{\mathbb{C}} \to \overline{\mathbb{C}}$ such that $H_r(c, z) = h(c, z)$ for $z \in \cup(E_{j_k})$.

Let ζ be a point in E. Replacing E_i by $E_i \cup \{\zeta\}$ and repeating the previous construction we obtain a $(1 + |c|/r)/(1 - |c|/r)$-quasiconformal homeomorphism $\tilde{H}_r$ which coincides with $h(c, z)$ on $(\cup E_{i_k}) \cup \{\zeta\}$. But $z \mapsto H_r(c, z)$ and $z \mapsto \tilde{H}_r(c, z)$ are continuous everywhere and coincide on $\cup E_{i_k}$, hence on E. So $H_r(c, \zeta) = \tilde{H}_r(c, \zeta) = h(c, \zeta)$ for any $\zeta \in E$.

Now for any $z \neq 0, 1, \infty$, since $H_i(c, z) : \Delta \mapsto \overline{\mathbb{C}}$ are holomorphic and omit three points $0, 1, \infty$. So the functions $c \mapsto H_i(c, z)$ form a normal family. Any convergent subsequence $H_{i_k}(c, z)$ still has a holomorphic limit $H_r(c, z)$, thus $H_r(c, z) : \Delta_r \times \overline{\mathbb{C}} \mapsto \overline{\mathbb{C}}$ is a holomorphic motion which extends $h(c, z)$ on $\Delta_r \times \overline{\mathbb{C}}$.

Now we are ready to take the limit as $r \to 1$. For each $0 < r < 1$, let $H_r(c, z) : \Delta_r \times \overline{\mathbb{C}} \to \overline{\mathbb{C}}$ be a holomorphic motion such that $H_r = h$ on $\Delta_r \times E$. From Theorem 4.3, $H_r(c, \cdot)$ is $(1 + |c|/r)/(1 - |c|/r)$-quasiconformal for every c with $|c| \leq r$.

Take a sequence $Z = \{z_i\}_{i=1}^{\infty}$ of points in $\overline{\mathbb{C}}$ such that $\overline{Z} = \overline{\mathbb{C}}$, and assume 0, 1, and ∞ are not elements of Z. For each $i = 1, 2, \cdots$, $H_r(c, z_i) : \Delta_r \to \overline{\mathbb{C}}$ is holomorphic and omits $0, 1, \infty$. Thus

$$\{H_r(c, z_i), c \in \Delta_r\}_{0 < r < 1}$$

forms a normal family. We have a subsequence $r_n \to 1$ such that $H_{r_n}(c, z_i)$ tends to a holomorphic function $\tilde{H}(c, z_i)$ defined on Δ uniformly in the spherical metric for all $i = 1, 2, \cdots$. For a fixed $c \in \Delta$, $H_{r_n}(c, \cdot)$ are $(1 + |c|/r_n)/(1 - |c|/r_n)$-quasiconformal for all $r_n > |c|$. So $\{H_{r_n}(c, \cdot)\}_{r_n > |c|}$ is a normal family. Since $H_{r_n}(c, \cdot)$ fixes $0, 1, \infty$, there is a subsequence of $\{H_{r_n}(c, \cdot)\}$, which we still denote by $\{H_{r_n}(c, \cdot)\}$, that converges uniformly in the spherical metric to a $(1 + |c|)/(1 - |c|)$-quasiconformal homeomorphism $H(c, \cdot)$. Since $\tilde{H}(c, z_i) = H(c, z_i)$ for all $i = 1, 2, \cdots$, this implies that, for any fixed z_i, $H(c, z_i)$ is holomorphic. Thus $H(c, z) : \Delta \times Z \to \overline{\mathbb{C}}$ is a holomorphic motion.

For any $0 < r < 1$, $H(c, z)$ is $(1 + r)/(1 - r)$-quasiconformal for all c with $|c| \leq r$, it is α-Hölder continuous, that is,

$$d(H(c, z), H(c, z')) \leq Ad(z, z')^{\alpha} \quad \text{for all } z, z' \in \overline{\mathbb{C}} \quad \text{and for all } |c| \leq r,$$

where $d(\cdot, \cdot)$ is the spherical distance and where A and $0 < \alpha < 1$ depend only on r.

For any $z \in Z$ such that its spherical distances to 0, 1, ∞ are greater than $\epsilon > 0$, the map $H(c, z)$ is a holomorphic map on Δ, which omits the values 0, 1, and ∞. So $H(c, z)$ decreases the hyperbolic distance ρ_Δ on Δ and the hyperbolic distance $\rho_{0,1}$ on $\overline{\mathbb{C}} \setminus \{0, 1, \infty\}$. So we have a constant $B > 0$ depending only on r and ϵ such that

$$d(H(c, z), H(c', z)) \leq B|c - c'|$$

for all $|c|, |c'| \leq r$ and all $z \in Z$ such that spherical distances between them and 0, 1, and ∞ are greater than $\epsilon > 0$. Thus we get that

$$d(H(c, z), H(c', z')) \leq A\delta(z, z')^{\alpha} + B|c - c'|$$

for $|c|, |c'| \leq r$ and $z, z' \in Z$ such that their spherical distances from 0, 1, and ∞ are greater than $\epsilon > 0$. This implies that $H(c, z)$ is uniformly equicontinuous on $|c| \leq r$ and $\{z \in Z \mid d(z, \{0, 1, \infty\}) \geq \epsilon\}$. Therefore, its continuous extension $H(c, z)$ is holomorphic in c with $|c| \leq r$ for any $\{z \in \overline{\mathbb{C}} \mid d(z, \{0, 1, \infty\}) \geq \epsilon\}$. Letting $r \to 1$ and $\epsilon \to 0$, we get that $H(c, z)$ is holomorphic in $c \in \Delta$ for any $z \in \overline{\mathbb{C}}$. Thus $H(c, z) : \Delta \times \overline{\mathbb{C}} \to \overline{\mathbb{C}}$ is a holomorphic motion such that $H(c, z)|\Delta \times E = h(c, z)$. This completes the proof. □

6 The $|\epsilon \log \epsilon|$ continuity of a holomorphic motion

In this section we show how the $|\epsilon \log \epsilon|$ modulus of continuity for the tangent vector to a holomorphic motion can be derived directly from Schwarz's lemma. Then we go on to show how the Hölder continuity of the mapping $z \mapsto w(z) = h(c, z)$ with Hölder exponent $\frac{1-|c|}{1+|c|}$ follows from the $|\epsilon \log \epsilon|$ continuity of the tangent vectors to the curve $c \mapsto h(c, z)$. In particular, since any K-quasiconformal map $z \mapsto f(z)$ coincides with $z \mapsto h(c, z)$ where $K \le \frac{1+|c|}{1-|c|}$, we conclude that f satisfies a Hölder condition with exponent $1/K$.

Lemma 6.1 *Let $h(c, z)$ be a normalized holomorphic motion parametrized by Δ and with base point 0 and let $V(z)$ be the tangent vector to this motion at $c = 0$ defined by*

$$V(z) = \lim_{c \to 0} \frac{h(c, z) - z}{c}. \tag{6.1}$$

Then $V(0) = 0$, $V(1) = 0$ and $|V(z)| = o(|z|^2)$ as $z \to \infty$.

Proof Since $h(c, z)$ is normalized, $h(c, 0) = 0$ and $h(c, 1) = 1$ for every $c \in \Delta$, and therefore $V(0) = 0$ and $V(1) = 0$. Since $h(c, \infty) = \infty$ for every $c \in \Delta$ if we introduce the coordinate $w = 1/z$ and consider the motion $h_1(c, w) = 1/h(c, 1/w)$, we see that $h_1(c, 0) = 0$ for every $c \in \Delta$.

Put $p(c) = h(c, z)$ and if we think of z as a local coordinate for the Riemann sphere,

$$z \circ p(c) = z + cV^z(z) + o(c^2),$$

and in terms of the local coordinate $w = 1/z$,

$$w \circ p(c) = w + cV^w(w) + o(c^2).$$

Then $V^w(0) = 0$. Putting $g = w \circ z^{-1}$, the identity $g(z(p(c))) = w(p(c))$ yields

$$g'(z(p(0)))z'(p(0)) = w'(p(0)). \tag{6.2}$$

Since $g(z) = 1/z$, $g'(z) = -(1/z)^2$ and since

$$V^w(0) = 0, \quad \frac{d}{dc} w(p(c))|_{c=0} = V^w(w(p(0)))$$

and $V^w(w(p(c)))$ is a continuous function of c, the equation

$$V^z(z(p(c)))\frac{dw}{dz} = V^w(w(p(c)))$$

implies

$$\frac{V^z(z)}{z^2} \to 0$$

as $z \to \infty$. □

Let $\rho_{0,1}(z)$ be the infinitesimal form for the hyperbolic metric on $\overline{\mathbb{C}} \setminus \{0, 1, \infty\}$ and let $\rho_{\Delta}(z) = 2/(1 - |z|^2)$ be the infinitesimal form for the hyperbolic metric on Δ. For any four distinct points a, b, c and d, the cross ratio

$$g(c) = cr(h_c(a), h_c(b), h_c(c), h_c(d))$$

is a holomorphic function of $c \in \Delta$, and omitting the values $0, 1$ and ∞. Then by Schwarz's lemma,

$$\rho_{0,1}(g(c))|g'(c)| \leq \sigma_{\Delta}(c) = \frac{2}{1 - |c|^2}$$

and

$$\rho_{0,1}(g(0))|g'(0)| \leq 2. \tag{6.3}$$

But $|g'(0)|$ is equal to

$$|g(0)| \left| \frac{V(b) - V(a)}{b - a} - \frac{V(c) - V(b)}{c - b} + \frac{V(d) - V(c)}{d - c} - \frac{V(a) - V(d)}{a - d} \right|, \tag{6.4}$$

where $g(0) = cr(a, b, c, d) = \frac{(b-a)(d-c)}{(c-b)(a-d)}$.

Lemma 6.2 *If $V(b) = o(b^2)$ as $b \to \infty$, then*

$$\left(\frac{V(b) - V(a)}{b - a} - \frac{V(c) - V(b)}{c - b} \right) \to 0 \quad as \quad b \to \infty.$$

Proof

$$\left(\frac{V(b) - V(a)}{b - a} - \frac{V(c) - V(b)}{c - b} \right)$$

simplifies to

$$\frac{cV(b) - bV(c) - aV(b) - cV(a) + bV(a) + aV(c)}{(b - a)(c - b)}.$$

As $b \to \infty$ the denominator grows like b^2 but the numerator is $o(b^2)$. □

Theorem 6.3 *For any vector field V tangent to a normalized holomorphic motion and defined by (6.1), there exists a number C depending on R such that for any two complex numbers z_1 and z_2 with $|z_1| < R$ and $|z_2| < R$ and $|z_1 - z_2| < \delta$,*

$$|V(z_2) - V(z_1)| \le |z_2 - z_1|(2 + \frac{C}{\log \frac{1}{\delta}})(\log \frac{1}{|z_2 - z_1|}).$$

Proof By applying Lemma 6.2, inequality (6.3) and equation (6.4) to $a = z_1, b = z_2, c = 0, d = \infty$, we obtain $g(0) = \frac{z_2 - z_1}{z_2}$,

$$\left| \frac{V(b) - V(a)}{b - a} - \frac{V(c) - V(b)}{c - b} + \frac{V(d) - V(c)}{d - c} - \frac{V(a) - V(d)}{a - d} \right|$$

$$= \left| \frac{V(z_2) - V(z_1)}{z_2 - z_1} - \frac{V(z_2)}{z_2} \right|,$$

and

$$\rho_{0,1}\left(\frac{z_2 - z_1}{z_2} \right) \left| \frac{z_2 - z_1}{z_2} \right| \left| \frac{V(z_2) - V(z_1)}{z_2 - z_1} - \frac{V(z_2)}{z_2} \right| \le 2,$$

and so

$$\left| \frac{V(z_2) - V(z_1)}{z_2 - z_1} - \frac{V(z_2)}{z_2} \right| \le \frac{2}{\rho_{0,1}\left(\frac{z_2 - z_1}{z_2} \right) \left| \frac{z_2 - z_1}{z_2} \right|}. \tag{6.5}$$

Applying (6.3) and (6.4) again with $a = 0, b = 1, c = \infty, d = z_2$, we obtain

$$\rho_{0,1}(z_2)|z_2| \left| \frac{V(z_2)}{z_2} \right| \le 2,$$

and so

$$\frac{|V(z_2)|}{|z_2|} \le \frac{2}{\rho_{0,1}(z_2)|z_2|}, \tag{6.6}$$

and this together with (6.5) implies

$$\left| \frac{V(z_2) - V(z_1)}{z_2 - z_1} \right| \le \frac{2}{\rho_{0,1}\left(\frac{z_2 - z_1}{z_2} \right) \left| \frac{z_2 - z_1}{z_2} \right|} + \frac{2}{\rho_{0,1}(z_2)|z_2|}. \tag{6.7}$$

To finish the proof we need the following lemma, a form of which appeared in [L, page 40]. We adapted similar ideas to prove the following version, which is sufficient for the proof of Theorem 6.3.

Lemma 6.4 *If* $0 < |z| < 1$, *then*

$$\rho_{0,1}(z) \geq \frac{1}{|z| \left(4 + \log 4 + \log \frac{1}{|z|}\right)}. \tag{6.8}$$

Proof From Agard's formula [Ag] (note that $\rho_{0,1}$ has the curvature -1),

$$\rho_{0,1}(z) = \left(\frac{1}{2\pi} \int\int_{\mathbb{C}} \left|\frac{z(z-1)}{\zeta(\zeta-1)(\zeta-z)}\right| d\xi d\eta\right)^{-1}.$$

Since the smallest value of $\rho_{0,1}(z)$ on the circle $|z| = 1$ occurs at $z = -1$, we see that

$$\min_{|z|=1} \rho_{0,1}(z) = \left(\frac{1}{2\pi} \int\int_{\mathbb{C}} \left|\frac{1}{(\zeta-1)(\zeta)(\zeta+1)}\right| d\xi d\eta\right)^{-1}.$$

The infinitesimal form of the Poincaré metric on the punctured disk $\Delta_r^* = \{z \in \mathbb{C} \mid 0 < |z| < r\}$ of radius r is

$$\rho_r(z) = \frac{1}{|z| \left[\log r + \log \frac{1}{|z|}\right]}. \tag{6.9}$$

Note that $\rho_r(z)$ takes the constant value $\frac{1}{\log r}$ along $|z| = 1$. Thus, if we choose r so that $\log r$ is no less than

$$\frac{1}{2\pi} \int\int_{\mathbb{C}} \left|\frac{1}{(\zeta-1)(\zeta)(\zeta+1)}\right| d\xi d\eta \tag{6.10}$$

then

$$\rho_r(z) \leq \rho_{0,1}(z) \text{ for all } z \text{ with } |z| = 1. \tag{6.11}$$

Our next objective is to show that the same inequality

$$\rho_r(z) \leq \rho_{0,1}(z) \tag{6.12}$$

holds for all z with $|z| < \delta$ when δ is sufficiently small. To prove (6.12) we will need to assume that $\log r$ is no smaller than $4 + \log 4$. Since numerical calculation shows that $4 + \log 4$ is larger than (6.10), in the final result the number $4 + \log 4$ is the number we require for the result in (6.8). In [KL] Keen and Lakic obtain an improved lower bound by showing that inequality (6.8) is still true if $4 + \log 4$ is replaced by (6.10). This improvement is unimportant for our purposes.

In [A1, pages 17–18] Ahlfors shows that

$$\rho_{0,1}(z) \geq \frac{|\zeta'(z)|}{|\zeta(z)|} \frac{1}{4 + \log \frac{1}{|\zeta(z)|}} \tag{6.13}$$

for $|z| \leq 1$ and $|z| \leq |z - 1|$, where ζ maps the complement of $[1, +\infty]$ conformally onto the unit disk, origins corresponding to each other and symmetry with respect to the real axis being preserved. ζ satisfies

$$\frac{\zeta'(z)}{\zeta(z)} = \frac{1}{z\sqrt{1-z}}, \tag{6.14}$$

$$\zeta(z) = \frac{\sqrt{1-z} - 1}{\sqrt{1-z} + 1} = \frac{z}{(\sqrt{1-z} + 1)^2} \tag{6.15}$$

with Re $\sqrt{1-z} > 0$, and

$$|\zeta(z)| \to \frac{|z|}{4} \tag{6.16}$$

as $z \to 0$.

We now show that there is a number $\delta > 0$ such that if $|z| < \delta$, then

$$\frac{|\zeta'|}{|\zeta|} \frac{1}{[4 + \log \frac{1}{|\zeta|}]} \geq \frac{1}{|z|[\log r + \frac{1}{|z|}]}.$$

From (6.14) this is equivalent to showing that

$$|\sqrt{1-z}|(4 + \log \frac{1}{|\zeta|}) \leq \log r + \log \frac{1}{|z|},$$

which is equivalent to

$$|\sqrt{1-z}|(4 + \log 4)$$

$$\leq \log r + \left\{ \left(\log \frac{1}{|z|} \right) \left((1 - |\sqrt{1-z}|) \left(\frac{\log \frac{1}{|\zeta|} - \log 4}{\log \frac{1}{|z|}} \right) \right) \right\}. \tag{6.17}$$

From (6.16)

$$\left(\frac{\log \frac{1}{|\zeta|} - \log 4}{\log \frac{1}{|z|}} \right)$$

approaches 1 as $z \to 0$ and the expression in the curly brackets on the right-hand side of (6.17) approaches zero. Thus, in order to prove (6.12), it suffices to assume

$$\log r > 4 + \log 4.$$

We have so far established that $\rho_{0,1}(z) \geq \rho_r(z)$ on the unit circle and on any circle $|z| = \delta$ for sufficiently small δ. To complete the proof of the lemma we observe that since both metrics $\rho_{0,1}(z)$ and $\rho_r(z)$ have constant curvatures equal to -1, if we denote the Laplacian by

$$\Delta = \left(\frac{\partial}{\partial x}\right)^2 + \left(\frac{\partial}{\partial y}\right)^2,$$

then

$$-\rho_{0,1}^{-2}\Delta \log \rho_{0,1} = -1 \text{ and } -\rho_r^{-2}\Delta \log \rho_r = -1.$$

Therefore,

$$\Delta(\log \rho_{0,1} - \log \rho_r) = \rho_{0,1}^2 - \rho_r^2 \qquad (6.18)$$

throughout the annulus $\{z : \delta \le |z| \le 1\}$. The minimum of $\rho_{0,1}/\rho_r$ in this annulus occurs either at a boundary point or an interior point. If it occurs at an interior point, then $\Delta \log(\rho_{0,1}/\rho_r) \ge 0$ at that point and so $\rho_{0,1}/\rho_r \ge 1$ at that point. But if it occurs on the boundary, we also have $\rho_{0,1}/\rho_r \ge 1$ at that point. So in either case

$$\rho_{0,1}/\rho_r \ge 1$$

at that point, and therefore

$$\rho_{0,1} \ge \rho_r$$

throughout the annulus provided $\log r$ is any number larger than $4+\log 4$. But this implies the previous inequality is also true when $4+\log 4 = \log r$, and this completes the proof of the lemma. $\qquad \square$

From (6.7) and this lemma we obtain

$$|V(z_2)-V(z_1)| \le |z_2-z_1|\left(\left|\frac{V(z_2)}{z_2}\right| + 2\log r + 2\log|z_2| + 2\log\frac{1}{|z_2 - z_1|}\right).$$

Therefore to prove the theorem we must show that for $\epsilon = \frac{C}{\log(1/\delta)}$

$$\left|\frac{V(z_2)}{z_2}\right| + 2\log r + 2\log|z_2| + 2\log\frac{1}{|z_2 - z_1|} \le (2+\epsilon)\log\frac{1}{|z_2 - z_1|}.$$

This is equivalent to showing that

$$\left|\frac{V(z_2)}{z_2}\right| + 2\log r + 2\log|z_2| \le \epsilon\log\frac{1}{|z_2 - z_1|}.$$

If $|z_2| < 1$, from (6.6) and Lemma 6.4, we have

$$\rho_{0,1}(z_2) \ge \frac{1}{|z_2|(\log r + \log\frac{1}{|z_2|})},$$

and

$$\frac{|V(z_2)|}{|z_2|} \le 2\log r + 2\log\frac{1}{|z_2|}.$$

Hence

$$\left|\frac{V(z_2)}{z_2}\right| + 2\log r + 2\log|z_2| \le 4\log r.$$

If $1 \le |z_2| \le R$, then since $|\frac{V(z_2)}{z_2}| + 2\log|z_2|$ is a continuous function, it is bounded by a number M_1, so

$$\left|\frac{V(z_2)}{z_2}\right| + 2\log r + 2\log|z_2| \le M_1 + 2\log r.$$

The constant $C = M_1 + 2\log r$ does not depend on δ and

$$\left|\frac{V(z_2)}{z_2}\right| + 2\log r + 2\log|z_2| \le C$$

for any $|z_2| \le R$. Thus, putting $\epsilon = C/\frac{1}{\log(1/\delta)}$, we obtain

$$|V(z_2) - V(z_1)| \le |z_2 - z_1|(2 + \epsilon)\left(\log\frac{1}{|z_2 - z_1|}\right).$$

$\square$

Applying the same argument at a variable value of c we obtain the following result.

Theorem 6.5 *Suppose $0 < r < 1$ and $R > 0$. If $|c| \le r$, $|z_1(c)| \le R$, $|z_2(c)| \le R$ and $|z_2(c) - z_1(c)| < \delta$, then*

$$|V(z_2(c)) - V(z_1(c))| \le \frac{2 + \epsilon}{1 - |c|^2}|z_2(c) - z_1(c)|\log\frac{1}{|z_2(c) - z_1(c)|}, \quad (6.19)$$

where $\epsilon \le \frac{M}{\log(1/\delta)}$ and $\delta \ge |z_1(0) - z_2(0)|$. Moreover, there is a constant C such that

$$|z_2(c) - z_1(c)| \le C \cdot |z_2 - z_1|^{\frac{1-|c|}{1+|c|}}.$$

Proof Equation (6.19) follows by the same calculations we have just completed. To prove the second inequality, put $s(c) = |z_2(c) - z_1(c)|$ and assume $0 < |c| < 1$. Then (6.19) yields

$$s'(c) \le \frac{2 + \epsilon}{1 - |c|^2}s(c)\log\frac{1}{s(c)}.$$

So

$$-(\log\frac{1}{s(c)})' \le \frac{2+\epsilon}{1-|c|^2}\log\frac{1}{s(c)}$$

and

$$-(\log(\log\frac{1}{s(c)}))' \le \frac{2+\epsilon}{1-|c|^2}.$$

By integration,

$$-\log(\log\frac{1}{s(c)})\Big|_0^c \le -\frac{2+\epsilon}{2}\log\frac{1-|c|}{1+|c|}\Big|_0^{|c|}$$

and

$$\log\log(\frac{1}{s(c)}) - \log\log(\frac{1}{s(0)}) \ge \log\left(\frac{1-|c|}{1+|c|}\right)^{1+\frac{\epsilon}{2}}.$$

Since $\log x$ is increasing,

$$\frac{\log\frac{1}{s(c)}}{\log\frac{1}{s(0)}} \ge \left(\frac{1-|c|}{1+|c|}\right)^{1+\frac{\epsilon}{2}},$$

$$\log s(c) \le \left(\frac{1-|c|}{1+|c|}\right)^{1+\frac{\epsilon}{2}}\log s(0)$$

and

$$s(c) \le s(0)^{(\frac{1-|c|}{1+|c|})^{1+\frac{\epsilon}{2}}}.$$

Putting $s = s(0)$ and $\alpha = \frac{1-|c|}{1+|c|}$, we wish to show that

$$s^{\alpha^{1+\epsilon}} \le Cs^\alpha \text{ or equivalently that } s^{(\alpha^{1+\epsilon}-\alpha)} \le C. \qquad (6.20)$$

This is equivalent to showing that

$$\alpha(\alpha^\epsilon - 1)\log s \le \log C$$

or that

$$\alpha(\exp(\frac{M}{\log(1/s)}\log\alpha) - 1)\log s \le \log C.$$

Since $0 < \alpha < 1$ and since we may assume $s < e^{-1}$, by using the inequality $e^x - 1 \le xe^{x_0}$ for $0 \le x \le x_0$, we see that it suffices to choose C so that

$$\alpha\frac{M}{\log(1/s)}\log(1/\alpha)e^{M\log\alpha}\log(1/s) = \alpha M\log(1/\alpha)e^{M\log\alpha} \le \log C.$$

$\square$

The idea for the proof of Theorem 6.5 is suggested but not worked out in [GK].

7 Kobayashi's metrics

Suppose $\mathcal{N}$ is a connected complex manifold over a complex Banach space. Let $\mathcal{H} = \mathcal{H}(\Delta, \mathcal{N})$ be the space of all holomorphic maps from Δ into $\mathcal{N}$. For p and q in $\mathcal{N}$, let

$$d_1(p, q) = \log \frac{1 + r}{1 - r},$$

where r is the infimum of the nonnegative numbers s for which there exists $f \in \mathcal{H}$ such that $f(0) = p$ and $f(s) = q$. If no such $f \in \mathcal{H}$ exists, then $d_1(p, q) = \infty$.

Let

$$d_n(p, q) = \inf \sum_{i=1}^{n} d_1(p_{i-1}, p_i)$$

where the infimum is taken over all chains of points $p_0 = p, p_1, ..., p_n = q$ in $\mathcal{N}$. Obviously, $d_{n+1} \leq d_n$ for all $n > 0$.

Definition 7.1 (Kobayashi's metric) *The Kobayashi pseudo-metric* $d_K = d_{K,\mathcal{N}}$ *is defined as*

$$d_K(p, q) = \lim_{n \to \infty} d_n(p, q), \quad p, q \in \mathcal{N}.$$

In general, it is possible that d_K is identically equal to 0, which is the case for example if $\mathcal{N} = \mathbb{C}$.

Another way to describe d_K is the following. Let the Poincaré metric on the unit disk Δ be given by

$$\rho_\Delta(z, w) = \log \frac{1 + \frac{|z - w|}{|1 - \bar{z}w|}}{1 - \frac{|z - w|}{|1 - \bar{z}w|}}, \quad z, w \in \Delta.$$

Then d_K is the largest pseudo-metric on $\mathcal{N}$ such that

$$d_K(f(z), f(w)) \leq \rho_\Delta$$

for all z and $w \in \Delta$ and for all holomorphic maps f from Δ into $\mathcal{N}$. The following is a consequence of this property.

Proposition 7.2 *Suppose $\mathcal{N}$ and $\mathcal{N}'$ are two complex manifolds and $F : \mathcal{N} \to \mathcal{N}'$ is holomorphic. Then*

$$d_{K,\mathcal{N}'}(F(p), F(q)) \leq d_{K,\mathcal{N}}(p, q).$$

Lemma 7.3 *Suppose $\mathcal{B}$ is a complex Banach space with norm $\|\cdot\|$. Let $\mathcal{N}$ be the unit ball of $\mathcal{B}$ and let d_K be the Kobayashi's metric on $\mathcal{N}$. Then*

$$d_K(0, \mathbf{v}) = \log \frac{1 + \|\mathbf{v}\|}{1 - \|\mathbf{v}\|} = 2 \tanh^{-1} \|\mathbf{v}\|, \quad \forall \, \mathbf{v} \in \mathcal{N}.$$

Proof Pick a point $\mathbf{v}$ in $\mathcal{N}$. The linear function $f(c) = c\mathbf{v}/\|\mathbf{v}\|$ maps the unit disk Δ into the unit ball $\mathcal{N}$, and takes $\|\mathbf{v}\|$ into $\mathbf{v}$, and 0 into $\mathbf{0}$. Therefore

$$d_K(0, \mathbf{v}) \leq \rho_\Delta(0, \|\mathbf{v}\|),$$

where ρ_Δ is the Kobayashi's metric on Δ (it coincides with the Poincaré metric on Δ).

On the other hand, by the Hahn-Banach theorem, there exists a continuous linear function L on $\mathcal{N}$ such that $L(\mathbf{v}) = \|\mathbf{v}\|$ and $\|L\| = 1$. Thus, L maps $\mathcal{N}$ into the unit disk Δ, and so

$$d_K(0, \mathbf{v}) \geq \rho_\Delta(0, \|\mathbf{v}\|).$$

Therefore,

$$d_K(0, \mathbf{v}) = \rho_\Delta(0, \|\mathbf{v}\|) = \log \frac{1 + \|\mathbf{v}\|}{1 - \|\mathbf{v}\|} = 2 \tanh^{-1} \|\mathbf{v}\|.$$

$\square$

8 Teichmüller's and Kobayashi's metrics on $T(R)$

Assume R is a Riemann surface conformal to Δ/Γ where Γ is a discontinuous, fixed point free group of hyperbolic isometries of Δ. Let $\mathcal{M} = \mathcal{M}(\Gamma)$ be the unit ball of the complex Banach space of all $\mathcal{L}^\infty$ functions defined on Δ satisfying the Γ-invariance property:

$$\mu(\gamma(z)) \frac{\overline{\gamma'(z)}}{\gamma'(z)} = \mu(z) \tag{8.1}$$

for all z in Δ and all γ in Γ. An element $\mu \in \mathcal{M}$ is called a Beltrami coefficient on R. Points of the Teichmüller space $T = T(R)$ are represented by equivalence classes of Beltrami coefficients $\mu \in \mathcal{M}$. Two Beltrami coefficients $\mu, \nu \in \mathcal{M}$ are in the same Teichmüller equivalence class if the quasiconformal self maps f^μ and f^ν which preserve Δ and which are normalized to fix $0, i$ and -1 on the boundary of the unit disk coincide at all boundary points of the unit disk.

Definition 8.1 (Teichmüller's metric) *For two elements $[\mu]$ and $[\nu]$ of $T(R)$, Teichmüller's metric is equal to*

$$d_T([\mu], [\nu]) = \inf \log K(f^\mu \circ (f^\nu)^{-1}),$$

where the infimum is over all μ and ν in the equivalence classes $[\mu]$ and $[\nu]$, respectively. In particular,

$$d_T(0, [\mu]) = \log \frac{1 + k_0}{1 - k_0},$$

where k_0 is the minimal value of $\|\mu\|_\infty$, where μ ranges over the Teichmüller class $[\mu]$.

Lemma 8.2 *Let d_K and d_T be Kobayashi's and Teichmüller's metrics of $T(R)$. Then $d_K \leq d_T$.*

Proof Let a Beltrami coefficient μ satisfying (8.1) be extremal in its class and $\|\mu\|_\infty = k$. This is possible because by normal families argument every class possesses at least one extremal representative. By the definition of Teichmüller's metric

$$d_T(0, [\mu]) = \log \frac{1 + k}{1 - k}.$$

For such a μ, let $g(c) = [c\mu/k]$. Then $g(c)$ is a holomorphic function of c for $|c| < 1$ with values in the Teichmüller space $T(R)$, $g(0) = 0$ and $g(k) = [\mu]$. Hence

$$d_K(0, [\mu]) \leq d_1(0, [\mu]) \leq d_T(0, [\mu]).$$

Now the right translation mapping $\alpha([f^\mu]) = [f^\mu \circ (f^\nu)^{-1}]$ is biholomorphic, so it is an isometry in Kobayashi's metric. We also know that it is an isometry in Teichmüller's metric. Therefore, the inequality

$$d_K([\nu], [\mu]) \leq d_1([\nu], [\mu]) \leq d_T([\nu], [\mu])$$

holds for an arbitrary pair of points $[\mu]$ and $[\nu]$ in the Teichmüller space $T(R)$.   □

In order to describe holomorphic maps into $T(R)$ we will use the Bers' embedding by which $T(R)$ is realized as a bounded domain in the Banach space $\mathcal{B}(R)$ of equivariant cusp forms. Here $\mathcal{B}(R)$ consists of the functions φ holomorphic in Δ^c for which

$$\sup_{z \in \Delta^c} \{|(|z|^2 - 1)^2 |\varphi(z)|\} < \infty$$

and for which

$$\varphi(\gamma(z))(\gamma'(z))^2 = \varphi(z) \text{ for all } \gamma \in \Gamma.$$

We assume Γ is a Fuchsian covering group such that Δ/Γ is conformal to R. For any Beltrami differential μ supported on Δ, we let w^μ be the quasiconformal self-mapping of $\overline{\mathbb{C}}$ which fixes $1, i$ and -1 and which has Beltrami coefficient μ in Δ and Beltrami coefficient identically equal to zero in Δ^c. Let w^μ restricted to Δ^c be equal to the Riemann mapping g^μ. Then g^μ has the following properties:

a) g^μ fixes the points $1, i$ and -1,
b) $g^\mu(\partial\Delta)$ is a quasiconformal image of the circle $\partial\Delta$,
c) g^μ is univalent and holomorphic in Δ^c,
d) $g^\mu \circ \gamma \circ (g^\mu)^{-1}$ is equal to a Möbius transformation $\tilde{\gamma}$, for all γ in Γ, and
e) g^μ determines and is determined uniquely by the corresponding point in $T(R)$.

The Bers' embedding maps the Teichmüller equivalence class of μ to the Schwarzian derivative of g^μ where the Schwarzian derivative of a C^3 function g is defined by

$$S(g) = \left(\frac{g''}{g'}\right)' + \frac{1}{2}\left(\frac{g''}{g'}\right)^2.$$

In the next section we use this realization of the complex structures to prove that $d_T \leq d_K$.

9 The lifting problem

Let Φ be the natural map from the space $\mathcal{M}$ of Beltrami differentials on R onto $T(R)$ and let f be a holomorphic map from the unit disk into $T(R)$ with $f(0)$ equal to the base point of $T(R)$. The lifting problem is the problem of finding a holomorphic map $\tilde{f}$ from Δ into $\mathcal{M}$, such that $\tilde{f}(0) = 0$ and $\Phi \circ \tilde{f} = f$.

In this section we prove the theorem of Earle, Kra and Krushkal [EKK] which says that the lifting problem always has a solution. We follow their technique which relies on proving an equivariant version of Slodkowski's extension theorem and then going on to show that the positive solution to the lifting problem implies $d_T \leq d_K$ for every Riemann surface that has a nontrivial Teichmüller space with complex structure.

Theorem 9.1 (An equivariant version of Slodkowski's extension theorem) *Let $h(c, z)$ be a holomorphic motion of $\Delta^c = \overline{\mathbb{C}} \setminus \Delta$ parametrized by Δ and with base point 0 and let Γ be a torsion-free group of Möbius transformations mapping Δ^c onto itself. Suppose for each $\gamma \in \Gamma$ and $c \in \Delta$ there is a Möbius transformation $\tilde{\gamma}_c$ such that*

$$h(c, \gamma(z)) = \tilde{\gamma}_c(h(c, z)), \ \forall\, z \in \Delta^c.$$

Then $h(c, z)$ can be extended to a holomorphic motion $H(c, z)$ of $\overline{\mathbb{C}}$ parametrized by Δ and with base point 0 in such a way that

$$H(c, \gamma(z)) = \tilde{\gamma}_c(H(c, z))$$

holds for $\gamma \in \Gamma$, $c \in \Delta$ and $z \in \overline{\mathbb{C}}$.

Proof Observe that $\tilde{\gamma}_c$ is uniquely determined for all $c \in \Delta$ because Δ^c contains more than two points. To extend $h(c, z)$ to Δ, start with a point $w \in \Delta$. By Theorem 1.1, the motion $h(c, z)$ can be extended to a holomorphic motion (still denote it as $h(c, z)$) of the closed set $\Delta^c \cup \{w\}$. Furthermore, we may extend it to the orbit of w using the Γ-invariant property:

$$h(t, \gamma(w)) = \tilde{\gamma}_c(h(c, w)),$$

for all $\gamma \in \Gamma$. Since every $\gamma \in \Gamma$ is fixed point free on Δ, the motion $h(c, z)$ is well defined and satisfies the Γ-invariant property for all $c \in \Delta$ and all z in the set

$$E = \{\gamma(w) : \gamma \in \Gamma\} \cup (\overline{\mathbb{C}} \setminus \Delta).$$

So we only need to show that $h(c, z)$ is a holomorphic motion of E. Observe first that $h(0, z) = z$ since $\tilde{\gamma}_0 = \gamma$ for all $\gamma \in \Gamma$. To show $h(c, z)$ is injective for all fixed $c \in \Delta$, suppose $h(c, z_1) = h(c, z_2)$ for some $c \in \Delta$.

Since $h(c, z)$ is injective on $\Delta^c \cup \{w\}$, we may assume that $z_1 = g(w)$ for some $g \in \Gamma$. By the Γ-invariant property,

$$h(c, w) = (\tilde{g}_c)^{-1}(h(c, z_1)).$$

Thus,

$$h(c, w) = (\tilde{g}_c)^{-1}(h(c, z_2)) = h(c, g^{-1}(z_2)),$$

and we conclude that z_2 belongs to the Γ-orbit of w. Let $z_2 = \beta(w)$ for some $\beta \in \Gamma$. Then

$$h(c, w) = \tilde{\gamma}_c(h(c, w)),$$

where $\gamma = g^{-1} \circ \beta$. Therefore $h(c, w)$ is a fixed point of $\tilde{\gamma}_c$. On the other hand, since γ is a hyperbolic Möbius transformation, $\tilde{\gamma}_c$ is also hyperbolic, so unless $\tilde{\gamma}_c$ is the identity, it can only fix points on the set $h(c, \partial\Delta)$. Hence γ is the identity map and $z_1 = z_2$.

Finally, we will show that $l : c \to h(c, z)$ is holomorphic for any fixed $z \in E$. We may assume $z = g(w)$, $g \in \Gamma \setminus \{identity\}$. Then $l(c) = h(c, g(w)) = \tilde{g}_c(h(c, w))$. Since $c \to h(c, w)$ is holomorphic and $\tilde{g}_c$ is a Möbius transformation, it is enough to prove the map $k : c \to \tilde{g}_c(\zeta)$ is holomorphic for any fixed ζ. Applying the Γ-invariant property to the three points $0, 1, \infty$, we obtain

$$\tilde{g}_c(0) = h(c, g(0)),$$

$$\tilde{g}_c(1) = h(c, g(1)),$$

$$\tilde{g}_c(\infty) = h(c, g(\infty)).$$

The right-hand sides of these three equations are holomorphic, so the maps $c \mapsto \tilde{g}_c(0)$, $c \mapsto \tilde{g}_c(1)$ and $c \mapsto \tilde{g}_c(\infty)$ are holomorphic. Since $\tilde{g}_c$ is a Möbius transformation, $k : c \to \tilde{g}_c(\zeta)$ is holomorphic.

Therefore, we have extended $h(c, z)$ to a holomorphic motion of

$$\Delta^c \cup \{the \ \Gamma \ orbit \ of \ z\}.$$

By repeating this extension process to a countable set of points whose Γ orbits are dense in Δ, we obtain the extension $H(c, z)$ of $h(c, z)$ with the property that

$$H(c, \gamma(z)) = \tilde{\gamma}_c(H(c, z))$$

for all $\gamma \in \Gamma$, $c \in \Delta$ and $z \in \overline{\mathbb{C}}$. $\square$

This equivariant version of Slodkowski's extension theorem leads almost immediately to the following lifting theorem.

Theorem 9.2 (The lifting theorem) *If $f : \Delta \to T(R)$ is holomorphic, then there exists a holomorphic map $\tilde{f} : \Delta \to \mathcal{M}$ such that*

$$\Phi \circ \tilde{f} = f.$$

If $\mu_0 \in \mathcal{M}$ and $\Phi(\mu_0) = f(0)$, we can choose $\tilde{f}$ such that $\tilde{f}(0) = \mu_0$.

Proof By using the translation mapping α of the Teichmüller space given by

$$\alpha([w^\mu]) = [w^\mu \circ (w^\nu)^{-1}],$$

we may assume $f(0) = 0$. For each $c \in \Delta$, let $g(c, \cdot)$ be a meromorphic function whose Schwarzian derivative is $f(c)$. Then on $\overline{\mathbb{C}} \setminus \Delta$ the map $g(c, \cdot)$ is injective, and we can specify $g(c, \cdot)$ uniquely by requiring that it fix $1, i$ and -1. Thus $g(0, z) = z$. It is easy to verify that

$$g(c, z) : \Delta \times (\overline{\mathbb{C}} \setminus \Delta) \to \overline{\mathbb{C}}$$

is a holomorphic motion. For every $\gamma \in \Gamma$ and $c \in \Delta$, there exists a Möbius transformation $\tilde{\gamma}_c$ such that

$$g(c, \gamma(z)) = \tilde{\gamma}_c(g(c, z)).$$

Using the equivalent version of Slodkowski's extension theorem, we extend g to a Γ-invariant holomorphic motion (still denote it as g) of $\overline{\mathbb{C}}$. For each $c \in \Delta$, let $\tilde{f}(c)$ be the complex dilatation

$$\tilde{f}(c) = \frac{g_{\bar{z}}}{g_z}.$$

Then the Γ-invariant property of g implies that $\tilde{f}(c) \in \mathcal{M}$. From Theorem 1.1 in Section 1, we know that $\tilde{f}(c)$ is a holomorphic function of c. By the definition of the Bers embedding, $\Phi(\tilde{f}(c))$ is the Schwarzian derivative of g. So $\Phi(\tilde{f}(c)) = g(c)$. $\qquad\square$

Now we will use the lifting theorem to show that the Teichmüller metric and Kobayashi's metric of $T(R)$ coincide.

Lemma 9.3 *Suppose $\mathcal{M}$ is the unit ball in the space of essentially* *bounded Beltrami differentials on a Riemann surface R. Let d_K be the Kobayashi's metric on $\mathcal{M}$. Then*

$$d_K(\mu, \nu) = 2 \tanh^{-1} \left\| \frac{\mu - \nu}{1 - \overline{\nu}\mu} \right\|_\infty$$

for all μ and ν in $\mathcal{M}$.

Proof From Lemma 7.3, for any $\nu \in \mathcal{M}$,

$$d_K(0, \nu) = 2 \tanh^{-1} \|\nu\|_\infty .$$

Observe the function defined by

$$\lambda \to \frac{\nu - \lambda}{1 - \overline{\nu}\lambda}$$

is a biholomorphic self map of $\mathcal{M}$. Therefore

$$d_K(\mu, \nu) = 2 \tanh^{-1} \left\| \frac{\mu - \nu}{1 - \overline{\nu}\mu} \right\|_\infty .$$

$\square$

Theorem 9.4 ([R], [G1], [G2]) *The Teichmüller's and Kobayashi's metrics of $T(R)$ coincide.*

Proof In Lemma 8.2 we already showed that $d_K \leq d_T$, so we only need to prove $d_K \geq d_T$. Choose a holomorphic map $f : \Delta \to T(R)$ so that $f(0) = 0$ and $f(c) = [\mu]$ for some $c \in \Delta$. Then the lifting theorem implies there exists a holomorphic map $\tilde{f} : \Delta \to \mathcal{M}$ so that

$$\Phi(\tilde{f}(c)) = f(c) = [\mu].$$

So

$$d_K(0, \tilde{f}(c)) \leq \rho_\Delta(0, c).$$

By Lemma 9.3 and definition of Teichmüller metric,

$$d_T(0, [\mu]) \leq d_K(0, \tilde{f}(c)).$$

Therefore,

$$d_T(0, [\mu]) \leq \rho_\Delta(0, c).$$

Taking the infimum over all such f, we have

$$d_T(0, [\mu]) \leq d_K(0, [\mu]).$$

Hence $d_T \leq d_K$. $\qquad\qquad\qquad\qquad\qquad\qquad\qquad\qquad$ $\square$

References

[Ag] Agard, S. Distortion of quasiconformal mappings. *Ann. Acad. Sci. Fenn.* **413** (1968), 1–7.

[A1] Ahlfors, L. V. *Conformal Invariants: Topics in Geometric Function Theory*, McGraw-Hill, New York, 1973.

[A2] Ahlfors, L. V. *Lectures on Quasiconformal Mapping*, University Lecture Series. **38**, *Amer. Math. Soc.*, 2006.

[AB] Ahlfors, L. V. and Bers, L. Riemann's mapping theorem for variable metrics. *Annals of Math.* **72** (1961), 385–404.

[AM] Astala, K. and Martin, G. Papers on analysis, a volume dedicated to Olli Martio on the occasion of his 60th birthday. *Report. Univ. Jyvaskyla* **211** (2001), 27–40.

[BR] Bers, L. and Royden, H. Holomorphic families of injections. *Acta Math.* **157** (1986), 259–286.

[C] Chirka, E. M. On the extension of holomorphic motions. *Doklady Akademii Nauk* **397** (2004), 37–40.

[CR] Chirka, E. M. and Rosay, J.-P. On the extension of holomorphic motions. *Ann. Pol. Math.* **70** (1998), 43–77.

[CH] Courant, R. and Hilbert, D. *Methods of Mathematical Physics*, Wiley-Interscience, New York, 1963.

[D] Douady, A. Prolongements de meuvements holomorphes (d'apres Slodkowski et autres). *Asterisque* **227** (1995), 7–20.

[EKK] Earle, C. J., Kra, I. and Krushkal, S. L. Holomorphic motions and Teichmüller spaces. *Trans. Am. Math. Soc.* **343**(2) (1994), 927–948.

[G1] Gardiner, F. P. Approximation of infinite dimensional Teichmüller spaces. *Trans. Amer. Math. Soc.* **282**(1) (1984), 367–383.

[G2] Gardiner, F. P. *Teichmüller Theory and Quadratic Differentials*, John Wiley & Sons, New York, 1987.

[GK] Gardiner, F. P. and Keen, L. Holomorphic motions and quasi-Fuchsian manifolds. *Contemp. Math., AMS* **240** (1998), 159–174.

[GL] Gardiner, F. P. and Lakic, N. *Quasiconformal Teichmüller Theory*, AMS, Providence, RI, 2000.

[KL] Keen, L. and Lakic, N. *Hyperbolic Geometry from a Local Viewpoint*, Cambridge University Press, Cambridge, 2007.

[LV] Lehto, O. and Virtanen, K. I. *Quasiconformal Mapping in the Plane*, Springer-Verlag, New York and Berlin, 1965.

[L] Li, Z. *Introduction to Complex Analysis* [In Chinese], Peking University Press, Beijing, 2004.

[MSS] Màñé, R. M., Sad, P. and Sullivan, D. On the dynamics of rational maps. *Ann. Éc. Norm. Sup.* **96** (1983), 193–217.

[R] Royden, H. Automorphisms and isometries of Teichmüller space. *Advances in the Theory of Riemann Surfaces* (1971), 369–384.

[S] Slodkowski, Z. Holomorphic motions and polynomial hulls. *Proc. Amer. Math. Soc.* **111** (1991), 347–355.

[ST] Sullivan, D. P. and Thurston, W. P. Extending holomorphic motions. *Acta Math.* **157** (1986), 243–257.

Cutting sequences and palindromes

Jane Gilman[1]
Rutgers University
gilman@andromeda.rutgers.edu

Linda Keen[2]
Lehman College and Graduate Center, CUNY
linda.keen@lehman.cuny.edu

1 Introduction

In this paper we discuss several more or less well-known theorems about primitive and palindromic words in two generator free groups. We describe a geometric technique that ties all of these theorems together and gives new proofs of all but the last of them, which is an enumerative scheme for palindromic words. This geometric approach and the enumerative scheme will be useful in applications. These applications will be studied elsewhere [GK3].

The main object here is a two generator free group which we denote by $G = \langle A, B \rangle$.

Definition 1.1 *A word $W = W(A, B) \in G$ is* primitive *if there is another word $V = V(A, B) \in G$ such that W and V generate G. V is called a* primitive associate *of W and the unordered pair W and V is called a pair of primitive associates.*

We remark that if W, V is a pair of primitive associates then both WV and WV^{-1} are primitive and $W, WV^{\pm 1}$ and $V, WV^{\pm 1}$ are both primitive pairs.

Definition 1.2 *A word $W = W(A, B) \in G$ is a* palindrome *if it reads the same forward and backward.*

In [GK1] we found connections between a number of different forms of primitive words and pairs of primitive associates in a two generator free group. These were obtained using both algebra and geometry. The theorems that we discuss, Theorems 2.9, 2.10 and 2.11, can be found in [GK1] and Theorem 2.13 can be found in Piggott [P]. Theorem 2.14, the enumeration scheme, along with another proof of Theorem 2.13 can be found in [GK2].

There are several different geometric objects that can be associated to two generator free groups; among them are the punctured torus, the three holed sphere and the genus two handlebody. Here we focus on the punctured torus and use "cutting sequences" for simple curves to obtain proofs of Theorems 2.9, 2.10 and 2.13.

A similar treatment can be made using the three holed sphere. Looking at the technique developed in Vidur Malik's thesis [M] for the three holed sphere representation of two generator groups we first noticed that the palindromes and products of palindromes were inherent in the geometry. The concept of a geometric center of a primitive word was inherent in his work. We thank him for his insight.

2 Notation and Definitions

In this section we establish the notation and give the definitions needed to state five theorems and we state them. Note that our statements of these theorems gather together results from several places into one theorem. Thus, for example, a portion of the statements in Theorem 2.11 appears in [KS] while another portion appears in [GK1].

We assume that the reader is familiar with standard group theory terminology as found in Magnus-Karass-Solitar [MKS], but we review some terms here. We give fuller details of well known terms related to continued fraction expansions and Farey sequences as we work intensely with these.

A word $W = W(A, B) \in G$ is an expression $A^{n_1} B^{m_1} A^{n_2} \cdots B^{m_r}$ for some set of $2r$ integers $n_1, ..., n_r, m_1, ..., m_r$.

A word $X_1 X_2 X_3 \cdots X_{n-1} X_n$ in G is *freely reduced* if each X_i is a generator of G and $X_i \neq X_{i+1}^{-1}$. It is *cyclically reduced* if it is freely reduced and $X_1 \neq X_n^{-1}$. We note that given a cyclically reduced word $X_1 X_2 X_3 \cdots X_{n-1} X_n$, a *cyclic permutation* of the word is a word of the form $X_r X_{r+1} \cdots X_n X_1 X_2 X_3 \cdots X_{r-1}$ where $1 \leq r \leq n$. A cyclic permutation of such a word is also cyclically reduced and is conjugate to the initial word.

Our ultimate aim is to characterize the exponents that occur in

$W = W(A, B)$ when W is primitive. These will be called *primitive exponents*. Before we do that we associate to each rational $p/q \in \mathbb{Q}$ $q \neq 0$, where p and q are relatively prime integers, a set of words which we denote by $\{W_{p/q}\}$ as follows:

Definition 2.1 *If $p/q \geq 1$ set*

$$\{W_{p/q}\} = \{B^{n_0} AB^{n_1} AB^{n_2} \ldots AB^{n_q}\} \tag{2.1}$$

where $n_0 \geq 0$, $n_i > 0, i = 1, \ldots, q$ and $p = \sum_{i=1}^q n_i$.

If $p/q \leq -1$, then replace A by A^{-1} in (2.1) and if $|p/q| < 1$, interchange both A and B and p and q in (2.1) so that after the interchanges $q = \sum_{i=1}^p n_i$.

Note that because each of the A exponents have the same sign, and each of the B exponents have the same sign each word in $\{W_{p/q}\}$ must be cyclically reduced and thus, there are no shorter words in its conjugacy class. Also note that in the union of the sets $\{W_{p/q}\}$ over all p/q, we never have both a word and its inverse.

After we recall some elementary number theory in the next section, we will be able to state Theorem 2.11 which gives necessary and sufficient conditions on the exponents n_i so that a particular word $W_{p/q} \in \{W_{p/q}\}$ is primitive.

Using the notation [ ] for the greatest integer function, one immediate corollary of Theorem 2.11 is

Corollary 2.2 *If $W_{p/q} \in \{W_{p/q}\}$ is primitive, then the primitive exponents satisfy $n_j = [|p/q|]$ or $n_j = [|p/q|] + 1$, $0 < j \leq p$.*

Another immediate corollary of Theorem 2.11 is

Corollary 2.3 *The primitive word $W_{p/q} \in \{W_{p/q}\}$ is unique up to cyclic permutation and inverse.*

In what follows, we will pick unique representatives for elements in the set $\{W_{p/q}\}$ using iteration. There are three iteration schemes, which will be termed the W-iteration, the V-iteration and the E-iteration schemes. The W-iteration scheme is patterned on Farey sequences. We will show: We can characterize pairs of primitive associates as follows:

Corollary 2.4 *Two primitive words $W_{p/q}$ and $W_{r/s}$ that occur in the W-iteration scheme are a pair of primitive associates if and only if $|ps - qr| = 1$.*

We will see that this corollary also holds for the V- and E-iteration schemes.

2.1 Farey Arithmetic

In the rest of this section we work only with rationals $r/s \in [0, \infty]$ where we assume $r, s \geq 0$, but not simultaneously 0, and that they are relatively prime, which we denote by $(r, s) = 1$. We use the notation $1/0$ to denote the point at infinity. We use $\mathbb{Q}^+$ to denote these rational numbers and we think of them as points on the real axis in the extended complex plane.

To state the second theorem, we need the concept of Farey addition for fractions in $\mathbb{Q}^+$.

Definition 2.5 *If $\frac{p}{q}, \frac{r}{s} \in \mathbb{Q}^+$, the* Farey sum *is*

$$\frac{p}{q} \oplus \frac{r}{s} = \frac{p+r}{q+s}.$$

Two fractions are called Farey *neighbors or simply* neighbors *if $|ps - qr| = 1$; the corresponding words are also called neighbors.*

Note that the Farey neighbors of $1/0$ are the rationals $n/1$. If $\frac{p}{q} < \frac{r}{s}$ then it is a simple computation to see that

$$\frac{p}{q} < \frac{p}{q} \oplus \frac{r}{s} < \frac{r}{s}$$

and that both pairs of fractions

$$(\frac{p}{q}, \frac{p}{q} \oplus \frac{r}{s}) \text{ and } (\frac{p}{q} \oplus \frac{r}{s}, \frac{r}{s})$$

are neighbors if $(p/q, r/s)$ are.

We can create the diagram for the Farey tree by marking each fraction by a point on the real line and joining each pair of neighbors by a semi-circle orthogonal to the real line in the upper half plane. The points $n/1$ are joined to their neighbor $1/0$ by vertical lines. The important thing to note here is that because of the properties above none of the semi-circles or lines intersect in the upper half plane. To simplify the exposition when we talk about a point or a vertex, we also mean the word corresponding to that rational number. Each pair of neighbors together with their Farey sum form the vertices of a curvilinear or hyperbolic triangle and the interiors of two such triangles are disjoint. Together the set of these triangles forms a tessellation of the hyperbolic plane which is known as the Farey tree.

Fix any point ζ on the positive imaginary axis above i. Given a fraction, $\frac{p}{q}$, there is a hyperbolic geodesic γ from ζ to $\frac{p}{q}$ that intersects a minimal number of these triangles.

Definition 2.6 *The* Farey level *or the* level *of* p/q, $Lev(p/q)$, *is the number of triangles traversed by* γ.

Note that the curve (line) γ joining ζ to either $0/1$ or $1/0$ does not cross any triangle so these rationals have level 0. The geodesic joining ζ to $1/1$ intersects only the triangle with vertices $1/0, 0/1$ and $1/1$ so the level of $1/1$ is 1. Similarly the level of $n/1$ is n.

We emphasize that we now have two different and independent orderings on the rational numbers: the ordering as rational numbers and the ordering by level. That is, given $\frac{p}{q}$ and $\frac{r}{s}$, we might, for example, have as rational numbers $\frac{p}{q} \leq \frac{r}{s}$, but $Lev(\frac{r}{s}) \leq Lev(\frac{p}{q})$. If we say one rational is larger or smaller than the other, we are referring to the standard order on the rationals. If we say one rational is higher or lower than the other, we are referring to the *levels* of the fractions.

Definition 2.7 *We determine a* Farey sequence *for* $\frac{p}{q}$ *inductively by choosing the new vertex of the next triangle in the sequence of triangles traversed by* γ.

The Farey sequence for $\frac{8}{5}$ is shown in Figure 2.1.

Given p/q, we can find the smallest and largest rationals m/n and r/s that are its neighbors. These also have the property that they are the only neighbors with lower level. That is, as rational numbers $m/n < p/q < r/s$ and the levels satisfy $Lev(m/n) < Lev(p/q)$ and $Lev(r/s) < Lev(p/q)$, and if u/v is any other neighbor $Lev(u/v) > Lev(p/q)$. Thinking of the Farey diagram as a tree whose nodes are the triangles of the diagram leads us to define

Definition 2.8 *The smallest and the largest neighbors of the rational* p/q *are the* parents *of* p/q.

Note that we can tell which parent r/s is smaller (respectively larger) than p/q by the sign of $rq - ps$.

Farey sequences are related to continued fraction expansions of

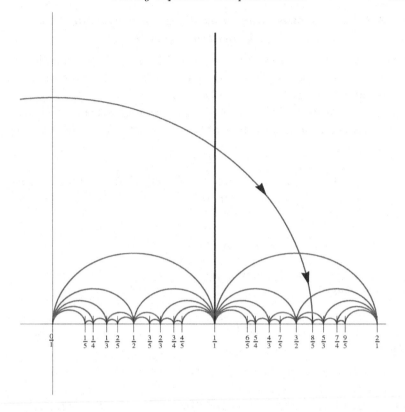

Fig. 2.1. Farey sequence for $8/5$ is $1/1, 2/1, 3/2, 5/3, 8/5$

fractions (see for example, [HW]). In particular, write

$$\frac{p}{q} = a_0 + \cfrac{1}{a_1 + \cfrac{1}{a_2 + \cfrac{1}{a_3 + \cdots + \frac{1}{a_k}}}} = [a_0, \ldots, a_k]$$

where $a_0 \geq 0$, $a_j > 0$, $j = 1, \ldots, k$.

The level of p/q can be expressed in terms of the continued fraction expansion by the formula

$$Lev(p/q) = \sum_{j=0}^{k} a_j.$$

The parents of p/q have continued fractions

$$[a_0, \ldots, a_{k-1}] \text{ and } [a_0, \ldots, a_{k-1}, a_k - 1].$$

2.2 *Farey Words, Continued Fraction Expansions and Algorithmic Words*

In this section we relate primitive words with their associated rationals p/q to Farey sequences. We tacitly assume $p/q \in \mathbb{Q}^+$; for negative p/q we use the transformation rules in Definition 2.1.

The next theorem gives a recursive enumeration scheme for primitive words using Farey sequences of rationals. This is the W-iteration scheme.

Theorem 2.9 (The W-iteration scheme) ([GK1], [KS]) *The primitive words in $G = \langle A, B \rangle$ can be enumerated inductively by using Farey sequences as follows: set*

$$W_{0/1} = A, \quad W_{1/0} = B.$$

Given p/q, consider its Farey sequence. Let $\frac{m}{n}$ and $\frac{r}{s}$ be its parents labeled so that

$$\frac{m}{n} < \frac{p}{q} < \frac{r}{s}.$$

Then

$$W_{\frac{p}{q}} = W_{\frac{m}{n} \oplus \frac{r}{s}} = W_{r/s} \cdot W_{m/n}.$$

A pair $W_{p/q}, W_{r/s}$ is a pair of primitive associates if and only if $\frac{p}{q}, \frac{r}{s}$ are neighbors, that is, $|ps - qr| = 1$.

Because a product of primitive associates is always primitive, it is clear that the words we obtain in the theorem are primitive. As we will see when we give the proofs of the theorems, the words $W_{p/q}$ in the theorem belong to the set $\{W_{p/q}\}$ we defined above so the notation is consistent.

We note that the two products $W_{m/n} \cdot W_{r/s}$ and $W_{r/s} \cdot W_{m/n}$ are always conjugate in G. In this W-iteration scheme we always choose the product where the larger subscript comes first. The point is that in order for the scheme to work the choice has to be made consistently. We emphasize that $W_{p/q}$ always denotes the word obtained using this enumeration scheme.

The other enumeration schemes are the V-scheme where the words come from the algorithm discussed below and the E-enumeration scheme that gives the words in a palindromic form or a canonical product palindromic form. While it might seem that we have introduced more notation here than necessary, we have done so because we want to emphasize the different ways in which the primitive words can arise.

The $W_{p/q}$ words can be rewritten using the entries in the continued fraction expansion instead of the primitive exponents. Since this is the form in which the words arise in the $PSL(2,\mathbb{R})$ discreteness algorithm [G1], [G2], [J], [M], the rewritten form is also known as the *algorithmic form* of the primitive words.

The algorithm begins with a pair of generators (X_0, Y_0) for a subgroup of $PSL(2,\mathbb{R})$ and runs through a sequence of primitive pairs of generators. At each step the algorithm either replaces a generating pair (X, Y) with a new generating pair (XY, Y), called a "linear step" or it replaces a generating pair with the pair (Y, XY) called a "Fibonacci step". There can be many linear steps between Fibonacci steps; the replacement pair after N linear steps is $(X^N Y, X)$. Each Fibonacci step, however, occurs between two linear steps. The algorithm thus generates new pairs of primitive associates at each step, and as we will see, the entries of the continued fraction are visible in the form of the words generated.

The main point of the discreteness algorithm, however, is not only to generate new pairs of primitive generators but to decide whether a particular pair of matrices generates a discrete group. Therefore, at each step a "stopping condition" is tested for and when it is reached, the algorithm prints out either: *the group is discrete* or *the group is not discrete*. The pair of generators given by the algorithm when it stops are called the *stopping generators*. Associated to any implementation of the algorithm is a sequence of integers, the F-sequence, or Fibonacci sequence, which gives the number of linear steps that occur between consecutive Fibonacci steps at the point the algorithm stops. The F-sequence is usually denoted by $[a_1, ..., a_k]$. The stopping generators can be determined from the F-sequence.

The algorithm can be run backwards from the stopping generators when the group is discrete and free, and any primitive pair can be obtained from the stopping generators using the backwards F-sequence. The F-sequence of the algorithm is used in [G1] and [J] to determine the computational complexity of the algorithm. In [G1] it is shown that most forms of the algorithm run in polynomial time and in [J] it is shown that all forms do.

The entries in the F-sequence are also called the *algorithmic exponents*. In [GK1] it is shown that they are also the entries in the continued fraction expansion of the rational corresponding to that primitive word determined by the algorithm — hence the same notation. Some of the theorems below are stated in [GK1] with the term F-sequence instead of the equivalent term, the continued fraction expansion.

The next theorem exhibits the primitive words with the continued fraction expansion exponents in their most concise form.

If $p/q = [a_0, \ldots, a_k]$, set $\frac{p_j}{q_j} = [a_0, \ldots, a_j]$, $j = 0, \ldots, k$. These fractions, called the *approximants* of p/q, can be computed recursively from the continued fraction for p/q as follows:

$$p_0 = a_0, q_0 = 1 \text{ and } p_1 = a_0 a_1 + 1, q_1 = a_1$$

$$p_j = a_j p_{j-1} + p_{j-2}, \ q_j = a_j q_{j-1} + q_{j-2}, \quad j = 2, \ldots, k.$$

Theorem 2.10 ([GK1]) *If $[a_0, \ldots, a_k]$ is the continued fraction expansion of p/q, the primitive word $W_{p/q}$ can be written inductively using the continued fraction approximants $p_j/q_j = [a_0, \ldots, a_j]$.*
   Set

$$W_{0/1} = A, \ W_{1/0} = B \ and \ W_{1/1} = BA.$$

For $j = 1, \ldots, k$ if $p_{j-2}/q_{j-2} > p/q$ set

$$W_{p_j/q_j} = W_{p_{j-2}/q_{j-2}} \left(W_{p_{j-1}/q_{j-1}} \right)^{a_j}$$

and set

$$W_{p_j/q_j} = \left(W_{p_{j-1}/q_{j-1}} \right)^{a_j} W_{p_{j-2}/q_{j-2}}$$

otherwise.

The relationship between the Farey sequence and the approximants of p/q is that the Farey sequence contains the approximating fractions as a subsequence. The points of the Farey sequence between $\frac{p_j}{q_j}$ and $\frac{p_{j+1}}{q_{j+1}}$ have continued fraction expansions

$$[a_0, a_1, \ldots, a_j + 1], [a_0, a_1, \ldots, a_j + 2], \ldots, [a_0, a_1, \ldots, a_j + a_{j+1} - 1].$$

As real numbers, the approximants are alternately larger and smaller than $\frac{p}{q}$. The number a_j counts the number of times the new endpoint in the Farey sequence lies on one side of the old one. Theorem 2.10 is thus an easy consequence of Theorem 2.9.

We have an alternative recursion scheme which reflects the form of the words as they show up in the algorithm; this is not quite the same as the form in Theorem 2.10. Words at linear steps are formed as in

Theorem 2.10 but words that occur at Fibonacci steps are slightly different; they are a cyclic permutation of the words in Theorem 2.10. We use the notation V for these words. To define the V-scheme, assume without loss of generality that $p/q > 1$ and write $p/q = [a_0, \ldots, a_k]$. By assumption $a_0 > 0$. Set $V_{-1} = B$ and $V_0 = W_{p_0/q_0} = AB^{a_0}$. Then for $j = 1, \ldots, k$, set

$$V_j = V_{j-2}[V_{j-1}]^{a_j}.$$

In the V words we see the entries of the continued fraction occurring as exponents. When we expand the word as a product of A's and B's, we see the primitive exponents. The next theorem gives us formulas for the primitive exponents in terms of the entries of the continued fraction expansion and hence necessary and sufficient conditions to recognize primitive words in $\{W_{p/q}\}$.

The notation here is a bit cumbersome because there are several indices to keep track of. When we expand the word V_j as a product of A's and B's, we need to keep track both of the index j and the primitive exponents of V_j so we denote them $n_i(j)$.

Theorem 2.11 ([GK1], [M]) *Given $p/q > 1$, expand the words V_j into*

$$V_j = B^{n_0(j)} AB^{n_1(j)} \ldots AB^{n_{t_j}(j)}$$

for $j = 0, \ldots, k$. Then $V_k \in \{W_{p/q}\}$; it is primitive and its exponents are related to the continued fraction of $p/q = [a_0, \ldots, a_k]$ or the F-sequence as follows:

If $j = 0$, then $t_0 = 1 = q_0$, $n_0 = 0$ and $n_1(0) = a_0$.

If $j = 1$, then $t_1 = a_1 = q_1$, $n_0(1) = 1$ and $n_i(1) = a_0$, $i = 1, \ldots, t_1$.
If $j = 2$ then $t_2 = a_2 a_1 + 1 = q_2$, $n_0(2) = 0$ and for $i = 1, \ldots, t_2$, $n_i(2) = a_0 + 1$ if $i \equiv 1 \mod t_2$ and $n_i(j) = a_0$ otherwise.
For $j > 2, \ldots, k$,

- *$n_0(j) = 0$ if j is even and $n_0(j) = 1$ if j is odd.*
- *$t_j = q_j$.*
- *For $i = 1 \ldots t_{j-2}$, $n_i(j) = n_i(j-2)$.*
- *For $i = t_{j-2} + 1 \ldots t_j$, $n_i(j) = a_0 + 1$ if $i \equiv 1 + t_{k-2} \mod t_{k-1}$ and $n_i(j) = a_0$ otherwise.*

Moreover any word in $\{W_{p/q}\}$ whose exponents, up to cyclic permutation, satisfy the above conditions is primitive.

There is an ambiguity because $[a_0, \ldots, a_k] = [a_0, \ldots, a_k - 1, 1]$ and these continued fractions give different words. One, however, is a cyclic permutation of the other.

Using the recursion formulas for the exponents in the theorem, we obtain a new proof of the following corollary which was originally proved in [BS]. We omit the proof as it is a fairly straightforward induction argument on the Farey level.

Corollary 2.12 ([BS]) *In the expression for a primitive word $W_{p/q}$, for any integer m, $0 < m < p$, the sums of any m consecutive primitive exponents n_i differ by at most ± 1.*

We note that the authors in [BS] also prove that this exponent condition is also sufficient for a word to be primitive.

The following theorem was originally proved in [OZ] and in [P] and [KR].

Theorem 2.13 ([GK2], [KR], [OZ], [P]) *Let $G = \langle A, B \rangle$ be a two generator free group. Then any primitive element $W \in G$ is conjugate to a cyclic permutation of either a palindrome in A, B or a product of two palindromes. In particular, if the length of W is $p + q$, then, up to cyclic permutation, W is a palindrome if and only if $p + q$ is odd and is a product of two palindromes otherwise.*

We note that this can be formulated equivalently using the parity of pq which is what we do below.

In the pq odd case, the two palindromes in the product can be chosen in various ways. We will make a particular choice in the next theorem.

2.3 E-Enumeration

The next theorem, proved in [GK2], gives yet another enumeration scheme for primitive words, again using Farey sequences. The new scheme to enumerate primitive elements is useful in applications, especially geometric applications. These applications will be studied elsewhere [GK3]. Because the words we obtain are cyclic permutations of the words $W_{p/q}$, we use a different notation for them; we denote them as $E_{p/q}$.

Theorem 2.14 (The E-iteration scheme) ([GK2]) *The primitive elements of a two generator free group can be enumerated recursively using* *their Farey sequences as follows. Set*

$$E_{0/1} = A, \ E_{1/0} = B, \ \text{and } E_{1/1} = BA.$$

Given p/q with parents $m/n, r/s$ such that $m/n < r/s$,

- *if pq is odd, set $E_{p/q} = E_{r/s} E_{m/n}$ and*
- *if pq is even, set $E_{p/q} = E_{m/n} E_{r/s}$. In this case $E_{p/q}$ is the unique palindrome cyclically conjugate to $W_{p/q}$.*

$E_{p/q}$ and $E_{p'/q'}$ are primitive associates if and only if $pq' - qp' = \pm 1$.

Note that when pq is odd, the order of multiplication is the same as in the enumeration scheme for $W_{p/q}$ but when pq is even, it is reversed. This theorem says that if pq is even, $E_{p/q}$ is the unique palindrome cyclically conjugate to $W_{p/q}$. If pq is odd, then $E_{p/q}$ is a product of a uniquely determined pair of palindromes.

In this new enumeration scheme, Farey neighbors again correspond to primitive pairs but the elements of the pair $(W_{p/q}, W_{p'/q'})$ are not necessarily conjugate to the elements of the pair $(E_{p/q}, E_{p'/q'})$ by the same element of the group. That is, they are not necessarily conjugate as pairs.

3 Cutting Sequences

We represent G as the fundamental group of a punctured torus and use the technique of *cutting sequences* developed by Series (see [Ser], [KS], [N]) as the unifying theme. This representation assumes that the group G is a discrete free group. Cutting sequences are a variant on Nielsen boundary development sequences [N]. In this section we outline the steps to define cutting sequences.

- It is standard that $G = \langle A, B \rangle$ is isomorphic to the fundamental group of a punctured torus S. Each element of G determines a free homotopy class of curves on S. The primitive elements are free homotopy classes of simple curves that do not simply go around the puncture. Primitive pairs are classes of simple closed curves with a single intersection point.
- Let $\mathcal{L}$ be the lattice of points in $\mathbb{C}$ of the form $m + ni$, $m, n \in \mathbb{Z}$ and let $\mathcal{T}$ be the corresponding lattice group generated by $a = z \mapsto z + 1, b = z \mapsto z + i$. The (unpunctured) torus is homeomorphic to the quotient $\mathbb{T} = \mathbb{C}/\mathcal{T}$. The horizontal lines map to longitudes and the vertical lines to meridians on $\mathbb{T}$.

The punctured torus is homeomorphic to the quotient of the plane punctured at the lattice, $(\mathbb{C} \setminus \mathcal{L})/\mathcal{T}$. Any curve in $\mathbb{C}$ whose endpoints are identified by the commutator $aba^{-1}b^{-1}$ goes around a puncture and is no longer homotopically trivial.

- The simple closed curves on $\mathbb{T}$ are exactly the projections of lines joining pairs of lattice points (or lines parallel to them). These are lines $L_{p/q}$ of rational slope p/q. The projection $l_{p/q}$ consists of p longitudinal loops and q meridional loops. We assume that p and q are relatively prime; otherwise the curve has multiplicity equal to the common factor.

  For the punctured torus, any line of rational slope, not passing through the punctures projects to a simple closed curve and any simple closed curve, not enclosing the puncture, lifts to a curve freely homotopic to a line of rational slope.

- Note that, in either case, if we try to draw the projection of $L_{p/q}$ as a simple curve, the order in which we traverse the loops on $\mathbb{T}$ (or S) matters. In fact there is, up to cyclic permutation and reversal, only one way to draw the curve. We will find this way using cutting sequences. This is the content of Theorem 2.11. Below, we assume we are working on $\mathbb{T}$.

- Choose as fundamental domain (for S or $\mathbb{T}$) the square D with corners (puncture points) $\{0, 1, 1 + i, i\}$. Label the bottom side B and the top side $\bar{B}$; label the left side A and the right side $\bar{A}$. Note that the transformation a identifies A with $\bar{A}$ and b identifies B with $\bar{B}$. Use the translation group to label the sides of all copies of D in the plane.

- Assume for simplicity that $p/q \geq 0$. We will indicate how to modify the discussion otherwise. Choose a fundamental segment of the line $L_{p/q}$ and pick one of its endpoints as initial point. It passes through $p + q$ copies of the fundamental domain. Call the segment in each copy a strand.

  Because the curve is simple, there will either be "vertical" strands joining the sides B and $\bar{B}$, or "horizontal" strands joining the sides A and $\bar{A}$, but not both.

  Call the segments joining a horizontal and vertical side corner strands. There are four possible types of corner strands: from left to bottom, from left to top, from bottom to right, from top to right. If all four types were to occur, the projected curve would be trivial on $\mathbb{T}$. There cannot be only one or three different types of corner strands because the curve would not close up. Therefore the only corner strands occur on one pair of opposite corners and there are an equal number on each corner.

- Traversing the fundamental segment from its initial point, the line goes through or "cuts" sides of copies of D. We will use the side labeling to define a *cutting sequence* for the segment. Since each side belongs to two copies it has two labels. We have to pick one of these labels in a consistent way. As the segment passes through, there is the label from the copy it leaves and the label from the copy it enters. We always choose the label from the copy it enters. Note that the cyclic permutation depends on the starting point.

- If $p/q > 1$, the resulting cutting sequence will contain p B's and q A's and there will be $p - q$ horizontal strands and q corner strands; if $0 \le p/q < 1$, the resulting cutting sequence will contain q B's, p A's and there will be $q - p$ vertical strands and p corner strands. If $p/q < 0$ we either replace A by $\bar{A}$ or B by $\bar{B}$. We identify the cutting sequence with the word in G interpreting the labels A, B and the generators and the labels $\bar{A}, \bar{B}$ as their inverses.

- Given an arbitrary word $W = A^{m_1} B^{n_1} A^{m_2} B^{n_2} \ldots A^{m_p} B^{n_p}$ in G, we form a cutting sequence for it by drawing strands from the word through successive copies of D. We then draw translates of the resulting strands by elements of the lattice group so that they all lie in the fundamental domain D. The strands in D are all disjoint up to homotopy if and only if the word is primitive.

Let us illustrate with three examples. In the first two, we draw the cutting sequences for the fractions $\frac{p}{q} = \frac{1}{1}$ and $\frac{p}{q} = \frac{3}{2}$. In the third, we construct the cutting sequence for the word $A^2 B^3$. See Figures 3.1 and 3.2.

- A fundamental segment of $L_{1/1}$ can be chosen to begin at a point on the left (A) side and pass through D and the adjacent copy above D. There will be a single corner strand connecting the A side to a $\bar{B}$ side and another connecting a B side to an $\bar{A}$ side.

  To read off the cutting sequence begin with the point on A and write A. Then as we enter the next (and last) copy of D we have a B side. The word is thus AB.

  Had we started on the bottom, we would have obtained the word BA.

- A fundamental segment of $L_{3/2}$ passes through 5 copies of the fundamental domain. (See Figure 3.2.) There is one "vertical" segment joining a B and a $\bar{B}$, two corner segments joining an A and a $\bar{B}$ and

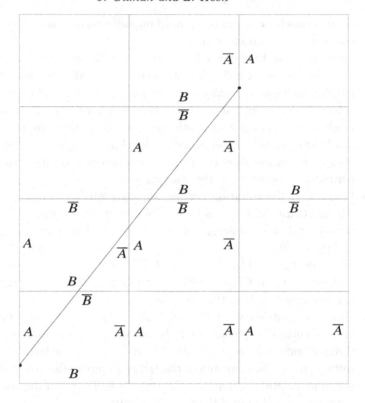

Fig. 3.1. The cutting sequence construction

two joining the opposite corners. Start on the left side. Then, depending on where on this side we begin we obtain the word *ABABB* or *ABBAB*.

If we start on the bottom so that the vertical side is in the last copy we encounter we get *BABAB*.

- To see that the word *AABBB* cannot correspond to a simple loop, draw a vertical line of length 3 and join it to a horizontal line of length 2. Translate it one to the right and one up. Clearly the translate intersects the curve and projects to a self-intersection on the torus. This will happen whenever the horizontal segments are not separated by a vertical segment.

Another way to see this is to try to draw a curve with two meridian loops and three longitudinal loops on the torus. You will easily find that if you try to connect them arbitrarily the strands will cross on 𝕋,

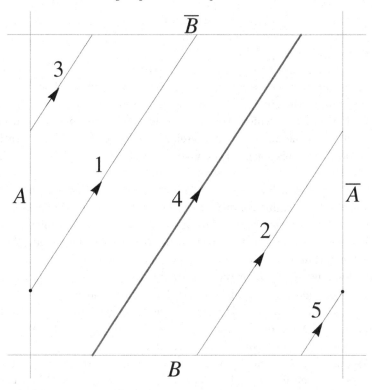

Fig. 3.2. The cutting sequence in a single fundamental domain

but if you use the order given by the cutting sequence they will not. Start in the middle of the single vertical strand and enter a letter every time you come to the beginning of a new strand. We get $BABAB$.

- Suppose $W = B^3 A^2$. To draw the cutting sequence, begin on the bottom of the square and, since the next letter is B again, draw a vertical strand to a point on the top and a bit to the right. Next, since we have a third B, in the copy above D draw another vertical strand to the top and again go a bit to the right. Now the fourth letter is an A so we draw a corner strand to the right. Since we have another A we need to draw a horizontal strand. We close up the curve with a last corner strand from the left to the top.

Because we have both horizontal and vertical strands, the curve is not simple and the word is not primitive.

4 Proofs

Proof of Theorem 2.11.

A word is primitive if and only if its cutting sequence defines a curve with no intersecting strands in the fundamental domain. The word corresponding to the cutting sequence is determined up to cyclic permutation, depending on which strand is chosen as the first one, and inverse, depending on the direction. It therefore corresponds to a line of rational slope p/q. This means that given a word in $\{W_{p/q}\}$ we can check whether it is primitive by drawing its cutting sequence.

We start with a line of slope p/q and find its cutting sequence. This process will determine conditions on the form of the corresponding word that will be necessary and sufficient for the word to be primitive.

The cases $p/q = 0/1, 1/0$ are trivial. We suppose first that $p/q \geq 1$ and treat that case. The other cases follow in the same way, either interchanging A and B or replacing B by $\overline{B}$.

Set $p/q = [a_0, \ldots, a_k]$. Since $p/q > 1$ we know that $a_0 > 0$. Note that there is an ambiguity in this representation; we have $[a_0, \ldots, a_k - 1, 1] = [a_0, \ldots, a_k]$. We can eliminate this by assuming $a_k > 1$. With this convention, the parity of k is well defined.

Assume first k is even; choose as starting point the lowest point on an A side. The line $L_{p/q}$ has slope at least 1 so there will be at least one vertical strand and no horizontal strands. Because there are no horizontal strands, we must either go up or down; assume we go up. The first letter in the cutting sequence is A and since the strand must be a corner strand, the next letter is B. As we form the cutting sequence we see that because there are no horizontal strands, no A can be followed by another A. Because we always go up or to the right, no strand begins with $\overline{B}$. Because we started at the lowest point on A, the last strand we encounter before we close up must start at the rightmost point on a B side. Since there are $p + q$ strands, this means the sequence, and hence the word belongs to $\{W_{p/q}\}$. Since we begin with an A, $n_0 = 0$ and the word looks like

$$AB^{n_1}AB^{n_2}A\ldots B^{n_q}, \quad \sum_{i=1}^{q} n_i = p$$

with all $n_i > 0$. If we use the translation group to put all the strands into one fundamental domain, the endpoints of the strands on the sides are ordered. We see that if we are at a point on the B side, the next time we come to the B side we are at a point that is q to the right $\mod(p)$.

Let us see exactly what the exponents are. Since we began with the lowest point on the left, the first B comes from the qth strand on the bottom. There are p strands on the bottom; the first (leftmost) $p - q$ strands are vertical and the last q are corner strands. Since we move to the right p strands at a time, we can do this $a_0 = [p/q]$ times. The word so far is AB^{a_0}.

At this point we have a corner strand so the next letter will be an A. Define r_1 by $p = a_0 q + r_1$. The corner strand ends at the right endpoint $r_1 + 1$ from the bottom and the corresponding corner strand on the A side joins with the $(q - r_1)^{th}$ vertical strand on the bottom. We again move to the right q strands at a time, a_0 times, while $a_0 q - r_1 > p - q$. After repeating this some n times, $a_0 q - r_1 \leq p - q$. This number, n, will satisfy $q = r_1 n + r_2$ and $r_2 < r_1$. Notice that this is the first step of the Euclidean algorithm for the greatest common denominator of p and q and it generates the continued fraction coefficients at each step. Thus $n = a_1$ and the word at this point is $[AB^{a_0}]^{a_1}$. Since we are now at a corner strand, the next letter is an "extra" B. We repeat the sequence we have already obtained a_3 times where $r_1 = a_3 r_2 + r_3$ and $r_3 < r_2$. The word at this point is $[AB^{a_0}]^{a_1} B[[AB^{a_0}]^{a_1}]^{a_3}$ which is the word V_3 in the V-recursion scheme.

We continue in this way. We see that the Euclidean algorithm tells us that each time we have an extra B, the sequence up to that point repeats as many times as the next a_i entry in the continued fraction expansion of p/q. When we come to the last entry a_k, we have used all the strands and are back to our starting point. The exponent structure is thus forced on us by the number p/q and the condition that the strands not intersect.

If we had intersecting strands we might have a horizontal strand violating the condition on the A exponents. If a vertical strand intersected either another vertical strand or a corner strand, or if two corner strands intersected, the points on the top or bottom would not move along q at a time and the exponents wouldn't satisfy the rules above.

If k is odd, we begin the process at the rightmost bottom strand and begin the word with B and obtain the recursion.

Note that had we chosen a different starting point we would have obtained a cyclic permutation of $W_{p/q}$, or, depending on the direction, its inverse.

For $0 < p/q < 1$ we have no vertical strands and we interchange the roles of A and B. We use the continued fraction $q/p = [a_0, \ldots, a_k]$ and argue as above, replacing "vertical" by "horizontal".

For $p/q < 0$, we replace A or B by $\bar{A}$ or $\bar{B}$ as appropriate.

Thus, we have proved that if we have a primitive word its exponents satisfy the conditions in Theorem 2.11 up to cyclic permutation. In addition, if the exponents n_i of a word $W \in \{W_{p/q}\}$, or some cyclic permutation of it, do not satisfy the exponent conditions, the strands of its cutting sequence must either intersect somewhere or they do not close up properly and the word is not primitive. The conditions are therefore both necessary and sufficient for the word to be primitive and Theorem 2.11 follows.   □

It is obvious that the only primitive exponents that can occur are a_0 and $a_0 + 1$. This gives the simple necessary conditions of Corollary 2.2.
□

For the proof of Corollary 2.4 we want to see when two primitive words $W_{p/q}$ and $W_{r/s}$ are associates. Note the vectors joining zero with $m+ni$ and $r+si$ generate the lattice $\mathcal{L}$ if and only if $|ps-qr| = 1$; or equivalently, if and only if $(p/q, r/s)$ are neighbors. Fundamental segments of lines $L_{p/q}, L_{r/s}$ correspond to primitive words in $\{W_{p/q}\}$ and $\{W_{r/s}\}$. If we choose words $W_{p/q}$ and $W_{r/s}$ and draw their cutting sequences they both start at a point closest to the left bottom corner. Connecting the strands for each word we see that they become generators for the lattice and hence associates if and only if $(p/q, r/s)$ are neighbors.   □

Proof of Theorems 2.9 and 2.10. Although Theorems 2.9 and 2.10 can be deduced from the proof above, we give independent proofs.

The theorems prescribe a recursive definition of a primitive word associated to a rational p/q. We assume $p/q \geq 0$ and that m/n and r/s are parents with

$$\frac{m}{n} < \frac{p}{q} < \frac{r}{s}.$$

We need to show that if we draw the strands for the cutting sequences of $W_{m/n}$ and $W_{r/s}$ in the same diagram, then the result is the cutting sequence of the product.

Note again that if $r/s, m/n$ are neighbors, the vectors joining zero with $m + ni$ and $r + si$ generate the lattice $\mathcal{L}$. Draw a fundamental segment $s_{m/n}$ for $W_{m/n}$ joining 0 to $m+ni$ and a fundamental segment $s_{r/s}$ for $W_{r/s}$ joining $m + ni$ to $(m + r) + (n + s)i$. The straight line s joining 0 to $(m + r) + (n + s)i$ doesn't pass through any of the lattice points because by the neighbor condition $rn - sm = 1$, $s_{m/n}$ and $s_{r/s}$ generate the lattice. We therefore get the same cutting sequence whether we follow $s_{m/n}$ and $s_{r/s}$ in turn or follow the straight line s. This means that the cutting sequence for $W_{p/q}$ is the concatenation of the cutting sequences of $W_{r/s}$ and $W_{m/n}$ which is what we had to show.

If $p/q < 0$, reflect the lattice in the imaginary axis. This corresponds to applying the rules in Definition 2.1.

We note that proving Theorem 2.10 is just a matter of notation.

□

Proof of Theorem 2.13.

We prove the theorem for $0 < p/q < 1$. The other cases follow as above by interchanging the roles of A and B or replacing B by $\overline{B}$. The idea is to choose the starting point correctly.

Suppose pq is even. We want to show that there is a unique cyclic permutation of $W_{p/q}$ that is a palindrome.

Draw a line of slope p/q. By assumption, there are horizontal but no vertical strands and $p - q > 0$ must be odd. This implies that in a fundamental segment there are an odd number of horizontal strands. In particular, if we pull all the strands of a fundamental segment into one copy of D, one of the horizontal strands is the middle strand. Choose the fundamental segment for the line in the lattice so that it is centered about this middle horizontal strand.

To form the cutting sequence for the corresponding word W, begin at the right endpoint of the middle strand and take as initial point the leftpoint that it corresponds to. Now go to the other end of the middle strand on the left and take as initial point the rightpoint that it corresponds to and form the cutting sequence for a word V. By the symmetry, since we began with a middle strand, V is W with all the A's replaced by $\overline{A}$'s and all the B's replaced by $\overline{B}$'s. Since $V = W^{-1}$, we see that W must be a palindrome. Moreover, since it is the cutting sequence of a fundamental segment of the line of slope p/q, it must be a cyclic permutation of $W_{p/q}$.

Note that since we began with a horizontal strand, the first letter in the sequence is an A and, since it is a palindrome, so is the last letter.

When $q/p > 1$, there are horizontal and no vertical strands, and there is a middle vertical strand. This time we choose this strand and go up and down to see that we get a palindrome. The first and last letters in this palindrome will be B.

If $p/q < 0$, we argue as above but either A or B is replaced by respectively $\overline{A}$ or $\overline{B}$.

□

Heuristic for the enumeration, Theorem 2.14.

The proof of the enumeration theorem, Theorem 2.14, involves purely algebraic manipulations and can be found in [GK2]. We do not reproduce it here but rather give a heuristic geometric idea of the enumeration and the connection with palindromes that comes from the $PSL(2, \mathbb{R})$ discreteness algorithm [G1], [G3].

Note that the absolute value of the trace of an element $X \in PSL(2, \mathbb{R})$, $|\text{trace}(X)|$, is well defined. Recall that X is elliptic if $|\text{trace}(X)| < 2$ and hyperbolic if $|\text{trace}(X)| > 2$. As an isometry of the upper half plane, each hyperbolic element has an invariant geodesic called its *axis*. Each point on the axis is moved a distance $l(X)$ towards one endpoint on the boundary. This endpoint is called the attractor and the distance can be computed from the trace by the formula $\cosh \frac{l(X)}{2} = \frac{1}{2}|\text{trace}(X)|$. The other endpoint of the axis is a repeller.

For convenience we use the unit disk model and consider elements of $PSL(2, \mathbb{R})$ as isometries of the unit disk. In the algorithm one begins with a representation of the group where the generators A and B are (appropriately ordered) hyperbolic isometries of the unit disk. The algorithm applies to any non-elementary representation of the group where the representation is not assumed to be free or discrete. The axes of A and B may be disjoint or intersect. We illustrate the geometric idea using intersecting axes.

If the axes of A and B intersect, they intersect in unique point p. In this case one does not need an algorithm to determine discreteness or non-discreteness as long as the multiplicative commutator, $ABA^{-1}B^{-1}$, is not an elliptic isometry. However, the geometric steps used in determining discreteness or non-discreteness in the case of an elliptic commutator still make sense. We think of the representation as being that of a punctured torus group when the group is discrete and free.

Normalize at the outset so that the translation length of A is smaller than the translation length of B, the axis of A is the geodesic joining -1 and 1 with attracting fixed point 1 and the axis of B is the line joining $e^{i\theta}$ and $-e^{i\theta}$. This makes the point p the origin. Replacing B by its inverse if necessary, we may assume the attracting fixed point of B is $e^{i\theta}$ and $-\pi/2 < \theta \leq \pi/2$.

The geometric property of the palindromic words is that their axes all pass through the origin.

Suppose $(p/q, p'/q')$ is a pair of neighbors with pq and $p'q'$ even and $p/q < p'/q'$. The word $W_{r/s} = W_{p'/q'}W_{p/q}$ is not a palindrome or conjugate to a palindrome. Since it is a primitive associate of both $W_{p'/q'}$ and $W_{p/q}$ the axis of $Ax_{W_{r/s}}$ intersects each of the axes $Ax_{W_{p/q}}$ and $Ax_{W_{p'/q'}}$ in a unique point; denote these points by $q_{p/q}$ and $q_{p'/q'}$ respectively. Thus, to each triangle, $(p/q, r/s, p'/q')$ we obtain a triangle in the disk with vertices $(0, q_{p/q}, q_{p'/q'})$.

The algorithm provides a method of choosing a next neighbor and next associate primitive pair so that at each step the longest side of the triangle is replaced by a shorter side. The procedure stops when the sides are as short as possible. Of course, it requires proof to see that this procedure will stop and thus will actually give an algorithm.

There is a similar geometric description of the algorithm and palindromes in the case of disjoint axes.

Notes

1 This work was partially supported by the Rutgers Research Council, Yale University and the National Science Foundation.
2 This work was partially supported by PSC-CUNY.

Bibliography

[BS] Buser, P. and Semmler, K.-D. The geometry and spectrum of the one-holed torus. *Comment. Math. Helv.* **63**(2) (1988), 259–274.

[CMZ] Cohen, P., Metzler, W. and Zimmermann, B. What does a basis of $F(a, b)$ look like? *Math. Ann.* **257**(4) (1981), 435–445.

[G3] Gilman, Jane. Alternate discreteness tests. *Birman Conf. Proceedings, AMS/IP Studies in Advanced Mathematics* **24** (2001), 41–47.

[G1] Gilman, Jane. Two-generator discrete subgroups of $PSL(2, \mathbf{R})$. *Memoirs of the AMS* **561** (1995).

[G2] Gilman, Jane. Algorithms, complexity and discreteness criteria in $PSL(2, \mathbb{C})$. *Journal D'Analyse Mathematique* **73** (1997), 91–114.

[GK1] Gilman, Jane and Keen, Linda. Word sequence and intersection numbers. *Contemporary Math.* **311**(2002), 331–349.

[GK2] Gilman, Jane and Keen, Linda. Enumerating palindromes and primitives in rank two free groups. Submitted.

[GK3] Gilman, Jane and Keen, Linda. Discreteness Criteria and the Hyperbolic Geometry of Palindromes. *Conformal Geometry and Dynamics* **13** (2009), 76–90.

[HW] Hardy, G.H. and Wright, E. M. *An Introduction to the Theory of Numbers*, Clarendon Press, Oxford, 1938.

[J] Jiang, Yichang. Complexity of the Fuchsian group discreteness algorithm. *Ann. Acad. Sci. Fenn.* **26**. (2001), 375–390.

[KR] Kassel, Christian and Reutenauer, Christophe. Sturmian morphisms, the braid group B_4, Christoffel words and bases of F_2. *Ann. Mat. Pura Appl.* *(4)* **186**(2) (2007), 317–339.

[KS] Keen, Linda and Series, Caroline. Pleating coordinates for the Maskit embedding of Teichmüller space for a punctured torus. *Topology* **32**(4) (1993), 719–749.

[MKS] Magnus, Wilhelm, Karass, Abraham and Solitar, Donald. *Combinatorial Group Theory*, John Wiley & Sons, New York, 1966.

[M] Malik, Vidur. Curves generated on surfaces by the Gilman-Maskit algorithm, PhD Thesis, Rutgers University, Newark, NJ, 2007.

[N] Nielsen, Jacob. Untersuchengen zur Topolie der geschlossenen zweiseitigen Flächen, I, II III. *Acta Math.* **50** (1927), 189–358; **53** (1927), 1–76; **58** (1932), 87–167.

[OZ] Osborne, R. P. and Zieschang, H. Primitives in the free group on two generators. *Invent. Math.* **63**(1) (1981), 17–24.

[P] Piggott, Adam. Palindromic primitives and palindromic bases in the free group of rank two. *J. Algebra* **304**(1) (2006), 359–366.

[S] Series, Caroline. The geometry of Markoff numbers. *Math. Intelligencer* **7**(3) (1985), 20–29.

[W] Wright, David J. Searching for the cusp, in *Spaces of Kleinian Groups*, LMS Lecture Notes **329**, Cambridge University Press, Cambridge, 2004, pp. 1–36.

On a Schottky problem for the singular locus of $\mathcal{A}_5$

Víctor González-Aguilera [1]

Departamento de Matemática
Universidad Técnica Federico Santa María
victor.gonzalez@usm.cl

Abstract

Let Sing$\mathcal{M}_g$ be the singular locus of the moduli space of compact connected Riemann surfaces of genus g and Sing$\mathcal{A}_g$ be the singular locus of the moduli space of principally polarized abelian varieties of dimension g. In this survey: we review the main analogies and differences among the results on automorphisms of Riemann surfaces and principally polarized abelian varieties, we describe and compare the sublocus of Sing$\mathcal{A}_5$ corresponding to Jacobians of Riemann surfaces with the sublocus corresponding to intermediate Jacobians of smooth cubic threefolds and we solve a Schottky problem for the 0-dimensional singular locus of $\mathcal{A}_5$.

1 Introduction

Let S be a compact connected Riemann surface of genus g, Aut(S) its group of biholomorphic automorphisms and $\mathcal{M}_g$ the Riemann moduli space, whose points are conformal equivalence classes of compact connected Riemann surfaces of genus g. By a famous theorem of Hurwitz Aut(S) is of order at most $84(g-1)$. The moduli space $\mathcal{M}_g$ can be realized as the quotient of the Teichmüller space $\mathcal{T}_g$ under the action of the Teichmüller modular group Mod(S), this way $\mathcal{M}_g$ is a normal complex analytic space of dimension $3g-3$. Since the work of Rauch and Oort [Ra], [O1], for $g > 3$ the singular points of $\mathcal{M}_g$ correspond to conformal classes of Riemann surfaces $[S]$ such that Aut(S) is non-trivial. The description of $\mathcal{M}_2$ and its only one singular point is due to Igusa [I]; if $[S] \in \mathcal{M}_3$ is non-hyperelliptic, $[S]$ corresponds to a singular point if and only if Aut$(S) \neq \{I\}$, however if $[S] \in \mathcal{M}_3$ is hyperelliptic then $[S]$ is a singular point if and only if Aut$(S) \neq \mathbb{Z}/2\mathbb{Z}$.

217

In 1966, [H1], W. J. Harvey studied cyclic groups of conformal auto-morphisms of a compact Riemann surface, he determined the minimum genus g of $[S]$ for which $\mathrm{Aut}(S)$ contains a cyclic subgroup of order n and later in 1971, [H2], in a pioneering work, using classes of Fuchsian groups G of a fixed isomorphism type, he described the sublocus of $\mathcal{T}_g$ which consists of those points which are fixed under the action of the Teichmüller modular group $\mathrm{Mod}(S)$. Also later [GH] in a joint work, G. González-Díez and Harvey proved that certain irreducible subvari-eties of the singular locus of $\mathcal{M}_g$ with a fixed prescribed symmetry are unirational.

Let $\mathrm{Sing}\mathcal{M}_g$ be the locus of (curves) compact connected Riemann surfaces of genus g with non-trivial automorphisms; in 1987 M. Cornalba [C] described (for any g) its irreducible components, also in [K] and [G] the authors gave results mainly concerning the isolated points of $\mathrm{Sing}\mathcal{M}_g$.

Let (X, H) be a principally polarized abelian variety (p.p.a.v.) of di-mension g (over $\mathbb{C}$), the biholomorphic automorphisms of the complex torus X could be a non-finite group, nevertheless the group of automor-phisms of X that preserves the polarization H, denoted by $\mathrm{Aut}_H(X)$ is a finite group. We will denote by $\mathcal{A}_g$ the moduli space of p.p.a.v. of dimension g, $\mathcal{A}_g$ is a non-smooth, non-compact analytic space of di-mension $\frac{g(g+1)}{2}$, its singular locus, denoted by $\mathrm{Sing}\mathcal{A}_g$ corresponds to points $[(X, H)] \in \mathcal{A}_g$ such that $\mathrm{Aut}_H(X) \neq \{I, -I\}$, [O2]. The subva-riety $\mathrm{Sing}\mathcal{A}_g$ is reducible, its irreducible components had been partially determined (for any g) in [GMZ1] and its isolated points in [GMZ2].

For any compact connected Riemann surface S of genus g its asso-ciated Jacobian variety $(J(S), \Theta)$ is a p.p.a.v. of dimension g, then a period map $p : \mathcal{M}_g \longrightarrow \mathcal{A}_g$ is determined. Let S and S' be compact con-nected Riemann surfaces of genus g; the well known theorem of Torelli (for Riemann surfaces) asserts that S and S' are isomorphic if and only if its Jacobians $(J(S), \Theta)$ and $(J(S'), \Theta')$ are isomorphic as p.p.a.v. If $g \geq 4$, then $\dim\mathcal{M}_g < \dim\mathcal{A}_g$ and the question that arises is how to characterize the p.p.a.v. that are isomorphic (as p.p.a.v.) to the Jaco-bian of Riemann surfaces. This is the famous Schottky problem; there have been many approaches to this difficult problem and several general solutions, an update can be found in [D]. It is also a natural question to decide when the p.p.a.v. in $\mathrm{Sing}\mathcal{A}_g$ are Jacobians of Riemann surfaces, that is, a restricted Schottky problem for p.p.a.v. with automorphisms.

This survey consists of two sections, in the first we review the main analogies and differences among the results on automorphisms of

Riemann surfaces and the automorphisms of p.p.a.v. Most of the results of that section are relatively well known, and our contribution consists in to set in an explicit form their analogies and differences. The moduli space $\mathcal{A}_5$ contains a 12-dimensional sublocus corresponding to Jacobians of Riemann surfaces S of genus 5, denoted by $\mathcal{J}\mathcal{S}_5$, and a 10-dimensional sublocus corresponding to intermediate Jacobians of smooth cubic hypersurfaces of $\mathbb{P}^4$, denoted by $\mathcal{J}\mathcal{H}_5$. In the second section we describe the irreducible components of the previous subloci, apply some general results from our joint research work with J. M. Muñoz-Porras and A. G. Zamora in a concrete analytic way for this particular case and we solve a concrete Schottky problem for the 0-dimensional components of $\mathrm{Sing}\mathcal{A}_g$.

Professor Harvey participated in the first Iberoamerican Congress in Geometry in Olmué, Chile, during January 1998, from that time he generously and enthusiastically helped the group of Mexican and Chilean mathematicians that began to work around Riemann surfaces and Algebraic Geometry, then for me it is an honour and a great pleasure to contribute with this survey in Professor Harvey's Fest schrift.

2 Automorphisms

Let S be a compact connected Riemann surface of genus g, $\varphi \in \mathrm{Aut}(S)$ an automorphism of prime order p and $\langle \varphi \rangle$ the cyclic group generated by φ,

$$\pi_\varphi : S \to X = S/\langle \varphi \rangle$$

is a branched analytic covering of degree p over a Riemann surface X of genus γ, the ramification points q_1, q_2,..., q_n of π_φ are exactly the images of the fixed points of φ and all the ramification indexes are p, eventually φ can act on S without fixed points (étale covering).

Let $H^0(S, \Omega_S^{q,0})$ be the $\mathbb{C}$-vector space of holomorphic q-differential forms on S, we denote by $H^0_\varphi(S, \Omega_S^{q,0})$ the subspace of holomorphic φ-invariant q-differential forms. The following proposition is well known [Far].

Proposition 2.1

   (i) $\dim(H^0_\varphi(S, \Omega_S^{1,0})) = \mathrm{genus}(X) = \gamma$.

  (ii) *Either $p \leq g+1$ or $p = 2g+1$; in the first case, $n = 4$ and in the second case, $n = 3$.*

 (iii) $\dim(H^0_\varphi(S, \Omega_S^{2,0})) = 3\gamma - 3 + n$.

The automorphism φ induces an invertible $\mathbb{C}$-linear map

$$\varphi^* : H^0(S, \Omega_C^{1,0}) \to H^0(S, \Omega_C^{1,0}).$$

The statement (i) of the proposition gives the genus of the quotient Riemann surface X as the dimension of the $\mathbb{C}$-subspace of the φ^*-invariant 1-differential forms on S.

The statement (ii) of the proposition gives an upper bound for the prime order of an automorphism of S and also states that the only prime order automorphisms of S of order bigger than g, are of order $g + 1$ or $2g + 1$, when these numbers are primes.

Every branched cyclic covering $\pi : S \to X$ of prime order p is defined by an invertible sheaf $\mathcal{L}$ and a divisor $D = a_1 q_1 + a_2 q_2 + \cdots + a_n q_n$ on X such that $\mathcal{L}^p \cong \mathcal{O}_S(D)$, with $1 \le a_i \le p$, where $\{q_1, q_2, ..., q_n\}$ are the branch points of π and the degree of D is a multiple of p, if we denote by:

$$S(p, \gamma; a_1, a_2, ..., a_n) \subset \mathcal{M}_g$$

the sublocus corresponding to Riemann surfaces S of genus g with an automorphism φ of prime order p such that

$$\pi_\varphi : S \to X = S/\langle \varphi \rangle$$

is a branched analytic covering of degree p over a Riemann surface X of genus γ, and $D = a_1 q_1 + a_2 q_2 + \cdots + a_n q_n$ is a divisor on X. Eventually $n = 0$ (étale covering), in which case we will denote the sublocus by $S(p, \gamma; -)$. Then the statement (iii) of the proposition gives the dimension of the family $S(p, \gamma; a_1, a_2, ..., a_n)$.

Now we will state and prove an analogous proposition to the previous one, in the case of p.p.a.v.

A principally polarized abelian variety of dimension g (over $\mathbb{C}$), is a pair (X, H), where $X \cong V/L$ is a g-dimensional complex torus and H is a principal Riemann form on X. Also in a more contemporary form, a polarized abelian variety can be considered as a pair $(X, \mathcal{L}_X)$, where $\mathcal{L}_X$ is a positive definite line bundle on X. If

$$c_1 : H^1(X, \mathcal{O}_X^*) \longrightarrow H^2(X, \mathbb{Z}) \cong \mathrm{Alt}^2(L, \mathbb{Z})$$

is the first Chern class, the connection with the classical point of view is given by the requirement that $c_1(\mathcal{L}_X) \cong \mathrm{Im}(H)$, where $\mathrm{Im}(H)$ denotes the imaginary part of the principal Riemann form H.

The group $\{\varphi \in \mathrm{Aut} X \mid \varphi^*(H) \cong H\}$ is denoted by $\mathrm{Aut}_H(X)$ and is called the automorphisms group of (X, H); for each $\varphi \in \mathrm{Aut}_H(X)$ we will denote by $\varphi_c \in GL_{\mathbb{C}}(V)$ its complex representation and by $\varphi_r \in GL_{\mathbb{Z}}(L)$ its rational representation, a basic result is that the rational representation φ_r is the direct sum of φ_c and its complex conjugate $\overline{\varphi_c}$. We denote by Ξ_{φ} the set of eigenvalues of φ_c and by $|\,\Xi_{\varphi}\,|$ its cardinal. The set of points $x \in X$ such that $\varphi(x) = x$, denoted by $\mathrm{Fix}(\varphi)$, is a closed algebraic subgroup of the complex torus X. The set of l-torsion points of X, denoted by X_l, is a subgroup of X isomorphic to $(\mathbb{Z}/l\mathbb{Z})^{2g}$.

The automorphisms group of a g-dimensional complex torus X could be a non-finite group, but $\mathrm{Aut}_H(X)$ is a finite group as $\mathrm{Aut}(S)$. In fact any $\varphi \in \mathrm{Aut}_H(X)$ preserves the real and imaginary part of the Hermitian form H giving a representation

$$\rho : \mathrm{Aut}_H(X) \longrightarrow Sp(2g, \mathbb{R})$$

where $Sp(2g, \mathbb{R})$ denotes the real symplectic group. As the real part of H is symmetric, $\mathrm{Aut}_H(X)$ is a compact subgroup of $Sp(2g, \mathbb{R})$, and since $\varphi(L) \subseteq L$, we get a representation:

$$\rho : \mathrm{Aut}_H(X) \longrightarrow Sp(2g, \mathbb{Z})$$

therefore $\mathrm{Aut}_H(X)$ is compact and discrete, then is a finite group.

Let $\varphi \in \mathrm{Aut}_H(X)$ be an automorphism of order p (p is a prime number). We can assume that φ_c acting in the complex tangent space T_0X is given by a diagonal matrix

$$\mathrm{diag}(\xi^{i_0}; n_0, \xi^{i_1}; n_1, ..., \xi^{i_k}; n_k)$$

where ξ is the primitive root of the unity of order p, each integer $i_0 < i_1 < \cdots < i_k$ belongs to the set $\{0, 1, 2, ..., p-1\}$, the integers n_j denote the multiplicity of the eigenvalue ξ^{i_j} and satisfy $n_0 + n_1 + \cdots + n_k = g$.

Let $\rho : \mathrm{Aut}_H(X) \longrightarrow GL_{\mathbb{C}}(T_0X)$ be a faithful representation; we will denote by $\mathcal{A}_g(\rho, p) \subset \mathcal{A}_g$ the family of p.p.a.v. of dimension g such that each $[(X, H)] \in \mathcal{A}_g(\rho, p)$ admits an automorphism φ of order p such that its complex representation is given by:

$$\rho(\varphi) = \varphi_c = \mathrm{diag}(\xi^{i_0}; n_0, \xi^{i_1}; n_1, ..., \xi^{i_k}; n_k)$$

We begin by recalling and proving a basic lemma [GMZ1].

Lemma 2.2 *Let (X, H) be an abelian variety of dimension g, $\varphi \in \mathrm{Aut}_H(X)$, $\varphi^p = I$ with p prime and such that n_i, $i \in \{1, 2, ..., p-1\}$, are the dimensions of the eigenspaces associated to the eigenvalues ξ^i. Then there exists a non-negative integer r such that $n_i + n_{p-i} = r$ for* *all $i \in \{1, 2, ..., \frac{p-1}{2}\}$ and*

$$g = n_0 + r\frac{(p-1)}{2}$$

Proof We denote by $\langle \varphi \rangle$ the cyclic group of order p generated by φ, consider the integral representation:

$$\rho_r : \langle \varphi \rangle \to GL_{\mathbb{Z}}(L)$$

and let $R(x)$ be the characteristic polynomial of φ_r; as φ_r is the direct sum of φ_c and its conjugate $\overline{\varphi_c}$:

$$R(x) = (x-1)^{2n_0} \prod_{i=1}^{p-1} (x - \xi^i)^{n_i + n_{p-i}}$$

we have a decomposition of the $\langle \varphi_r \rangle$-module L,

$$L = L_0 \oplus L_1$$

where L_0 is the $\mathbb{Z}$-module of $\langle \varphi_r \rangle$-invariants; the polynomial

$$Q(x) = \prod_{i=1}^{p-1} (x - \xi^i)^{n_i + n_{p-i}}$$

must be the characteristic polynomial of φ_r acting on L_1, but the minimal polynomial of the action of φ_r on L_1 is

$$x^p + x^{p-1} + \cdots + x + 1$$

thus

$$Q(x) = (x^p + x^{p-1} + \cdots + x + 1)^r$$

and there exists a natural number r such that $n_i + n_{p-i} = r$ for all $i \neq 0$ and

$$g = n_0 + r\frac{(p-1)}{2}$$

$\square$

For the prime p we have:

$$p = 1 + \frac{2(g - n_0)}{r}, \qquad 0 \leq n_0 \leq (g-1), \qquad r \geq 1$$

The maximal value of p is attained when $n_0 = 0$ and $r = 1$, thus we have that $p = 2g + 1$. If a p.p.a.v. of dimension g admits an automorphism φ of prime order p with Fix(φ) finite and $g < p$, it follows from the previous equality that $p = g + 1$.

The group $\mathrm{Fix}(\varphi)$ is a closed analytic subgroup of X whose dimension is equal to the multiplicity of 1 as an eigenvalue of φ_c [BL]. As $\mathrm{Fix}(\varphi)$ is finite, using the holomorphic Lefschetz fixed point formula, we can determine the order of $\mathrm{Fix}(\varphi)$; furthermore $\mathrm{Fix}(\varphi)$ is a subgroup of the p-torsion points of the complex torus X.

From the previous lemma we obtain the following proposition, that is the analogue to the statement (ii) of proposition 12:

Proposition 2.3 *Let (X, H) be a p.p.a.v. of dimension g and φ an automorphism of prime order p with $\mathrm{Fix}(\varphi)$ finite and $g < p$ then $p = g + 1$ or $p = 2g + 1$. In the first case the order of $\mathrm{Fix}(\varphi)$ is p^2 and in the second case its order is p.*

Remark 2.4 *Let $\mathcal{H}_g = \{Z \in M_{g \times g}(\mathbb{C}) \mid {}^t Z = Z, \mathrm{Im} Z > 0\}$ be the Siegel space and $Sp(2g, \mathbb{Z})$, the symplectic modular group, acts on $\mathcal{H}_g$ as follows:*

$$Sp(2g, \mathbb{Z}) \times \mathcal{H}_g \to \mathcal{H}_g \qquad (M, Z) \to M \star Z = (AZ + B)(CZ + D)^{-1}$$

where

$$M = \begin{pmatrix} A & B \\ C & D \end{pmatrix}$$

is a matrix in $Sp(2g, \mathbb{Z})$.

The group $Sp(2g, \mathbb{Z})$ acts properly and discontinuously on $\mathcal{H}_g$, the quotient $\mathcal{A}_g \cong \mathcal{H}/Sp(2g, \mathbb{Z})$ with its natural quotient structure is a normal complex analytic space of dimension $\frac{g(g+1)}{2}$ [BL], that parametrizes the isomorphism classes of p.p.a.v. of dimension g.

The family $\mathcal{A}_g(\rho, p) \subset \mathcal{A}_g$ can be described from an algebraic point of view [GMZ1], but in the spirit of this survey we will follow an analytic point of view.

Let (X, H) be a p.p.a.v. of dimension g. There is a choice of a basis $\{v_1, ..., v_g\}$ of the tangent space $T_0 V$ and of a basis $\{\alpha_1, ..., \alpha_g, \beta_1, ..., \beta_g\}$ of L in such a way that the basis $\{v_1, ..., v_g\}$ in terms of the basis $\{\alpha_1, ..., \alpha_g, \beta_1, ..., \beta_g\}$ is given by a matrix $\Omega_X = (I_{g \times g}, Z_X)$, where $Z_X \in \mathcal{H}_g$. The matrix Z_X is called a period matrix for (X, H).

If φ is an automorphism of prime order p of (X, H), $\rho(\varphi) = \varphi_c$ induce an element $\varphi_s \in Sp(2g, \mathbb{Z})$ in such a way that Z_X is fixed under the action of φ_s on $\mathcal{H}_g$. The family $\mathcal{A}_g(\rho, p)$ can be described locally at (X, H) as the subvariety of period matrices fixed under the action of φ_s.

The following proposition, that is the analogue to the statement (iii) of proposition 12 can be obtained, [GMZ1].

Proposition 2.5 *For the family* $\mathcal{A}_g(\rho,p) \subset \mathcal{A}_g$ *we have:*

$$\dim\mathcal{A}_g(\rho,p) = \frac{n_0(n_0+1)}{2} + \sum_{i=1}^{\frac{(p-1)}{2}} n_i n_{p-i}$$

The study of compact connected Riemann surfaces S of genus g admitting an automorphism of maximal prime order $p = 2g+1$ goes back to Lefschetz, and was developed by several researchers, [Le], [Ll], [K], [G]. Those Riemann surfaces arise as branched covers of $\mathbb{P}^1$ of degree p, branched at exactly 3 points. Nowadays those surfaces are called Lefschetz surfaces or Lefschetz curves, they have zero moduli and their number of equivalence classes is given in the following proposition.

Proposition 2.6 *The number of equivalence classes of compact connected Riemann surfaces* $[S]$ *admitting an automorphism of prime order* $p = 2g+1$ *is given by:*

$$\frac{(p+1)}{6}, \qquad 2g+1 = p \equiv 2 \pmod 3$$

$$\frac{(p+5)}{6}, \qquad 2g+1 = p \equiv 1 \pmod 3$$

Some of the previous classes $[S]$ could be contained in other components of $\text{Sing}\mathcal{M}_g$, the number of isolated classes had been calculated for example in [G].

Proposition 2.7 *The number of equivalence classes of compact connected Riemann surfaces* $[S]$ *admitting an automorphism of prime order* $p = 2g+1$, *that are isolated singular points of* $\mathcal{M}_g$ *is given by:*

$$\frac{(p-5)}{6}, \qquad 2g+1 = p \equiv 2 \pmod 3$$

$$\frac{(p-7)}{6}, \qquad 2g+1 = p \equiv 1 \pmod 3$$

Remark 2.8 *The equivalence classes of compact connected Riemann surfaces* $[S]$ *of genus* g *admitting an automorphism of prime order* $p = g+1$ *are 1-dimensional families of* $\text{Sing}\mathcal{M}_g$, *their number was determined in [U], and it is* $\frac{g(g+2)}{24}$ *if* $p > 3$.

Now we present a result for p.p.a.v. that is analogous to the proposition 15.

Let (X, H) be a p.p.a.v. of dimension g and φ an automorphism of prime order p, with $p = 2g + 1$. In a suitable basis of $T_0 X$ we can assume that the complex representation of the automorphism is given by $\rho(\varphi) = \mathrm{diag}(\xi^{i_1}, ..., \xi^{i_g})$ where ξ is a primitive root of the unity of order $p = 2g + 1$. By proposition 14 the family $\mathcal{A}_g(\rho, p) \subset \mathcal{A}_g$ is zero dimensional; furthermore for the set $C = \{i_1, ..., i_g\}$ we have:

$$C \cap (-C) = \emptyset, \qquad C \cup (-C) = \mathbb{F}_p^*$$

and (X, H) is isomorphic to a p.p.a.v. $(\mathbb{Q}(\xi_p) \otimes_{\mathbb{Q}} \mathbb{R}, C)$ of complex multiplication type [BL]. The Galois group of the extension: $\mathbb{Q} \to \mathbb{Q}(\xi_p)$ acts on the different types of complex multiplications.

The components $\mathcal{A}_g(\rho, 2g + 1) \subset \mathrm{Sing}\mathcal{A}_g$ have been characterized in [GMZ2].

Proposition 2.9 *Let* (X, H) *be a p.p.a.v. of dimension g admitting an automorphism φ of prime order $p = 2g + 1$ and ξ a primitive p-root of the unity, then*

(i) *The components $\mathcal{A}_g(\rho, 2g + 1)$ are 0-dimensional.*

(ii) *The p.p.a.v. (X, H) is of complex multiplication type.*

(iii) *A component $\mathcal{A}_g(\rho, 2g + 1)$ is an isolated point if and only if the Galois group $\mathrm{Gal}(\mathbb{Q}(\xi)/\mathbb{Q})$ acts on the different complex multiplication types with trivial stabilizer.*

Also the number of 0-dimensional different components of $\mathrm{Sing}\mathcal{A}_g$ can be calculated [RR], [BB]. This number is related, in a subtle way, to properties of the representation of φ over the ring of integers of the field $\mathbb{Q}(\xi)$.

Concerning the beautiful upper bound of Hurwitz for $\mathrm{Aut}(S)$, it is well known that it is attained for infinitely many values of g, [M]; the first two values of g being 3, with Klein's quartic, and 7 with Macbeath's curve. For a p.p.a.v. (X, H) the group $\mathrm{Aut}_H(X)$ is a finite group, the author is unable to find in the literature a systematic analogous study for an upper bound of $\mathrm{Aut}_H(X)$. Nevertheless we outline and discuss shortly some of our partial knowledge concerning this problem.

Recall that X_l is the subgroup of l-torsion points of a p.p.a.v. (X, H) of dimension g. This subgroup is a free $\mathbb{Z}/l\mathbb{Z}$-module of rank $2g$. It is well known that for any subgroup G of $\mathrm{Aut}_H(X)$ and any $l \geq 3$ there exists a faithful representation $G \longrightarrow GL_{\mathbb{Z}/l\mathbb{Z}}(X_l)$. That representation allows to obtain some rudimentary upper bound for the order of the group $\mathrm{Aut}_H(X)$.

It is possible to construct p.p.a.v. (X, H) of dimension g with $\mathrm{Aut}_H(X)$ of large order, in fact: for each $g \geq 2$ there are 1-dimensional families of p.p.a.v. (X, H) of dimension g such that $\mathrm{Aut}_H(X)$ is isomorphic to the Weyl groups of the irreducible roots system: A_g, B_g, C_g, D_g. For $g \in \{2, 4, 6, 7, 8\}$ there are 1-dimensional families of p.p.a.v. (X, H) of dimension g such that $\mathrm{Aut}_H(X)$ is isomorphic to the Weyl groups of the irreducible roots system: G_2, F_4, E_6, E_7 and E_8 [CGR]. In particular, for each $g \geq 3$, if $\mathrm{Aut}_H(X)$ is isomorphic to the Weyl group of the irreducible root system A_g, that provides a family of p.p.a.v. of dimension g with an automorphism group of order $(g+1)!$. These abelian varieties are irreducibles (as p.p.a.v.), but they are isomorphic (as abelian varieties) to the product of elliptic curves [GR].

Moreover in the case $g = 4$, following the method developed in [GR], it would be possible to construct p.p.a.v. (X, H) of dimension 4 with $\mathrm{Aut}_H(X)$ of order 240 (A_4), 284 (B_4, C_4), and 1152 (D_4, F_4). Nevertheless the following example due to Beauville takes place:

We consider the elliptic curve E_i with endomorphism ring $\mathbb{Z}[i]$, and $X \cong (E_i)^4$. The $\mathbb{Z}[i]$-module $L \cong \mathbb{Z}[i]^4$ is a free $\mathbb{Z}$-module of rank 8. In order to get a Riemann form H on X, we consider as its real part the quadratic canonical form associated to the root system E_8 and as a complex structure an element $J \in W(E_8)$ (Weyl group) with $J^2 = -I$, in this way we get a p.p.a.v. (X, H) of dimension 4 and $\mathrm{Aut}_H(X) \cong \{\alpha \in \mathrm{Aut}(X)/\alpha \circ J = J \circ \alpha\}$ is a group of order 46080.

Remark 2.10 *Any symplectic representation of a finite group G comes from an integer representation, thus an upper bound for the order of finite subgroups of $GL_{\mathbb{Z}}(2g)$ could help to get an upper bound for $\mathrm{Aut}_H(X)$. With regard to this bound W. Feit, [F1], [F2], has announced that $(n!)2^n$ is an upper bound for the order of a finite subgroup $G \subset GL_{\mathbb{Z}}(n)$, except for $n = 2, 4, 6, 7, 8, 9$ and 10 for which some exceptional Weyl groups provide the counterexamples. Nevertheless: an upper bound for $\mathrm{Aut}_H(X)$ is closely related to the upper bound of the order of finite subgroups of $Sp(2g, \mathbb{Z})$.*

3 Sing$\mathcal{M}_5$ and Sing$\mathcal{A}_5$

In this section we study the Schottky problem for some components of Sing$\mathcal{A}_5$, the choice of $g = 5$ is motivated by the fact that $\mathcal{A}_5$ contains also intermediate Jacobians of smooth cubic threefolds. We begin by describing the components of $\mathrm{Sing}\mathcal{M}_5$ and the families of smooth cubic threefolds with automorphisms of prime order. Next, following an analytic point of view, we describe $\mathrm{Sing}\mathcal{A}_5$ and compare the three different subloci and we solve a particular Schottky problem for the 0-dimensional components of $\mathrm{Sing}\mathcal{A}_5$.

If $S(p, \gamma; a_1, a_2, ..., a_n) \subset \mathcal{M}_g$ is the subloci previously defined in section 2, it can be deduced from the work of Cornalba [C] that:

Theorem 3.1 *The irreducible components of* $\mathrm{Sing}\mathcal{M}_5$ *denoted by*

$$S(p, \gamma; a_1, a_2, ..., a_n)$$

are given in the following list:

p	$S(p, \gamma; a_1, a_2, \ldots, a_n)$	d
2	$S(2, 0; 12\{1\})$	*9*
2	$S(2, 1; 8\{1\})$	*8*
2	$S(2, 2; 4\{1\})$	*7*
2	$S(2, 3; -)$	*6*
3	$S(3, 0; 5\{1\}, 2\{2\})$	*4*
3	$S(3, 1; 2\{1\}, 2\{2\})$	*4*
11	$S(11, 0; 1, 2, 8)$	*0*

where p denotes the prime order, d denotes the dimension of the family and $l\{k\}$ means $l, l, ..., l$ of length k.

By proposition 12 a Riemann surface S of genus $g = 5$ can admit an automorphism of prime order p, with $p \in \{2, 3, 5, 11\}$. If S of genus five admits an automorphism φ of prime order $p = 5$, from the Riemann-Hurwitz formula we have that $\pi_\varphi : S \to E = S/\langle\varphi\rangle$ is a branched analytic covering of degree 5 over an elliptic curve E with 2 branched points, then it follows from corollary 1, in [C] that the corresponding 2-dimensional family is contained in another component of $\mathrm{Sing}\mathcal{M}_5$.

It can be also remarked that it is possible to give an explicit description of those subvarieties, using the fact that the canonical curve of genus $g = 5$ is the intersection of three quadrics in $\mathbb{P}^4$. Also recalling that there exists a period map $p_S : \mathcal{M}_5 \longrightarrow \mathcal{A}_5$, we denote by $\mathcal{JM}_5$ the image of $\mathcal{M}_5$ under p_S.

Some p.p.a.v. of dimension 5 can arise as intermediate Jacobians of smooth cubic hypersurfaces of $\mathbb{P}^4$; for the sake of completeness we begin recalling the construction of its intermediate Jacobians [CG].

Let $\mathbb{P}^4$ be the 4-dimensional complex projective space with homogeneous coordinates $(x_1, ..., x_5)$ and X a smooth cubic hypersurface of $\mathbb{P}^4$ (a smooth cubic threefold), let $H^q(X)$ be the image of $H^q(X, \mathbb{Z})$ in the de Rham group of closed q-forms modulo exact q-forms and Ω_X^p the sheaf of holomorphic differential p-forms on X, as $H^0(X, \Omega_X^3) \cong \{0\}$, in the Hodge decomposition:

$$H^3(X) \otimes \mathbb{C} \cong H^2(X, \Omega_X^1) \oplus H^1(X, \Omega_X^2)$$

We project the subgroup $H^3(X)$ of $H^3(X) \otimes \mathbb{C}$ into the subspace $W \cong H^2(X, \Omega_X^1)$, the image is a lattice U_X in W and there is an isomorphism: $\rho : H^3(X) \to U_X$ such that for any $\alpha, \beta \in H^3(X)$:

$$\int_X \alpha \wedge \beta = 2\mathrm{Re} \int_X \rho(\alpha) \wedge \overline{\rho(\beta)}$$

Now there is a non-degenerate Hermitian form P_X in W defined by the formula:

$$P_X(\omega_1, \omega_2) = 2i \int_X \omega_1 \wedge \overline{\omega_2}$$

As $\mathrm{Im}P_X$ correspond under the isomorphism ρ to the cup pairing on $H^3(X)$, the alternating bilinear form $\mathrm{Im}P_X$ is unimodular in U_X, then P_X is a principal Riemann form and we have a 5-dimensional principally polarized abelian variety:

$$\mathcal{J}(X) \cong (W/U_X, P_X)$$

called the intermediate Jacobian of X. It can be remarked that P_X determine a non-degenerate divisor Θ on W/L_X such that $\dim \mid \Theta \mid = 0$.

We will denote by $\mathcal{V}(3, \mathbb{P}^4)$ the 10-dimensional moduli space of smooth cubic hypersurfaces of $\mathbb{P}^4$, then we obtain a period map

$$p_T : \mathcal{V}(3, \mathbb{P}^4) \longrightarrow \mathcal{A}_5$$

We will denote by $\mathcal{JH}_5$ the image of $\mathcal{V}(3, \mathbb{P}^4)$ under the map p_T.

Also we recall the Torelli theorem [CG].

Theorem 3.2 *Let X and Y be smooth cubic hypersurfaces of $\mathbb{P}^4$, $\mathcal{J}(X)$*

and $\mathcal{J}(Y)$ their associated intermediate Jacobians, then X and Y are isomorphic if and only if $\mathcal{J}(X)$ and $\mathcal{J}(Y)$ are isomorphic as p.p.a.v.

Let F be a homogeneous form of degree 3 in 5 variables, $V(F)$ the cubic hypersurface of $\mathbb{P}^4$ defined by F and φ an automorphism of $V(F)$. The families of forms admitting an automorphism of prime order p can be obtained from the Magister thesis of A. Liendo [Li].

Theorem 3.3 *Let $X = V(F)$ be a non-singular cubic hypersurface which admits an automorphism φ of order p prime, then after a linear change of variables which diagonalizes φ, F is given in the following list*

$\mathcal{V}_p^j(\sigma_1,\sigma_2,\sigma_3,\sigma_4,\sigma_5;d)$	F
$\mathcal{V}_2^1(0,0,0,0,1;7)$	$x_5^2 L_1(x_1,x_2,x_3,x_4) + L_3(x_1,x_2,x_3,x_4)$
$\mathcal{V}_2^2(0,0,0,1,1;6)$	$x_1 L_2(x_4,x_5) + x_2 M_2(x_4,x_5)$ $+x_3 N_2(x_4,x_5) + L_3(x_1,x_2,x_3)$
$\mathcal{V}_3^1(0,0,0,0,1;4)$	$L_3(x_1,x_2,x_3,x_4) + x_5^3$
$\mathcal{V}_3^2(0,0,0,1,1;1)$	$L_3(x_1,x_2,x_3) + M_3(x_4,x_5)$
$\mathcal{V}_3^3(0,0,0,1,2;4)$	$L_3(x_1,x_2,x_3) + x_4 x_5 L_1(x_1,x_2,x_3) + x_4^3 + x_5^3$
$\mathcal{V}_3^4(0,0,1,1,2;2)$	$L_3(x_1,x_2) + M_3(x_3,x_4) + x_5^3$ $+L_1(x_1,x_2)M_1(x_3,x_4)x_5$
$\mathcal{V}_5^1(0,1,2,3,4;2)$	$a_1 x_1^3 + a_2 x_1 x_2 x_5 + a_3 x_1 x_3 x_4 + a_4 x_2^2 x_4$ $+a_5 x_2 x_3^2 + a_6 x_3 x_5^2 + a_7 x_4^2 x_5$
$\mathcal{V}_{11}^1(1,3,4,5,9;0)$	$x_1^2 x_5 + x_4^2 x_1 + x_2^2 x_4 + x_2 x_3^2 + x_3 x_5^2$

A generator of $\langle\varphi\rangle$ is given by

$$(x_1,x_2,x_3,x_4,x_5) \rightarrow (\xi^{\sigma_1} x_1, \xi^{\sigma_2} x_2, \xi^{\sigma_3} x_3, \xi^{\sigma_4} x_4, \xi^{\sigma_5} x_5)$$

where ξ is a primitive root of order p. Each family of hypersurfaces is denoted by $\mathcal{V}_p^j(\sigma_1,\sigma_2,\sigma_3,\sigma_4,\sigma_5;d)$, where p denotes the order of φ and d the dimension of the family. The subscripts in the homogeneous forms L_i, M_i, N_i indicate their degree.

It can be remarked that there exist some inclusions between the families of the previous list:

Every generic element of the family $\mathcal{V}_3^3$ is invariant under the permutation of coordinates given by: $(x_4 x_5)$, this family is contained in the family $\mathcal{V}_2^1$.

Every generic element of the 2-dimensional family $\mathcal{V}_5^1$ is invariant under the permutation of coordinates given by: $(x_2 x_5)(x_3 x_4)$, this family is contained in the family $\mathcal{V}_2^2$.

The 0-dimensional family $\mathcal{V}_{11}^1$ (Klein's cubic threefold) admit as automorphism group, a group isomorphic to $PSL_2(\mathbb{F}_{11})$, [A]. From the representation of $PSL_2(\mathbb{F}_{11})$ in $GL(5,\mathbb{C})$ and Torelli's theorem for smooth cubic threefolds, it follows that $\mathcal{V}_{11}^1$ belongs to the families $\mathcal{V}_2^2$, $\mathcal{V}_3^4$ and $\mathcal{V}_5^1$.

Remark 3.4 *The family $\mathcal{V}_3^1$ correspond to triple covers of $\mathbb{P}^3$ branched over a cubic surface $Y \subset \mathbb{P}^3$. In [ACT1], using the Hodge structures associated to those covers, it was proved that the moduli space of stable cubic surfaces is isomorphic to a quotient of the complex hyperbolic 4-space, by the action of a specific discrete subgroup.*

Now we study Sing$\mathcal{A}_5$, by proposition 13 we must search for components $\mathcal{A}_5(p,\alpha)$ with the primes $p = 2$, $p = 3$, $p = 5$ and $p = 11$. Beginning with the prime $p = 11$, from the corollary of proposition 17 it follows:

Proposition 3.5 *There exist exactly 4 isomorphism classes of p.p.a.v. (X_i, H_i, C_i), $i \in \{1,...,4\}$ of dimension 5, such that $\mathbb{Z}_{11} \subseteq$ Aut$_{H_i}(X_i, C_i)$. For each class we can choose a representant C_i for the complex multiplication class given by:*

$$C_1 = \{\xi, \xi^2, \xi^3, \xi^4, \xi^5\} \quad C_2 = \{\xi, \xi^2, \xi^3, \xi^4, \xi^6\}$$
$$C_3 = \{\xi, \xi^2, \xi^3, \xi^5, \xi^7\} \quad C_4 = \{\xi, \xi^3, \xi^9, \xi^5, \xi^4\}$$

where ξ is a primitive root of the unity of order 11.

In the proposition 17, it can be verified that the isotropy subgroups H_{C_1}, H_{C_2} and H_{C_3} of the Galois group Gal$(\mathbb{Q}(\xi_{11})/\mathbb{Q})$ are trivial, then we have that:

$$(X_1, H_1, C_1) \quad (X_2, H_2, C_2) \quad (X_3, H_3, C_3)$$

are isolated points in Sing$\mathcal{A}_5$.

Now we can solve the following particular Schottky problem for 0-dimensional components of Sing$\mathcal{A}_5$.

Problem: Which of the p.p.a.v. (X_i, H_i, C_i), $i \in \{1, ..., 4\}$ of dimension 5 are Jacobian of curves or intermediate Jacobian of smooth cubic hypersurfaces?

Proposition 3.6

(i) X_1 *is isomorphic to the Jacobian of the hyperelliptic curve:*

$$y^2 = \prod_{j=1}^{11} (x - \xi^j)$$

(ii) X_2 *is isomorphic to the Jacobian of the Lefschetz curve:*

$$y^{11} = x^2 (x - 1)^8$$

(iii) X_4 *is isomorphic to the intermediate Jacobian of the Klein cubic threefold*

$$x_0^2 x_1 + x_1^2 x_2 + x_2^2 x_3 + x_3^2 x_4 + x_4^2 x_0 = 0$$

(iv) X_3 *is not a Jacobian of curves and it is not an intermediate Jacobian of a smooth cubic hypersurface.*

Proof From proposition 15, it follows that there exist two 0-dimensional components of $\mathrm{Sing}\mathcal{M}_5$: a hyperelliptic Riemann surface and a Lefschetz curve both admitting an automorphism φ of prime order $p = 11$, a calculation of the action of the corresponding automorphisms on the complex vector spaces of holomorphic 1-differential forms, and the Torelli theorem give that its Jacobians are isomorphic as p.p.a.v. to (X_1, H_1, C_1) and (X_2, H_2, C_2) respectively.

From theorem 31 and Torelli's theorem for smooth cubic threefold, it follows that X_4 is isomorphic to the intermediate Jacobian of the Klein cubic threefold.

We suppose that (X_3, H_3, C_3) is isomorphic to the intermediate Jacobian of a smooth cubic threefold $V = V(F)$, and $\varphi \in GL(5, \mathbb{C})$ such that $\varphi = \mathrm{diag}(\xi, \xi^2, \xi^3, \xi^5, \xi^7)$. An elementary calculation shows that $V(F)$ is non-smooth for any φ-invariant cubic form F. $\square$

We consider $(X_4, H_4, C_4) \in \mathcal{A}_5(\theta_4, 11)$, where

$$\theta_4 : \mathrm{Aut}_{H_4}(X_4) \to GL_{\mathbb{C}}(T_0 X_4)$$

with $\theta_4(\varphi) = \varphi_c = \mathrm{diag}(\xi, \xi^3, \xi^9, \xi^5, \xi^4)$. The element $\sigma \in \mathrm{Gal}(\mathbb{Q}(\xi_{11})/\mathbb{Q})$ given by

$$\sigma\{\xi^{k_1}, \xi^{k_2}, \xi^{k_3}, \xi^{k_4}, \xi^{k_5}\} = \{\xi^{3k_1}, \xi^{3k_2}, \xi^{3k_3}, \xi^{3k_4}, \xi^{3k_5}\}$$

fixes C_4 (as a set), σ acts on the $\mathbb{Q}$-vector space $\mathbb{Q}(\xi_p) \otimes \mathbb{R}$ as the permutation $n = (12345)$, giving an automorphism of (X_4, H_4, C_4) of order 5 acting with eigenvalues $\{1, \zeta, \zeta^2, \zeta^3, \zeta^4\}$, where ζ is a primitive 5-root of the unity: if $(\eta)(n) = n_c = \mathrm{diag}(1, \zeta, \zeta^2, \zeta^3, \zeta^4)$, then (X_4, H_4, C_4) is contained in a family $\mathcal{A}_5(\eta, 5)$, from proposition 14 this family is 3-dimensional.

We will describe explicitly that family: $n \oplus {}^t n^{-1}$ determines a symplectic representation η_s, the period matrices given by:

$$\mathrm{Fix}(\eta_s) = \begin{pmatrix} \tau_{11} & \tau_{12} & \tau_{13} & \tau_{13} & \tau_{12} \\ \tau_{12} & \tau_{11} & \tau_{12} & \tau_{13} & \tau_{13} \\ \tau_{13} & \tau_{12} & \tau_{11} & \tau_{12} & \tau_{13} \\ \tau_{13} & \tau_{13} & \tau_{12} & \tau_{11} & \tau_{12} \\ \tau_{12} & \tau_{13} & \tau_{13} & \tau_{12} & \tau_{11} \end{pmatrix}$$

correspond to a subvariety of the Siegel space $\mathcal{H}_5$ representing the family $\mathcal{A}_5(\eta, 5)$ locally at (X_4, H_4, C_4).

Let $(X, H) \in \mathcal{A}_5(\eta, 5)$ and μ be the permutation of coordinates of $T_0 X \cong \mathbb{C}^5$ given by (15)(24), $\mu_s = \mu \oplus {}^t \mu^{-1} \in Sp(10, \mathbb{Z})$, the period matrices given by:

$$\mathrm{Fix}(\mu_s) = \begin{pmatrix} \mu_{11} & \mu_{12} & \mu_{13} & \mu_{14} & \mu_{15} \\ \mu_{12} & \mu_{22} & \mu_{23} & \mu_{24} & \mu_{14} \\ \mu_{13} & \mu_{23} & \mu_{33} & \mu_{23} & \mu_{13} \\ \mu_{14} & \mu_{24} & \mu_{23} & \mu_{22} & \mu_{12} \\ \mu_{15} & \mu_{14} & \mu_{13} & \mu_{12} & \mu_{11} \end{pmatrix}$$

correspond to a subvariety of the Siegel space $\mathcal{H}_5$ representing the family $\mathcal{A}_5(\mu, 5)$ locally at (X, H).

The subvariety $\mathrm{Fix}(\mu_s)$ is 9-dimensional and contains $\mathrm{Fix}(\eta_s)$, thus we have the following inclusions of components:

$$\mathcal{A}_5(\theta_4, 11) \subset \mathcal{A}_5(\eta, 5) \subset \mathcal{A}_5(\mu, 2)$$

where

$$\dim_{\mathbb{C}} \mathcal{A}_5(\theta_4, 11) = 0 \qquad \dim_{\mathbb{C}} \mathcal{A}_5(\eta, 5) = 3 \qquad \dim_{\mathbb{C}} \mathcal{A}_5(\mu, 2) = 9$$

We consider in $\mathrm{Sing} \mathcal{A}_5$ the components admitting automorphisms φ of prime order $p = 3$ such that $\mathrm{Fix}(\varphi)$ is finite, the equality:

$$g = n_0 + r \frac{(p-1)}{2}$$

force $n_0 = 0$, $r = 5$ and $n_1 + n_2 = 5$, thus the characteristic polynomial of φ_r acting on L is $Q(x) = (x^2 + x + 1)^5$. If ω is a primitive root of order 3 we can assume that ϕ_r acting in the complex tangent space T_0X is given by a diagonal matrix $\mathrm{diag}(\omega, ..., \omega, \omega^2, ..., \omega^2)$ where ω has multiplicity m and ω^2 has multiplicity n, with $m + n = g$.

In order to analyse this case we provide an analytic proof, using the Eichler trace formula of a general proposition due to Zarhin [Z].

Proposition 3.7 *Let* (X, H) *be a p.p.a.v. of dimension* g *and* φ *an automorphism of order 3, such that* $\rho(\varphi) = \varphi_c$ *acting in the complex tangent space* T_0X *is given by a diagonal matrix* $\mathrm{diag}(\omega, ..., \omega, \omega^2, ..., \omega^2)$, *where* ω *has multiplicity* m *and* ω^2 *has multiplicity* n, *with* $m + n = g$.

  (i) *We have that* $\dim\mathcal{A}_g(\rho, 3) = m \cdot n$.

 (ii) *If* $3 \mid m - n \mid > (g + 2)$ *then* (X, H) *cannot be the Jacobian of a Riemann surface of genus* g *with canonical polarization.*

Proof The statement (i) follows from proposition 14.

We will supose that (X, H) is the Jacobian of a compact connected Riemann surface S of genus g; $(X, H) \cong (J(S), \Theta)$. By the Torelli theorem there exists an automorphism $\alpha \in \mathrm{Aut}(S)$ that induces φ or $-\varphi$, if α induces $-\varphi$, α^4 induces φ, thus we can assume that α induces an automorphism α^* in the g-dimensional complex vector space of holomorphic 1-differential forms $H^0(S, \Omega_S^{1,0})$, the representation $\alpha \to \alpha^*$ is faithful and α^* has the same spectrum as φ.

We denote by

$$\pi : S \to S/\langle \alpha \rangle \cong X$$

the branched analytic covering of degree 3 determined by the cyclic group generated by α, their ramification points are exactly the images of the fixed points of α and all the ramification indexes are 3. We denote by $H^0_\alpha(S, \Omega_S^{1,0})$ the subspace of holomorphic α-invariant 1-differential forms, we have that

$$\dim(H^0_\alpha(S, \Omega_S^{1,0})) = \mathrm{genus}(X) = 0$$

If t is the number of fixed points of α by the Riemann-Hurwitz formula we have that $t = g + 2$.

As α and φ have the same trace we have that:

$$\mathrm{Trace}(\alpha) = m\omega + n\omega^2$$

Also by the Eichler trace formula:

$$\text{Trace}(\alpha) = 1 + \sum_{j=1}^{t} \frac{\omega^{\nu_j}}{1 - \omega^{\nu_j}}$$

where the ν_j are one or two.

We have:

$$|\,\text{Trace}(\alpha) - 1\,| \leq \sum_{j=1}^{t} \frac{1}{|\,1 - \omega^{\nu_j}\,|} = \frac{t}{|\,1 - \omega^{\nu_j}\,|}$$

$$|\,\text{Trace}(\alpha) - 1\,| \leq \frac{g+2}{\sqrt{3}}$$

but

$$|\,\text{Trace}(\alpha) - 1\,|^2 = \frac{(m+n+2)^2 + 3(m-n)^2}{4} = \frac{(g+2)^2 + 3(m-n)^2}{4}$$

$$\frac{(g+2)^2 + 3(m-n)^2}{4} \leq \frac{(g+2)^2}{3}$$

thus

$$3\,|\,m - n\,| \leq (g+2)$$

and we obtain the statement (ii). □

Let $(X, H) \in \mathcal{A}_5$, $\varphi \in \text{Aut}_H(X)$ such that $\rho(\varphi) = \text{diag}(\omega, \omega, \omega, \omega, \omega^2)$ and $\psi \in \text{Aut}_H(X)$ such that $\varrho(\psi) = \text{diag}(\omega, \omega, \omega, \omega^2, \omega^2)$. The 4-dimensional family $\mathcal{A}_5(\rho, 3) \subset \text{Sing}\mathcal{A}_5$ contains all the intermediate Jacobians of the family $\mathcal{V}_3^1$ of the theorem 31, and by the previous proposition cannot contain Jacobians of Riemann surfaces of genus 5. The 6-dimensional family $\mathcal{A}_5(\varrho, 3) \subset \text{Sing}\mathcal{A}_5$ contains all the Jacobians of the 4-dimensional family $\mathcal{S}(3, 0; 5\{1\}, 2\{2\})$ of the theorem 29.

Remark 3.8 *It is well known, that in general the sublocus of Jacobians of Riemann surfaces $\mathcal{J}\mathcal{M}_5$ and the sublocus of intermediate Jacobians of smooth cubic threefolds $\mathcal{J}\mathcal{H}_5$ do not intersect in $\mathcal{A}_5$. In fact if S is non-hyperelliptic, each singular point of the theta divisor Θ of $(J(S), \Theta)$ is quadratic, but if V is a smooth cubic threefold the theta divisor Θ of $(J(V), \Theta)$ has a singular point of multiplicity 3.*

We end by considering in $\text{Sing}\mathcal{A}_5$ the components admitting automorphisms of prime order $p = 2$, in that case [Re], we have 12 non-conjugated

Reiner involutions of $Sp(10, \mathbb{Z})$, given by the matrices U_i con

$$U_i = \begin{pmatrix} \mu_i & 0 \\ 0 & {}^t\mu_i \end{pmatrix}$$

where $\mu_i \in Gl(5, \mathbb{Z})$. As the 12 non-conjugated Reiner involutions include the identity, minus the identity and 4 diagonal involutions, that gives reducible p.p.a.v. In an explicit form the 6 non-trivial non-conjugated Reiner involutions of $Sp(10, \mathbb{Z})$, are given by

$$\mu_1 = \begin{pmatrix} 1 & 0 & 0 & 0 & 0 \\ 1 & -1 & 0 & 0 & 0 \\ 0 & 0 & -1 & 0 & 0 \\ 0 & 0 & 0 & -1 & 0 \\ 0 & 0 & 0 & 0 & -1 \end{pmatrix} \qquad \mu_2 = \begin{pmatrix} 1 & 0 & 0 & 0 & 0 \\ 1 & -1 & 0 & 0 & 0 \\ 0 & 0 & 1 & 0 & 0 \\ 0 & 0 & 0 & 1 & 0 \\ 0 & 0 & 0 & 0 & 1 \end{pmatrix}$$

$$\mu_3 = \begin{pmatrix} 1 & 0 & 0 & 0 & 0 \\ 1 & -1 & 0 & 0 & 0 \\ 0 & 0 & -1 & 0 & 0 \\ 0 & 0 & 0 & -1 & 0 \\ 0 & 0 & 0 & 0 & 1 \end{pmatrix} \qquad \mu_4 = \begin{pmatrix} 1 & 0 & 0 & 0 & 0 \\ 1 & -1 & 0 & 0 & 0 \\ 0 & 0 & -1 & 0 & 0 \\ 0 & 0 & 0 & 1 & 0 \\ 0 & 0 & 0 & 0 & 1 \end{pmatrix}$$

$$\mu_5 = \begin{pmatrix} 1 & 0 & 0 & 0 & 0 \\ 1 & -1 & 0 & 0 & 0 \\ 0 & 0 & 1 & 0 & 0 \\ 0 & 0 & 1 & -1 & 0 \\ 0 & 0 & 0 & 0 & -1 \end{pmatrix} \qquad \mu_6 = \begin{pmatrix} 1 & 0 & 0 & 0 & 0 \\ 1 & -1 & 0 & 0 & 0 \\ 0 & 0 & 1 & 0 & 0 \\ 0 & 0 & 1 & -1 & 0 \\ 0 & 0 & 0 & 0 & 1 \end{pmatrix}$$

If we denote by $\mathcal{A}_5(\mu_i, 2)$ the family of p.p.a.v. of dimension 5 admitting as automorphism the involution μ_i, it would be interesting to characterize explicitly, in each case, the Jacobians of the families $\mathcal{S}(2, 0; 12\{1\})$, $\mathcal{S}(2, 1; 8\{1\})$, $\mathcal{S}(2, 2; 4\{1\})$, $\mathcal{S}(2, 3; -)$ and the Jacobians of the families $\mathcal{V}_2^1$, $\mathcal{V}_2^2$ that are contained in the families $\mathcal{A}_5(\mu_i, 2)$.

Remark 3.9 *A recent result of Casalaina-Martin and Friedmann [CF] characterizes the intermediate Jacobians of cubic threefolds as those five-dimensional p.p.a.v. whose theta divisor has a unique singular point, which is of multiplicity three. Also recently X. Roulleau proved that X_3 is not the intermediate Jacobian of a smooth cubic surface.*

Notes

1 Partially supported by Fondecyt Project N 1080030

References

[A] Adler, A. On the automorphism group of a certain cubic threefold. *American Journal of Maths.* **100**(6) (1978), 1275–1280.

[ACT1] Allcock, D., Carlson, J. A. and Toledo, D. The complex hyperbolic geometry of the moduli space of cubic surfaces. *Journal of Algebraic Geometry* **11** (2002), 659–724.

[ACT2] Allcock, D., Carlson, J. A. and Toledo, D. The moduli space of cubic threefolds as a ball quotient. arXiv:math.AG/0608287 v.1, August 2006.

[BB] Bennama,. H. and Bertin, J. Varietés abéliennes avec un automorphisms d'ordre premier. *Manuscripta Math.* **94** (1997), 409–425.

[BL] Birkenhake, Ch. and Lange, H. *Complex Abelian Varieties*, Grundlehren der mathematischen Wissenschaften **302**, Springer, 2003.

[CGR] Carocca, A., González-Aguilera, V. and Rodríguez R. E: Weyl groups and abelian varieties. *Journal of Group Theory* **9**(2) (2006), 265–282.

[CF] Casalaina-Martin, S. and Friedmann, R. Cubic threefolds and abelian varieties of dimension five. *Journal of Algebraic Geometry* **14** (2005), 295–326.

[CG] Clemens, C. H. and Griffiths, P. H. The intermediate jacobian of the cubic threefold. *Annals of Maths.* **95** (1972), 281–356.

[C] Cornalba, M. On the locus of curves with automorphisms. *Annali di Matematica Pura ed Applicatta* **149** (1987), 135–151.

[D] Debarre, O. *The Schottky Problem: An Update*, Math. Sci. Res. Inst. Publ. **28**, Cambridge University Press, 1995, pp. 57–64.

[F-K] Farkas, H.M. and Kra, I. *Riemann Surfaces*, Graduate Texts in Mathematics **71**, Springer-Verlag, Berlin, 1991.

[F1] Feit, W. Private correspondence, September 5, 1994.

[F2] Feit, W. MathReview. MR1318996 (95k:20075), MathSciNet, AMS.

[G] Gómez González, E. Irreducible components and isolated points in the branch locus of the moduli space of smooth curves. *Boletín de la Sociedad Matemática Mexicana* **2**(3) (1996), 115–128.

[GMZ1] González-Aguilera, V., Muñoz-Porras, J. M. and R. Zamora, A. G. On the irreducible components of the singular locus of $\mathcal{A}_g$. *Journal of Algebra* **240** (2001), 230–250.

[GMZ2] González-Aguilera, V., Muñoz-Porras, J. M. and R. Zamora, A. G. On the 0-dimensional irreducible components of the singular locus of $\mathcal{A}_g$. *Archiv der Mathematik* **84** (2006), 298–303.

[GR] González-Aguilera, V. and Rodríguez, R. E. Families of irreducible principally polarized abelian varieties isomorphic to a product of elliptic curves. *Proceedings of the AMS* **128**(3) (2000), 629–636.

[GH] González-Díez, G. and Harvey, W. J. Moduli of Riemann surfaces with symmetry, in *Discrete Groups and Geometry*, London Math. Soc. Lecture Notes **173**, Cambridge University Press, 1992, pp. 75–94.

[H1] Harvey, W. J. Cyclic groups of automorphism of a compact Riemann surface. *Quart. J. Math. Oxford* **17**(2) (1966), 86–97.

[H2] Harvey, W. J. On branch loci in Teichmüller space. *Trans. Amer. Math. Soc.* **153**(2) (1971), 387–399.

[I] Igusa, J. Arithmetic variety of moduli for genus two. *Ann. Math.* **72** (1960), 612–649.

[K] Kulkarni, R. S. Isolated points in the branch locus of the moduli space of compact Riemann surfaces. *Ann. Acad. Sci. Fenn.* **16** (1991), 71–81.

[Le] Lefschetz, S. *Selected Papers*, Chelsea, New York, 1971.

[Li] Liendo, A. Sobre hipersuperficies cúbicas de $\mathbb{P}^4$ y puntos de Eckardt, Tésis Magister en Matemática. U. Santa María, Valparaíso, Chile, 2005.

[Ll] Lloyd, E. K. Riemann surface transformation groups. *J. Combin. Theory Ser. A.* **13** (1972), 17–27.

[M] Macbeath, A. M. On a theorem of Hurwitz, *Proc. Glasgow. Math Soc.* **5** (1961), 90–96.

[O1] Oort, F. The singularities of the moduli schemes of curves. *J. of Number Theory* **1** (1969), 90–107.

[O2] Oort, F. Singularities of coarse moduli schemes. *Séminaire Dubreil* **16** (1976).

[Ra] Rauch, H. E. A transcendental view of the space of algebraic Riemann surfaces. *Bull. Amer. Math. Soc.* **71** (1965), 1–39.

[Re] Reiner, I. Automorphism of the symplectic modular group. *Transactions of the AMS* **80**(1) (1955), 35–50.

[RR] Riera, G. and Rodríguez, R. E. Riemann surfaces and abelian varieties with an automorphism of prime order. *Duke Mathematical Journal.* **69** (1993), 199–217.

[U] Urzua, G. C. Riemann surfaces of genus g with an automorphism of prime order and $p > g$. *Manuscripta Mathematica* **121** (2006), 169–189.

[Z] Zarhin, Y. G. Cubic surfaces and cubic threefolds, jacobians and intermediate jacobians. arXiv: math/0610138v3 math.AG. 18 April 2007.

Non-special divisors supported on the branch set of a p-gonal Riemann surface

Gabino González-Diez [1]

Departamento de Matemáticas, Universidad Autónoma de Madrid
gabino.gonzalez@uam.es

Abstract

A compact Riemann surface S is called *cyclic p-gonal* if it possesses an automorphism τ of order p such that the quotient $S/\langle\tau\rangle$ has genus zero. It is well known that if p is a prime number and $Q_1, \ldots, Q_r \in S$ are the fixed points of τ then S has genus $g = \frac{p-1}{2}(r-2)$.

In this article we find a criterion to decide when a divisor of the form $D = Q_1^{d_1} \cdots Q_r^{d_r}$, with $\sum d_i = g$, is non-special.

The criterion is very easy to apply in practice since it only depends on the arithmetic of the local rotation numbers of τ at the points Q_i and the multiplicities of these points on the divisor D, i.e. the integers d_i.

Knowledge of the set of non-special divisors supported on the ramification set seems to be essential in all attempts to extend the classical Thomae formulae, which apply to hyperelliptic (i.e. 2-gonal) Riemann surfaces, to the case of p-gonal ones.

Notation. *Throughout this paper we use the following notation. Given an integer $n \in \mathbb{Z}$ we shall denote by $\overline{n} \in \{0, 1, \ldots, p-1\}$ and $[n] \in \mathbb{Z}/p\mathbb{Z}$ its remainder and its residue class modulo p, respectively; thus, we have $[\overline{n}] = [n]$.*

1 Introduction and statement of the main result

Among the many ways in which a compact Riemann surface S of genus $g \geq 1$ can be described are the *algebraic equation*

$$F(x, y) = \sum a_{ij} x^i y^j = 0$$

satisfied by any pair of meromorphic functions x and y generating the function field of S (*Riemann's existence theorem*) and the *Jacobian of* S defined as

$$J(S) = \frac{\mathbb{C}^g}{\mathbb{Z}^g \oplus \mathbb{Z}^g \cdot \Omega}$$

where $\Omega = (\int_{B_j} \omega_i)$ is a $g \times g$ matrix, called the *period matrix*, whose entries are the B-periods of the basis of holomorphic 1-forms $\omega_i, i = 1, \ldots, g$ which is dual to a chosen symplectic basis $\{A_j, B_j, j = 1, \ldots, g\}$ of the first homology group $H_1(S, \mathbb{Z})$ (*Torelli's theorem*).

One may therefore hope to express the coefficients a_{ij} of a certain algebraic equation for S in terms of *Riemann's theta* function $\theta(z, \Omega)$ evaluated at a suitable finite collection of points $z = a + b \cdot \Omega; a, b \in \mathbb{R}^g$ (*theta constants*) and, conversely, to obtain these theta constants, usually denoted $\theta \begin{bmatrix} a \\ b \end{bmatrix} (0, \Omega)$, as a function of the coefficients a_{ij}.

This correspondence between the algebraic and the transcendental moduli theory of Riemann surfaces works very well at a theoretical level. But, of course, it is not clear at all how to materialize these ideas when an arbitrary Riemann surface is given.

However, it can be satisfactorily achieved for hyperelliptic Riemann surfaces. In fact the formulae performing this relationship go back to the work of Frobenius [Fro] and Thomae [T1], [T2]. We refer to [FK], chapter VII.4. for a modern account of this correspondence in one direction (*expressing the branch points as functions of the periods*) and to [Mum] and [EF], in the other one (*Thomae's formula*).

Several authors have generalized these formulae to certain families of cyclic p-gonal Riemann surfaces with $p > 2$. The reader may consult the articles [Far], [G], [GH], [K] and [Mat] for formulae expressing the branch points of p-gonal surfaces as functions of the periods and [Ac], [BR], [EbF], [EF], [EG], [GH] and [Ko] for generalizations of Thomae formulae to several kinds of p-gonal Riemann surfaces. To obtain these identities one of the key points is to detect sufficiently many suitable degree g non-special divisors supported on the ramification locus of the structural p-gonal morphism $S \to \widehat{\mathbb{C}}$ of the p-gonal Riemann surface S.

The aim of this paper is to provide a criterion to check when a given divisor of this kind is special.

We recall that a compact Riemann surface S is called *cyclic p-gonal* if it possesses an automorphism τ of order p such that the quotient

$S/\langle\tau\rangle$ has genus zero and so the natural map $S \to S/\langle\tau\rangle \simeq \widehat{\mathbb{C}}$ provides a degree p, or p-gonal, morphism which ramifies at the points fixed by τ. Accordingly, the set of fixed points will be referred to as *the ramification (or branch) locus (or set)*.

Throughout this paper we shall assume that p is a prime positive integer.

It is well known (see e.g. [G], [H]) that such a Riemann surface admits an algebraic model of the form

$$y^p = (x - a_1)^{m_1} ... (x - a_r)^{m_r} \qquad (1.1)$$

where

- $\sum m_i = np$, for some positive integer n.
- $1 \le m_i \le p - 1$.
- The Riemann surface S consists of the affine points of the curve (1.1) plus p points at infinity.
- The cyclic group $\langle\tau\rangle$ is generated by the automorphism $\tau(x, y) = (x, e^{2\pi i/p} y)$.
- The full fixed point set of τ is $Fix(\tau) = \{Q_1 = (a_1, 0), \ldots, Q_r = (a_r, 0)\}$. The points at infinity get permuted by τ.
- The integer m_k is called the *rotation number* of τ at the point Q_k. The rotation number of τ at a fixed point Q is independent of the choice of the model (1.1) (because, locally, $\tau^{-1}(z) = e^{2\pi i m_k/p} \cdot z$).
- The genus of S is $g = \frac{p-1}{2}(r - 2)$.

Let D be an integral divisor of degree g, that is $D = P_1^{d_1} \cdots P_r^{d_r}$ with $d_i \ge 0$ and $\sum d_i = g$. Recall that D is said to be special if there is a non constant function f whose set of poles is bounded by D or, as it is usually written, $f \in L(D^{-1})$.

The significance of the special divisors can be explained as follows. Let us identify the set of integral divisor of degree g with the g-fold symmetric product $S^{(g)}$, then, after choosing a base point $Q \in S$, there is a holomorphic map, *the Abel-Jacobi map*, from S^g to $J(S)$ defined by

$$A(D) = \sum_{i=1}^{r} d_i \int_Q^{P_i} (\omega_1, \cdots, \omega_g) \in J(S)$$

It is a classical result that this map is a birational map which fails to be an isomorphism precisely at the special divisors.

It is a trivial fact that if one of the multiplicities d_i is bigger than or equal to p then D is special (see Proposition 2.1, *ii* below). Therefore we can assume from the start that $0 \le d_i < p$, for all $i = 1, \ldots, r$.

We can now state our criterion

Theorem 1.1

Let S be a compact Riemann surface and τ an automorphism of S of prime order p such that the quotient $S/\langle\tau\rangle$ has genus zero. Let $Fix(\tau) = \{Q_1, \ldots, Q_r\}$ be the fixed point set of τ and let us denote by m_k the rotation number of the point Q_k.

Then, for a divisor D of the form $D = Q_1^{d_1} \cdots Q_r^{d_r}$ with $0 \le d_i \le p-1$ and $\sum d_i = g$, the following four conditions are equivalent

(i) D is non-special.

(ii) $\sum_{i=1}^{r} \overline{d_i + m_i k} > g$ for every $k = 1, \ldots, p-1$.

(iii) $\sum_{i=1}^{r} \overline{d_i + m_i k} = g + p$ for every $k = 1, \ldots, p-1$.

(iv) $\sum_{i=1}^{r} \overline{d_i + m_i k} = g + p$ for $p-2$ integers $k \in \{1, \ldots, p-1\}$.

Moreover, if for a certain $k \in \{1, \ldots, p-1\}$ one has the inequality $\sum_{i=1}^{r} \overline{d_i + m_i k} \le g$, then the function

$$y^k \prod_{i=1}^{r} (x - a_i)^{s_{i,k}}, \quad with \quad s_{i,k} = \frac{\overline{d_i + m_i k} - (d_i + m_i k)}{p},$$

belongs to $L(D^{-1})$.

The proof of this theorem is carried out in section 2 while in section 3 we explicitly describe the set of non-special divisors supported on the ramification set for some particularly interesting cases. These include most of the families of p-gonal Riemann surfaces that appear in the literature in connection with problems involving theta constants.

Acknowledgement 1.2 *I would like to thank Hershel M. Farkas for kindly pointing out an error in my article [G] concerning the choice of a collection of non-special divisors (see 2.2). It was his observation* that made me initiate the search for a complete characterization of non-special divisors supported on the ramification set of p-gonal Riemann surfaces.

2 Proof of the criterion

We begin by recalling the most elementary facts and definitions relative to the group of divisors on a Riemann surface. For a detailed account the reader is referred to [FK].

A *divisor* $\mathcal{U}$ on S is a formal symbol

$$\mathcal{U} = P_1^{s_1} \cdots P_r^{s_r}$$

with $P_j \in S, s_j \in \mathbb{Z} \setminus \{0\}$. The subset $\{P_1 \cdots P_r\} \subset S$ is called the *support* of $\mathcal{U}$. At times we will also allow $s_j = 0$ which, of course, means that P_j does not belong to the support. The integer s_j is called *the multiplicity* of the point P_j in the divisor $\mathcal{U}$.

The set of divisors carries a structure of an abelian group under the obvious multiplicative law; the inverse of $\mathcal{U}$ being $\mathcal{U}^{-1} = P_1^{-s_1} \cdots P_r^{-s_r}$.

The divisor of a meromorphic function $f \in \mathcal{M}(S)$ is defined by

$$(f) = \frac{P_1^{d_1} \cdots P_r^{d_r}}{R_1^{l_1} \cdots R_r^{l_r}} = P_1^{d_1} \cdots P_r^{d_r} R_1^{-l_1} \cdots R_r^{-l_r}; \quad d_i, l_i \geq 0$$

where the numerator (resp. denominator) stands for the zero (resp. pole) set of f, multiplicities being taken into account.

Given two divisors $\mathcal{U} = P_1^{d_1} \cdots P_r^{d_r}$ and $\mathcal{B} = P_1^{l_1} \cdots P_r^{l_r}$ we say that $\mathcal{B} \geq \mathcal{U}$ if $d_i \geq l_i; i = 1, \ldots, r$.

We also need to introduce the $\mathbb{C}$-vector space

$$L(\mathcal{U}) = \{f \in \mathcal{M}(S) : (f) \geq \mathcal{U}\}$$

whose dimension we shall denote by $r(\mathcal{U})$.

Let D be an integral divisor of degree g, that is $D = P_1^{d_1} \cdots P_r^{d_r}$ with $d_i \geq 0$ and $\sum d_i = g$. We say that D is *special* if $r(D^{-1}) > 1$, that is if there is a non constant function f whose set of poles is bounded by D.

The following proposition collects a list of well known facts relative to the function field of a p-gonal Riemann surface.

Proposition 2.1 *Let S be the Riemann surface defined by equation (1.1) and let $\mathcal{M}(S)$ be its function field. The following properties hold*
(i) $\mathcal{M}(S)$ is generated by the coordinate functions x and y.

(ii) The divisors of the functions $x - a_i$ and y are

$$(x - a_i) = \frac{Q_i^p}{\infty}, \text{ and } (y) = \frac{Q_1^{m_1}, \dots, Q_r^{m_r}}{\infty^n}$$

where ∞ stands for the integral divisor of degree p supported on the p points that S possesses at infinity.

(iii) The divisor of a meromorphic function of the form

$$f = f(x) = \prod_{i=1}^{r} (x - a_i)^{s_i} \prod_{j} (x - b_j)^{t_j}; \quad s_i, t_j \in \mathbb{Z}, \ a_i \neq b_j$$

is

$$(f) = \prod_{i=1}^{r} Q_i^{p s_i} \cdot \prod_{j} B_j^{t_j} \cdot \infty^{-(\sum s_i + \sum t_j)}$$

where B_j is the divisor of degree p given by:

$$B_j = \prod_{k=1}^{p} (b_j, e^{2\pi k/p} \sqrt[p]{\prod_i (b_j - a_i)^{m_i}})$$

Definition 2.2 *Let D be a divisor of the form $D = Q_1^{d_1} \cdots Q_r^{d_r}$ with $0 \leq d_i \leq p - 1$ and $\sum d_i = g$. Associated to D we define the following objects*

(i) A collection of integers $s_{i,k}; i = 1, \dots, r, k = 0, 1, \dots, p - 1$ defined by

$$s_{i,k} = \frac{\overline{d_i + m_i k} - (d_i + m_i k)}{p} \tag{2.1}$$

(ii) A collection of meromorphic functions $f_k = f_k(x)$, $k = 1, \dots, p - 1$ defined by

$$f_k = \prod_{i=1}^{r} (x - a_i)^{s_{i,k}} \tag{2.2}$$

In proving Theorem 1.1 the following simple lemma will be useful

Lemma 2.3 *Let $U = Q_1^{s_1} \cdots Q_r^{s_r}$ be a divisor supported in the ramification locus. Then we have the following direct sum decomposition*

$$L(U) = (L(U) \cap \mathbb{C}(x)) \oplus (L(U) \cap \mathbb{C}(x)y) \oplus \cdots \oplus (L(U) \cap \mathbb{C}(x)y^{p-1})$$

Proof Let $\tau^* : \mathcal{M}(S) \to \mathcal{M}(S)$ denote the map induced by τ on the function field. This is a linear automorphism with (distinct) eigenvalues $1, e^{2\pi i/p}, ..., e^{2\pi i(p-1)/p}$. Therefore $\mathcal{M}(S)$ decomposes as the direct sum of its corresponding eigenspaces, namely

$$\mathcal{M}(S) = \mathbb{C}(x) \oplus \mathbb{C}(x)y \oplus \cdots \oplus \mathbb{C}(x)y^{p-1}$$

Now, as by hypothesis $\tau(U) = U$, $L(U)$ is an invariant $\mathbb{C}$-linear subspace, hence we have a corresponding diagonal decomposition for $L(U)$ as required. □

For the divisor $U = D^{-1}$, D as in Theorem 1.1, Lemma 2.3 implies the following crucial observation.

Corollary 2.4 *Let D be as in Theorem 1.1. Then D is a special divisor if and only if there is a function $h \in L(D^{-1})$ of the form $h = f(x)y^k$, for some $f(x) \in \mathbb{C}(x)$ and some $k \geq 1$.*

Proof By Lemma 2.3 if there is a non constant function $h \in L(D^{-1})$ then there is a non constant function $h_k \in L(D^{-1})$ of the form $h_k = f(x)y^k$. Thus we only have to rule out the case in which $h = h_0 = f(x)$. So, let us write $h = f$ as in Proposition 2.1, *iii*). Then, if $h \in L(D^{-1})$, we must have

$$\prod_{i=1}^{r} Q_i^{ps_i} \cdot \prod_{i=1}^{r} B_j^{t_j} \cdot \infty^{-(\sum s_i + \sum t_j)} \geq Q_1^{-d_1} \cdots Q_r^{-d_r}$$

From here we infer the following inequalities
1) $ps_i \geq -d_i$, hence $s_i \geq 0$
2) $t_j \geq 0$ and
3) $-(\sum s_i + \sum t_j) \geq 0$, hence $s_i = t_j = 0$.
Thus, h would be a constant function. □

Lemma 2.5 *Let $f = \prod_{i=1}^{r}(x - a_i)^{s_i} \prod_j (x - b_j)^{t_j}$ as in Proposition 2.1, iii) and D as in Theorem 1.1. Then $fy^k \in L(D^{-1})$ if and only if the following numerical conditions hold*
(i) $t_j \geq 0$
(ii) $ps_i \geq -d_i - m_i k$
(iii) $-(\sum s_i + \sum t_j) \geq nk$

Proof $(fy^k) = \prod_{i=1}^{r} Q_i^{ps_i} \cdot \prod_{i=1}^{r} B_j^{t_j} \cdot \infty^{-(\sum s_i + \sum t_j)} \left(\dfrac{Q_1^{m_1 k}, ..., Q_r^{m_r k}}{\infty^{nk}} \right) =$

$\prod_{i=1}^{r} B_{j}^{t_j} \prod_{i=1}^{r} Q_{i}^{ps_i + m_i k} \infty^{-(nk + \sum s_i + \sum t_j)}$. This divisor is bigger than $D^{-1} = Q_1^{-d_1} \cdots Q_r^{-d_r}$ if and only if conditions (i), (ii) and (iii) hold. $\quad\square$

Corollary 2.6 *There exists $f \in \mathbb{C}(x)$ such that $fy^k \in L(D^{-1})$ if and only if there are integers $s_i \in \mathbb{Z}$ satisfying the following conditions*
(i) $ps_i \geq -d_i - m_i k$
(ii) $\sum s_i \leq -nk$
Moreover, in this situation f can be chosen to be $f = \prod_{i=1}^{r} (x - a_i)^{s_i}$.

Proof If such integers s_i exist then one can take $f = \prod_{i=1}^{r} (x - a_i)^{s_i}$ (Lemma above with $t_j = 0$).

Conversely, if $f = \prod_{i=1}^{r} (x - a_i)^{s_i} \prod_j (x - b_j)^{t_j}$ is a function such that $fy^k \in L(D^{-1})$ then conditions (i) and (iii) of the previous lemma imply condition (ii) of this corollary. $\quad\square$

2.1 Proof of Theorem 1.1

2.1.1 (i) $\Leftrightarrow$ (ii)

By Corollary 2.4 it is enough to prove the following

Proposition 2.7 *Let $1 \leq k \leq p-1$. Then there exists $h_k \in \mathbb{C}(x)$ such that $h_k(x)y^k \in L(D^{-1})$ if and only if $\sum \overline{d_i + m_i k} \leq g$.*

Proof This is an easy application of Corollary 2.6. Suppose that $\sum_{i=1}^{r} \overline{d_i + m_i k} \leq g$ for some $k \in \{1, ..., p-1\}$. Then we would have $p\sum s_{i,k} = \sum \overline{d_i + m_i k} - \sum d_i - npk = \sum \overline{d_i + m_i k} - g - npk \leq -pnk$. On the other hand, by definition, $ps_{i,k} \geq -d_i - m_i k$. Thus, the two conditions in Corollary 2.6 for h_k to exist hold.

Conversely, if such function h_k exists then, by Corollary 2.6, there must be integers $\widetilde{s}_{i,k} \in \mathbb{Z}$ such that $p\widetilde{s}_{i,k} \geq -d_i - m_i k$ and $p\sum \widetilde{s}_{i,k} \leq -pnk$. Therefore we have $p(\widetilde{s}_{i,k} - s_{i,k}) \geq -d_i - m_i k - ps_{i,k} = -d_i - m_i k - (\overline{d_i + m_i k} - d_i - m_i k) = -\overline{d_i + m_i k} > -p$; in other words $\widetilde{s}_{i,k} \geq s_{i,k}$. From here we infer that $\sum \overline{d_i + m_i k} - g - pnk = \sum ps_{i,k} \leq \sum p\widetilde{s}_{i,k} \leq -pnk$ hence $\sum \overline{d_i + m_i k} - g \leq 0$ as wanted. $\quad\square$

The final statement in Theorem 1.1 follows at once from the final statement in Corollary 2.6.

$$2.1.2 \ (ii) \Leftrightarrow (iii)$$

Clearly $[\overline{\sum(d_i + m_i k)}] = [\sum d_i + \sum m_i k] = [g + np] = [g]$, hence (ii) implies that

$$\sum_{i=1}^{r} \overline{d_i + m_i k} = g + pN_k \quad \text{for some positive integer } N_k \qquad (2.3)$$

It remains to be seen that, for each k, $N_k = 1$. This obviously follows from the following lemma

Lemma 2.8 *For any divisor D such that*

$$D = Q_1^{d_1} \cdots Q_r^{d_r} \ \text{with } 0 \le d_i \le p-1 \ \text{and} \ \sum d_i = g$$

we have

$$\sum_{k=1}^{p-1} \sum_{i=1}^{r} \overline{d_i + m_i k} = (p-1)(g+p)$$

Proof For each $d = 0, 1, \ldots, p-1$ and $m = 1, \ldots, p-1$ we denote by $x_{d,m}$ the number of ramification points Q_i with rotation number m and multiplicity d on D. Before we proceed we observe that

$$\sum_{m=1}^{p-1} \sum_{d=0}^{p-1} x_{d,m} = r, \ \text{the total number of ramification points}$$

and

$$\sum_{m=1}^{p-1} \sum_{d=0}^{p-1} x_{d,m} \cdot d = g, \ \text{the degree of the divisor } D.$$

We then have

$$\sum_{k=1}^{p-1}\sum_{i=1}^{r}\overline{d_i+m_ik} = \sum_{k=1}^{p-1}\sum_{m=1}^{p-1}\sum_{d=0}^{p-1}x_{d,m}\cdot\overline{d+mk}$$

$$= \sum_{m=1}^{p-1}\sum_{d=0}^{p-1}x_{d,m}\left(\sum_{k=1}^{p-1}\overline{d+mk}\right)$$

$$= \sum_{m=1}^{p-1}\sum_{d=0}^{p-1}x_{d,m}\left(\sum_{l=1}^{d-1}l+\sum_{l=d+1}^{p-1}l\right)$$

$$= \sum_{m=1}^{p-1}\sum_{d=0}^{p-1}x_{d,m}\left(\frac{p(p-1)}{2}-d\right)$$

$$= r\frac{p(p-1)}{2}-g = (r-2)\frac{p(p-1)}{2}+p(p-1)-g$$

$$= pg+p(p-1)-g = (p-1)(g+p)$$

$\square$

2.1.3 (iii) ⇔ (iv)

This clearly follows from the lemma above.

2.2 Corrigenda

Lemma 2.3 above is the correct version of Lemma 2.3 in my paper [G]. That lemma, as it stands, is incorrect. As a result the choice of non-special divisors D_i, V_i made in Proposition 2.1 of [G] with the property that D_i/V_i is the p-root of the divisor of the function $(x-a_i)/(x-a_r)$, i.e. such that $D_i/V_i = Q_i/Q_r$, should be modified so as to meet the criterion established in Theorem 1.1 above. This can be readily done, for instance, for the families of p-gonal Riemann surfaces

$$y^p = (x-a_1)\cdots(x-a_{np})$$

defined by the rotation data $(1,\ldots,1)$, and

$$y^p = (x-a_1)\cdots(x-a_n)\left((x-a_{n+1})\cdots(x-a_{2n})\right)^{p-1}$$

corresponding to the rotation data $(1,\ldots,1,p-1,\ldots,p-1)$, whose non-special divisors have been listed in section 3 (see Proposition 3.3 and Corollary 3.7).

As a general description of the set of non-special divisors which is valid for families corresponding to arbitrary rotation data $m_1,\ldots,m_r$ looks unmanageable (see Remark 3.8), the choices of D_i and V_i will have to be made for each family individually. Clearly, Theorem 1.1 above is the right tool to do that since it enables us to produce the list of non-special divisors of any given family (perhaps through a simple computer program, if p is large). Alternatively, one can set $D_i = Q_iU, V_i = Q_rU$ where U is a degree $(g-1)$ integral divisor such that D_i and V_i are non-special; a choice of such U can be made, since the degree g special divisors on S are a subvariety of $S^{(g)}$ of codimension at least 2 (see [K], [Mum]). Note, however, that this general result does not guarantee that the divisor U can be always chosen with support in the ramification set (in fact, see Example 3.2); the consequence being that the points in $J(S)$ which – via the Abel-Jacobi map – correspond to the divisors D_i and V_i may not be points of order p of the abelian variety $J(S)$ as it would be the case otherwise.

3 Applications

3.1 A couple of interesting examples

Theorem 1.1 provides an easy recipe to check if an explicitly given divisor D is special. Let us see how it works in two concrete examples.

Example 3.1 *(Klein's Riemann surface of genus 3)*

$$y^7 = (x - a_1)(x - a_2)^2(x - a_3)^4$$

1) Let us check if the divisor $D = Q_1^2Q_2 = Q_1^2Q_2Q_3^0$ is non-special.
We have to compute the integers $\overline{d_i + m_ik}$ for $i = 1, 2, 3$ and $k = 1, \ldots, 6$. In this case we have $p = 7, d_1 = 2, d_2 = 1, d_3 = 0$ and $m_1 = 1, m_2 = 2, m_3 = 4$. We get the following table of values

	$\overline{d_i + m_i}$	$\overline{d_i + m_i2}$	$\overline{d_i + m_i3}$	$\overline{d_i + m_i4}$	$\overline{d_i + m_i5}$	$\overline{d_i + m_i6}$
$i = 1$	3	4	5	6	0	1
$i = 2$	3	5	0	2	4	6
$i = 3$	4	1	5	2	6	3
$\sum \overline{d_i + m_ik}$	10	10	10	10	10	10

Thus, for all $k = 1, \ldots, 6$ we obtain $\sum \overline{d_i + m_i k} > 3 = g$ and therefore $D = Q_1^2 Q_2$ is not special.

2) Let us look now at the divisor $D = Q_1^3 = Q_1^3 Q_2^0 Q_3^0$.

When we compute the integers $\overline{d_i + m_i 4}$ we find

$$\overline{d_1 + m_1 4} = \overline{3 + 1 \cdot 4} = 0$$
$$\overline{d_2 + m_2 4} = \overline{0 + 2 \cdot 4} = 1$$
$$\overline{d_3 + m_3 4} = \overline{0 + 4 \cdot 4} = 2$$

hence $\sum \overline{d_i + m_i 4} = 3$ and so our criterion tells us that $D = Q_1^3$ is special. But not only that, the criterion also gives us a non-constant function whose pole divisor is bounded by D, namely the function

$$f_k y^k = (x - a_1)^{s_{1,4}} (x - a_2)^{s_{2,4}} (x - a_3)^{s_{3,4}} y^4 = (x - a_1)^{-1} (x - a_2)^{-1} (x - a_3)^{-2} y^4$$

whose divisor is $Q_1^{-3} Q_2 Q_3^2$.

3) It turns out that the integral divisors supported on the branch locus are precisely

$$D = Q_1^2 Q_2, \ Q_1 Q_3^2, \ Q_1 Q_2 Q_3, \ Q_2^2 Q_3$$

Example 3.2 *(the 7-gonal hyperelliptic Riemann surface of genus 3)*

$$y^7 = (x - a_1)(x - a_2)(x - a_3)^5$$

We want to find out which divisors of the form $D = Q_1^{d_1} Q_2^{d_2} Q_3^{d_3}$, with $d_1 + d_2 + d_3 = g$, are non-special, thus we have to compute the integers $\overline{d_i + m_i k}$ for $i = 1, 2, 3$ and $k = 1, \ldots, 6$. In this case we have $p = 7, m_1 = 1, m_2 = 1, m_3 = 5$ and $d_1 + d_2 + d_3 = 3$, hence $d_i \leq 3$.

For D to be non-special the following table of values must hold

	$\overline{d_i + m_i}$	$\overline{d_i + m_i 2}$	$\overline{d_i + m_i 3}$	$\overline{d_i + m_i 4}$	$\overline{d_i + m_i 5}$	$\overline{d_i + m_i 6}$
$i = 1$	$\overline{d_1 + 1}$	$\overline{d_1 + 2}$	$\overline{d_1 + 3}$	$\overline{d_1 + 4}$	$\overline{d_1 + 5}$	$\overline{d_2 + 6}$
$i = 2$	$\overline{d_2 + 1}$	$\overline{d_2 + 2}$	$\overline{d_2 + 3}$	$\overline{d_2 + 4}$	$\overline{d_2 + 5}$	$\overline{d_2 + 6}$
$i = 3$	$\overline{d_3 + 5}$	$\overline{d_3 + 3}$	$\overline{d_3 + 1}$	$\overline{d_3 + 6}$	$\overline{d_3 + 4}$	$\overline{d_3 + 2}$
$\sum \overline{d_i + m_i k}$	10	10	10	10	10	10

From the first column we infer that, in that case, $d_3 = 0, 1$. On the other hand, the 4-th column rules out the value $d_3 = 0$, hence we must have $d_3 = 1$. But then, the 5-th column shows that $d_i \neq 1, i = 1, 2$. The conclusion is that in this case we only have four non-special divisors, namely

$$Q_1^2 Q_3, Q_2^2 Q_3, Q_2^3 \text{ and } Q_1^3$$

3.2 The case of equal rotation numbers

Let us consider the case in which all rotation numbers m_i agree. In fact, by replacing the automorphism τ by a suitable power τ^k we can assume that $m_i = 1$ for every i. That is, we can assume that S is a Riemann surface of genus $g = \frac{p-1}{2}(np - 2)$ given by an equation of the form

$$y^p = (x - a_1) \cdots (x - a_{np}) \qquad (3.1)$$

Let D be a divisor of degree g supported on the branch locus. We can write it in the form

$$D = D_0^0 \cdot D_1 \cdot D_2^2 \cdot D_3^3 \cdots D_{p-2}^{p-2} \cdot D_{p-1}^{p-1}$$

where

- $D_0 = Q_1 \cdots Q_{r_0}$ (and so $Q_1, \ldots, Q_{r_0}$ are the points which are omitted)
- $D_1 = Q_{r_0+1} \cdots Q_{r_0+r_1}$
- $D_2 = Q_{r_0+r_1+1} \cdots Q_{r_0+r_1+r_2}$
- $D_3 = Q_{r_0+r_1+r_2+1} \cdots Q_{r_0+r_1+r_2+r_3}$

• $\cdots$

• $D_{p-1} = Q_{r_0 + \cdots + r_{p-2} + 1} \cdots Q_{r_0 + \cdots + r_{p-2} + r_{p-1}}$

We have the following relations among the non-negative integers r_i

$$r_0 + r_1 + r_2 + \cdots + r_{p-2} + r_{p-1} = pn$$

and

$$0 \cdot r_0 + 1 \cdot r_1 + 2 \cdot r_2 + \cdots + (p-2) \cdot r_{p-2} + (p-1) \cdot r_{p-1} = g$$

Let us now work out the integers $\overline{d_i + m_i k} = \overline{d_i + k}$; the last equality because we are now assuming that each m_i equals 1.

For $k = 1$ we get

$$\overline{d_i + m_i \cdot 1} = \begin{cases} \overline{0+1} & = & 1 & \text{(for the } r_0 \text{ points in } D_0) \\ \overline{1+1} & = & 2 & \text{(for the } r_1 \text{ points in } D_1) \\ \overline{2+1} & = & 3 & \text{(for the } r_1 \text{ points in } D_2) \\ \cdots & = & \cdots & \cdots \\ \overline{(p-3)+1} & = & p-2 & \text{(for the } r_{p-3} \text{ points in } D_{p-3}) \\ \overline{(p-2)+1} & = & p-1 & \text{(for the } r_{p-2} \text{ points in } D_{p-2}) \\ \overline{(p-1)+1} & = & 0 & \text{(for the } r_{p-1} \text{ points in } D_{p-1}) \end{cases}$$

According to our criterion for D to be non-special we need to have

$$1 \cdot r_0 + 2 \cdot r_1 + 3 \cdot r_2 + \cdots + (p-1) \cdot r_{p-2} + 0 \cdot r_{p-1} = g + p$$

For $k = 2$ we get

$$\overline{d_i + m_i 2} = \begin{cases} \overline{0+2} & = & 2 & \text{(for the } r_0 \text{ points in } D_0) \\ \overline{1+2} & = & 3 & \text{(for the } r_1 \text{ points in } D_1) \\ \overline{2+2} & = & 4 & \text{(for the } r_1 \text{ points in } D_2) \\ \cdots & = & \cdots & \cdots \\ \overline{(p-3)+2} & = & p-1 & \text{(for the } r_{p-3} \text{ points in } D_{p-3}) \\ \overline{(p-2)+2} & = & 0 & \text{(for the } r_{p-2} \text{ points in } D_{p-2}) \\ \overline{(p-1)+2} & = & 1 & \text{(for the } r_{p-1} \text{ points in } D_{p-1}) \end{cases}$$

As in the previous case for D to be non-special we need to have

$$2 \cdot r_0 + 3 \cdot r_1 + 4 \cdot r_2 + \cdots + 0 \cdot r_{p-2} + 1 \cdot r_{p-1} = g + p$$

We proceed in the same manner to obtain the corresponding equations for all $k = 1, \ldots, p-1$. For instance, the last one is

$$(p-1) \cdot r_0 + 0 \cdot r_1 + 1 \cdot r_2 + \cdots + (p-3) \cdot r_{p-2} + (p-2) \cdot r_{p-1} = g + p$$

This way we obtain a linear system consisting of a $p+1$ linear equation in the unknowns $r_0, r_1, \ldots, r_{p-1}$. To solve it we employ the Gaussian elimination algorithm. The matrix corresponding to this linear system is

$$A = \left(\begin{array}{cccccc|c} 1 & 1 & 1 & \cdots & 1 & 1 & np \\ 0 & 1 & 2 & \cdots & p-2 & p-1 & g \\ 1 & 2 & 3 & \cdots & p-1 & 0 & g+p \\ 2 & 3 & 4 & \cdots & 0 & 1 & g+p \\ \cdots & \cdots & \cdots & \cdots & \cdots & \cdots & \cdots \\ p-2 & p-1 & 0 & \cdots & p-4 & p-3 & g+p \\ p-1 & 0 & 1 & \cdots & p-3 & p-2 & g+p \end{array} \right)$$

Now, obvious elementary operations among the rows of the matrix A give the matrix

$$A_1 = \left(\begin{array}{cccccc|c} 1 & 1 & 1 & \cdots & 1 & 1 & np \\ 0 & 1 & 2 & \cdots & p-2 & p-1 & g \\ 1 & 1 & 1 & \cdots & 1 & 1-p & p \\ 1 & 1 & 1 & \cdots & 1-p & 1 & 0 \\ \cdots & \cdots & \cdots & \cdots & \cdots & \cdots & \cdots \\ 1 & 1 & 1-p & \cdots & 1 & 1 & 0 \\ 1 & 1-p & 1 & \cdots & 1 & 1 & 0 \end{array} \right)$$

and then

$$A_2 = \left(\begin{array}{cccccc|c} 1 & 1 & 1 & \cdots & 1 & 1 & np \\ 0 & 1 & 2 & \cdots & p-2 & p-1 & g \\ 0 & 0 & 0 & \cdots & 0 & p & np-p \\ 0 & 0 & 0 & \cdots & p & 0 & np \\ \cdots & \cdots & \cdots & \cdots & \cdots & \cdots & \cdots \\ 0 & 0 & p & \cdots & 0 & 0 & np \\ 0 & p & 0 & \cdots & 0 & 0 & np \end{array} \right)$$

and from here

$$A_3 = \left(\begin{array}{cccccc|c} 1 & 0 & 0 & \cdots & 0 & 0 & np-(p-1)n+1 \\ 0 & 0 & 0 & \cdots & 0 & 0 & g-\frac{p(p-1)n}{2}+p-1 \\ 0 & 0 & 0 & \cdots & 0 & 1 & n-1 \\ 0 & 0 & 0 & \cdots & 1 & 0 & n \\ \cdots & \cdots & \cdots & \cdots & \cdots & \cdots & \cdots \\ 0 & 0 & 1 & \cdots & 0 & 0 & n \\ 0 & 1 & 0 & \cdots & 0 & 0 & n \end{array} \right)$$

Now the solutions of our linear system can be read off the matrix A_3.

We find that

$$r_0 = n + 1, r_1 = n, \ldots, r_{p-2} = n, r_{p-1} = n - 1$$

In other words we have proved the following result

Proposition 3.3

Let S be a p-gonal Riemann surface whose branch locus consists of pn ramification points all of them having the same rotation number. Then, a degree g integral divisor D supported on the branch locus is non-special if and only if D is of the form

$$(R_1 \cdots R_n) \cdots (R_{(p-3)n+1} \cdots R_{(p-2)n})^{p-2} (R_{(p-2)n+1} \cdots R_{(p-1)n-1})^{p-1}$$

where $\{R_1 \cdots R_{(p-1)n-1}\}$ is any collection of $np - (n+1)$ ramification points.

Example 3.4 *(hyperelliptic case, $p = 2$)*

$$y^2 = (x - a_1) \cdots (x - a_{2(g+1)})$$

In this case the non-special divisors are those of the form $D = Q_{i_1} \cdots Q_{i_g}$.

Example 3.5 *(the trigonal case, $p = 3$)*

$$y^3 = (x - a_1) \cdots (x - a_{3n}) \tag{3.2}$$

The non-special divisors supported on the branch locus are those of the form

$$(Q_{i_1} \cdots Q_{i_n})(Q_{i_{n+1}} \cdots Q_{i_{2n-1}})^2$$

In [EF] Thomae type formulae for Riemann surfaces of the form (3.2), with $n = 2$, have been investigated. In that case the relevant divisors take the form $(Q_{i_1} Q_{i_2}) Q_{i_3}^2$.

3.3 The case of two different rotation angles

Let us consider the Riemann surface S given by an equation of the form

$$y^p = (x - a_1) \cdots (x - a_n) ((x - a_{n+1}) \cdots (x - a_{2n}))^{p-1} \tag{3.3}$$

Its genus is

$$g = (p - 1)(n - 1)$$

Let D be a divisor of degree g supported on the branch locus. Let r_d

(resp. t_d) be the number of points with rotation number 1 (resp. $p - 1$) and multiplicity $0 \le d \le p - 1$ on D. We have

$$
\begin{cases}
\sum_{d=0}^{p-1} r_d = n \\[2mm]
\sum_{d=0}^{p-1} t_d = n \\[2mm]
\sum_{d=0}^{p-1} dr_d + \sum_{d=0}^{p-1} dt_d = g
\end{cases}
\tag{3.4}
$$

With this notation Theorem 1.1 tells us that a necessary and sufficient condition for our divisor D to be non-special is

$$
\sum_{d=0}^{p-1} \overline{(d + k)} r_d + \sum_{d=0}^{p-1} \overline{(d + (p-1)k)} t_d = g + p \quad \text{for every } k = 1, ..., p - 1
\tag{3.5}
$$

All these $p+2$ conditions can be assembled together in a linear system which in Gaussian terminology can be represented as follows

$$
\left(
\begin{array}{ccccccc|c}
1 & 1 & \cdots & 1 & 0 & 0 & \cdots & 0 & n \\
0 & 0 & \cdots & 0 & 1 & 1 & \cdots & 1 & n \\
0 & 1 & \cdots & p-1 & 0 & 1 & \cdots & p-1 & g \\
1 & 2 & \cdots & 0 & p-1 & 0 & \cdots & p-2 & g+p \\
2 & 3 & \cdots & 1 & p-2 & p-1 & \cdots & p-3 & g+p \\
3 & 4 & \cdots & 2 & p-3 & p-2 & \cdots & p-4 & g+p \\
\cdots & \cdots & \cdots & \cdots & \cdots & \cdots & \cdots & \cdots \\
p-3 & p-2 & \cdots & p-4 & 3 & 4 & \cdots & 2 & g+p \\
p-2 & p-1 & \cdots & p-3 & 2 & 3 & \cdots & 1 & g+p \\
p-1 & 0 & \cdots & p-2 & 1 & 2 & \cdots & 0 & g+p
\end{array}
\right)
$$

Now, keeping fixed the first three rows and subtracting from each of the remaining rows the previous one we obtain

$$\begin{pmatrix}
1 & 1 & \cdots & 1 & 0 & 0 & \cdots & 0 & 0 & \vdots & n \\
0 & 0 & \cdots & 0 & 1 & 1 & \cdots & 1 & 1 & \vdots & n \\
0 & 1 & \cdots & 0 & 1 & 2 & \cdots & p-2 & p-1 & \vdots & g \\
1 & 1 & \cdots & 1-p & p-1 & -1 & \cdots & -1 & -1 & \vdots & p \\
1 & 1 & \cdots & 1 & -1 & p-1 & \cdots & -1 & -1 & \vdots & 0 \\
1 & 1 & \cdots & 1 & -1 & -1 & \cdots & -1 & -1 & \vdots & 0 \\
\cdots & \cdots & \cdots & \cdots & \cdots & \cdots & \cdots & \cdots & \cdots & \vdots & \cdots \\
1 & 1 & \cdots & -1 & -1 & -1 & \cdots & -1 & -1 & \vdots & 0 \\
1 & 1 & \cdots & 1 & -1 & -1 & \cdots & -1 & -1 & \vdots & 0 \\
1 & 1-p & \cdots & 1 & -1 & -1 & \cdots & p-1 & -1 & \vdots & 0
\end{pmatrix}$$

The next operations we perform are as follows: we leave the first two rows untouched, we erase the third row and subtract the first row from the remaining ones. This way we obtain

$$\begin{pmatrix}
1 & 1 & \cdots & 1 & 0 & 0 & \cdots & 0 & 0 & \vdots & n \\
0 & 0 & \cdots & 0 & 1 & 1 & \cdots & 1 & 1 & \vdots & n \\
0 & 0 & \cdots & -p & p-1 & -1 & \cdots & -1 & -1 & \vdots & p-n \\
0 & 0 & \cdots & 0 & -1 & p-1 & \cdots & -1 & -1 & \vdots & -n \\
0 & 0 & \cdots & 0 & -1 & -1 & \cdots & -1 & -1 & \vdots & -n \\
\cdots & \cdots & \cdots & \cdots & \cdots & \cdots & \cdots & \cdots & \cdots & \vdots & \cdots \\
0 & 0 & \cdots & 0 & -1 & -1 & \cdots & -1 & -1 & \vdots & -n \\
0 & 0 & \cdots & 0 & -1 & -1 & \cdots & -1 & -1 & \vdots & -n \\
0 & -p & \cdots & 0 & -1 & -1 & \cdots & p-1 & -1 & \vdots & -n
\end{pmatrix}$$

Now we first multiply the first row by p and the second one by (-1). Then, we add to the first row all the other p rows. We get

$$\begin{pmatrix}
p & 0 & \cdots & 0 & 0 & 0 & \cdots & 0 & -p & \vdots & p \\
0 & 0 & \cdots & 0 & -1 & -1 & \cdots & -1 & -1 & \vdots & -n \\
0 & 0 & \cdots & -p & p-1 & -1 & \cdots & -1 & -1 & \vdots & p-n \\
0 & 0 & \cdots & 0 & -1 & p-1 & \cdots & -1 & -1 & \vdots & -n \\
0 & 0 & \cdots & 0 & -1 & -1 & \cdots & -1 & -1 & \vdots & -n \\
\cdots & \cdots & \cdots & \cdots & \cdots & \cdots & \cdots & \cdots & \cdots & \vdots & \cdots \\
0 & 0 & \cdots & 0 & -1 & -1 & \cdots & -1 & -1 & \vdots & -n \\
0 & 0 & \cdots & 0 & -1 & -1 & \cdots & -1 & -1 & \vdots & -n \\
0 & -p & \cdots & 0 & -1 & -1 & \cdots & p-1 & -1 & \vdots & -n
\end{pmatrix}$$

In our last step we subtract the second row from each of the rows below it. We finally obtain

$$
\begin{pmatrix}
p & 0 & \cdots & 0 & 0 & 0 & \cdots & 0 & 0 & -p & \vline & p \\
0 & 0 & \cdots & 0 & -1 & -1 & \cdots & -1 & -1 & -1 & \vline & -n \\
0 & 0 & \cdots & -p & p & 0 & \cdots & 0 & 0 & 0 & \vline & p \\
0 & 0 & \cdots & 0 & 0 & p & \cdots & 0 & 0 & 0 & \vline & 0 \\
0 & 0 & \cdots & 0 & 0 & 0 & \cdots & 0 & 0 & 0 & \vline & 0 \\
\cdots & \cdots & \cdots & \cdots & \cdots & \cdots & \cdots & \cdots & \cdots & \cdots & \vline & \cdots \\
0 & 0 & \cdots & 0 & 0 & 0 & \cdots & 0 & 0 & 0 & \vline & 0 \\
0 & 0 & \cdots & 0 & 0 & 0 & \cdots & p & 0 & 0 & \vline & 0 \\
0 & -p & \cdots & 0 & 0 & 0 & \cdots & 0 & p & 0 & \vline & 0
\end{pmatrix}
$$

Thus, we infer the following relations among the unknowns r_i, t_i

$$
\begin{cases}
r_0 & + & -t_{p-1} & = & 1 \\
-r_{p-1} & + & t_0 & = & 1 \\
-r_{p-2} & + & t_1 & = & 0 \\
-r_{p-3} & + & t_2 & = & 0 \\
\cdots & \cdots & \cdots & \cdots & \cdots \\
-r_{p-k} & + & t_{k-1} & = & 0 \\
\cdots & \cdots & \cdots & \cdots & \cdots \\
-r_2 & + & t_{p-3} & = & 0 \\
-r_1 & + & t_{p-2} & = & 0
\end{cases}
\Leftrightarrow
\begin{cases}
t_0 & = & 1 + r_{p-1} \\
t_1 & = & r_{p-2} \\
t_2 & = & r_{p-3} \\
t_3 & = & r_{p-4} \\
\cdots & \cdots & \cdots \\
t_k & = & r_{p-k-1} \\
\cdots & \cdots & \cdots \\
t_{p-2} & = & r_1 \\
t_{p-1} & = & r_0 - 1
\end{cases}
\tag{3.6}
$$

This allows a full description of the set of non-special divisors.

Proposition 3.6 *Let S be any p-gonal Riemann surface whose ramification set consists of $n \geq 2$ ramification points of rotation number 1 and n ramification points of rotation number $p-1$. Then, a degree g integral divisor D supported on the branch locus is non-special if and only if D is of the form*

$$D = (T_1 \cdots T_{r_0-1})^{p-1}$$
$$(R_{r_0} \cdots R_{r_0+r_1-1})(T_{r_0} \cdots T_{r_0+r_1-1})^{p-2}$$
$$(R_{r_0+r_1} \cdots R_{r_0+r_1+r_2-1})^2 (T_{r_0+r_1} \cdots T_{r_0+r_1+r_2-1})^{p-3}$$
$$\cdots$$
$$(R_{r_0+r_1+\cdots+r_{p-2}} \cdots R_{r_0+r_1+\cdots+r_{p-1}-1})^{p-1}$$

where $(r_0, r_1, \ldots, r_{p-1})$ is any p-tuple of non-negative integers such that

$$r_0 \geq 1, \quad \sum_{d=0}^{p-1} r_d = n$$

and $\{T_1, \ldots, T_{r_1+\cdots+r_{p-2}-1}\}$ (resp. $\{R_{r_0} \cdots R_{r_0+r_1+\cdots+r_{p-1}-1}\}$) is any choice of $n - r_{p-1} - 1$ (resp. $n - r_0$) ramification points with rotation number $p - 1$ (resp. 1).

In particular we have

Corollary 3.7 *Let S be as in Proposition 3.6.*

(i) The set of non-special divisors on S supported on the ramification set such that all points in the support have multiplicity $p - 1$ consists of the following divisors

$$D = (R_1 \cdots R_a)^{p-1}(T_1 \cdots T_{n-1-a})^{p-1}$$

where $0 \leq a \leq n - 1$ and $R_1 \cdots R_a$ (resp. $T_1 \cdots T_{n-1-a}$) are arbitrary distinct points with rotation number 1 (resp. $p - 1$).

(ii) The set of non-special divisors on S supported on the ramification set such that the multiplicity at all points but one of its support is $p - 2$ consists of the following divisors

$$D = R_1^{p-k-1}(R_2 \cdots R_{b+1})^{p-1}(T_1 \cdots T_{n-2-b})^{p-1}T_{n-b-1}^k$$
$$\left(resp. \quad V = T_1^{p-k-1}(T_2 \cdots T_{b+1})^{p-1}(R_1 \cdots R_{n-2-b})^{p-1}R_{n-b-1}^k \right)$$

where $1 \leq k \leq p - 2$, $0 \leq b \leq n - 2$ and the R_i's (resp. the T_j's) are arbitrary distinct points with rotation number 1 (resp. $p - 1$).

In [EG] Thomae type formulae for Riemann surfaces of the form (3.3) have been investigated. The non-special divisors used to obtain those formulae are among the ones described in Corollary 3.7(i) and Corollary 3.7(ii) with $k = 1$.

Remark 3.8 *An explicit description of the set of non-special divisors for arbitrary families, although doable, seems to be too much involved to be of any use. For instance, already in the case of the general curve with two arbitrary rotation numbers*

$$y^p = (x - a_1) \cdots (x - a_{n_1})\left((x - a_{n_1+1}) \cdots (a_{n_1+n_2})\right)^m \ ; \quad n_1 + n_2 m = np$$

the simple formulae given in (3.6) to express the relation between the r_i's and the t_j's take now the following rather involved form

$$\begin{cases} r_0 & = \sum_{r=m}^{p-1} t_r & + \quad n - n_2 + 1 \\[2ex] r_k & = \sum_{r=m}^{p-1} t_{\overline{km+r}} & + \quad n - n_2 \quad if \ k \neq 0, p-1 \\[2ex] r_{p-1} & = \sum_{r=m}^{p-1} t_{\overline{-m+r}} & + \quad n - n_2 - 1 \end{cases}$$

Notes

1 Research partially supported by the MCE research project MTM2006-01859

References

[Ac] Accola, R. Riemann Surfaces, Theta Functions and Abelian Automorphism Groups. *Lecture Notes in Mathematics* vol. 483 Springer-Verlag.

[BR] Bershadsky, M. and Radul, A. Fermionic fields on Z_N-curves. *Comm. Math. Phys.* **116**(4) (1988), 689–700.

[EbF] Ebin, D. G. and Farkas, H. M. Thomae's Formula for Z_N curves, to appear in *Journal d'Analyse Mathématique.*

[EF] Eisenmann, A. and Farkas, H. M. An elementary proof of Thomae's formulae *Online Journal of Analytic Combinatorics*, (3) (2008).

[EG] Enolski, V. Z. and Grava, T. Thomae type formulae for singular Z_N curves. *Lett. Math. Phys.* **76**(2–3) (2006), 187–214.

[Far] Farkas, H. M. Generalizations of the λ functions. *Israel Mathematical Conference Proceedings* **9** (1996), 231–239.

[FK] Farkas, H. M. and Kra, I. *Riemann Surfaces.* Graduate Texts in Mathematics **71**, Springer-Verlag, New York and Berlin, 1980.

[Fro] Frobenius, F. G. Uber die constanten Factoren der Thetareihen. *Crelle's Journal,* **98** (1885).

[G] González-Díez, G. Loci of curves which are prime Galois coverings of $\mathbb{P}^1$. *Proc. London Math. Soc. (3)* 62(3) (1991), 469–489.

[GH] González-Díez, G. and Harvey, W. J. Moduli of Riemann surfaces with symmetry, in Discrete Groups and Geometry, London Math. Soc. Lecture Note Ser. **173**, Cambridge University Press, Cambridge, 1992, pp. 75–93.

[H] Harvey, W. J. On branchi loci in Teichmüller space. *Trans. Amer. Math. Soc.* **153** (1971), 387–399.

[Ko] Kopeliovich, Y. Non-vanishing divisors for general cyclic covers. (preprint)

[K] Kuribayashi, A. On the generalized Teichmuller spaces and differential equations. *Nagoya Math. J.* **64** (1976), 97–115.

[Mat] Matsumoto, K. Theta constants associated with the cyclic triple coverings of the complex projective line branching at six points. *Publ. Res. Inst. Math. Sci.* **37**(3) (2001), 419–440.

[Mum] Mumford, D. *Tata Lectures on Theta I and II*, Birkhauser, 1983.

[N] Nakayashiki, A. On the Thomae formula for Z_N curves. *Publ. Res. Inst. Math. Sci.* **33**(6) (1997), 987–1015.

[P] Piponi, D. A generalization of Thomae's formula for cyclic covers of the sphere, thesis, supervised by W. J. Harvey, 1993, *King's College London*.

[T1] Thomae, J. Bestimmung von dlg $\vartheta(0, \ldots, 0)$ durch die Klassenmoduln. *Crelle's Journal* **66** (1866), 92–96.

[T2] Thomae, J. Beitrag zur bestimug von $\theta(0, \ldots, 0)$ durch die Klassenmoduln algebraischer Funktionen. *Crelle's Journal* **71** (1870).

A note on the lifting of automorphisms

Rubén A. Hidalgo [1]

Departamento de Matemática, Universidad Técnica Federico Santa María
ruben.hidalgo@usm.cl

Bernard Maskit

Department of Mathematics, Stony Brook University
bernie@math.sunysb.edu

1 Introduction

The goal of this paper is to study uniformizations of surfaces and orbifolds (either Riemann or Klein). There is of course a well developed theory of regular coverings based on the correspondence with subgroups of the fundamental group. This theory can be applied to branched regular coverings by removing the discrete set of branch points and their preimages. There is however no known extension of the theory of regular coverings to include folded coverings; this is the case where the covering group contains elements with real co-dimension one sets of fixed points. In this paper we take a first step towards laying a foundation for the study of such coverings.

We start with some necessary definitions. Let S be an analytically finite Riemann surface or orbifold; then S_0 denotes the subsurface obtained by removing the (finitely many) orbifold points. A *marked system of loops* on S is a collection $\{(u_1, k_1), \ldots, (u_n, k_n)\}$ where $u_1, \ldots, u_n$ are n homotopically distinct pairwise disjoint simple loops on S_0 and $k_1, \ldots, k_n$ are positive integers. We require further that, for each orbifold point $z \in S$, there is an index i, so that k_i is the order of z, and, in an appropriate parameter disc, u_i bounds a disc on S containing z and no other orbifold point. Let $P : \widetilde{S_0} \to S_0$ be a regular planar covering of S_0, that is, $\widetilde{S_0}$ is planar. The marked system of loops, $\{(u_1, k_1), \ldots, (u_n, k_n)\}$, *determines* this covering if the following holds.

(1) Every lift to $\widetilde{S_0}$ of each of the multi-loops $w_i = u_i^{k_i}$, $i = 1, \ldots, n$, is a simple loop.

(2) Every loop in $\widetilde{S_0}$ is freely homotopic to a product of the liftings of the loops, $w_1 = u_1^{k_1}, \ldots, w_n = u_n^{k_n}$.

We remark that this definition can be restated as follows. After choosing base-points, the regular planar covering $P : \widetilde{S}_0 \to S_0$ has a defining subgroup $N \subset \pi_1(S_0)$. The marked system of loops, $\{(u_1, k_1), \ldots, (u_n, k_n)\}$, determines such a planar covering if and only if $u_i^{k_i} \in N$, $i = 1, \ldots, n$, and N is equal to the normal closure of the set of loops $\{u_1^{k_1}, \ldots, u_n^{k_n}\}$.

The following theorem, which is an easy consequence of a result in [M1], provides a technique for constructing planar regular coverings of orientable orbifolds. In fact, the planarity theorem [M1] asserts that, up to homeomorphism, every such planar covering can be constructed in this manner.

Theorem 1.1 ([M1]) *Let S be an analytically finite Riemann surface or orbifold, and let $\{(u_1, k_1), \ldots, (u_n, k_n)\}$ be a marked system of loops on S. This marked system of loops determines a regular covering $P :$ $\widetilde{S}_0 \to S_0$, where $\widetilde{S}_0$ is a plane domain.*

In the above, we note that each orbifold point is surrounded by a simple loop that, when raised to the appropriate power, lifts to a loop, at each possible starting point. It then easily follows that there are well defined preimages of each of the orbifold points on S. That is, there is a plane domain $\widetilde{S}$ containing $\widetilde{S}_0$, obtained by filling in the points above the orbifold points of S. In this case, we obtain a branched regular covering $P : \widetilde{S} \to S$, where, as above, $\widetilde{S}$ is a plane domain and the branching occurs exactly at the orbifold points of the orbifold S and with the same branching order.

Let S be a Riemann surface, let $P : \widetilde{S} \to S$ be a possibly branched regular covering of S and let $h : S \to S$ be a homeomorphism. We say that h *lifts* to $\widetilde{S}$ if there is a homeomorphism $\widehat{h} : \widetilde{S} \to \widetilde{S}$ so that $P\widehat{h} = hP$. If H is a group of homeomorphisms of S, where every $h \in H$ lifts to $\widehat{S}$, then we say that H *lifts* to $\widetilde{S}$. We note that, in this case, if G is the group of deck transformations of $P : \widetilde{S} \to S$, then the set of lifts of H forms a new group of homeomorphisms of $\widetilde{S}$ containing G as a normal subgroup.

In the case that the above covering is branched, so that we can regard S as an orbifold, we remove from $\widetilde{S}$ the set of points where the deck group does not act freely to obtain the subsurface $\widetilde{S}_0$, and we let S_0 be its projection. We note that, in this case, since we require the covering map to be conformal, a homeomorphism of S lifts to $\widetilde{S}$, if and only if the same homeomorphism, when restricted to S_0, lifts to $\widetilde{S}_0$.

2 Main Topological Result

We next state the main result of this paper, giving conditions for a group of automorphisms to lift to a given possibly branched planar covering. This theorem can be viewed as a fixed point theorem on the collection of marked systems of loops on the orbifold S.

Given an automorphism h of S, we say that h *preserves* the marked system of loops, $\{(u_1, k_1), \ldots, (u_n, k_n)\}$, if, for each $i = i, \ldots, n$, there is a j, $1 \leq j \leq n$, so that $h(u_i) = u_j$ and $k_j = k_i$.

The group of automorphisms H *preserves* a marked system of loops if every element of H preserves the marked system of loops. In this case, we also say that the marked system of loops is *invariant* under H.

Theorem 2.1 *Let S be an analytically finite Riemann surface, and let $P : \widetilde{S} \to S$ be a possibly branched planar regular covering, with covering group G. Let S_0 be S with the set of orbifold points removed, and let $\widetilde{S_0}$ be its preimage. Let H be a group of conformal and anti-conformal homeomorphisms acting on S. Then H lifts to a group of homeomorphisms of $\widetilde{S}$ normalizing G if and only if there is a marked system of loops that both determines this covering and is invariant under H.*

Proof In what follows, we will use the natural (hyperbolic, parabolic or elliptic) metric on the Riemann surface S_0. We remark that it is well known that a homeomorphism h of S_0 lifts to $\widetilde{S_0}$ if and only if $h_*(N) = N$, where $N \subset \pi_1(S_0)$ is the defining subgroup for the regular covering $P : \widetilde{S_0} \to S_0$, and h_* is the induced map on the fundamental group.

First suppose that there is a determining marked system of loops, $\{(u_1, k_1), \ldots, (u_n, k_n)\}$ that is preserved by H. The statement that the marked system determines the covering is equivalent to the statement that N is the smallest normal subgroup of $\pi_1(S_0)$ containing the loops $u_1^{k_1}, \ldots, u_n^{k_n}$. Since every $h \in H$ preserves this set of loops, every $h \in H$ preserves N, and so lifts to $\widetilde{S_0}$. Since $\widetilde{S_0}$ is obtained from $\widetilde{S}$ by the deletion of a discrete set of points, the lift of h, which is either directly or inversely conformal, extends to all of $\widetilde{S}$.

Next assume that H lifts to $\widetilde{S}$. Of necessity, every element of H preserves the set of orbifold points on S, with their orders; that is, an element $h \in H$ maps an orbifold point of order k to a perhaps different orbifold point of the same order k. We surround each of these orbifold points with a small loop, and note that this loop, when raised to an appropriate power, lifts to a loop at each possible starting point. Since H is finite, it is easy to ensure that this marked system of loops is preserved by H.

Since there is nothing further to prove if $\widetilde{S}$ is simply connected, we can assume that it is not.

Let W be a shortest geodesic on $\widetilde{S}_0$, and let $w = P(W)$. It is well known that W is necessarily a simple loop.

Proposition 2.2 *The loop w is a simple geodesic loop raised to some power.*

Proof Since W is a geodesic, so is w. Write $w = w(t)$, $0 \leq t \leq 1$. If w were not a power of a simple geodesic, then there would be two points, t_1 and t_2 with $w(t_1) = w(t_2)$ and, further, looking at the action on the tangent space, we would also have $w'(t_1) \neq \pm w'(t_2)$. There are then two distinct lifts of w starting at any point lying over $w(t_1) = w(t_2)$. Since $\widetilde{S}_0$ is planar, these two lifts have a second point of intersection. Cutting and pasting these two loops at these two points of intersection either yields at least one homotopically non-trivial loop that is shorter than W, or yields a homotopically non-trivial loop, whose length is equal to that of W, and which has a corner. In the latter case, the corresponding geodesic will be shorter than W. $\qquad\square$

Proposition 2.3 *Let $J \subset H$ be the stabilizer of w in H. Then w, as a subset of S, is precisely invariant under J in H.*

Proof Suppose there is some $h \in H$, with $h(w) \cap w \neq \emptyset$ and $h(w) \neq w$. Since h is conformal, it is an isometry, so $h(w)$ is a geodesic of the same length as w. Since h lifts to $\widetilde{S}$, $h(w)$ lifts to a geodesic loop W' in $\widetilde{S}_0$. Then, since $\widetilde{S}_0$ is planar, W and W' have at least two points of intersection. As in the proof above (see also [M3, pp. 252–253]), we cut and paste these two geodesics to produce a homotopically non-trivial geodesic on $\widetilde{S}_0$ that is shorter than W, which is impossible. $\qquad\square$

We now finish the proof of Theorem 2.1. If the set of loops, $\{H(w)\}$, together with the small loops about the orbifold points, determines the covering $P : \widetilde{S}_0 \to S$, then we are finished. If not, then let $w_{11}, \ldots, w_{1n}$ be an enumeration of the distinct translates of w. Note that there is a single integer k_1 so that the loops $w_{1i}^{k_1}$, $i = 1, \ldots, n_1$, all lift to loops at every possible starting point; we can also assume that this integer k_1 is minimal.

Now assume that we have found an H-invariant marked system of loops, which we label as $(w_{11}, k_1), \ldots, (w_{jn_j}, k_j)$, so that each $w_{im}^{k_i}$ lifts to a loop, k_i the minimal positive integer with such a property, but that this marked system of loops does not determine the covering. Let $\widetilde{\mathcal{W}}$ denote the set of liftings of these loops raised to appropriate powers; this is a set of simple disjoint loops on $\widetilde{S}_0$. Also, let $\mathcal{W}$ denote the set of loops on S_0 that are the projections of the loops in $\widetilde{\mathcal{W}}$. Then there is some homotopically non-trivial loop W on $\widetilde{S}_0$, which, as an element of $\pi_1(\widetilde{S}_0)$, does not lie in the normal closure of the loops in $\mathcal{W}$.

We next observe that we can choose W to be disjoint from all the loops in $\widetilde{\mathcal{W}}$. Look at the set of points of intersection of W with the various loops in $\widetilde{\mathcal{W}}$. Since $\widetilde{S}_0$ is planar, there are two such points on W that are both points of intersection with the same loop $U \in \widetilde{\mathcal{W}}$, where the two points of intersection occur with opposite orientation. Choose one of the two arcs of U lying between these two points, and cut and paste W at these two points of intersection, so as to obtain two new loops, where each has an arc parallel to this same arc of U, but with opposite orientations, and each contains exactly one of the two arcs of W cut by these two points of intersection. Then the only points of intersection of these two new loops with $\widetilde{\mathcal{W}}$ are exactly the original points of intersection of U with $\widetilde{\mathcal{W}}$, except that the two points where we have cut and pasted are no longer points of intersection. It follows that each of these two loops has fewer points of intersection with the loops of $\widetilde{\mathcal{W}}$ than does W. Since the product of these two loops, with appropriate choice of base-point, is homotopic to W, at least one of them does not lie in the normal closure of the loops of $\widetilde{\mathcal{W}}$. Continuing inductively, we arrive at a new loop, which we again call W, that is disjoint from all the loops of $\widetilde{\mathcal{W}}$, and does not lie in its normal closure.

In addition to assuming that W is disjoint from all the loops in $\widetilde{\mathcal{W}}$, and does not lie in its normal closure, we can also assume that W is a shortest geodesic with these properties. We now apply the argument of Proposition 2.2 to conclude that $w = P(W)$ is a power of a simple geodesic loop.

Since the set of loops, $\mathcal{W}$, is H-invariant, and $u = P(U)$ is disjoint from all the loops of $\mathcal{W}$, the sets of loops $\{H(u)\}$ and $\mathcal{W}$ are disjoint. The argument of Proposition 2.3 shows that the set of loops $\{H(u)\}$ is precisely invariant under H; we already know that the set of loops, $\mathcal{W}$, is precisely invariant under H. Since these two sets of loops are disjoint, we conclude that their union is an H-invariant set of simple disjoint loops. We also observe that the normal closure of the homotopy classes of this enlarged set of loops raised to powers is properly contained in N.

Since we have assumed that S is analytically finite, the above process must end after a finite number of steps. $\qquad\square$

3 Koebe Groups

Theorem 2.1 is stated in purely topological terms; we now restate it in terms of Kleinian groups and extended Kleinian groups.

Let $\mathbb{M}$ be the group of Möbius transformations and let $\widehat{\mathbb{M}}$ be a group generated by $\mathbb{M}$ and complex conjugation, $j(z) = \bar{z}$; the orientation preserving half of $\widehat{\mathbb{M}}$ is precisely $\mathbb{M}$. The transformations in $\widehat{\mathbb{M}} - \mathbb{M}$ are called *extended Möbius transformations*. If $G \subset \widehat{\mathbb{M}}$, then we set $G^{+} = \mathbb{M} \cap G$. A discrete group $G \subset \mathbb{M}$ is, as usual, called a *Kleinian group*; and a discrete group $G \subset \widehat{\mathbb{M}}$ containing at least one extended Möbius transformation, is called an *extended Kleinian group*.

A *function group* (respectively, *extended function group*) is a pair (G, Δ), where G is a finitely generated Kleinian group (respectively, extended Kleinian group) and $\Delta \subset \Omega(G)$ is an invariant component under the action of G. One special case of function groups are the *Koebe groups*; these are geometrically finite function groups where every component of the set of discontinuity other than the invariant component is a Euclidean (circular) disc.

A parabolic element g in a function group G, acting on a hyperbolic invariant component Δ, is *accidental* if, in terms of the hyperbolic metric on Δ, there is a geodesic that is invariant under g. We note that this geodesic becomes a loop if we adjoin the parabolic fixed point to it.

A *uniformization* of a Riemann surface or orbifold S is a triple $(\Delta, G, P : \Delta \to S)$ where (G, Δ) is a function group and $P : \Delta \to S = \Delta/G$ is the natural quotient map. In the special case that the Kleinian group $G \subset \mathbb{M}$ acts freely (i.e., S is a Riemann surface), then $P : \Delta \to S$ is a regular covering; otherwise, every element of G has only isolated fixed points, and $P : \Delta \to S$ is a branched regular covering.

It was shown by Maskit [M2] that, if $P : \tilde{S} \to S$ is a perhaps branched regular covering of an analytically finite Riemann surface S, then there is a unique Koebe group G, with invariant component Δ, where G contains no accidental parabolic elements, so that the covering defined by this Koebe group is conformally equivalent to the given covering. It was further shown in [M2] that if G is a Koebe group, with invariant

component Δ, uniformizing S without accidental parabolic transformations, and if h is a directly conformal homeomorphism of S that lifts to
Δ, then any such lift of h lies in $\mathbb{M}$. This was extended by Haas [Ha], who
showed that if h is an orientation-reversing conformal homeomorphism
of S that lifts to Δ, then any such lift of h lies in $\widehat{\mathbb{M}}$.

Combining the above with Theorem 2.1, we obtain the following.

Corollary 3.1 *Let G be a Koebe group with invariant component Δ,
where G acts without accidental parabolic transformations. Let S_0 be
$S = P(\Delta) = \Delta/G$ with its finite number of orbifold points removed,
and let Δ_0 be its preimage. Let H be a group of conformal and anti-
conformal automorphisms of S. Then the Koebe group G can be extended
to a subgroup $\widetilde{G} \subset \widehat{\mathbb{M}}$ so that $\Delta/\widetilde{G} = S/H$ if and only if there is a
marked system of loops on S_0 that both determines the uniformization
$P : \Delta_0 \to S_0$ and is invariant under H.*

We remark that the above can be extended to the following situation.
Let G be a Koebe group uniformizing the Riemann surface or orbifold
S, where G contains accidental parabolic transformations. As above, let
S_0 be S with the orbifold points removed, and let Δ_0 be the preimage
of S_0. Then [M3], there is a *determining* set of simple disjoint loops
$w_1, \ldots, w_n$ on S_0, with corresponding "integers", $k_1, \ldots, k_n$, where each
k_i is either a positive integer or the symbol ∞, where $k_i = \infty$ if and only
if the lift of w_i determines an accidental parabolic transformation.

Theorem 3.2 *Let G be a Koebe group with invariant component Δ,
and let $S = \Delta/G$. Let S_0 be $S = P(\Delta) = \Delta/G$ with its finite set of
orbifold points removed, and let Δ_0 be its preimage. Let H be a group
of conformal and anti-conformal automorphisms of S. Then the Koebe
group G can be extended to a subgroup $\widetilde{G} \subset \widehat{\mathbb{M}}$ so that $\Delta/\widetilde{G} = S/H$ if
and only if there is a set of loops on S_0, with corresponding "integers",
that determines the uniformization $P : \Delta_0 \to S_0$ and that is invariant
under H.*

Proof We first look at the case that there is a determining set of loops
that is invariant under H. Then, looking only at those loops with finite
integers, we see that H lifts to a group of conformal and anti-conformal
homeomorphisms of Δ that normalize G. Then, looking at the loops
that determine accidental parabolic elements, we see that every element
of H preserves the set of accidental parabolic elements. The techniques

of Maskit [M2] and Haas [Ha] now show that the lifted group is in fact a subgroup of $\widehat{M}$.

In the other direction, we now assume that H lifts to a subgroup of $\widehat{M}$. We start with a shortest geodesic, on S_0, whose lift determines an accidental parabolic element; it is necessarily simple. Using cut and paste as above, this geodesic is necessarily precisely invariant under its stabilizer in H. Also, since H lifts to a subgroup of $\widehat{M}$, the lift of every H-translate of this geodesic also determines an accidental parabolic element. If there are further accidental parabolic elements, then a shortest geodesic whose lift determines such an accidental parabolic element is necessarily simple and disjoint from the previously found geodesics determining accidental parabolic elements. Hence, after a finite number of steps, we will have found a set of simple disjoint geodesics whose lifts determine all accidental parabolic elements.

We conclude the argument with the observation that if W is a geodesic on Δ_0, and W intersects some of the geodesics determining accidental parabolic elements, then we can cut and paste to produce a new set of geodesics on Δ_0, all of which are disjoint from all geodesics determining accidental parabolic elements. Hence we can proceed as above to produce the desired set of determining simple disjoint loops on S_0. $\square$

Notes

1   The first author was partially supported by Project Fondecyt 1070271 and UTFSM 12.08.01

References

[Ha] Haas, A. Linearization and mappings onto pseudocircle domains. *Trans. Amer. Math. Soc.* **282** (1984), 415–429.

[M1] Maskit, B. A theorem on planar covering surfaces with applications to 3-manifolds. *Ann. of Math. (2)* **81** (1965), 341–355.

[M2] Maskit, B. On the classification of Kleinian groups: I. Koebe groups. *Acta Math.* **135** (1975), 249–270.

[M3] Maskit, B. *Kleinian Groups*, Springer-Verlag, Heidelberg, 1988.

Simple closed geodesics of equal length on a torus

Greg McShane

Institut Fourier, Université de Grenoble I
greg.mcshane@gmail.com

Hugo Parlier

Department of Mathematics, University of Toronto
hugo.parlier@epfl.ch

Abstract

Starting with a classical conjecture of Frobenius on solutions of the Markoff cubic, we are led, via the work of Harvey Cohn, to explore the multiplicities of lengths of simple geodesics on surfaces. We indicate recent progress on this and related questions stemming from the work of Schmutz Schaller. As an illustration we compare the cases of multiplicities on euclidean and hyperbolic once-punctured tori; in the euclidean case basic number theory gives a complete understanding of the spectrum. We explain an elementary construction using iterated Dehn twists that gives useful information about the lengths of simple geodesics in the hyperbolic case. In particular it shows that the marked simple length spectrum satisfies a rigidity condition: knowing just the order in the marked simple length spectrum is enough to determine the surface up to isometry. These results are special cases of a more general result [MP].

1 Introduction

The length spectrum of a hyperbolic surface is defined as the set of lengths of closed geodesics counted with multiplicities, and has been studied extensively in its relationship with the Laplace operator of a surface. A natural subset of the length spectrum is the *simple length spectrum*: the set of lengths of simple closed geodesics counted with multiplicities. This set is more naturally related to Teichmüller space and the mapping class group. In the particular case of the one-holed torus, we shall explore the following question: when, how often and on what type of subsets of Teichmüller space can two distinct simple closed geodesics be of equal length?

The origins of this question can be traced back to Frobenius who conjectured that any solution (a, b, c) of the *Markoff cubic*

$$a^2 + b^2 + c^2 - 3abc = 0 \qquad (1.1)$$

admits infinitely many solutions (a, b, c) in positive integers, and such a triple (a, b, c) is called a *Markoff triple*. Frobenius was led to conjecture that a Markoff triple is uniquely determined by $\max\{a, b, c\}$. (The conjecture is generally called the *Markoff uniqueness conjecture*.) Making the change of variable, $(x, y, z) = (3a, 3b, 3c)$ the Markoff cubic becomes

$$x^2 + y^2 + z^2 - xyz = 0. \qquad (1.2)$$

By work of Fricke and others, given a once-punctured hyperbolic torus $\mathbb{M}$ and α, β, γ a triple of simple closed curves, meeting pairwise in a single point, then

$$(2 \cosh \frac{\ell_M(\alpha)}{2}, 2 \cosh \frac{\ell_M(\beta)}{2}, 2 \cosh \frac{\ell_M(\gamma)}{2}),$$

where $\ell_M(.)$ is the hyperbolic length, is a solution to (1.2). By work of Harvey Cohn [C1] and others, a solution over the integers corresponds to the lengths of a triple in the so-called modular torus $\mathbb{M}$. The modular torus is the unique hyperbolic torus with a single cusp as boundary which is conformally equivalent to the flat hexagonal torus. Stated otherwise, it is the only once-punctured torus with an isometry group of maximal order (the order is 12). It is called the modular torus because it can be seen as the quotient of $\mathbb{H}$ by a subgroup of index 3 of $PSL(2, \mathbb{Z})$. Frobenius' conjecture on Markoff triples is in fact equivalent to the following conjecture on the modular torus:

Given any two simple closed geodesics of $\mathbb{M}$, there is an isometry of $\mathbb{M}$ which takes one to the other.

This property of having an isometry between any two simple closed geodesics of equal length on a torus will be called the *Markoff uniqueness property*.

There are a number of partial results which lend weight to the conjecture of Frobenius, notably:

Theorem 1.1 (Baragar [Bar], Button [But], Schmutz-Schaller [S1]) *A Markoff number is unique if it is a prime power or 2 times a prime power.*

And:

Theorem 1.2 (Zhang [Z]) *A Markoff number c is unique if one of $3c + 2$ and $3c + 2$ is a prime power, 4 times a prime power, or 8 times a prime power.*

Zhang's proof is elementary and relies on a clever study of congruences. Unfortunately for a geometer, this leads one to think that the solution of the Markoff uniqueness conjecture is outside the scope of classical geometry. We think that our study of Schmutz-Schaller's conjecture gives further weight to this point of view.

Let us call *simple multiplicity* of a torus the maximum multiplicity which appears in the simple length spectrum. Another rephrasing of the Markoff unicity conjecture is that the modular torus has simple multiplicity equal to 6. Schmutz-Schaller [S2] made the following generalization of the Markoff uniqueness conjecture:

All once-punctured tori have simple multiplicity at most 6.

Let us now consider the Teichmüller space $\mathcal{T}$ of all hyperbolic tori with either geodesic or cusp boundary. The main result we would like to present is the following:

Theorem 1.3 *The set of hyperbolic tori $\mathcal{N}eq$ with all simple closed geodesics of distinct length is Baire dense in $\mathcal{T}$. Conversely, the set $\mathcal{N}eq$ contains no arcs, and as such is totally disconnected.*

The theorem is in fact true for any Teichmüller space [MP]. Here we present only a proof of the converse (which is in fact the interesting part).

As mentioned above, the modular torus is the unique once-punctured torus with an isometry group of order 12, but it is not the only one-*holed* torus (tori with either cusp *or* geodesic boundary). In fact, such tori represent a connected dimension 1 subset of $\mathcal{T}$ which we shall denote $\mathcal{T}^*$. The techniques used to prove the above theorem can be used to show the following:

Theorem 1.4 ([MP]) *The set of one-holed tori with multiplicity at least* 12 *is dense in* T^*.

Thus Schmutz-Schaller's conjecture cannot be generalized to one-holed tori.

Note that theorem 1.3 implies that most tori *do* have the Markoff uniqueness property. However, knowing whether a particular torus has this property is in general a difficult question. An analogy can be made with the case of transcendental real numbers. Although most real numbers are transcendental, given a particular real number, proving that it is transcendental is often a very difficult question, for example we know that $\zeta(3)$ is irrational but we do not know whether it is transcendental.

This note is organized as follows. We begin by showing theorem 1.3 in the case of flat tori. Sections 3 to 5 are dedicated to the proof of theorem 1.3 in the case of hyperbolic tori. In the last section, we discuss bounds on simple multiplicity. First, we show that there are flat tori with unbounded multiplicity. Finally, we end our exposition by presenting certain tori which do have the Markoff uniqueness property, and thus multiplicity bounded by 6.

2 The flat torus

Our general approach is to study the sets of Teichmüller space where two simple closed geodesics are of equal length. In the case of flat tori, these sets are straightforward to characterize.

Recall that Riemann's Uniformization Theorem tells us that every flat or euclidean torus $\mathbb{T}^2$ is obtained as a quotient of its universal cover $\mathbb{C}$ by the group of deck transformations Γ, which is isomorphic to $\mathbb{Z} \oplus \mathbb{Z}$. The lift to $\mathbb{C}$ of a closed geodesic on $\mathbb{T}^2$ is a straight line $L \subset \mathbb{C}$ invariant by some cyclic subgroup of deck transformations $\langle z \mapsto z + \omega \rangle$. It follows that the length of the geodesic is equal to the translation length of $z \mapsto z + \omega$, that is $|\omega|$. Note that, if $c \in \mathbb{C}$ then $L + c/\langle z \mapsto z + \omega \rangle$ is a closed geodesic, freely homotopic to the original geodesic and of the same length. We identify Γ with the fundamental group of $\mathbb{T}^2$ and note that, contrary to strictly negatively curved spaces, there are infinitely many closed geodesics in each free homotopy class. Thus, in order to make sense of multiplicity in the spectrum, we choose the unique geodesic in the (free) homotopy class which passes through the base point of $\mathbb{T}^2$.

In fact, given a flat torus $\mathbb{T}^2$ there exists τ, $\mathrm{Im}(\tau) > 0$ such that $\mathbb{C}/\Gamma$ is conformally equivalent to $\mathbb{T}^2$ where Γ is generated by the translations $z \mapsto z + 1, z \mapsto z + \tau$. Teichmüller space of flat tori can be seen as $\mathbb{H}$ in the following way. Consider a torus obtained by quotienting $\mathbb{C}$ by $\mathbb{T}^2$ generated by two complex translations $z \mapsto z + 1$ and $z \mapsto z + \tau$ with $\mathrm{Im}(\tau) > 0$. Teichmüller space can be seen as the space of deformations of such a torus by letting the parameter τ vary. As we exclude singular tori, we only let τ live in $\mathbb{H}$. Up to homothety, we have described all possible flat tori, and of course a bit more. By the uniformization theorem, we've also described all smooth tori up to *conformal* equivalence. To obtain the *Moduli space* of smooth tori, that is the set of tori up to *conformal equivalence*, one takes our set of flat tori and quotients by homothety. This corresponds to quotienting $\mathbb{H}$ by $PSL(2,\mathbb{Z})$, the mapping class group in this instance. The resulting space is the *modular surface* and has an orbifold structure with three singular points. The modular surface is a rather deep first example of a moduli space and is a very useful source of natural questions one might want to ask for a moduli space in general.

Now simple closed geodesics on a flat torus (up to free homotopy) are naturally associated to rational numbers (union infinity) in the following fashion. Consider the square torus, i.e., when $\tau = i$. Now consider a line in $\mathbb{C}$ of slope σ: clearly the line projects to a simple closed geodesic if and only if $\sigma \in \mathbb{Q}$ (or if the line is vertical, we say the line is of slope $\infty = \frac{1}{0}$). We are interested in *primitive* curves, meaning curves that are not the n-iterate of another curve. Thus up to free homotopy, each simple closed geodesic is described by a unique element of $\mathbb{Q} \cup \infty$.

Consider $\frac{p}{q}, \frac{r}{s} \in \mathbb{Q} \cup \infty$ distinct. The set of $\mathbb{H}$ where their associated simple closed geodesics are equal is the set where τ satisfies

$$|p\tau + q| = |r\tau + s|. \tag{2.1}$$

A straightforward calculation shows this set to be the Poincaré geodesic between endpoints $\frac{q-s}{r-p}$ and $\frac{q+s}{r+p}$.

Conversely, between any given pair of distinct rationals $\frac{a}{b}, \frac{a'}{b'}$, the Poincaré geodesic $[a/b, a'/b']$ between them is the set of $\mathbb{H}$ where two simple closed geodesics are of equal length. To show this, consider the map $z \mapsto -\bar{z}$. It preserves the rationals and fixes $[0, \infty] = \{z \in \mathbb{C} \mid \Re z = 0\}$. For any rational $\frac{m}{n}$ the curves of slope $\frac{m}{n}$ and $-\frac{m}{n}$ have the same length at $\tau \in [0, \infty]$. One maps $[0, \infty]$ onto any geodesic $[a/b, a'/b']$ using $PSL(2, \mathbb{R})$, which is transitive on pairs of rationals and thus finds a pair of curves which are of equal length on $[a/b, a'/b']$.

Using this characterization, we can now show the following.

Theorem 2.1 *The set of flat tori $\mathcal{N}eq$ with all simple closed geodesics of distinct length is Baire dense. Conversely, the set $\mathcal{N}eq$ contains no arcs, and as such is totally disconnected.*

The first statement follows from the fact that rationals are countable and the second from the fact that any two distinct points in $\mathbb{H}$ are separated by a Poincaré geodesic between rationals.

Unfortunately in the case of hyperbolic tori, the set of tori with two simple closed geodesics of equal length is not so easy to characterize.

3 Hyperbolic tori

Let us recall a few definitions and facts from the theory of surfaces; all this is available in a more detailed treatment in either [A], [Bea] or [Bus]. Throughout M will denote a surface with constant curvature -1 and we shall insist that M is *complete* with respect to this metric (although we will only be concerned by what happens inside the *Nielsen core* of the surface). This means that M is *locally modeled* on the hyperbolic plane $\mathbb{H}^2$ and there is a natural covering map $\pi : \mathbb{H} \to M$. By $\mathcal{T}$ we mean the Teichmüller space of M, meaning the space of marked complete hyperbolic structures on M. The signature of a surface M will be denoted (g, n) where M is homeomorphic to a surface of genus g with n simple closed boundary curves. In this article, we are interested in hyperbolic tori with one boundary component, which we will consider to be either a cusp or a simple closed geodesic (surfaces of signature $(1, 1)$).

(It is worth noting however, that in fact all of our arguments either apply to, or can easily be made to fit, the case of tori with a cone angle.)

Let us recall a few facts about curves on surfaces (see [Bus] or [CB] for details). Firstly, a *simple curve* is a curve which has no self intersections. A curve is said to be *essential* if it bounds neither a disc nor a punctured disc (or an annulus). For each free homotopy class which contains an essential simple loop, there is a unique geodesic representative.

There is a natural function, $\ell : \mathcal{T} \times$ essential homotopy classes $\to \mathbb{R}^+$, which takes the pair $M, [\alpha]$ to the length $\ell_M(\alpha)$ of the geodesic in the homotopy class $[\alpha]$ (measured in the Riemannian metric on M). It is an abuse, though common in the literature, to refer merely to *the length of the geodesic α* (rather than, more properly, the length of the geodesic in the appropriate homotopy class). Using length functions one can describe Teichmüller space. In the case of surfaces of signature $(1,1)$, Teichmüller space is, topologically, $\mathbb{R}^+ \times \mathbb{R}^2$. The first parameter corresponds to boundary length, and the other two correspond to an interior (or essential) simple closed geodesic in the following way. One can think of the first parameter as being the length of the simple closed geodesic (thus formally it lies in $\mathbb{R}^{+,*}$) and the second is a twist parameter, a real valued parameter which tells you how the simple closed geodesic is pasted together to get a torus. (These are the Fenchel-Nielsen parameters.) Note that these parameters are not homogeneous in nature. In the next section, we give a set of homogeneous parameters for the Teichmüller space of one-holed tori.

4 A projectively injective map

In the case of one-holed tori, we will make essential use of the following lemma. Recall that a projectively injective map is a map f such that $f(x) = \lambda f(y)$ for $\lambda \in \mathbb{R}$ implies that $x = y$.

Lemma 4.1 *There are four interior simple closed curves α, β, γ, and δ of a one-holed torus such that the map $\varphi : M \mapsto (\ell_M(\alpha), \ell_M(\beta), \ell_M(\gamma), \ell_M(\delta))$ is projectively injective.*

Proof Let M be a one-holed torus and let α, β, γ, and δ be the simple closed curves as in figure 4.1.

We've chosen our curves as follows. We begin by choosing any α and β that intersect once. It is not difficult to see that given α and β, there are exactly two curves (γ and δ) that intersect both α and β exactly once.

The curves γ and δ intersect twice. Now the remarkable fact about the geodesic representatives of simple closed curves on a one-holed torus is that they pass through exactly two of the three Weierstrass points of the torus in diametrically opposite points. In the case of the curves α, β, γ, and δ, their intersection points are all necessarily Weierstrass points. Therefore they can be seen in the universal cover as in figure 4.1.

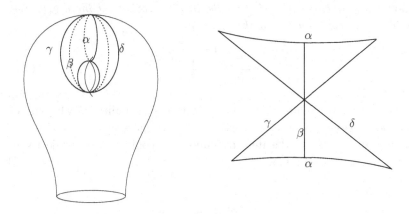

Fig. 4.1. The one-holed torus with four interior geodesics and the four curves seen in the universal cover

In fact the lengths of α, β, and γ determine a unique point in the Teichmüller space of one-holed tori. One can show this by recovering the Fenchel-Nielsen parameters from the three lengths (this is done in detail in [BS] for instance). However, up to a multiplicative constant, they do not (otherwise the real dimension of the Teichmüller space would be 2 and not 3). For this we need the curve δ. What we need to prove is that if we have two one-holed tori M_1 and M_2 in Teichmüller space with

$$(\ell_{M_1}(\alpha), \ell_{M_1}(\beta), \ell_{M_1}(\gamma), \ell_{M_1}(\delta)) = \lambda(\ell_{M_2}(\alpha), \ell_{M_2}(\beta), \ell_{M_2}(\gamma), \ell_{M_2}(\delta))$$

for some $\lambda \in \mathbb{R}$, then $\lambda = 1$ and then, by what precedes, $M_1 = M_2$. Figure 4.1 shows four hyperbolic triangles. Consider the two bottom ones. The side lengths of the bottom left triangle are $\frac{\ell(\alpha)}{2}$, $\frac{\ell(\beta)}{2}$, and $\frac{\ell(\gamma)}{2}$. The side lengths of the bottom right triangle are $\frac{\ell(\alpha)}{2}$, $\frac{\ell(\beta)}{2}$, and $\frac{\ell(\delta)}{2}$. The bottom intersection point between α and β forms two angles depending on the surface M, say $\theta_1(M)$ and $\theta_2(M)$ such that $\theta_1 + \theta_2 = \pi$. Suppose without loss of generality that $\lambda \geq 1$. Now if for M_1 the triangle lengths are equal to a, b, c and d, the triangle lengths for M_2 are $\lambda a, \lambda b, \lambda c$ and λd. This implies that $\theta_1(M_1) \leq \theta_1(M_2)$ as well as $\theta_2(M_1) \leq \theta_2(M_2)$, equality occurring only if $\lambda = 1$. As $\theta_1 + \theta_2$ is always equal to π, this concludes the proof. $\qquad\square$

We will make essential use of the following corollary to this lemma.

Corollary 4.2 *Let $M_1, M_2 \in \mathcal{T}$ be distinct tori, then there exist two interior simple curves γ_1 and γ_2 such that*

$$\frac{\ell_{M_1}(\gamma_1)}{\ell_{M_1}(\gamma_2)} \neq \frac{\ell_{M_2}(\gamma_1)}{\ell_{M_2}(\gamma_2)}. \tag{4.1}$$

Proof Suppose inequality (4.1) was in fact an equality for all pairs of simple curves γ_1, γ_2, thus in particular for all pairs in the set of curves α, β, γ and δ. Then the map of lemma 4.1 could not be projectively injective, a contradiction. $\qquad\square$

5 Dehn twisting

Here we show the second part of theorem 1.3 for one-holed tori, namely that a path between two distinct points of $\mathcal{T}$ contains a surface with two simple closed geodesics of equal length.

Let M_1 and M_2 be two distinct points of $\mathcal{T}$. Corollary 4.2 guarantees the existence of two simple closed geodesics who satisfy inequality (4.1) above. For the remainder of this section, these two curves shall be denoted α and β.

Given these curves, we can choose a pair of curves $\tilde{\alpha}$ and $\tilde{\beta}$, such that $\mathrm{int}(\alpha, \tilde{\alpha}) = \mathrm{int}(\beta, \tilde{\beta}) = 1$. Consider the two families of curves $\{\alpha_k\}_{k \in \mathbb{N}}$, $\{\beta_k\}_{k \in \mathbb{N}}$ obtained by performing k right Dehn twists of $\tilde{\alpha}$, resp. $\tilde{\beta}$, around α, resp. β. These two families satisfy the following lemma.

Lemma 5.1 *For each $k \in \mathbb{N}$, and for any surface M we have*

(i) $\mathrm{int}(\alpha, \alpha_k) = 1$, $\mathrm{int}(\beta, \beta_k) = 1$,
(ii) $k\ell_M(\alpha) - \ell_M(\tilde{\alpha}) < \ell_M(\alpha_k) \leq k\ell_M(\alpha) + \ell_M(\tilde{\alpha})$,
(iii) $k\ell_M(\beta) - \ell_M(\tilde{\beta}) < \ell_M(\beta_k) \leq k\ell_M(\beta) + \ell_M(\tilde{\beta})$.

Proof The first statement is obvious and the last two follow by lifting to the universal cover $\mathbb{H}$ and by applying the triangle inequality to the geodesics.   □

For a surface M, set $B_i(M) := \{\beta_k : \ell_M(\beta_k) \leq \ell_M(\alpha_i)\}$. Our aim is to calculate the ratio $\ell_M(\alpha)/\ell_M(\beta)$ from the asymptotic formula in lemma 5.1.

Proposition 5.2 *With the notation above:*

$$\frac{\sharp B_i(M)}{i} \longrightarrow \frac{\ell_M(\alpha)}{\ell_M(\beta)}.$$

Proof As M is fixed, we set $B_i := B_i(M)$ and $\ell := \ell_M$.

By lemma 5.1 we have

$$\sharp B_i \leq \sharp\{k : 2k\ell(\beta) - \ell(\beta_0) \leq 2i\ell(\alpha) + \ell(\alpha_0)\},$$

and

$$\sharp B_i \geq \sharp\{k : 2k\ell(\beta) + \ell(\beta_0) \leq 2i\ell(\alpha) - \ell(\alpha_0)\}.$$

It follows that

$$i\frac{\ell(\alpha)}{\ell(\beta)} - \frac{\ell(\beta_0)}{2\ell(\beta)} - \frac{\ell(\alpha_0)}{2\ell(\beta)} \leq \sharp B_i \leq i\frac{\ell(\alpha)}{\ell(\beta)} + \frac{\ell(\alpha_0)}{2\ell(\beta)} + \frac{\ell(\beta_0)}{2\ell(\beta)}.$$

The statement of the proposition is immediate.   □

We can now establish the following.

Corollary 5.3 *There exist two simple closed geodesics α_k and $\beta_{\tilde{k}}$ such that*

$$\ell_{M_1}(\beta_{\tilde{k}}) > \ell_{M_1}(\alpha_k)$$

and

$$\ell_{M_2}(\beta_{\tilde{k}}) > \ell_{M_2}(\alpha_k).$$

In particular, the marked order in lengths of simple closed geodesics determines a unique surface in $\mathcal{T}$.

Proof Recall that α and β satisfy inequality (4.1). Applying proposition 5.2 to M_1 and M_2, we see that there is an integer k such that $\sharp B_k(M_1) \neq \sharp B_k(M_2)$. In particular, there exists a $\tilde{k}$ such that α_k and $\beta_{\tilde{k}}$ satisfy the desired inequalities.   □

Now $M \mapsto \ell_M(\alpha_k) - \ell_M(\beta_{\bar{k}})$ is a continuous function, so applying the intermediate value theorem to the arc $\mathcal{A}$ between the points M_1 and M_2, yields the existence of a surface $Z \in \mathcal{A}$ so that $\ell_Z(\alpha_k) = \ell_Z(\beta_{\bar{k}})$. This establishes the second part of theorem 1.3.

Remark 5.4 *In fact one can show something stronger than corollary 5.3, namely that for any given $M_1 \neq M_2$ and any integer N, there exists a set of simple closed geodesics $\alpha_k, \beta_{\bar{k}+1}, \dots, \beta_{\bar{k}+N}$ such that*

$$\ell_{M_1}(\beta_{\bar{k}+i}) > \ell_{M_1}(\alpha_k) \ and$$
$$\ell_{M_2}(\beta_{\bar{k}+i}) > \ell_{M_2}(\alpha_k)$$

for all $i \in \{1, \dots, N\}$. The proof goes as follows. In the proof of corollary 5.3, we used the fact that $\sharp B_k(M_1) \neq \sharp B_k(M_2)$. Suppose by contradiction that $|\sharp B_k(M_1) - \sharp B_k(M_2)|$ was bounded by some constant for all k. The limit of the ratios from proposition 5.2 would be then the same for both M_1 and M_2, a contradiction.

6 Bounds on multiplicity of simple closed geodesics

Although most surfaces have all simple multiplicities equal to 1, it is an open question as to whether or not hyperbolic surfaces with unbounded simple multiplicity exist. The remark at the end of the last section shows why it might be difficult to prove that simple multiplicity is always bounded. For the full length spectrum, a theorem of Randol [R], based on a construction of Horowitz, shows that multiplicity is *always* unbounded. In the particular case of tori with a single cusp, Schmutz-Schaller [S2] conjectured that all simple multiplicities of once-punctured tori are bounded by 6. He also notes that, to the best of his knowledge, one does not know a surface for which we are sure that simple multiplicity is bounded. After having shown why multiplicities can be unbounded in the case of flat tori, we shall give examples of hyperbolic tori for which we are sure that multiplicities are bounded.

6.1 The multiplicity of the spectrum of a flat torus

In this section we give a short account of unboundedness of multiplicities in the length spectrum of a flat or euclidean torus. Our exposition is based on elementary number theory, and we concentrate on only the two "most symmetric" such tori, though a more thorough knowledge of class field theory [C2] might allow more cases to be treated.

Consider a flat torus $\mathbb{T}^2$. As explained in Section 2, there exists $\tau, \operatorname{Im} \tau > 0$ such that $\mathbb{C}/\Gamma$ is conformally equivalent to $\mathbb{T}^2$ where Γ is generated by the translations $z \mapsto z + 1, z \mapsto z + \tau$. For certain values of τ one can compute the multiplicities of the numbers which appear in the length spectrum by studying the ring of integers $\mathbb{Z}[\tau]$ of a quadratic field $\mathbb{Q}(\tau)$. Throughout we will assume that $\mathbb{Z}[\tau]$ is a unique factorization domain, that is every $\omega \in \mathbb{Z}[\tau]$ factors as $uq_1 q_2 \dots q_n$ where u is a unit and q_i are irreducible elements of $\mathbb{Z}[\tau]$ and this factorization is unique up to permutation of q_i and multiplication by the units of $\mathbb{Z}[\tau]$. Whenever $\mathbb{Z}[\tau]$ is a euclidean domain e.g. $\tau = i, \sqrt{-2}, \sqrt{-3}, \sqrt{-7}, \sqrt{-11}$ then it is a unique factorization domain, the former condition being easier to verify [C2].

We restrict our attention to τ such that τ is a *quadratic irrational* that is it satisfies a quadratic with integer coefficients

$$\tau^2 + B\tau + C = 0,$$

since, for such τ, the ring $\mathbb{Z}[\tau]$ embeds in $\mathbb{C}$ as a lattice and there is an isomorphism of abelian groups

$$\mathbb{Z}[\tau] \to \Gamma, \ x + \tau y \mapsto (z \mapsto z + x + \tau y),$$

where, as above, Γ denotes the group of deck transformations of $\mathbb{T}^2$. We are interested primarily in $\tau = i, \frac{-1+\sqrt{-3}}{2}$ as the resulting torus, $\mathbb{C}/\Gamma$, is respectively the square torus and the regular hexagonal (or modular) torus.

By convention, the *norm* of $\omega \in \mathbb{Z}[\tau]$ is defined to be $\omega\bar{\omega}$; this is an integer and it is evidently the square of the translation length of $z \mapsto z + \omega$. For example when $\tau = i$ the norm of $x + \tau y \in \mathbb{Z}[\tau]$ is just $x^2 + y^2$ and when $\tau = \frac{-1+\sqrt{-3}}{2}$ the norm is $x^2 + xy + y^2$. Now a prime $p \neq 2$ can be written as a sum of squares $x^2 + y^2$, $x, y \in \mathbb{N}$ if and only if p is congruent to 1 modulo 4 and it can be written as $x^2 + xy + y^2$, $x, y \in \mathbb{N}$ if and only if it is congruent to 1 modulo 3. It is a celebrated theorem of Dirichlet that there are infinitely many primes in any arithmetic progression and so there are infinitely many congruent to 1 modulo 4 and to 1 modulo 3. Such a prime p admits a factorization $p = (x + \tau y)(x + \bar{\tau} y)$ where $x + \tau y, x + \bar{\tau} y$ are irreducible elements of the ring of $\mathbb{Z}[\tau]$. By Dirichlet's theorem we may choose n such distinct primes $p_k \in \mathbb{N}$, $1 \leq k \leq n$, let $a_k \in \mathbb{Z}[i]$, $p_k = a_k \bar{a}_k$ and let N denote their product. Now N factorizes over $\mathbb{Z}[\tau]$ and

$$N = (a_1 \bar{a}_1)(a_2 \bar{a}_2) \ldots (a_n \bar{a}_n).$$

Consider the set $R_N \subset \mathbb{Z}[\tau]$ of the form $c_1 c_2 \ldots c_n$ where $c_k \in \{a_k, \bar{a}_k\}$. Note that the norm of each element of R_N is N. It is easy to check, using the fact that $\mathbb{Z}[\tau]$ is a unique factorization domain, that R_N contains exactly 2^{n-1} distinct elements. Note further that if $c_1 c_2 \ldots c_n \in R_N$ and $c_1 c_2 \ldots c_n = x + iy$ then x, y are coprime integers, for otherwise there is a prime p that divides x, y hence $x + \tau y$, now as the c_i are irreducible p factors as

$$p = u c_{i_1} \ldots c_{i_l}$$

for some unit $u \in \mathbb{Z}[\tau]$. Considering the norms of both sides of the above one has

$$p^2 = p_{i_1} \ldots p_{i_l}$$

which contradicts the hypothesis that the p_i were distinct. The set of deck transformation $z \mapsto z + \omega, \omega \in R_N$ yields a set of pairwise non-homotopic, primitive, simple closed geodesics of length $\sqrt{N}$ on the torus.

6.2 Hyperbolic tori

Given a hyperbolic structure on a surface M, not necessarily of finite volume, then the holonomy of the metric gives a representation of the fundamental group into the group of isometries of the hyperbolic plane $PSL(2, \mathbb{R})$. In fact, since there is no 2-torsion, one can lift this representation $\hat{\rho} : \pi_1 \to SL(2, \mathbb{R})$ and for any element $\gamma \in \pi_1$

$$2 \cosh(\ell_M(\gamma)) = \operatorname{tr} \hat{\rho}(\gamma).$$

In the case of the once-punctured torus one obtains a representation of the free group on two generators $\langle \alpha, \beta \rangle$ into $SL(2, \mathbb{R})$.

Theorem 6.1 (Fricke, Horowitz, Keen) *Let* A, B *be matrices in* $SL(2, \mathbb{C})$. *If* W *is a word in* A, B *then there is a polynomial* $P_W \in \mathbb{Z}[x, y, z]$ *such that*

$$\operatorname{tr} W = P_W(\operatorname{tr} A, \operatorname{tr} B, \operatorname{tr} AB).$$

A celebrated construction of Horowitz [H], see [Bus] and [R] for details, yields pairs of words W, W' such that W, W', W^{-1}, W'^{-1} are pairwise inconjugate but $P_W = P_{W'}$. However, the Horowitz construction cannot be applied to W representing a simple closed curve (see [MP]):

Fact: If W represents a simple closed curve and $P_W = P_{W'}$ then W' is conjugate to W or W^{-1}. Thus, we remark that if W, W' represent simple closed curves such that W, W', W^{-1}, W'^{-1} are pairwise inconjugate then $P_W - P_{W'}$ is a non-zero element of $\mathbb{Z}[x, y, z]$.

An immediate corollary of this remark is that, given $\lambda > 2$ a transcendental real number and A, B, AB such that $\operatorname{tr} A = \operatorname{tr} B = \operatorname{tr} AB = \lambda$, then the quotient $\mathbb{H}/\langle A, B \rangle$ is a hyperbolic one-holed torus which satisfies Markoff uniqueness.

Acknowledgements: The authors thank the referee for his suggestions and the editors for putting together this book for Bill.

References

[A] Abikoff, William. *The Real Analytic Theory of Teichmüller Space*, Lecture Notes in Mathematics **820**, Springer, Berlin, 1980.

[Bar] Baragar, Arthur. On the unicity conjecture for Markoff numbers. *Canad. Math. Bull.* **39**(1) (1996), 3–9.

[Bea] Beardon, Alan F. *The Geometry of Discrete Groups*, Graduate Texts in Mathematics **91**, Springer-Verlag, New York, 1995. Corrected reprint of the 1983 original.

[Bus] Buser, Peter. *Geometry and Spectra of Compact Riemann Surfaces*, Progress in Mathematics **106**, Birkhäuser Boston Inc., Boston, MA, 1992.

[BS] Buser, P. and Semmler, K.-D. The geometry and spectrum of the one-holed torus. *Comment. Math. Helv.* **63**(2) (1988), 259–274.

[But] Button, J. O. The uniqueness of the prime Markoff numbers. *J. London Math. Soc. (2)* **58**(1) (1998), 9–17.

[CB] Casson, Andrew J. and Bleiler, Steven A. *Automorphisms of Surfaces after Nielsen and Thurston*, London Mathematical Society Student Texts. **9**, Cambridge University Press, Cambridge, 1988.

[C1] Cohn, Harvey. Representation of Markoff's binary quadratic forms by geodesics on a perforated torus. *Acta Arith.* **18** (1971), 125–136.

[C2] Cohn, Harvey. *A Classical Invitation to Algebraic Numbers and Class Fields*, Springer-Verlag, New York, 1978. With two appendices by Olga Taussky: "Artin's 1932 Göttingen lectures on class field theory" and "Connections between algebraic number theory and integral matrices", Universitext.

[H] Horowitz, Robert D. Characters of free groups represented in the two-dimensional special linear group. *Comm. Pure Appl. Math.* **25** (1972), 635–649.

[MP] McShane, Greg and Parlier, Hugo. Multiplicities of simple closed geodesics and hypersurfaces in Teichmüller space. *Geom. Topol.* **12**(4) (2008), 1883–1919.

[R] Randol, Burton. The length spectrum of a Riemann surface is always of unbounded multiplicity. *Proc. Amer. Math. Soc.* **78**(3) (1980), 455–456.

[S1] Schmutz, Paul. Systoles of arithmetic surfaces and the Markoff spectrum. *Math. Ann.* **305**(1) (1996), 191–203.

[S2] Schmutz Schaller, Paul. Geometry of Riemann surfaces based on closed geodesics. *Bull. Amer. Math. Soc. (N.S.)* **35**(3) (1998), 193–214.

[Z] Zhang, Ying. Congruence and uniqueness of certain Markoff numbers. *Acta Arith.* **128**(3) (2007), 295–301.

On extensions of holomorphic motions
—a survey

Sudeb Mitra [1]

Queens College of the City University of New York
sudeb.mitra@qc.cuny.edu

Abstract

In this survey article, we discuss some of the main results on extensions of holomorphic motions. Our emphasis is on holomorphic motions over infinite dimensional parameter spaces. The Teichmüller space of a closed set E in the Riemann sphere is a universal parameter space for holomorphic motions of that set. This universal property is exploited throughout the article.

Notation. In this paper we will use the following notations: $\mathcal{C}$ for the complex plane, $\widehat{\mathcal{C}}$ for the Riemann sphere, and Δ for the open unit disk $\{z \in \mathcal{C} : |z| < 1\}$.

1 Introduction

Motivated by the study of dynamics of rational maps, Mañé, Sad, and Sullivan (in [MSS]) defined a holomorphic motion of a set E in $\widehat{\mathcal{C}}$, to be a curve $\phi_t(z)$ defined for every z in E and for every t in Δ, such that:

(i) $\phi_0(z) = z$ for all z in E,

(ii) $z \mapsto \phi_t(z)$ is injective as a function from E to $\widehat{\mathcal{C}}$, for each fixed t in Δ, and

(iii) $t \mapsto \phi_t(z)$ is holomorphic for $|t| < 1$, and for each fixed z in E.

As t moves in the unit disk, the set $E_t = \phi_t(E)$ moves in $\widehat{\mathcal{C}}$. We think of t as the complex time-parameter for the motion. As Gardiner and Keen remark in the introduction in [GK], although E may start out as smooth as a circle and although the points of E move holomorphically, for every $t \neq 0$, $E_t = \phi_t(E)$ can be an interesting fractal with fractional Hausdorff dimension. Julia sets of rational maps and

283

limit sets of Kleinian groups are almost always fractals. The famous λ-lemma of Mañé, Sad, and Sullivan (in [MSS]) showed the surprising fact that injectivity and holomorphic dependence automatically lead to quasiconformality. Since then, holomorphic motions have found interesting applications in Teichmüller theory, dynamical systems, Kleinian groups, and geometric function theory.

The fundamental paper of Bers and Royden ([BR]) showed the deep relationship between holomorphic motions and Teichmüller spaces. They used results from Teichmüller theory to prove some theorems about holomorphic motions. Later, the paper [EKK] reversed this approach; using an important fact on holomorphic motions they obtained new results about Teichmüller spaces. These two papers established the remarkable relationship between the study of holomorphic motions and the theory of Teichmüller spaces. In these papers, the authors used classical Teichmüller spaces to study holomorphic motions over the open unit disk. We shall consider motions over more general parameter spaces (see Definition 2.2 below).

A central topic in the study of holomorphic motions is the question of extensions. Let V be a connected complex manifold with a basepoint. If E is a proper subset of $\widetilde{E}$ and $\phi : V \times E \to \widehat{C}$, $\widetilde{\phi} : V \times \widetilde{E} \to \widehat{C}$ are two maps, we say that $\widetilde{\phi}$ extends ϕ if $\widetilde{\phi}(x, z) = \phi(x, z)$ for all (x, z) in $V \times E$. If $\phi : V \times E \to \widehat{C}$ is a holomorphic motion, a natural question is whether there exists a holomorphic motion $\widetilde{\phi} : V \times \widehat{C} \to \widehat{C}$ that extends ϕ. For $V = \Delta$, important results were obtained in [BR] and in [ST]. A complete affirmative answer was given by Slodkowski in [Sl1]. He showed that any holomorphic motion of E over Δ can be extended to $\widehat{C}$.

In this survey article, we give a broad overview of the fundamental results on extensions of holomorphic motions, including some of the most recent advances. We study holomorphic motions over a connected complex Banach manifold. For that, we need to discuss some properties of more general Teichmüller spaces. We define "universal holomorphic motions" and discuss some applications to extensions of holomorphic motions.

Acknowledgement. I want to thank the referee for his valuable suggestions that helped me to improve an earlier version of this article. I also want to thank the organizers of the Harvey Conference (Anogia, 2007) for giving me the opportunity to participate in that wonderful event.

2 Basic definitions

Definition 2.1 *A complex-valued function $w = f(z)$ defined in a region Ω in $\mathcal{C}$ is called a quasiconformal mapping if it is a sense-preserving homeomorphism of Ω onto its image and its complex distributional derivatives*

$$w_z = \frac{1}{2}\left(\frac{\partial f}{\partial x} - i\frac{\partial f}{\partial y}\right) \text{ and } w_{\bar{z}} = \frac{1}{2}\left(\frac{\partial f}{\partial x} + i\frac{\partial f}{\partial y}\right)$$

are Lebesgue measurable locally square integrable functions on Ω that satisfy the inequality $|w_{\bar{z}}| \leq k|w_z|$ almost everywhere in Ω, for some real number k with $0 \leq k < 1$.

If $w = f(z)$ is a quasiconformal mapping defined on the region Ω then the function w_z is known to be nonzero almost everywhere on Ω. Therefore the function

$$\mu_f = \frac{w_{\bar{z}}}{w_z}$$

is a well-defined L^∞ function on Ω, called the *complex dilatation* or the *Beltrami coefficient* of f. The L^∞ norm of every Beltrami coefficient is less than one.

The positive number

$$K(f) = \frac{1 + \|\mu_f\|_\infty}{1 - \|\mu_f\|_\infty}$$

is called the *dilatation* of f. We say that f is K-quasiconformal if f is a quasiconformal mapping and $K(f) \leq K$.

We call a homeomorphism of $\widehat{\mathcal{C}}$ *normalized* if it fixes the points 0, 1, and ∞.

Let $M(\mathcal{C})$ denote the open unit ball of the complex Banach space $L^\infty(\mathcal{C})$. Then, for each μ in $M(\mathcal{C})$, there exists a unique normalized quasiconformal homeomorphism of $\widehat{\mathcal{C}}$ onto itself that has Beltrami coefficient μ; we denote this quasiconformal map by w^μ. Furthermore, for every fixed $z \in \widehat{\mathcal{C}}$, the map $\mu \mapsto w^\mu(z)$ of $M(\mathcal{C})$ into $\widehat{\mathcal{C}}$ is holomorphic. The basepoint of $M(\mathcal{C})$ is the zero function. The reader is referred to the book [Ah] and the paper [AB] for basic facts on quasiconformal mappings.

Definition 2.2 *Let V be a connected complex manifold with a basepoint x_0, and let E be a subset of $\widehat{\mathcal{C}}$. A holomorphic motion of E over V is a map $\phi\colon V \times E \to \widehat{\mathcal{C}}$ that has the following three properties:*

(a) $\phi(x_0, z) = z$ for all z in E,

(b) the map $\phi(x, \cdot) \colon E \to \widehat{C}$ is injective for each x in V, and

(c) the map $\phi(\cdot, z) \colon V \to \widehat{C}$ is holomorphic for each z in E.

We will sometimes write $\phi(x, z)$ as $\phi_x(z)$ for x in V and z in E.

We say that V is the *parameter space* of the holomorphic motion ϕ.

We will always assume that ϕ is a *normalized* holomorphic motion; i.e. 0, 1, and ∞ belong to E and are fixed points of the map $\phi_x(\cdot)$ for every x in V.

Definition 2.3 *Let V and W be connected complex manifolds with base-points, and f be a basepoint preserving holomorphic map of W into V. If ϕ is a holomorphic motion of E over V its pullback by f is the holomorphic motion*

$$f^*(\phi)(x, z) = \phi(f(x), z) \qquad \forall (x, z) \in W \times E \qquad (2.1)$$

of E over W.

An example: Define $\Psi_{\widehat{C}} \colon M(\mathcal{C}) \times \widehat{C} \to \widehat{C}$ as follows:

$$\Psi_{\widehat{C}}(\mu, z) = w^\mu(z)$$

where w^μ is the unique normalized quasiconformal homeomorphism of $\widehat{C}$ onto itself that has the Beltrami coefficient μ. Theorem 11 in [AB] implies that $\Psi_{\widehat{C}}$ is a holomorphic motion. We will refer to this example several times in this article.

In [MSS], Mañé, Sad, and Sullivan proved the famous λ-lemma. That can be stated as follows:

The λ-lemma. *If ϕ is a holomorphic motion of E over Δ, then for each λ in Δ, the map $z \mapsto \phi(\lambda, z)$, for z in E, extends to a quasiconformal homeomorphism of $\widehat{C}$ onto itself.*

3 The Teichmüller space of a closed set E

From now on, we will assume that E is a closed subset of $\widehat{C}$ and that $0, 1, \infty \in E$.

Definition 3.1 *Two normalized quasiconformal self-mappings f and g of $\widehat{C}$ are said to be E-equivalent if and only if $f^{-1} \circ g$ is isotopic to the identity rel E. The Teichmüller space $T(E)$ is the set of all E-equivalence classes of normalized quasiconformal self-mappings of $\widehat{C}$.*

The basepoint of $T(E)$ is the E-equivalence class of the identity map.

The following analytic description of $T(E)$ as a quotient space of $M(\mathcal{C})$ will be more useful for our purpose.

Define the quotient map

$$P_E : M(\mathcal{C}) \to T(E)$$

by setting $P_E(\mu)$ equal to the E-equivalence class of w^μ. Clearly, P_E maps the basepoint of $M(\mathcal{C})$ to the basepoint of $T(E)$.

In his doctoral dissertation ([Li]), G. Lieb proved that $T(E)$ is a complex Banach manifold such that the projection map P_E from $M(\mathcal{C})$ to $T(E)$ is a holomorphic split submersion. This result is also proved in [EM].

The Teichmüller distance $d_M(\mu, \nu)$ between μ and ν on $M(\mathcal{C})$ is defined by

$$d_M(\mu, \nu) = \tanh^{-1} \left\| \frac{\mu - \nu}{1 - \bar{\mu}\nu} \right\|_\infty.$$

The *Teichmüller metric* on $T(E)$ is the quotient metric

$$d_{T(E)}(s, t) = \inf\{d_M(\mu, \nu) : \mu \text{ and } \nu \in M(\mathcal{C}), P_E(\mu) = s \text{ and } P_E(\nu) = t\}$$

for all s and t in $T(E)$.

Let w be a normalized quasiconformal self-mapping of $\widehat{\mathcal{C}}$, and let $\widehat{E} = w(E)$. By definition, the *allowable map* g from $T(\widehat{E})$ to $T(E)$ maps the $\widehat{E}$-equivalence class of f to the E-equivalence class of $f \circ w$ for every normalized quasiconformal self-mapping f of $\widehat{\mathcal{C}}$.

Proposition 3.2 *The allowable map* $g \colon T(\widehat{E}) \to T(E)$ *is biholomorphic. If μ is the Beltrami coefficient of w, then g maps the basepoint of $T(\widehat{E})$ to the point $P_E(\mu)$ in $T(E)$.*

See Proposition 7.20 in [EM] or Proposition 6.7 in [M1]. The map g is also called the *geometric isomorphism* induced by the quasiconformal map w. See [EGL2] for other biholomorphic maps between the spaces $T(E)$.

The following fact will be very useful in our discussion.

Proposition 3.3 *There is a continuous basepoint preserving map s from $T(E)$ to $M(\mathcal{C})$ such that $P_E \circ s$ is the identity map on $T(E)$.*

For a complete proof we refer the reader to Proposition 7.22 in [EM] (or Proposition 6.3 in [M1]).

Since $M(\mathcal{C})$ is contractible, we conclude:

Corollary 3.4 *The space $T(E)$ is contractible.*

Here is an outline for the construction of $s(t)$ for t in $T(E)$. It is given in [M2]. We include that discussion here, for the reader's convenience. Choose an extremal μ in $M(\mathcal{C})$ such that $P_E(\mu) = t$. We set $s(t) = \mu$ in E. Let X be a connected component of $\widehat{\mathcal{C}} \setminus E$. On X, $s(t)$ is defined as follows. Choose a holomorphic universal cover $\pi : \Delta \to X$. Lift μ to Δ and let $\widetilde{\mu} = \pi^*(\mu)$ (the lift of μ). If $\pi(\zeta) = z$ we have

$$\widetilde{\mu}(\zeta) = \mu(z)\frac{\overline{\pi'(\zeta)}}{\pi'(\zeta)}.$$

Let $\widetilde{w} : \Delta \to \Delta$ be a quasiconformal map whose Beltrami coefficient is $\widetilde{\mu}$, and let $h : \partial\Delta \to \partial\Delta$ be the boundary homeomorphism. Let $w : \Delta \to \Delta$ be the barycentric extension of h and $\widetilde{\nu}$ be the Beltrami coefficient of w. (Barycentric extensions were studied in [DE].) Then, $\widetilde{\nu}$ is the lift of a uniquely determined L^∞ function ν on X. We set $s(t) = \nu$ in X. Then $\|\widetilde{\mu}\|_\infty = \|\mu|X\|_\infty \leq k := \|\mu\|_\infty$; so,

$$\|s(t)|X\|_\infty = \|\nu\|_\infty \leq c(k)$$

by Proposition 7 in [DE], where $c(k)$ depends only on k and $0 \leq c(k) < 1$. Since X is any connected component of $\widehat{\mathcal{C}} \setminus E$, we conclude that $\|s(t)\|_\infty \leq \max(k, c(k))$.

When E is a finite set, the Teichmüller space $T(E)$ is a classical object. Write $W = \widehat{\mathcal{C}} \setminus E$. There is a natural identification of $T(E)$ with the classical Teichmüller space $Teich(W)$ (i.e. the Teichmüller space of the sphere with punctures at the points of E), which we describe below.

Recall that two quasiconformal mappings f and g with domain W belong to the same Teichmüller class if and only if there is a conformal map h of $f(W)$ onto $g(W)$ such that the self-mapping $g^{-1} \circ h \circ f$ of W is isotopic to the identity rel the boundary of W. The Teichmüller space $Teich(W)$ is the set of Teichmüller classes of quasiconformal mappings with domain W.

The Teichmüller class of f depends only on its Beltrami coefficient, which is a function μ in the open unit ball $M(W)$ of the complex Banach space $L^\infty(W)$. The standard projection Φ of $M(W)$ onto

$Teich(W)$ maps μ to the Teichmüller class of any quasiconformal map whose domain is W and whose Beltrami coefficient is μ. The base-points of $M(W)$ and $Teich(W)$ are 0 and $\Phi(0)$ respectively. It is well-known that $Teich(W)$ is a complex Banach manifold and that $\Phi : M(W) \to Teich(W)$ is a holomorphic split submersion. See [G], [GL], [IT], [N] for basic results in Teichmüller theory.

We can define a map θ from $T(E)$ into $Teich(W)$ by setting $\theta(P_E(\mu))$ equal to the Teichmüller class of the restriction of w^μ to W. It is easy to see that this map θ is biholomorphic (see Example 3.1 in [M1] for the details). This gives a canonical identification of $T(E)$ with the classical Teichmüller space $Teich(W)$ (for finite sets E).

When $E = \widehat{C}$, the space $T(\widehat{C})$ consists of all the normalized quasi-conformal self-mappings of $\widehat{C}$, and the map $P_{\widehat{C}}$ from $M(C)$ to $T(\widehat{C})$ is bijective. We use it to identify $T(\widehat{C})$ biholomorphically with $M(C)$.

4 Universal holomorphic motion of a closed set E

Definition 4.1 *The universal holomorphic motion Ψ_E of E over $T(E)$ is defined as follows:*

$$\Psi_E(P_E(\mu), z) = w^\mu(z) \text{ for } \mu \in M(C) \text{ and } z \in E.$$

The definition of P_E in §3 guarantees that Ψ_E is well-defined. It is a holomorphic motion since P_E is a holomorphic split submersion and $\mu \mapsto w^\mu(z)$ is a holomorphic map from $M(C)$ to $\widehat{C}$ for every fixed z in $\widehat{C}$. This holomorphic motion is "universal" in the following sense:

Theorem 4.2
   Let $\phi : V \times E \to \widehat{C}$ be a holomorphic motion. If V is simply connected, then there exists a unique basepoint preserving holomorphic map $f : V \to T(E)$ such that $f^*(\Psi_E) = \phi$.

For a proof see Section 14 in [M1]. This universal property has found several interesting applications; we will outline them in this article. It is also worth noting that all the three main theorems in [BR] can be deduced from this universal property.

Theorem 4.2 becomes very explicit when $E = \widehat{C}$. Recall from §3 that $T(\widehat{C})$ and $M(C)$ are identified biholomorphically by the map

$P_{\widehat{\mathcal{C}}} : M(\mathcal{C}) \to T(\widehat{\mathcal{C}})$. The pullback $\widetilde{\Psi}_{\widehat{\mathcal{C}}}$ of $\Psi_{\widehat{\mathcal{C}}}$ to $M(\mathcal{C})$ by $P_{\widehat{\mathcal{C}}}$ satisfies

$$\widetilde{\Psi}_{\widehat{\mathcal{C}}}(\mu, z) = \Psi_{\widehat{\mathcal{C}}}(P_{\widehat{\mathcal{C}}}(\mu), z) = w^{\mu}(z), \qquad (\mu, z) \in M(\mathcal{C}) \times \widehat{\mathcal{C}}.$$

Thus, the universal holomorphic motion of $\widehat{\mathcal{C}}$ becomes the map

$$\Psi_{\widehat{\mathcal{C}}}(\mu, z) = w^{\mu}(z) \tag{4.1}$$

from $M(\mathcal{C}) \times \widehat{\mathcal{C}}$ to $\widehat{\mathcal{C}}$.

In particular, if $\phi : V \times \widehat{\mathcal{C}} \to \widehat{\mathcal{C}}$ is a holomorphic motion and V is simply connected, Theorem 4.2 provides a unique basepoint preserving holomorphic map $f : V \to M(\mathcal{C})$ such that

$$\phi(x, z) = \Psi_{\widehat{\mathcal{C}}}(f(x), z) = w^{f(x)}(z)$$

for all x in V and z in $\widehat{\mathcal{C}}$. In other words, $f(x)$ is the Beltrami coefficient of the quasiconformal map $\phi_x(z)$ for each x in V.

Remark 4.3 *The map $f : V \to M(\mathcal{C})$ that sends x in V to the Beltrami coefficient of $\phi_x(z)$ is well-defined for every holomorphic motion $\phi : V \times \widehat{\mathcal{C}} \to \widehat{\mathcal{C}}$. By Theorem 4.2, it is holomorphic when V is simply connected. It follows readily that f is holomorphic for every connected complex manifold V. See Theorem 4 in [E2]. For $V = \Delta$, see Theorem 2 in [BR].*

We end this section by stating a fact that will be very useful in our discussion. Suppose E_1 and E_2 are closed subsets of $\widehat{\mathcal{C}}$ such that E_1 is a proper subset of E_2, and as usual, $0, 1, \infty$ belong to E_1. If μ is in $M(\mathcal{C})$, then the E_2-equivalence class of w^{μ} is contained in the E_1-equivalence class of w^{μ}. Therefore, there is a well-defined "forgetful map" p_{E_2, E_1} from $T(E_2)$ to $T(E_1)$ such that $P_{E_1} = p_{E_2, E_1} \circ P_{E_2}$. It is easy to see that this forgetful map is a basepoint preserving holomorphic split submersion. We also have the universal holomorphic motions $\Psi_{E_1} : T(E_1) \times E_1 \to \widehat{\mathcal{C}}$ and $\Psi_{E_2} : T(E_2) \times E_2 \to \widehat{\mathcal{C}}$.

Lemma 4.4 *Let V be a connected complex Banach manifold with basepoint x_0 and let f and g be basepoint preserving holomorphic maps from V into $T(E_1)$ and $T(E_2)$, respectively. Then $p_{E_2, E_1} \circ g = f$ if and only if $g^*(\Psi_{E_2})$ extends $f^*(\Psi_{E_1})$.*

See Proposition 13.1 in [M1] for a complete proof.

The identification of $T(\widehat{\mathcal{C}})$ with $M(\mathcal{C})$ then gives us:

Corollary 4.5 *Let V be as above and let f and g be basepoint preserving holomorphic maps from V to $T(E)$ and $M(\mathcal{C})$ respectively. Then $P_E \circ g = f$ if and only if $g^*(\Psi_{\widehat{\mathcal{C}}})$ extends $f^*(\Psi_E)$.*

5 Some applications of Slodkowski's theorem

In his fundamental paper [Sl1], Slodkowski proved that if $\phi : \Delta \times E \to \widehat{\mathcal{C}}$ is a holomorphic motion, then there exists a holomorphic motion $\widetilde{\phi} :$ $\Delta \times \widehat{\mathcal{C}} \to \widehat{\mathcal{C}}$ such that $\widetilde{\phi}$ extends ϕ. See also the papers [AM], [Ch], [Do] and the book [Hu2] for proofs of this theorem.

We list below two applications of Slodkowski's theorem.

Proposition 5.1 *If $f : \Delta \to T(E)$ is a basepoint preserving holomorphic map, there is a holomorphic map $F : \Delta \to M(\mathcal{C})$ such that $P_E \circ F = f$.*

Proof Let $f^*(\Psi_E)$ be denoted by ϕ. Then, $\phi : \Delta \times E \to \widehat{\mathcal{C}}$ is a holomorphic motion. By Slodkowski's theorem, there exists a holomorphic motion $\widetilde{\phi} : \Delta \times \widehat{\mathcal{C}} \to \widehat{\mathcal{C}}$ such that $\widetilde{\phi}$ extends ϕ. By Theorem 4.2, there exists a unique basepoint preserving holomorphic map $F : \Delta \to M(\mathcal{C})$ such that $F^*(\Psi_{\widehat{\mathcal{C}}}) = \widetilde{\phi}$. Since $F^*(\Psi_{\widehat{\mathcal{C}}})$ extends $f^*(\Psi_E)$, it follows by Corollary 4.5 that $P_E \circ F = f$. $\qquad\qquad\square$

Remark 5.2 *We do not need the holomorphic map f to be basepoint preserving. If f is any holomorphic map from Δ into $T(E)$ and μ is any point in $M(\mathcal{C})$ such that $P_E(\mu) = f(0)$, we can use change of basepoints in $T(E)$ (see Proposition 3.2) to assume that f is a basepoint preserving map.*

Remark 5.3 *Proposition 5.1 was also proved in [EM] (see Proposition 7.27 in that paper). The proof we have given above is more direct.*

Remark 5.4 *It follows from the above proposition that the Teichmüller metric on $T(E)$ is the same as its Kobayashi metric. For more details we refer the reader to §7.18 in [EM].*

Another interesting application of Slodkowski's theorem is an easy proof of the universal property for holomorphic motions over the open unit disk.

Proposition 5.5 *Let $\phi : \Delta \times E \to \widehat{\mathcal{C}}$ be a holomorphic motion. Then* *there exists a unique basepoint preserving holomorphic map* $f : \Delta \to T(E)$ *such that* $f^*(\Psi_E) = \phi$.

Proof The uniqueness is straightforward.

Let $\phi : \Delta \times E \to \widehat{\mathcal{C}}$ be a holomorphic motion. By Slodkowski's theorem there exists a holomorphic motion $\widetilde{\phi} : \Delta \times \widehat{\mathcal{C}} \to \widehat{\mathcal{C}}$ such that $\widetilde{\phi}$ extends ϕ. By our discussion after Theorem 4.2, there exists a unique basepoint preserving holomorphic map $F : \Delta \to M(\mathcal{C})$ such that $F^*(\Psi_{\widehat{\mathcal{C}}}) = \widetilde{\phi}$, which means

$$\widetilde{\phi}(x, z) = w^{F(x)}(z)$$

for all $(x, z) \in \Delta \times \widehat{\mathcal{C}}$.

Define $f : \Delta \to T(E)$ as $f = P_E \circ F$. Then, clearly, f is holomorphic and is basepoint preserving. Also, we have

$$f^*(\Psi_E)(x, z) = \Psi_E(f(x), z) = \Psi_E(P_E(F(x)), z) = w^{F(x)}(z) = \phi(x, z)$$

for $(x, z) \in \Delta \times E$. $\square$

Remark 5.6 *This was proved independently by Lieb in his thesis [Li] without using Slodkowski's theorem.*

Remark 5.7 *Slodkowski's theorem has important applications in dynamics, holomorphic motions, and Teichmüller theory. In §8.2, we discuss some applications in Teichmüller theory. See [EGL1], [J], [Ma], and [Sl2] for some other interesting applications.*

Remark 5.8 *Slodkowski's theorem does not hold for more general parameter spaces. An explicit example is given in Appendix two in [EM] for a holomorphic motion over a domain in $\mathcal{C}^2$. In that example the set E is an infinite set. We give below an example of a holomorphic motion of a finite set over a simply connected parameter space for which Slodkowski's theorem does not hold. This was given in [JM]. We include this for the reader's convenience.*

An example. Let $E = \{0, 1, \infty, \zeta_1, ..., \zeta_n\}$, where $\zeta_i \neq \zeta_j$ for $i \neq j$ and n is a positive integer ≥ 2. Consider the universal holomorphic motion $\Psi_E : T(E) \times E \to \widehat{\mathcal{C}}$. By our discussion in §3, $T(E)$ and the classical Teichmüller space $Teich(\widehat{\mathcal{C}} \setminus E)$ are canonically identified. Consider the identity map $i : T(E) \to T(E)$. Obviously, $i^*(\Psi_E) = \Psi_E$. Suppose Ψ_E extends to a holomorphic motion $\widehat{\phi} : T(E) \times \widehat{\mathcal{C}} \to \widehat{\mathcal{C}}$. Then, there exists a basepoint preserving holomorphic map $F : T(E) \to M(\mathcal{C})$ such that

$F^*(\Psi_{\widehat{C}}) = \widehat{\phi}$. Since $\widehat{\phi}$ extends $i^*(\Psi_E)$, it follows by Corollary 4.5 that $P_E \circ F = i$. That means, the map P_E has a holomorphic section F. Since $T(E)$ and $Teich(\widehat{C} \setminus E)$ are naturally identified, this is impossible by Earle's theorem in [E1]. This proves that for the finite set E given above, the universal holomorphic motion $\Psi_E : T(E) \times E \to \widehat{C}$ cannot be extended to a holomorphic motion of $\widehat{C}$.

Definition 5.9 *A holomorphic motion* $\phi : V \times E \to \widehat{C}$ *is called maximal if there exists no set $\widehat{E}$ where E is a proper subset of $\widehat{E}$ such that ϕ can be extended to a holomorphic motion of $\widehat{E}$.*

Remark 5.10 *The universal holomorphic motion* $\Psi_E : T(E) \times E \to \widehat{C}$ *for a finite set E discussed in the above example is a maximal holomorphic motion. This was mentioned in [Do], and the details are given in [M4]. The proof is easy, using some results in [EK] (also [Hu1]). We want to emphasize that the non-extendability of the above holomorphic motion to a holomorphic motion of $\widehat{C}$ does not need the deep results in [Hu1] (which study the nonexistence of holomorphic sections of the universal Teichmüller curve).*

Remark 5.11 *See [E2] for other interesting examples of maximal holomorphic motions.*

6 Some applications of universal holomorphic motions

In this section we review some extension theorems that follow from the universal property of Ψ_E (Theorem 4.2).

Let E_0 be any subset of $\widehat{C}$, not necessarily closed; as usual, we assume that $0, 1, \infty \in E_0$. Let E be the closure of E_0. Using Theorem 4.2, we prove the following

Theorem 6.1 *Let* $\phi : V \times E_0 \to \widehat{C}$ *be a holomorphic motion where V is a connected complex Banach manifold with a basepoint. There exists a holomorphic motion* $\widehat{\phi} : V \times E \to \widehat{C}$ *such that $\widehat{\phi}$ extends ϕ.*

For a proof see Theorem 2 in [JM].

Remark 6.2 *For holomorphic motions over Δ, this was proved in [MSS].*

The following theorem was proved in [M1].

Theorem 6.3 *Let* $\phi : V \times E \to \widehat{C}$ *be a holomorphic motion, where V is a simply connected complex Banach manifold with basepoint x_0. Then, for every $x \in V$, $\phi(x, \cdot)$ is the restriction to E of a quasiconformal self map of $\widehat{C}$ with dilatation not exceeding $exp(2\rho_V(x_0, x))$, where ρ_V denotes the Kobayashi pseudometric on V.*

The proof is an easy consequence of Theorem 4.2. There is a basepoint preserving holomorphic map $f : V \to T(E)$ such that $f^*(\Psi_E) = \phi$. Let $x \in V$ and choose an extremal μ in $M(\mathcal{C})$ such that $P_E(\mu) = f(x)$. We have $\phi(x, z) = w^\mu(z)$ for all z in E, and also

$$d_{T(E)}(f(x_0), f(x)) = d_M(0, \mu) = \tanh^{-1}(\|\mu\|_\infty) = \frac{1}{2} \log K$$

where K is the dilatation of the quasiconformal map w^μ.

By Remark 5.4, the Teichmüller metric on $T(E)$ is the same as its Kobayashi metric. Hence, we get

$$\frac{1}{2} \log K = d_{T(E)}(f(x_0), f(x)) \le \rho_V(x_0, x)$$

which gives $\log K \le (2\rho_V(x_0, x))$.

Remark 6.4 *If $V = \Delta$ and $x_0 = 0$, it follows from Theorem 6.3, that*

$$K \le \frac{1 + |x|}{1 - |x|}$$

which is Theorem 1 in [BR].

An application. Theorems 6.1 and 6.3 were used in [JM] to study holomorphic dynamical systems. As above, V is a simply connected complex Banach manifold with a basepoint x_0, and ρ_V denotes the Kobayashi pseudometric on V.

Theorem 6.5 *Suppose $R(x, z)$ is a hyperbolic family of holomorphic dynamical systems over V. Then, for any $x \in V$, the Julia set J_x is quasiconformally equivalent to J_{x_0} by a quasiconformal map of $\widehat{C}$ whose dilatation does not exceed $exp(2\rho_V(x, x_0))$. Moreover, J_x depends holomorphically on x over V.*

See Theorem 3 in [JM] for the details.

Generalized harmonic λ-lemma. Let Ω be a plane region whose complement in $\mathcal{C}$ contains at least two points. Following [BR] and [MS], we call a Beltrami coefficient μ on Ω *harmonic* if the function $\overline{\mu}\lambda^2$ on Ω is holomorphic, where $ds = \lambda(z)|dz|$ is the infinitesimal Poincaré metric on $\overline{\Omega} = \{z \in \mathcal{C} : \overline{z} \in \Omega\}$.

Recall from §3, the standard projection $P_E : M(\mathcal{C}) \to T(E)$. It can be shown that there is a domain W in $T(E)$ on which we can define the *Ahlfors-Weill section* $\mathcal{S}$, as a holomorphic map of W into $M(\mathcal{C})$ such that $P_E(\mathcal{S}(\tau)) = \tau$ for all τ in W. It is also true that μ in $M(\mathcal{C})$ belongs to the image of $\mathcal{S}$ if and only if $\mu|\Omega$ is harmonic for each component Ω of $\widehat{\mathcal{C}} \setminus E$. See §6 in [M1] for the details.

In the next theorem, V is a simply connected complex Banach manifold with a basepoint x_0, ρ_V is the Kobayashi metric on V and ρ_Δ is the Poincaré metric on Δ.

It can be shown that if f is a basepoint preserving holomorphic map of V into $T(E)$ and $\rho_V(x, x_0)$ is less than $\rho_\Delta(0, 1/3)$, then $f(x)$ must belong to W. See Lemma 15.2 in [M1] for a proof.

Theorem 6.6 *Let* $\phi : V \times E \to \widehat{\mathcal{C}}$ *be a holomorphic motion. There is a* **unique** *holomorphic motion* $\widehat{\phi} : U \times \widehat{\mathcal{C}} \to \widehat{\mathcal{C}}$, *where* $U = \{x \in V : \rho_V(x, x_0) < \rho_\Delta(0, 1/3)\}$, *such that* $\widehat{\phi}(x, z) = \phi(x, z)$ *for all* $(x, z) \in U \times E$ *and the Beltrami coefficient of the map* $z \mapsto \widehat{\phi}(x, z)$ *is harmonic on each component of* $\widehat{\mathcal{C}} \setminus E$ *for each* $x \in U$.

Because of the importance of this theorem we include an outline of the proof. By Theorem 4.2, there exists a basepoint preserving holomorphic map $f : V \to T(E)$ such that $f^*(\Psi_E) = \phi$. We noted above that f maps the set U into the domain W. It is easy to show that U is open and connected. So, the restriction of ϕ to $U \times E$ is a holomorphic motion (still called ϕ) over the connected complex manifold U (with the same basepoint).

Since f maps U into W, we can compose it with the Ahlfors-Weill holomorphic section $\mathcal{S} : W \to M(\mathcal{C})$, getting a basepoint preserving holomorphic map $\widehat{f} = \mathcal{S} \circ f$ from U to $M(\mathcal{C})$ such that $P_E \circ \widehat{f} = f$. By Corollary 4.5, $\widehat{\phi} = \widehat{f}^*(\Psi_{\widehat{c}})$ is a holomorphic motion of $\widehat{\mathcal{C}}$ over U which extends ϕ. For each x in U, the Beltrami coefficient of the map $z \mapsto \widehat{\phi}(x, z)$ is $\widehat{f}(x) = \mathcal{S}(f(x))$, and so, its restriction to each component of $\widehat{\mathcal{C}} \setminus E$ is harmonic. This shows the existence part.

For the uniqueness, suppose that $\widetilde{\phi}$ is any holomorphic motion of $\widehat{\mathcal{C}}$ over U having the required properties. By Theorem 4 of [E2] (also see Remark 4.3), there is a basepoint preserving holomorphic map

$h : U \rightarrow M(\mathcal{C})$ such that $\widetilde{\phi} = h^*(\Psi_{\widehat{\mathcal{C}}})$. Since $\widetilde{\phi}$ extends ϕ, it follows by Corollary 4.5 that $P_E \circ h = f$ in U. The Beltrami coefficient $h(x)$ of the map $z \mapsto \widetilde{\phi}(x, z)$ is harmonic in each component of $\widehat{\mathcal{C}} \setminus E$, and therefore, $h(x)$ is in the image of $\mathcal{S}$ for each x in U. Hence, we get

$$h(x) = \mathcal{S}(P_E(h(x))) = \mathcal{S}(f(x)) = \widehat{f}(x)$$

for all x in U. It follows that $\widetilde{\phi} = \widehat{\phi}$.

Remarks.

1. For $V = \Delta$ this is Theorem 3 in [BR]. We call this the Bers-Royden extension theorem.

2. In [ST], it was shown that there is some universal constant $a > 0$ such that a holomorphic motion of any set $X \subset \mathcal{C}$ over the parameter space Δ can be extended to a holomorphic motion of $\mathcal{C}$ over the parameter space Δ_a of radius a about 0.

3. In [Do], Douady observed that the Bers-Royden extension theorem can be generalized to holomorphic motions over open unit balls in complex Banach spaces. A proof was sketched by Sugawa in [Su].

4. The Bers-Royden extension theorem was called the *harmonic λ-lemma* by McMullen and Sullivan in [MS]. The uniqueness of the extension has special advantages. That was used in [MS] to study quasiconformal conjugacies.

5. For some other applications of the Bers-Royden extension theorem see the papers [B1], [B2], and [Ro].

6. Theorem 6.6 proves the harmonic λ-lemma in its fullest generality. The central idea in our proof is the universal property of the holomorphic motion Ψ_E.

7 Some other extension theorems

Let X be a connected Hausdorff space with a basepoint x_0. For any map $\phi : X \times E \rightarrow \widehat{\mathcal{C}}$, x in X, and any quadruplet a, b, c, d of points in E, let $\phi_x(a, b, c, d)$ denote the cross-ratio of the values $\phi(x, a)$, $\phi(x, b)$, $\phi(x, c)$, and $\phi(x, d)$. We will write $\phi(x, z)$ as $\phi_x(z)$ for x in X and z in E. So we have:

$$\phi_x(a, b, c, d) = \frac{(\phi_x(a) - \phi_x(c))(\phi_x(b) - \phi_x(d))}{(\phi_x(a) - \phi_x(d))(\phi_x(b) - \phi_x(c))} \tag{7.1}$$

for each x in X.

Let ρ be the Poincaré metric on $\widehat{C} \setminus \{0, 1, \infty\}$. In their paper [ST], Sullivan and Thurston introduced the following definition.

Definition 7.1 *A quasiconformal motion is a map* $\phi : X \times E \to \widehat{C}$ *of E over X such that*

(i) $\phi(x_0, z) = z$ *for all z in E, and*

(ii) *given any x in X and any $\epsilon > 0$, there exists a neighborhood U_x of x such that for any quadruplet of distinct points a, b, c, d in E, we have*

$$\rho\big(\phi_y(a, b, c, d), \phi_{y'}(a, b, c, d)\big) < \epsilon \quad \text{for all } y \text{ and } y' \text{ in } U_x.$$

We will assume that ϕ is a normalized quasiconformal motion; i.e. 0, 1, and ∞ belong to E and are fixed points of the map $\phi_x(\cdot)$ for every x in X.

Remark 7.2 *If $\phi : X \times E \to \widehat{C}$ is a quasiconformal motion, $\phi_x(a, b, c, d)$ is a well-defined point in $\widehat{C} \setminus \{0, 1, \infty\}$, and then it is obvious that for each x in X, the map $\phi_x : E \to \widehat{C}$ is injective.*

Remark 7.3 *Suppose $\phi : V \times E \to \widehat{C}$ is a holomorphic motion (where V is a connected complex manifold with a basepoint). For any quadruplet of points a, b, c, d in E, the map $x \mapsto \phi_x(a, b, c, d)$ from V into $\widehat{C} \setminus \{0, 1, \infty\}$ is holomorphic. Therefore, it is distance-decreasing with respect to the Kobayashi metrics on V and $\widehat{C} \setminus \{0, 1, \infty\}$. It easily follows that the map $\phi : V \times E \to \widehat{C}$ is also a quasiconformal motion.*

Remark 7.4 *Let V and W be connected Hausdorff spaces with basepoints, and f be a continuous basepoint preserving map of W into V. If ϕ is a quasiconformal motion of E over V its pullback by f is*

$$f^*(\phi)(x, z) = \phi(f(x), z) \qquad \forall (x, z) \in W \times E. \tag{7.2}$$

Then, $f^*(\phi)$ is a quasiconformal motion of E over W.

Theorem 7.5 *Let $\phi : V \times E \to \widehat{C}$ be a holomorphic motion where V is a simply connected complex Banach manifold with a basepoint x_0. Then there exists a quasiconformal motion $\widetilde{\phi} : V \times \widehat{C} \to \widehat{C}$ such that $\widetilde{\phi}$ extends ϕ. Furthermore, the quasiconformal motion $\widetilde{\phi} : V \times \widehat{C} \to \widehat{C}$ satisfies the following properties:*

(i) the map $\widetilde{\phi}$ is continuous,

(ii) for each x in V, the map $\widetilde{\phi}_x : \widehat{C} \to \widehat{C}$ is quasiconformal,

(iii) the Beltrami coefficient μ_x of $\widetilde{\phi}_x$ depends continuously with respect to x, and

(iv) the L^∞ norm of μ_x is bounded above by a number less than 1, that depends only on the Kobayashi distance from x to x_0, denoted by $\rho_V(x, x_0)$.

Here is an outline of the construction of the map $\widetilde{\phi}$.

By Theorem 4.2, there exists a unique basepoint preserving holomorphic map $f : V \to T(E)$ such that $f^*(\Psi_E) = \phi$. Recall from Proposition 3.3 the continuous basepoint preserving map $s : T(E) \to M(\mathcal{C})$ such that $P_E \circ s$ is the identity map on $T(E)$. (See the construction of the map s given in §3.) Define $\widetilde{f} : V \to M(\mathcal{C})$ as $\widetilde{f} = s \circ f$. Let $\widetilde{\phi} : V \times \widehat{\mathcal{C}} \to \widehat{\mathcal{C}}$ be defined by $\widetilde{\phi}(x, z) = w^{\widetilde{f}(x)}(z)$ for all x in V and for all z in $\widehat{\mathcal{C}}$. It is easy to see that $\widetilde{\phi}$ extends the holomorphic motion ϕ.

Next, recall from §4, the universal holomorphic motion $\Psi_{\widehat{\mathcal{C}}} : M(\mathcal{C}) \times \widehat{\mathcal{C}} \to \widehat{\mathcal{C}}$. By Remark 7.3, $\Psi_{\widehat{\mathcal{C}}}$ is a quasiconformal motion of $\widehat{\mathcal{C}}$. Since $\widetilde{f} : V \to M(\mathcal{C})$ is a basepoint preserving continuous map, $\widetilde{f}^*(\Psi_{\widehat{\mathcal{C}}}) : V \times \widehat{\mathcal{C}} \to \widehat{\mathcal{C}}$ is a quasiconformal motion (by Remark 7.4). Note that for all $(x, z) \in V \times \widehat{\mathcal{C}}$, we have

$$\widetilde{f}^*(\Psi_{\widehat{\mathcal{C}}})(x, z) = \Psi_{\widehat{\mathcal{C}}}(\widetilde{f}(x), z) = w^{\widetilde{f}(x)}(z) = \widetilde{\phi}(x, z).$$

Hence, $\widetilde{\phi} : V \times \widehat{\mathcal{C}} \to \widehat{\mathcal{C}}$ is a quasiconformal motion.

It is not hard to show that this map $\widetilde{\phi}$ satisfies the properties (i)–(iv) in the theorem. See [M4] for the details.

The following example illustrates the above theorem for the universal holomorphic motion $\Psi_E : T(E) \times E \to \widehat{\mathcal{C}}$.

An example. We know from §3, that $T(E)$ is simply connected. The map $\widetilde{\Psi}_E : T(E) \times \widehat{\mathcal{C}} \to \widehat{\mathcal{C}}$ defined by the formula

$$\widetilde{\Psi}_E(t, z) = w^{s(t)}(z), \qquad (t, z) \in T(E) \times \widehat{\mathcal{C}},$$

is a quasiconformal motion of $\widehat{\mathcal{C}}$ over $T(E)$, that extends the universal holomorphic motion $\Psi_E : T(E) \times E \to \widehat{\mathcal{C}}$, and satisfies the conditions (i)–(iv) of Theorem 7.5.

In fact, we have

$$\Psi_E(t, z) = \Psi_E(P_E(s(t)), z) = w^{s(t)}(z) = \widetilde{\Psi}_E(t, z)$$

for all $(t, z) \in T(E) \times E$. So $\widetilde{\Psi}_E$ extends Ψ_E. Next, note that the pullback $s^*(\Psi_{\widehat{\mathcal{C}}})$ is a quasiconformal motion of $\widehat{\mathcal{C}}$ over $T(E)$. Also,

$s^*(\Psi_{\widehat{C}})(t,z) = \Psi_{\widehat{C}}(s(t),z) = w^{s(t)}(z)$ for all t in $T(E)$ and z in $\widehat{C}$. There-fore the map $\widetilde{\Psi}_E$ is a quasiconformal motion. The continuity of $\widetilde{\Psi}_E$ follows from Lemma 17 of [AB], which says that $w^{\mu_n} \to w^{\mu}$ uniformly in the spherical metric if $\mu_n \to \mu$ in $M(\mathcal{C})$. Properties (ii) and (iii) are obvious. Property (iv) is an easy consequence of the construction of the continuous section $s : T(E) \to M(\mathcal{C})$, and the estimate of the L^∞ norm of $s(t)$ (for t in $T(E)$), given in that construction.

Definition 7.6 *Let X be a path-connected Hausdorff space with a base-point x_0. A normalized continuous motion of $\widehat{C}$ over X is a continuous map $\phi : X \times \widehat{C} \to \widehat{C}$ such that:*

(i) $\phi(x_0,z) = z$ for all z in $\widehat{C}$, and

(ii) for each x in X, the map $\phi(x,\cdot)$ is a homeomorphism of $\widehat{C}$ onto itself that fixes the points 0, 1, and ∞.

We will always assume that the continuous motion is normalized and we will write $\phi(x,\cdot)$ as $\phi_x(\cdot)$ for x in X.

In [M3] the following facts are proved.

Proposition 7.7 *Let V be a connected Hausdorff space with basepoint x_0. Let $\phi : V \times \widehat{C} \to \widehat{C}$ be a map such that $\phi(x_0,z) = z$ for all z in $\widehat{C}$, and 0, 1, and ∞ are fixed points of $\phi(x,\cdot)$ for every x in V. Then ϕ is a quasiconformal motion of $\widehat{C}$ if and only if it satisfies:*

(a) the map $\phi_x : \widehat{C} \to \widehat{C}$ is quasiconformal for each x in V, and

(b) the map from V to $M(\mathcal{C})$ that sends x to the Beltrami coefficient of ϕ_x for each x in V is continuous.

One direction is easy. The map $\Psi_{\widehat{C}}$ in Equation (4.1) (see §4) is a quasiconformal motion, by Remark 7.3. Let $\phi : V \times \widehat{C} \to \widehat{C}$ be a map that satisfies (a) and (b). For each x in V, let μ_x be the Beltrami coefficient of the quasiconformal map ϕ_x, and let $f : V \to M(\mathcal{C})$ be the continuous map $x \mapsto \mu_x$. Then

$$f^*(\Psi_{\widehat{C}})(x,z) = \Psi_{\widehat{C}}(\mu_x,z) = w^{\mu_x}(z) = \phi_x(z) = \phi(x,z)$$

for all x in V and z in $\widehat{C}$, and so ϕ is the pullback of the quasiconformal motion $\Psi_{\widehat{C}}$ by f. Therefore ϕ is a quasiconformal motion (by Remark 7.4).

The other direction is non-trivial. See [M3] for the details.

Proposition 7.8 *Every quasiconformal motion of $\widehat{C}$ is a continuous motion.*

Remark 7.9 *We want to* **emphasize** *that Theorem 7.5 is completely independent of Propositions 7.7 and 7.8. The quasiconformal motion $\widetilde{\phi} : V \times \widehat{C} \to \widehat{C}$ that we construct in Theorem 7.5 follows from the universal property of holomorphic motions over simply connected parameter spaces, and the construction of the continuous section $s : T(E) \to M(\mathcal{C})$. With the exception of the facts given in Remarks 7.3 and 7.4 (both of which are obvious),* **no other** *properties of quasiconformal motions are necessary. All the properties (i)–(iv) in Theorem 7.5 follow from the construction of the map $\widetilde{\phi}$.*

Proposition 7.10 *Let $\phi : V \times E \to \widehat{C}$ be a holomorphic motion where V is a connected complex Banach manifold with a basepoint x_0. Suppose $\widetilde{\phi} : V \times \widehat{C} \to \widehat{C}$ and $\widetilde{\psi} : V \times \widehat{C} \to \widehat{C}$ are two quasiconformal motions that extend ϕ. Then, for each x in V, the quasiconformal maps $\widetilde{\phi}_x$ and $\widetilde{\psi}_x$ are isotopic rel E.*

The proof needs the following easy lemma. Let B be a path-connected topological space and $\mathcal{H}(\widehat{C})$ be the group of homeomorphisms of $\widehat{C}$ onto itself, with the topology of uniform convergence in the spherical metric.

Lemma 7.11 *Let $h : B \to \mathcal{H}(\widehat{C})$ be a continuous map such that $h(t)(e) = e$ for all t in B and for all e in E. If $h(t_0)$ is isotopic to the identity rel E for some fixed t_0 in B, then $h(t)$ is isotopic to the identity rel E for all t in B.*

The proof is easy; it is Lemma 12.1 in [M1].

Proof of Proposition 7.10. Define maps f and g from V to $\mathcal{H}(\widehat{C})$ by $f(x)(z) = \widetilde{\phi}(x, z)$ and $g(x)(z) = \widetilde{\psi}(x, z)$ for x in V and z in $\widehat{C}$. By Proposition 7.8, both $\widetilde{\phi}$ and $\widetilde{\psi}$ are continuous. Therefore, by Theorem 5 in [Ar], the maps f and g are both continuous. Hence, the map $h : V \to \mathcal{H}(\widehat{C})$ defined by

$$h(x) = g(x)^{-1} \circ f(x)$$

for x in V, is also continuous. Clearly, $h(x_0)$ is the identity map of $\widehat{C}$, and for each x in V, $h(x)$ fixes E pointwise. Therefore, by Lemma 7.11, it follows that $h(x)$ is isotopic to the identity rel E, for each x in V.

The following result gives the most general extension theorem for holomorphic motions over connected parameter spaces.

Theorem 7.12 *Let $\phi : V \times E \to \widehat{C}$ be a holomorphic motion where V*

is a connected complex Banach manifold with a basepoint x_0. Then the following are equivalent.

(i) There exists a continuous motion $\widetilde{\phi} : V \times \widehat{C} \to \widehat{C}$ that extends ϕ.

(ii) There exists a quasiconformal motion $\widehat{\phi} : V \times \widehat{C} \to \widehat{C}$ that extends ϕ.

(iii) There exists a basepoint preserving holomorphic map $F : V \to T(E)$ such that $F^*(\Psi_E) = \phi$.

The implications (iii) implies (ii) and also (iii) implies (i) are sketched in the outline of the proof of Theorem 7.5. The implication (i) implies (ii) uses Proposition 7.7, and Proposition 7.8 shows that (ii) implies (i). The deepest part is to prove that (i) implies (iii) which is given in §5 in [M2]. We refer the reader to [M2] and [M3] for all details.

Remark 7.13 *Propositions 7.7 and 7.8 describe properties of quasiconformal motions of $\widehat{C}$. These properties are used in the proof of Theorem 7.12. If the parameter space of the holomorphic motion ϕ is simply connected, then the quasiconformal motion that extends ϕ has the additional property (iv) of Theorem 7.5.*

Here is an explicit example of a holomorphic motion of a set that cannot be extended to a continuous motion of $\widehat{C}$. I thank Clifford J. Earle for this example.

An example. Let $\Delta^* := \{z \in C : 0 < |z| < 1\}$ and choose some basepoint a in Δ^*. Let $E := \{0, 1, a, \infty\}$. Let $\phi(t, z) = z$ for all (t, z) in $\Delta^* \times \{0, 1, \infty\}$ and $\phi(t, a) = t$ for all t in Δ^*. Clearly, ϕ is a holomorphic motion of E over Δ^*. We show that ϕ cannot be extended to a continuous motion of $\widehat{C}$ over Δ^*. For, suppose $\widetilde{\phi}$ is such an extension. For each ζ in $C \setminus \{0\}$, let $\gamma_\zeta : [0, 2\pi] \to C \setminus \{0\}$ be the closed curve

$$\gamma_\zeta(\theta) = \widetilde{\phi}(ae^{i\theta}, \zeta)$$

for θ in $[0, 2\pi]$. Since $\widetilde{\phi}$ is a continuous motion, the winding number of γ_ζ about zero is a continuous function of ζ. But that winding number is zero when $\zeta = 1$ and one when $\zeta = a$.

Remark 7.14 *Because of Theorem 7.12, the holomorphic motion ϕ in the above example cannot be extended to a quasiconformal motion of $\widehat{C}$.*

Remark 7.15 *Note that, in the above example the Teichmüller space $T(E)$ is the Teichmüller space of the four-punctured sphere (which is the* *upper half plane). So, by Theorem 7.12, this gives an example of a holomorphic motion ϕ, for which there cannot exist a (basepoint preserving) holomorphic map $f : \Delta^* \to T(E)$ such that $f^*(\Psi_E) = \phi$.*

The following extension theorem is an easy consequence of Theorem 48. See [M3] for a proof.

Theorem 7.16 *Let V be a connected complex Banach manifold with a basepoint and let $\phi : V \times E \to \widehat{C}$ be a holomorphic motion. Suppose the restriction of ϕ to $V \times K$ extends to a continuous motion of $\widehat{C}$ whenever $\{0, 1, \infty\} \subset K \subset E$ and K is finite. Then, ϕ extends to a quasiconformal motion of $\widehat{C}$.*

8 Some applications

8.1 Some applications in geometric function theory

In March 1996, at the Oberwolfach conference, Dieter Gaier posed four problems about the dependence of conformal invariants on parameters, that were hard to solve using classical methods. Three of them were on the modules of ring domains or quadrilaterals. The general question can be stated as follows:

Suppose a quadrilateral or ring domain D in $\widehat{C}$ is varied by letting its boundary points depend analytically on some parameters. Is the module of D a real analytic function of the parameters?

Using universal holomorphic motions, it was shown in [EM] that the answer is positive if the parameter space is a connected complex manifold V and the boundary of D undergoes a holomorphic motion over V. In this section we state the basic results. For details, the reader is referred to the paper [EM].

Let V be a connected complex Banach manifold with a basepoint t_0, and let ϕ be a holomorphic motion of E over V. For each t in V there exists a quasiconformal map $\widetilde{\phi}_t$ of $\widehat{C}$ onto itself such that $\widetilde{\phi}_t(z) = \phi(t, z)$ for all z in E. One can show that, if D is a component of $\widehat{C} \setminus E$, the region $\widetilde{\phi}_t(D)$ depends only on t, ϕ, and D, not on the choice of $\widetilde{\phi}_t$. For t in V, we set $\phi_t(E) = \{\phi(t, z) : z \in E\}$, and we define $\phi_t(D)$ to be the component of $\widehat{C} \setminus \phi_t(E)$ such that $\phi_t(D) = \widetilde{\phi}_t(D)$ for any quasiconformal map $\widetilde{\phi}_t$ of $\widehat{C}$ onto itself satisfying $\widetilde{\phi}_t(z) = \phi(t, z)$ for all z in E. So, if D is a region in $\widehat{C}$ and E is its boundary, we get a well-defined region $\phi_t(D)$ for each t in V.

The region D in $\widehat{C}$ is called a *nondegenerate ring domain* if and only if it is conformally equivalent to the annulus

$$A(R) = \{z \in C : 1 < |z| < R\}$$

for some finite number $R > 1$. The number R is uniquely determined by D, and the number

$$mod(D) = \frac{1}{2\pi} \log(R)$$

is called the *module* of D.

Theorem 8.1 *Let E be the boundary of the nondegenerate ring domain D. If ϕ is a holomorphic motion of E over V, then the set $\phi_t(D)$ is a nondegenerate ring domain for each t in V, and the function $t \mapsto mod(\phi_t(D))$ is real analytic in V.*

See [EM] and also the Corollaries 1.4 and 1.5 there.

We recall the usual definition of a quadrilateral; it is a Jordan domain in $\widehat{C}$ with four distinguished points a, b, c, and d on its boundary (with a positive orientation). Let f be a conformal mapping of D onto a rectangle in C whose vertices are the points $f(a)$, $f(b)$, $f(c)$, and $f(d)$. By definition, the module of the quadrilateral $(D; a, b, c, d)$ is the ratio

$$mod(D; a, b, c, d) = \frac{|f(b) - f(a)|}{|f(c) - f(b)|}.$$

Theorem 8.2 *Let E be the boundary of the quadrilateral $(D; a, b, c, d)$, and let ϕ be a holomorphic motion of E over V. The function*

$$t \mapsto mod(\phi_t(D); \phi(t, a), \phi(t, b), \phi(t, c), \phi(t, d))$$

is real analytic in V.

See [EM] for a proof.

Let $\phi : V \times E \to \widehat{C}$ be a holomorphic motion, and let γ be a closed curve in the component D of $\widehat{C} \setminus E$. Let $\widetilde{\phi}$ and $\widetilde{\psi}$ be two quasiconformal motions of $\widehat{C}$ that extend ϕ. For each t in V, the regions $\widetilde{\phi}_t(D)$ and $\widetilde{\psi}_t(D)$ are equal, and the closed curves $\widetilde{\phi}_t(\gamma)$ and $\widetilde{\psi}_t(\gamma)$ are homotopic in $\widetilde{\phi}_t(D)$, by Proposition 7.10. If γ is a Poincaré geodesic in D, then the homotopy class of $\widetilde{\phi}_t(\gamma)$ in $\widetilde{\phi}_t(D)$ contains a unique Poincaré geodesic, which we denote by $\phi_t(\gamma)$. See the discussion in Corollary and Definition 2.4 of [EM] for details.

We now state the next application. See §5 in [EM] for a complete proof.

Theorem 8.3 *Let D be a region in $\widehat{C}$ whose boundary E contains at least three points, and let ϕ be a holomorphic motion of E over the simply connected parameter space V. For any closed Poincaré geodesic γ in D, the length of the closed Poincaré geodesic $\phi_t(\gamma)$ in $\phi_t(D)$ is a real analytic function of t in V.*

8.2 Some applications in Teichmüller theory

In [EKK] the following group-equivariant version of Slodkowski's theorem is proved.

Theorem 8.4 *Let K be a subset of $\widehat{C}$ that contains at least three points, and let G be a group of Möbius transformations that map K onto itself. Let $\phi : \Delta \times K \to \widehat{C}$ be a holomorphic motion. Suppose that for each $g \in G$ and $t \in \Delta$, there is a Möbius transformation $\theta_t(g)$ such that*

$$\phi(t, g(z)) = \theta_t(g)(\phi(t, z)) \qquad (8.1)$$

for all z in K. Then ϕ can be extended to a holomorphic motion $\widehat{\phi}$ of $\widehat{C}$ in such a way that (8.1) holds for all $g \in G$, $t \in \Delta$, and $z \in \widehat{C}$.

For a complete proof see §2 in [EKK] (also [GL], [Hu2]).

Let G be a Fuchsian group acting on the upper half plane $\mathcal{U}$. As usual, let $M(G)$ be the open unit ball of the complex Banach space of bounded measurable Beltrami differentials for G on $\mathcal{U}$ denoted by $L^\infty(\mathcal{U}, G)$. Let $Teich(G)$ be the Teichmüller space of G and let $\Phi : M(G) \to Teich(G)$ be the canonical projection. See [GL], [Hu2], or [N] for standard facts on Teichmüller theory.

Using Theorem 8.4, the authors prove the following theorem in [EKK].

Theorem 8.5 *If $f : \Delta \to Teich(G)$ is a holomorphic map, then there exists a holomorphic map $g : \Delta \to M(G)$ such that $\Phi \circ g = f$. If $\mu_0 \in M(G)$ and $\Phi(\mu_0) = f(0)$, we can choose g so that $g(0) = \mu_0$.*

This lifting theorem has several remarkable applications in Teichmüller theory. One immediate consequence is that the Teichmüller metric on $Teich(G)$ is the same as its Kobayashi metric. For other interesting applications, we refer the reader to the paper [EKK].

We would like to end this survey article with a short discussion of an extremely interesting application of the λ-lemma. For some other applications of the λ-lemma in Teichmüller theory see [Sh] and [Su].

Definition 8.6 *A simple holomorphic family* (W, π, B) *consists of a pair of connected complex manifolds* W *and* B, *with* B *simply connected, and a surjective holomorphic map* $\pi : W \to B$ *satisfying the following conditions. First, the map* π *has horizontally holomorphic local trivializations. That means there is an open covering* $\{B_\alpha\}$ *of* B, *a topological space* X, *and homeomorphisms* θ_α *(the local trivializations) of* $B_\alpha \times X$ *onto* $\pi^{-1}(B_\alpha)$ *such that* $\pi(\theta_\alpha(t, x)) = t$ *for all* $(t, x) \in B_\alpha \times X$ *and the map* $\theta_\alpha(\cdot, x) : B_\alpha \to W$ *is holomorphic for each* x *in* X. *Secondly, each fiber* $\pi^{-1}(t)$ *should be a Riemann surface.*

A *morphism* of simple holomorphic families (W_1, π_1, B_1) to (W_2, π_2, B_2) is a pair of holomorphic maps $f : B_1 \to B_2$ and $g : W_1 \to W_2$ such that g restricts to each fiber as a bijective map of $\pi_1^{-1}(t)$ to $\pi_2^{-1}(f(t))$ for each t in B_1. If f and g are biholomorphic, the morphism is called an *isomorphism*. An isomorphism is an automorphism if $B_1 = B_2$ and $W_1 = W_2$.

In [EF2], the following characterization of $Teich(\Gamma)$ is proved:

Theorem 8.7 *Let* Γ *be a Fuchsian group acting freely on* Δ, *and let* $X = \Delta/\Gamma$. *There is a simple holomorphic family* (W, π, B) *with the following properties:*

(i) if (W', π', B') *is a simple holomorphic family with some fiber* $(\pi')^{-1}(t)$ *biholomorphically equivalent to* X, *there is a morphism* (f, g) *of* (W', π', B') *to* (W, π, B);

(ii) the above morphism (f, g) *is unique up to an automorphism of* (W, π, B).

These properties determine (W, π, B) *up to isomorphism, and* B *is biholomorphically equivalent to* $Teich(\Gamma)$.

Property (ii) means that if (f_1, g_1) and (f_2, g_2) are two morphisms of (W', π', B') to (W, π, B), there is an automorphism (ϕ, ψ) of (W, π, B) such that $f_2 = \phi \circ f_1$ and $g_2 = \psi \circ g_1$.

By Theorem 3 in [DE], the Teichmüller space $Teich(\Gamma)$ is simply connected. It is well-known that Γ acts on the Bers fiber space $F(\Gamma)$, producing a complex manifold $V(\Gamma) = F(\Gamma)/\Gamma$ and a holomorphic map

306   S. Mitra

$\pi : V(\Gamma) \to Teich(\Gamma)$. See [EF1] or Chapter 5 in [N] for the details. In [EF1], it is shown that the map π has horizontally holomorphic local trivializations. Thus, $(V(\Gamma), \pi, Teich(\Gamma))$ is a simple holomorphic family. In [EF2] it is proved that $(V(\Gamma), \pi, Teich(\Gamma))$ satisfies (i) of Theorem 8.7, and in [EF1], the authors show that $(V(\Gamma), \pi, Teich(\Gamma))$ satisfies (ii) of Theorem 8.7.

The crucial point in their proof (in [EF2]) is the following important application of the λ-lemma.

Lemma 8.8 *Let* $\theta_\alpha : B_\alpha \times X \to \pi^{-1}(B_\alpha)$ *be a local trivialization of* (W, π, B) *defined over the connected open set* $B_\alpha \subset B$. *If the map* $\theta_\alpha(\cdot, x) : B_\alpha \to W$ *is holomorphic for each* x *in* X, *then the map* $\theta_\alpha(t, x) \mapsto \theta_\alpha(s, x)$ *from the fibers* $\pi^{-1}(t)$ *to* $\pi^{-1}(s)$ *is quasiconformal for any fixed* t *and* s *in* B_α.

See §5 in [EF2] for the details.

Remark 8.9 *Theorem 8.7 gives the universal property for all Teichmüller spaces. Its beauty is that it gives a characterization of Teichmüller spaces without mentioning quasiconformal mappings. Of course, lurking behind, are quasiconformal mappings, via the* λ-lemma.

Notes
1 Research partially supported by a PSC-CUNY grant.

Bibliography

[Ah] Ahlfors, L. V. *Lectures on Quasiconformal Mappings*, second edition, with additional chapters by C. J. Earle and I. Kra, M. Shishikura, J. H. Hubbard. American Mathematical Society, University Lecture Series **38** (2006).
[AB] Ahlfors, L. V. and Bers, L. Riemann's mapping theorem for variable metrics. *Ann. of Math.* **72** (1960), 385–404.
[Ar] Arens, R. Topologies for homeomorphism groups. *Amer. J. Math.* **68** (1946), 593–610.
[AM] Astala, K. and Martin, G. J. Holomorphic motions. Papers on Analysis; *Report. Univ. Jyväskylä* **83** (2001), 27–40.
[B1] Bers, L. On a theorem of Abikoff. *Ann. Acad. Sci. Fenn.* **AI10** (1985), 83–87.
[B2] Bers, L. Holomorphic families of isomorphisms of Möbius groups. *J. Math. Kyoto Univ.* **26** (1986), 73–76.
[BR] Bers, L. and Royden, H. L. Holomorphic families of injections. *Acta Math.* **157** (1986), 259–286.

[Ch] Chirka, E. M. On the extension of holomorphic motions. *Doklady Mathematics.* **70**(1) (2004), 516–519.

[Do] Douady, A. Prolongement de mouvements holomorphes [d'aprés Slodkowski et autres]. *Séminaire N. Bourbaki, Vol. 1993/1994, Astérique No. 227* (1995), Exp. No. 775, pp. 3, 7–20.

[DE] Douady, A. and Earle, C. J. Conformally natural extensions of homeomorphisms of the circle. *Acta Math.* **157** (1986), 23–48.

[E1] Earle, C. J. On holomorphic cross-sections in Teichmüller spaces. *Duke Math. J.* **33** (1969), 409–416.

[E2] Earle, C. J. Some maximal holomorphic motions. *Contemp. Math.* **211** (1997), 183–192.

[EF1] Earle, C. J. and Fowler, R. S. Holomorphic families of open Riemann surfaces. *Math. Ann.* **270** (1985), 249–273.

[EF2] Earle, C. J. and Fowler, R. S. A new characterization of infinite dimensional Teichmüller spaces. *Ann. Acad. Sci. Fenn.* **A110** (1985), 149–153.

[EGL1] Earle, C. J., Gardiner, F. P. and Lakic, N. Vector fields for holomorphic motions of closed sets. *Contemporary Mathematics* **211** (1997), 193–225.

[EGL2] Earle, C. J., Gardiner, F. P. and Lakic, N. Isomorphisms between generalized Teichmüller spaces. *Contemporary Mathematics* **240** (1999), 97–110.

[EK] Earle, C. J. and Kra, I. On holomorphic mappings between Teichmüller spaces, in *Contributions to Analysis*, Academic Press, New York, 1974, pp. 107–124.

[EKK] Earle, C. J., Kra, I. and Krushkal, S. L. Holomorphic motions and Teichmüller spaces. *Trans. Amer. Math. Soc.* **343** (1994), 927–948.

[EM] Earle, C. J. and Mitra, S. Variation of moduli under holomorphic motions. *Contemporary Mathematics* **256** (2000), 39–67.

[G] Gardiner, F. P. *Teichmüller Theory and Quadratic Differentials*, Wiley-Interscience, 1987.

[GK] Gardiner, F. P. and Keen, L. Conformal dynamical systems and Lipman Bers, in *Selected Works of Lipman Bers, Papers on Complex Analysis*, Part 2, American Mathematical Society, 1998.

[GL] Gardiner, F. P. and Lakic, N. *Quasiconformal Teichmüller Theory*, AMS Mathematical Surveys and Monographs, **76** (2000).

[Hu1] Hubbard, J. H. Sur les sections analytiques de la courbe universelle de Teichmüller. *Memoirs Amer. Math. Soc.* **166** (1976), 1–137.

[Hu2] Hubbard, J. H. *Teichmüller Theory and Applications to Geometry, Topology, and Dynamics – Volume I: Teichmüller Theory*, Matrix Editions, Ithaca, NY, 2006.

[IT] Imayoshi, Y. and Taniguchi, M. *An Introduction to Teichmüller Spaces*, Springer-Verlag, Tokyo, 1992.

[J] Jiang, Y. α-Asymptotically conformal fixed points and holomorphic motions, in *Complex Analysis and its Applications*, Proceedings of the 13th ICFIDCAA (2006), World Scientific Publishing Co., to appear.

[JM] Jiang, Y. and Mitra, S. Some applications of universal holomorphic motions. *Kodai Mathematical Journal* **30**(1) (2007), 85–96.

[Li] Lieb, G. Holomorphic motions and Teichmüller space, PhD dissertation, Cornell University, 1990.

[MS] McMullen, C. T. and Sullivan, D. P. Quasiconformal Homeomorphisms and Dynamics III: The Teichmüller space of a holomorphic dynamical system. *Adv. Math.* **135** (1998), 351–395.

[MSS] Mañé, R., Sad, P. and Sullivan, D. P. On the dynamics of rational maps. *Ann. Sci. École Norm. Sup.* **16** (1983), 193–217.

[Ma] Martin, G. The distortion theorem for quasiconformal mappings, Schottky's theorem and holomorphic motions. *Proc. Amer. Math. Soc.* **125** (1997), 1093–1103.

[M1] Mitra, S. Teichmüller spaces and holomorphic motions. *J. d'Analyse Math.* **81** (2000), 1–33.

[M2] Mitra, S. Extensions of holomorphic motions. *Israel Journal of Mathematics* **159** (2007), 277–288.

[M3] Mitra, S. Extensions of holomorphic motions to quasiconformal motions. *Contemporary Mathematics* **432** (2007), 199–208.

[M4] Mitra, S. An extension theorem for holomorphic motions over infinite dimensional parameter spaces, in *Complex Analysis and its Applications*, Proceedings of the 15th International Conference on Finite or Infinite Dimensional Complex Analysis and Applications, OCAMI Studies **2** (2008), 89–98.

[N] Nag, S. *The Complex Analytic Theory of Teichmüller Spaces*, Canadian Math. Soc. Monographs and Advanced Texts, Wiley-Interscience, 1988.

[Ro] Rodin, B. Behavior of the Riemann mapping function under complex analytic deformations of the domain. *Complex Variables Theory Appl.* **5** (1986), 189–195.

[Sh] Shiga, H. Characterization of quasi-disks and Teichmüller spaces. *Tôhoku Math. J.* **37** (1985), 541–552.

[Sl1] Slodkowski, Z. Holomorphic motions and polynomial hulls. *Proc. Amer. Math. Soc.* **111** (1991), 347–355.

[Sl2] Slodkowski, Z. Extensions of holomorphic motions. *Ann. Scuola Norm. Sup. Pisa Cl. Sci. (4)* **22** (1995), 185–210.

[Su] Sugawa, T. The Bers projection and the λ-lemma. *J. Math. Kyoto Univ.* **32** (1992), 701–713.

[ST] Sullivan, D. and Thurston, W. P. Extending holomorphic motions. *Acta Math.* **157** (1986), 243–257.

Complex hyperbolic quasi-Fuchsian groups

John R. Parker

Department of Mathematical Sciences, Durham University
j.r.parker@dur.ac.uk

Ioannis D. Platis [1]

Department of Mathematics, University of Crete
jplatis@math.uoc.gr

Abstract

A complex hyperbolic quasi-Fuchsian group is a discrete, faithful, type-preserving and geometrically finite representation of a surface group as a subgroup of the group of holomorphic isometries of complex hyperbolic space. Such groups are direct complex hyperbolic generalisations of quasi-Fuchsian groups in three dimensional (real) hyperbolic geometry. In this article we present the current state of the art of the theory of complex hyperbolic quasi-Fuchsian groups.

1 Introduction

The purpose of this paper is to outline what is known about the complex hyperbolic analogue of quasi-Fuchsian groups. Discrete groups of complex hyperbolic isometries have not been studied as widely as their real hyperbolic counterparts. Nevertheless, they are interesting to study and should be more widely known. The classical theory of quasi-Fuchsian groups serves as a model for the complex hyperbolic theory, but results do not usually generalise in a straightforward way. This is part of the interest of the subject.

Complex hyperbolic Kleinian groups were first studied by Picard at about the same time as Poincaré was developing the theory of Fuchsian and Kleinian groups. In spite of work by several other people, including Giraud and Cartan, the complex hyperbolic theory did not develop as rapidly as the real hyperbolic theory. So, by the time Ahlfors and Bers were laying the foundations for the theory of quasi-Fuchsian groups, complex hyperbolic geometry was hardly studied at all. Later, work of Chen and Greenberg and of Mostow on symmetric spaces led to a resurgence of interest in complex hyperbolic discrete groups. The basic theory of complex hyperbolic quasi-Fuchsian groups was laid out by Goldman and these foundations have been built upon by many other people.

There are several sources of material on complex hyperbolic geometry. The book of Goldman [Gol4] gives an encyclopedic source of many facts, results and proofs about complex hyperbolic geometry. The forthcoming book of Parker [P] is intended to give a gentler background to the subject, focusing on discrete groups. The book of Schwartz [S5] also gives a general introduction, but concentrates on the proof and application of a particular theorem. Additionally, most of the papers in the bibliography contain some elementary material but they often use different conventions and notation. Therefore we have tried to make this paper as self contained as possible, and we hope that it will become a useful resource for readers who want to begin studying complex hyperbolic quasi-Fuchsian groups.

This paper is organised as follows. We give a wide ranging introduction to complex hyperbolic geometry in Section 2. We then go on to discuss the geometry of complex hyperbolic surface group representations in Section 3, including the construction of fundamental domains. One of the most striking aspects of this theory is that, unlike the real hyperbolic case, there is a radical difference between the case where our surface has punctures or is closed (and without boundary). We discuss these two cases separately in Sections 4 and 5. Finally, in Section 6 we give some open problems and conjectures.

It is a great pleasure for us to present this survey as a contribution to a volume in honour of Bill Harvey. Bill's contributions to the classical theory of Teichmüller and quasi-Fuchsian spaces have been an inspiration to us and, more importantly, he is a good friend to both of us.

2 Complex hyperbolic space

2.1 Models of complex hyperbolic space

The material in this section is standard. Further details may be found in the books [Gol4] or [P]. Let $\mathbb{C}^{2,1}$ be the complex vector space of complex dimension 3 equipped with a non-degenerate, indefinite Hermitian form $\langle \cdot, \cdot \rangle$ of signature $(2,1)$, that is $\langle \cdot, \cdot \rangle$ is given by a non-singular 3×3 Hermitian matrix H with 2 positive eigenvalues and 1 negative eigenvalue. There are two standard matrices H which give different Hermitian forms on $\mathbb{C}^{2,1}$. Following Epstein, see [E], we call these the first and second Hermitian forms. Let $\mathbf{z}$ and $\mathbf{w}$ be the column vectors $[z_1, z_2, z_3]^t$ and $[w_1, w_2, w_3]^t$ respectively. The *first Hermitian form* is defined to be:

$$\langle \mathbf{z}, \mathbf{w} \rangle_1 = \mathbf{w}^* H_1 \mathbf{z} = z_1 \overline{w}_1 + z_2 \overline{w}_2 - z_3 \overline{w}_3$$

where H_1 is the Hermitian matrix:

$$H_1 = \begin{pmatrix} 1 & 0 & 0 \\ 0 & 1 & 0 \\ 0 & 0 & -1 \end{pmatrix}.$$

The *second Hermitian form* is defined to be:

$$\langle \mathbf{z}, \mathbf{w} \rangle_2 = \mathbf{w}^* H_2 \mathbf{z} = z_1 \overline{w}_3 + z_2 \overline{w}_2 + z_3 \overline{w}_1$$

where H_2 is the Hermitian matrix:

$$H_2 = \begin{pmatrix} 0 & 0 & 1 \\ 0 & 1 & 0 \\ 1 & 0 & 0 \end{pmatrix}.$$

There are other Hermitian forms which are widely used in the literature.

Given any two Hermitian forms of signature $(2,1)$ we can pass between them using a *Cayley transform*. This is not unique for we may precompose and postcompose by any unitary matrix preserving the relevant Hermitian form. For example, one may check directly that the following Cayley transform interchanges the first and second Hermitian forms:

$$C = \frac{1}{\sqrt{2}} \begin{pmatrix} 1 & 0 & 1 \\ 0 & \sqrt{2} & 0 \\ 1 & 0 & -1 \end{pmatrix}.$$

In what follows we shall use subscripts only if it is necessary to specify which Hermitian form to use. When there is no subscript, then the reader can use either of these (or her/his favourite Hermitian form on $\mathbb{C}^3$ of signature $(2,1)$).

Since $\langle \mathbf{z}, \mathbf{z} \rangle$ is real for all $\mathbf{z} \in \mathbb{C}^{2,1}$ we may define subsets V_-, V_0 and

V_+ of $\mathbb{C}^{2,1}$ by

$$V_- = \left\{ \mathbf{z} \in \mathbb{C}^{2,1} \,|\, \langle \mathbf{z}, \mathbf{z} \rangle < 0 \right\},$$
$$V_0 = \left\{ \mathbf{z} \in \mathbb{C}^{2,1} - \{0\} \,|\, \langle \mathbf{z}, \mathbf{z} \rangle = 0 \right\},$$
$$V_+ = \left\{ \mathbf{z} \in \mathbb{C}^{2,1} \,|\, \langle \mathbf{z}, \mathbf{z} \rangle > 0 \right\}.$$

We say that $\mathbf{z} \in \mathbb{C}^{2,1}$ is *negative, null* or *positive* if $\mathbf{z}$ is in V_-, V_0 or V_+ respectively. Motivated by special relativity, these are sometimes called time-like, light-like and space-like. Let $\mathbb{P} : \mathcal{C}^{2,1} \longmapsto \mathcal{CP}^2$ be the standard projection map. On the chart of $\mathbb{C}^{2,1}$ where $z_3 \neq 0$ this projection map is given by

$$\mathbb{P} : \begin{bmatrix} z_1 \\ z_2 \\ z_3 \end{bmatrix} \longmapsto \left(\; z_1/z_3 \quad z_2/z_3 \; \right) \in \mathbb{C}^2.$$

Because $\langle \lambda \mathbf{z}, \lambda \mathbf{z} \rangle = |\lambda|^2 \langle \mathbf{z}, \mathbf{z} \rangle$ we see that for any non-zero complex scalar λ the point $\lambda \mathbf{z}$ is negative, null or positive if and only if $\mathbf{z}$ is. It makes sense to describe $\mathbb{P}\mathbf{z} \in \mathbb{CP}^2$ as positive, null or negative.

Definition 2.1 *The projective model of complex hyperbolic space is defined to be the collection of negative lines in $\mathbb{CP}^2$ and its boundary is defined to be the collection of null lines. In other words $\mathbf{H}_{\mathbb{C}}^2$ is $\mathbb{P}V_-$ and $\partial \mathbf{H}_{\mathbb{C}}^2$ is $\mathbb{P}V_0$.*

We define the other two standard models of complex hyperbolic space by taking the section of $\mathbb{C}^{2,1}$ defined by $z_3 = 1$ and considering $\mathbb{P}V_-$ for the first and second Hermitian forms. In particular, we define the *standard lift* of $z = (z_1, z_2) \in \mathbb{C}^2$ to be $\mathbf{z} = [z_1, z_2, 1]^t \in \mathbb{C}^{2,1}$. It is clear that $\mathbb{P}\mathbf{z} = z$. Points in complex hyperbolic space will be those $z \in \mathbb{C}^2$ for which their standard lift satisfies $\langle \mathbf{z}, \mathbf{z} \rangle < 0$. Taking the first and second Hermitian forms, this gives two models of complex hyperbolic space which naturally generalise, respectively, the Poincaré disc and half plane models of the hyperbolic plane.

For the first Hermitian form we obtain $z = (z_1, z_2) \in \mathbf{H}_{\mathbb{C}}^2$ provided

$$\langle \mathbf{z}, \mathbf{z} \rangle_1 = z_1 \overline{z}_1 + z_2 \overline{z}_2 - 1 < 0$$

or, in other words $|z_1|^2 + |z_2|^2 < 1$.

Definition 2.2 *The unit ball model of complex hyperbolic space $\mathbf{H}_{\mathbb{C}}^2$ is*

$$\mathbb{B}^2 = \left\{ z = (z_1, z_2) \in \mathbb{C}^2 \; : \; |z_1|^2 + |z_2|^2 < 1 \right\}.$$

Its boundary $\partial \mathbf{H}^2_{\mathbb{C}}$ is the unit sphere

$$S^3 = \left\{ z = (z_1, z_2) \in \mathbb{C}^2 \ : \ |z_1|^2 + |z_2|^2 = 1 \right\}.$$

For the second Hermitian form we obtain $z = (z_1, z_2) \in \mathbf{H}^2_{\mathbb{C}}$ provided

$$\langle \mathbf{z}, \mathbf{z} \rangle_2 = z_1 + z_2 \bar{z}_2 + \bar{z}_1 < 0.$$

In other words $2 \Re(z_1) + |z_2|^2 < 0$.

Definition 2.3 *The Siegel domain model of complex hyperbolic space $\mathbf{H}^2_{\mathbb{C}}$ is*

$$\mathbb{S}^2 = \left\{ z = (z_1, z_2) \in \mathcal{C}^2 \ : \ 2 \Re(z_1) + |z_2|^2 < 0 \right\}.$$

Its boundary is the one point compactification of $\mathbb{R}^3$. It turns out (see Section 2.4) that this is naturally endowed with the group structure of the Heisenberg group $\mathfrak{N}$. That is $\partial \mathbf{H}^2_{\mathbb{C}} = \mathfrak{N} \cup \{\infty\}$ where

$$\mathfrak{N} = \left\{ z = (z_1, z_2) \in \mathbb{C}^2 \ : \ 2 \Re(z_1) + |z_2|^2 = 0 \right\}.$$

The standard lift of ∞ is the column vector $[1, 0, 0]^t \in \mathbb{C}^{2,1}$.

2.2 Bergman metric

The *Bergman metric* on $\mathbf{H}^2_{\mathbb{C}}$ is defined by

$$ds^2 = -\frac{4}{\langle \mathbf{z}, \mathbf{z} \rangle^2} \det \begin{bmatrix} \langle \mathbf{z}, \mathbf{z} \rangle & \langle d\mathbf{z}, \mathbf{z} \rangle \\ \langle \mathbf{z}, d\mathbf{z} \rangle & \langle d\mathbf{z}, d\mathbf{z} \rangle \end{bmatrix}.$$

Alternatively, it is given by the distance function ρ given by the formula

$$\cosh^2 \left(\frac{\rho(z, w)}{2} \right) = \frac{\langle \mathbf{z}, \mathbf{w} \rangle \langle \mathbf{w}, \mathbf{z} \rangle}{\langle \mathbf{z}, \mathbf{z} \rangle \langle \mathbf{w}, \mathbf{w} \rangle} = \frac{|\langle \mathbf{z}, \mathbf{w} \rangle|^2}{|\mathbf{z}|^2 |\mathbf{w}|^2}$$

where $\mathbf{z}$ and $\mathbf{w}$ in V_- are the standard lifts of z and w in $\mathbf{H}^2_{\mathbb{C}}$ and $|\mathbf{z}| = \sqrt{-\langle \mathbf{z}, \mathbf{z} \rangle}$.

Substituting the first Hermitian form in these formulae gives the following expressions for the Bergman metric and distance function for the unit ball model

$$\begin{aligned} ds^2 &= \frac{-4}{(|z_1|^2 + |z_2|^2 - 1)^2} \det \begin{pmatrix} |z_1|^2 + |z_2|^2 - 1 & \bar{z}_1 d z_1 + \bar{z}_2 d z_2 \\ z_1 d \bar{z}_1 + z_2 d \bar{z}_2 & |d z_1|^2 + |d z_2|^2 \end{pmatrix} \\ &= \frac{4 \left(|dz_1|^2 + |dz_2|^2 - |z_1 dz_2 - z_2 dz_1|^2 \right)}{\left(1 - |z_1|^2 - |z_2|^2 \right)^2} \end{aligned}$$

and

$$\cosh^2\left(\frac{\rho(z,w)}{2}\right) = \frac{|1 - z_1\overline{w}_1 - z_2\overline{w}_2|^2}{\left(1 - |z_1|^2 - |z_2|^2\right)\left(1 - |w_1|^2 - |w_2|^2\right)}.$$

Likewise, the Bergman metric on the Siegel domain is given by

$$ds^2 = \frac{-4}{(z_1 + |z_2|^2 + \overline{z}_1)^2} \det\left(\begin{array}{cc} z_1 + |z_2|^2 + \overline{z}_1 & d\,z_1 + \overline{z}_2 d\,z_2 \\ d\,\overline{z}_1 + z_2 d\,\overline{z}_2 & |d\,z_2|^2 \end{array}\right)$$

$$= \frac{-4(z_1 + |z_2|^2 + \overline{z}_1)|d\,z_2|^2 + 4\left|d\,z_1 + \overline{z}_2 d\,z_2\right|^2}{(z_1 + |z_2|^2 + \overline{z}_1)^2}.$$

The corresponding distance formula is

$$\cosh^2\left(\frac{\rho(z,w)}{2}\right) = \frac{|z_1 + \overline{w}_1 + z_2\overline{w}_2|^2}{\left(z_1 + \overline{z}_1 + |z_2|^2\right)\left(w_1 + \overline{w}_1 + |w_2|^2\right)}.$$

2.3 Isometries

Let $U(2,1)$ be the group of matrices that are unitary with respect to the form $\langle\cdot,\cdot\rangle$ corresponding to the Hermitian matrix H. By definition, each such matrix A satisfies the relation $A^*HA = H$ and hence $A^{-1} = H^{-1}A^*H$, where $A^* = \overline{A}^t$.

The group of holomorphic isometries of complex hyperbolic space is the *projective unitary group* $PU(2,1) = U(2,1)/U(1)$, where we make the natural identification $U(1) = \{e^{i\theta}I, \theta \in [0, 2\pi)\}$ and I is the 3×3 identity matrix. The full group of complex hyperbolic isometries is generated by $PU(2,1)$ and the conjugation map $(z_1, z_2) \longmapsto (\overline{z}_1, \overline{z}_2)$.

For our purposes we shall consider instead the group $SU(2,1)$ of matrices that are unitary with respect to H and have determinant 1. Therefore $PU(2,1) = SU(2,1)/\{I, \omega I, \omega^2 I\}$, where ω is a non real cube root of unity, and so $SU(2,1)$ is a 3-fold covering of $PU(2,1)$. This is analogous to the well known fact that $SL(2,\mathbb{C})$ is the double cover of $PSL(2,\mathbb{C}) = SL(2,\mathbb{C})/\{\pm I\}$. Since $SU(2,1)$ comprises 3×3 matrices rather than 2×2 matrices we obtain a triple cover rather than a double cover.

The usual trichotomy which classifies isometries of real hyperbolic spaces also holds here. Namely:

(i) an isometry is *loxodromic* if it fixes exactly two points of $\partial\mathbf{H}_\mathbb{C}^2$;
(ii) an isometry is *parabolic* if it fixes exactly one point of $\partial\mathbf{H}_\mathbb{C}^2$;
(iii) an isometry is *elliptic* if it fixes at least one point of $\mathbf{H}_\mathbb{C}^2$.

In Section 2.5 below we will give another geometrical interpretation of this trichotomy.

2.4 The Heisenberg group and the boundary of $\mathbf{H}^2_{\mathbb{C}}$

Each unipotent, upper triangular matrix in $SU(2,1)$ (with respect to the second Hermitian form) has the form

$$T_{(\zeta,v)} = \begin{bmatrix} 1 & -\sqrt{2}\,\overline{\zeta} & -|\zeta|^2 + iv \\ 0 & 1 & \sqrt{2}\zeta \\ 0 & 0 & 1 \end{bmatrix}$$

where $\zeta \in \mathbb{C}$ and $v \in \mathbb{R}$. There is a natural map $\phi : \mathbb{C} \times \mathbb{R} \longrightarrow SU(2,1)$ given by $\phi(\zeta, v) = T_{(\zeta,v)}$. The set of unipotent, upper triangular matrices in $SU(2,1)$ is a group under matrix multiplication. In order to make ϕ a homeomorphism, we give $\mathbb{C} \times \mathbb{R}$ the following group operation:

$$(\zeta, v) \cdot (\zeta', v') = (\zeta + \zeta', v + v') + 2\mathrm{Im}\,(\zeta\overline{\zeta'}),$$

that is the group structure of the *Heisenberg group* $\mathfrak{N}$.

Given a finite point z of $\partial\mathbf{H}^2_{\mathbb{C}}$ there is a unique unipotent, upper triangular matrix in $SU(2,1)$ taking $o = (0,0)$ to z. Therefore we may identify $\partial\mathbf{H}^2_{\mathbb{C}}$ with the one point compactification of the Heisenberg group. (This generalises the well known fact that the boundary of hyperbolic three-space is the one point compactification of $\mathbb{C}$.)

The identification is done as follows. If z is a finite point of the boundary (that is any point besides ∞), then its standard lift is

$$\mathbf{z} = \begin{bmatrix} z_1 \\ z_2 \\ 1 \end{bmatrix}, \quad 2\Re(z_1) + |z_2|^2 = 0.$$

We write $\zeta = z_2/\sqrt{2} \in \mathbb{C}$ and this condition becomes $2\Re(z_1) = -2|\zeta|^2$. Hence we may write $z_1 = -|\zeta|^2 + iv$ for $v \in \mathbb{R}$. That is for $(\zeta, v) \in \mathfrak{N}$:

$$\mathbf{z} = \begin{bmatrix} -|\zeta|^2 + iv \\ \sqrt{2}\zeta \\ 1 \end{bmatrix}.$$

The Heisenberg group is not abelian but is 2-step nilpotent. Therefore any point in $\mathfrak{N}$ of the form $(0,t)$ is central and the commutator of any two elements lies in the centre.

Geometrically, we think of the $\mathbb{C}$ factor of $\mathfrak{N}$ as being horizontal and the $\mathbb{R}$ factor as being vertical. We refer to $T(\zeta, v)$ as *Heisenberg trans-lation* by (ζ, v). A Heisenberg translation by $(0, t)$ is called *vertical translation* by t. It is easy to see the Heisenberg translations are ordi-nary translations in the horizontal direction and shears in the vertical direction. The fact that $\mathfrak{N}$ is nilpotent means that translating around a horizontal square gives a vertical translation, rather like going up a spiral staircase. We define *vertical projection* $\Pi : \mathfrak{N} \to \mathbb{C}$ to be the map $\Pi(\zeta, v) = \zeta$.

The Heisenberg norm is given by

$$\left|(\zeta, v)\right| = \left||\zeta|^2 - iv\right|^{1/2}.$$

This gives rise to a metric, the Cygan metric, on $\mathfrak{N}$ by

$$\rho_0\big((\zeta_1, v_1), (\zeta_2, v_2)\big) = \left|(\zeta_1, v_1)^{-1} \cdot (\zeta_2, v_2)\right|.$$

2.5 Geodesic submanifolds

There are no totally geodesic, real hypersurfaces of $\mathbf{H}_{\mathbb{C}}^2$, but there are two kinds of totally geodesic 2-dimensional subspaces of complex hyperbolic space (see Section 3.1.11 of [Gol4]). Namely:

(i) *complex lines* L, which have constant curvature -1, and
(ii) totally real *Lagrangian planes* R, which have constant curvature $-1/4$.

Every complex line L is the image under some element of $SU(2, 1)$ of the complex line L_1 where the first coordinate is zero:

$$L_1 = \Big\{ (z_1, z_2)^t \in \mathbf{H}_{\mathbb{C}}^2 \ : \ z_1 = 0 \Big\}.$$

The subgroup of $SU(2, 1)$ stabilising L_1 is thus the group of block diagonal matrices $S\big(U(1) \times U(1, 1)\big) < SU(2, 1)$. The stabiliser of ev-ery other complex line is conjugate to this subgroup. The restriction of the Bergman metric to L_1 is just the Poincaré metric on the unit disc with curvature -1.

Every Lagrangian plane is the image under some element of $SU(2, 1)$ of the *standard real Lagrangian plane* $R_{\mathbb{R}}$ where both coordinates are real:

$$R_{\mathbb{R}} = \Big\{ (z_1, z_2)^t \in \mathbf{H}_{\mathbb{C}}^2 \ : \ \mathrm{Im}\,(z_1) = \mathrm{Im}\,(z_2) = 0 \Big\}.$$

The subgroup of $SU(2,1)$ stabilising $R_{\mathbb{R}}$ consists of all matrices with real entries, that is $SO(2,1) < SU(2,1)$. The stabiliser of every other Lagrangian plane is conjugate to this subgroup. The restriction of the Bergman metric to $R_{\mathbb{R}}$ is the Klein-Beltrami metric on the unit disc with curvature $-1/4$.

We finally define two classes of topological circles, which form the boundaries of complex lines and Lagrangian planes respectively:

(i) the boundary of a complex line is called a $\mathbb{C}$-*circle* and

(ii) the boundary of a Lagrangian plane is called an $\mathbb{R}$-*circle*.

The complex conjugation map $\iota_{\mathbb{R}} : (z_1, z_2) \longmapsto (\overline{z}_1, \overline{z}_2)$ is an involution of $\mathbf{H}^2_{\mathbb{C}}$ fixing the standard real Lagrangian plane $R_{\mathbb{R}}$. It too is an isometry. Indeed any anti-holomorphic isometry of $\mathbf{H}^2_{\mathbb{C}}$ may be written as $\iota_{\mathbb{R}}$ followed by some element of $PU(2,1)$. Any Lagrangian plane may be written as $R = B(R_{\mathbb{R}})$ for some $B \in SU(2,1)$ and so $\iota = B\iota_{\mathbb{R}}B^{-1}$ is an anti-holomorphic isometry of $\mathbf{H}^2_{\mathbb{C}}$ fixing R.

Falbel and Zocca [FZ] have used involutions fixing Lagrangian planes to give the following characterisation of elements of $SU(2,1)$:

Theorem 2.4 (Falbel and Zocca [FZ]) *Any element A of* $SU(2,1)$ *may be written as $A = \iota_1 \circ \iota_0$ where ι_0 and ι_1 are involutions fixing Lagrangian planes R_0 and R_1 respectively. Moreover*

(i) *$A = \iota_1 \circ \iota_0$ is loxodromic if and only if R_0 and R_1 are disjoint;*

(ii) *$A = \iota_1 \circ \iota_0$ is parabolic if and only if R_0 and R_1 intersect in exactly one point of $\partial \mathbf{H}^2_{\mathbb{C}}$;*

(iii) *$A = \iota_1 \circ \iota_0$ is elliptic if and only if R_0 and R_1 intersect in at least one point of $\mathbf{H}^2_{\mathbb{C}}$.*

2.6 Cartan's angular invariant

Let z_1, z_2, z_3 be three distinct points of $\partial \mathbf{H}^2_{\mathbb{C}}$ with lifts $\mathbf{z}_1$, $\mathbf{z}_2$ and $\mathbf{z}_3$. Cartan's angular invariant [Car] is defined as follows:

$$\mathbb{A}(z_1, z_2, z_3) = \arg\big(-\langle \mathbf{z}_1, \mathbf{z}_2 \rangle \langle \mathbf{z}_2, \mathbf{z}_3 \rangle \langle \mathbf{z}_3, \mathbf{z}_1 \rangle\big).$$

The angular invariant is independent of the chosen lifts $\mathbf{z}_j$ of the points z_j. It is clear that applying an element of $SU(2,1)$ to our triple of points does not change the Cartan invariant. The converse is also true.

Proposition 2.5 (Goldman, Theorem 7.1.1 of [Gol4]) *Let* z_1, z_2, z_3 *and* z_1', z_2', z_3' *be triples of distinct points of* $\partial\mathbf{H}_\mathbb{C}^2$. *Then the Cartan invariants* $\mathbb{A}(z_1, z_2, z_3)$ *and* $\mathbb{A}(z_1', z_2', z_3')$ *are equal if and only if there exists an* $A \in SU(2,1)$ *so that* $A(z_j) = z_j'$ *for* $j = 1, 2, 3$. *Moreover,* A *is unique unless the three points lie on a complex line.*

The properties of $\mathbb{A}$ may be found in Section 7.1 of [Gol4]. In the next proposition we highlight some of them, see Corollary 7.1.3 and Theorem 7.1.4 on pages 213–4 of [Gol4].

Proposition 2.6 (Cartan [Car]) *Let* z_1, z_2, z_3 *be three distinct points of* $\partial\mathbf{H}_\mathbb{C}^2$ *and let* $\mathbb{A} = \mathbb{A}(z_1, z_2, z_3)$ *be their angular invariant. Then,*

   (i) $\mathbb{A} \in [-\pi/2, \pi/2]$;
   (ii) $\mathbb{A} = \pm\pi/2$ *if and only if* z_1, z_2 *and* z_3 *all lie on a complex line;*
   (iii) $\mathbb{A} = 0$ *if and only if* z_1, z_2 *and* z_3 *all lie on a Lagrangian plane.*

Geometrically, the angular invariant $\mathbb{A}$ has the following interpretation; see Section 7.1.2 of [Gol4]. Let L_{12} be the complex line containing z_1 and z_2 and let Π_{12} denote orthogonal projection onto L_{12}. The Bergman metric restricted to L_{12} is just the Poincaré metric with curvature -1. Consider the hyperbolic triangle Δ in L_{12} with vertices z_1, z_2 and $\Pi_{12}(z_3)$. Then the angular invariant $\mathbb{A}(z_1, z_2, z_3)$ is half the signed Poincaré area of Δ. That is, the area of Δ is $|2\mathbb{A}|$. The sign of $\mathbb{A}$ is determined by the order one meets the vertices of Δ when going around $\partial\Delta$ in a positive sense with respect to the natural orientation of L_{12}. If one meets the vertices in a cyclic permutation of z_1, z_2, $\Pi_{12}(z_3)$ then $\mathbb{A} > 0$; if one meets them in a cyclic permutation of z_2, z_1, $\Pi_{12}(z_3)$ then $\mathbb{A} < 0$. In the case where $\Pi_{12}(z_3)$ lies on the geodesic joining z_1 and z_2 then the triangle is degenerate and has area zero. In the case where z_3 lies on ∂L_{12} then Δ is an ideal triangle and has area π.

2.7 The Korányi-Reimann cross-ratio

Cross-ratios were generalised to complex hyperbolic space by Korányi and Reimann [KR1]. Following their notation, we suppose that z_1, z_2, z_3, z_4 are four distinct points of $\partial\mathbf{H}_\mathbb{C}^2$. Let $\mathbf{z}_1$, $\mathbf{z}_2$, $\mathbf{z}_3$ and $\mathbf{z}_4$ be corresponding lifts in $V_0 \subset \mathbb{C}^{2,1}$. The *complex cross-ratio* of our four points is defined to be

$$\mathbb{X} = [z_1, z_2, z_3, z_4] = \frac{\langle \mathbf{z}_3, \mathbf{z}_1 \rangle \langle \mathbf{z}_4, \mathbf{z}_2 \rangle}{\langle \mathbf{z}_4, \mathbf{z}_1 \rangle \langle \mathbf{z}_3, \mathbf{z}_2 \rangle}.$$

We note that $\mathbb{X}$ is invariant under $SU(2,1)$ and independent of the chosen lifts. More properties of the complex cross-ratio may be found in Section 7.2 of [Gol4].

By choosing different orderings of our four points we may define other cross-ratios. There are some symmetries associated to certain permutations, see Property 5 on page 225 of [Gol4]. After taking these into account, there are three cross-ratios that remain. Given four distinct points $z_1, \ldots, z_4 \in \partial \mathbf{H}^2_{\mathbb{C}}$, we define

$$\mathbb{X}_1 = [z_1, z_2, z_3, z_4], \quad \mathbb{X}_2 = [z_1, z_3, z_2, z_4], \quad \mathbb{X}_3 = [z_2, z_3, z_1, z_4]. \quad (2.1)$$

These three cross-ratios determine the quadruple of points up to $SU(2,1)$ equivalence; see Falbel [Fal] or Parker and Platis [PP2].

Proposition 2.7 (Falbel [Fal]) *Let $z_1, \ldots, z_4$ be distinct points of $\partial \mathbf{H}^2_{\mathbb{C}}$ with cross-ratios $\mathbb{X}_1$, $\mathbb{X}_2$, $\mathbb{X}_3$ given by (2.1). Let $z'_1, \ldots, z'_4$ be another set of distinct points of $\partial \mathbf{H}^2_{\mathbb{C}}$ so that*

$$
\begin{aligned}
\mathbb{X}_1 &= [z_1, z_2, z_3, z_4] = [z'_1, z'_2, z'_3, z'_4], \\
\mathbb{X}_2 &= [z_1, z_3, z_2, z_4] = [z'_1, z'_3, z'_2, z'_4], \\
\mathbb{X}_3 &= [z_2, z_3, z_1, z_4] = [z'_2, z'_3, z'_1, z'_4].
\end{aligned}
$$

Then there exists $A \in SU(2,1)$ so that $A(z_j) = z'_j$ for $j = 1, 2, 3, 4$.

In [Fal] Falbel has given a general setting for cross-ratios that includes both Korányi-Reimann cross-ratios and the standard real hyperbolic cross-ratio. The normalisation (2.1) is somewhat different than his. The three cross-ratios satisfy two real equations; in Falbel's normalisation, the analogous relations are given in Proposition 2.3 of [Fal]. In his general setting there are six cross-ratios that lie on a complex algebraic variety in $\mathbb{C}^6$. Our cross-ratios correspond to the fixed locus of an anti-holomorphic involution on this variety.

In a recent preprint Cunha and Gusevskii [CuGu] have pointed out a minor error in Proposition 37. This result is not true for certain configurations of points all lying on a complex line.

Proposition 2.8 (Parker and Platis [PP2]) *Let z_1, z_2, z_3, z_4 be any four distinct points in $\partial \mathbf{H}^2_{\mathbb{C}}$. Let $\mathbb{X}_1$, $\mathbb{X}_2$ and $\mathbb{X}_3$ be defined by (2.1). Then*

$$|\mathbb{X}_2| = |\mathbb{X}_1||\mathbb{X}_3|, \quad (2.2)$$

$$2|\mathbb{X}_1|^2 \Re(\mathbb{X}_3) = |\mathbb{X}_1|^2 + |\mathbb{X}_2|^2 + 1 - 2\Re(\mathbb{X}_1 + \mathbb{X}_2). \quad (2.3)$$

The set of points $\mathfrak{X}$ consisting of triples $(\mathbb{X}_1, \mathbb{X}_2, \mathbb{X}_3)$ is thus a real algebraic variety in $\mathbb{C}^3$, which we call the *cross-ratio variety*.

The cross-ratio variety $\mathfrak{X}$ defined in Proposition 2.8 appears naturally in the study of the space of configurations of four points in the unit sphere S^3. In the classical case, it is well known that a configuration of four points on the Riemann sphere is determined up to Möbius equivalence by their cross-ratio and the set F of equivalence classes of configurations is biholomorphic to $\mathbb{C} - \{0,1\}$ which may be identified to the set of cross-ratios of four pairwise distinct points on $\mathbb{CP}^1$ (see section 4.4 of [Bea]). In the complex hyperbolic setting, things are more complicated. Denote by $\mathbb{C}$ the set of configurations of four points in S^3. The group of holomorphic isometries of $\mathbf{H}_{\mathbb{C}}^2$ acts naturally on $\mathbb{C}$. Denote by $\mathcal{F}$ the quotient of $\mathbb{C}$ by this action. In [Fal], Falbel proved the following.

Proposition 2.9 (Falbel [Fal]) *There exists a CR map* $\pi : \mathcal{F} \to \mathbb{C}^3$ *such that its image is* $\mathfrak{X}$.

Now, there are plenty of analytical and geometrical structures on large subsets of $\mathcal{F}$, all inherited from the natural CR structure of the Heisenberg group. Namely (see [FP1]), there exists a complex structure and a (singular) CR structure of codimension 2. To transfer these results to $\mathfrak{X}$, Falbel and Platis strengthened Proposition (2.9) by proving that the map π is a CR-embedding. Thus they prove the following.

Theorem 2.10 (Falbel and Platis [FP1]) *Let* $\mathfrak{X}$ *be the cross-ratio variety. Then, except for some subsets of measure zero,* $\mathfrak{X}$

  (i) *is a 2-complex manifold and*
  (ii) *admits a singular CR structure of codimension 2.*

In [PP3] Parker and Platis explored the topology of the cross-ratio variety by defining global coordinates on $\mathfrak{X}$ using only geometrical tools.

3 Representations of surface groups

In what follows we fix an oriented topological surface Σ of genus g with p punctures; its Euler characteristic is thus $\chi = \chi(\Sigma) = 2-2g-p$. We suppose that $\chi < 0$. We denote by $\pi_1 = \pi_1(\Sigma)$ the fundamental group of Σ. A specific choice of generators for π_1 is called a *marking*. The collection of marked representations of π_1 into a Lie group G up to conjugation will be denoted by $\mathrm{Hom}(\pi_1, G)/G$. We shall consider $\mathrm{Hom}(\pi_1, G)/G$ endowed with the compact-open topology; this will enable us to make sense of what it means for two representations to be close. We remark that in the cases we consider, the compact-open topology is equivalent to the l^2-topology on the relevant matrix group.

Our main interest is in the case where $G = \mathrm{SU}(2,1)$, that is representation variety consisting of marked representations up to conjugation into the group of holomorphic isometries of complex hyperbolic space $\mathbf{H}_{\mathbb{C}}^2$. In the following section we review in brief the classical cases of spaces of marked representations of π_1 into G when G is $\mathrm{SL}(2,\mathbb{R})$ or $\mathrm{SL}(2,\mathbb{C})$. These spaces are the predecessors of the space we study here.

3.1 The motivation from real hyperbolic geometry

It is well known that if $\rho : \pi_1 \longrightarrow \mathrm{SL}(2,\mathbb{R})$ is a discrete and faithful representation of π_1, then $\rho(\pi_1)$ is called *Fuchsian*. In this case ρ is necessarily geometrically finite. Moreover, $\rho(\pi_1)$ is necessarily totally loxodromic when $p = 0$. If $p > 0$ then this condition would be replaced with type-preserving, which requires that an element of $\rho(\pi_1)$ is parabolic if and only if it represents a peripheral curve. The group $\mathrm{SL}(2,\mathbb{R})$ is a double cover of the group of orientation preserving isometries of the hyperbolic plane. The quotient of the hyperbolic plane by $\rho(\pi_1)$ naturally corresponds to a hyperbolic (as well as to a complex) structure on Σ.

Definition 3.1 *The collection of distinct, marked Fuchsian representations, up to conjugacy within* $\mathrm{SL}(2,\mathbb{R})$, *is the Teichmüller space of* Σ, *denoted* $\mathcal{T} = \mathcal{T}(\Sigma) \subset \mathrm{Hom}\big(\pi_1, \mathrm{SL}(2,\mathbb{R})\big)/\mathrm{SL}(2,\mathbb{R})$.

Among its many properties, Teichmüller space is:

- topologically a ball of real dimension $6g - 6 + 2p$,
- a complex Banach manifold, equipped with a Kähler metric (the Weil-Petersson metric) of negative holomorphic sectional curvature.

We now consider representations of π_1 to $\mathrm{SL}(2,\mathbb{C})$. We again require such a representation ρ to be discrete, faithful, geometrically finite and, if $p = 0$ (respectively $p > 0$) totally loxodromic (respectively type-preserving). We call these representations *quasi-Fuchsian*.

Definition 3.2 *The collection of distinct, marked quasi-Fuchsian representations, up to conjugation in* $\mathrm{SL}(2,\mathbb{C})$, *is called (real hyperbolic) quasi-Fuchsian space and is denoted by*

$$\mathcal{Q}_{\mathbb{R}} = \mathcal{Q}_{\mathbb{R}}(\Sigma) \subset \mathrm{Hom}\big(\pi_1, \mathrm{SL}(2,\mathbb{C})\big)/\mathrm{SL}(2,\mathbb{C}).$$

If ρ is a quasi-Fuchsian representation, then it corresponds to a three dimensional hyperbolic structure on an interval bundle over Σ. Real hyperbolic quasi-Fuchsian space $\mathcal{Q}_{\mathbb{R}}(\Sigma)$:

- may be identified with the product of two copies of Teichmüller space according to the Simultaneous Uniformization Theorem of Bers [Ber],
- it is a complex manifold of dimension $6g - 6 + 2p$ and it is endowed with a hyper-Kähler metric whose induced complex symplectic form is the complexification of the Weil-Petersson symplectic form on $\mathcal{T}(\Sigma)$; see [Pla1].

3.2 Complex hyperbolic quasi-Fuchsian space

Motivated by these two examples, one may consider representations of π_1 into $\mathrm{SU}(2,1)$ up to conjugation, that is $\mathrm{Hom}\big(\pi_1, \mathrm{SU}(2,1)\big)/\mathrm{SU}(2,1)$. A definition consistent with the ones in the classical cases would then be the following.

Definition 3.3 *A representation in* $\mathrm{Hom}\big(\pi_1, \mathrm{SU}(2,1)\big)/\mathrm{SU}(2,1)$ *is said to be complex hyperbolic quasi-Fuchsian if it is*

- *discrete,*
- *faithful,*
- *geometrically finite and*
- *type-preserving.*

We remark that in the case where $p = 0$ we may replace "geometrically finite" with "convex-cocompact" and we may replace "type-preserving" with "totally loxodromic".

Since the group $\mathrm{SU}(2,1)$ is a triple cover of the holomorphic isometry group of complex hyperbolic space $\mathbf{H}^2_{\mathbb{C}}$ it turns out that such a representation corresponds to a complex hyperbolic structure on a disc bundle over Σ.

There are two matters to clarify in Definition 3.3. The first has to do with the geometrical finiteness of a representation. The notion of geometrical finiteness in spaces with variable negative curvature generalises the one in spaces with constant negative curvature and has been studied by Bowditch in [Bow]. (We keep Bowditch's labels, in particular there is no condition F3.) These definitions require the extensions to complex hyperbolic space of several familiar notions from the theory of Kleinian groups. We refer to [Bow] for details of how this extension takes place.

In particular, core(M) is the quotient by Γ of the convex hull of the limit set Λ, that is the *convex core*; thin$_\epsilon$(M) is the part of M where the injectivity radius is less than ϵ, that is the ϵ-*thin part* and thick$_\epsilon$(M) is its complement. In [Bow] Bowditch explains that there is no condition F3 because that would involve finite sided fundamental polyhedra.

Theorem 3.4 (Bowditch [Bow]) *Suppose that* Γ *is a discrete subgroup of* $\mathrm{SU}(2,1)$. *Let* $\Lambda \subset \partial\mathbf{H}^2_\mathbb{C}$ *be the limit set of* Γ *and let* $\Omega = \partial\mathbf{H}^2_\mathbb{C} - \Lambda$ *be the domain of discontinuity of* Γ. *Let* $M_C(\Gamma)$ *denote the orbifold* $\left(\mathbf{H}^2_\mathbb{C} \cup \Omega\right)/\Gamma$. *Then the following conditions are equivalent and any group satisfying one of them will be called geometrically finite:*

F1. $M_C(\Gamma)$ has only finitely many topological ends, each of which is a parabolic end.

F2. Λ consists only of conical limit points and bounded parabolic fixed points.

F4. There exists $\epsilon > 0$ so that core(M) $\cap$ thick$_\epsilon$(M) is compact. Here ϵ is chosen small enough so that thin$_\epsilon$(M) is the union of cusps and Margulis tubes.

F5. There is a bound on the orders of every finite subgroup of Γ and there exists $\eta > 0$ so that the η neighbourhood of core(M) has finite volume.

The second issue to clarify in Definition 3.3 concerns type-preserving representations. As for Fuchsian groups, a type-preserving representation is automatically discrete. This follows from the following theorem of Chen and Greenberg. In spite of this, we include the hypothesis of discreteness first because it reinforces the connection with the classical definition and, secondly, because our usual method of showing that a representation is quasi-Fuchsian is to show that it is discrete. In particular we usually construct a fundamental polyhedron and use Poincaré's polyhedron theorem to verify the other criteria.

Theorem 3.5 (Chen and Greenberg [CG]) *Suppose that* Γ *is a nonelementary subgroup of* $\mathrm{SU}(2,1)$. *If the identity is not an accumulation point of elliptic elements of* Γ *then* Γ *is discrete.*

Note that in Chen and Greenberg's statement, Corollary 4.5.3 of [CG], they suppose that Γ has more than one limit point and the lowest dimensional, Γ-invariant, totally geodesic subspace of $\mathbf{H}^2_\mathbb{C}$ has even dimensions. Since the only odd dimensional, totally geodesic submanifolds of $\mathbf{H}^2_\mathbb{C}$ are geodesics, in our case this means that Γ does not fix a point of $\overline{\mathbf{H}^2_\mathbb{C}}$ and does not leave a geodesic invariant. Hence Γ is non-elementary. (Our statement does not hold for SU(3, 1), although Chen and Greenberg's does. This is because there are totally loxodromic, non-elementary, non-discrete subgroups of SO(3, 1) < SU(3, 1) preserving a copy of hyperbolic 3-space.)

This contrasts with the case of representations to SL(2, $\mathbb{C}$). In our definition of complex hyperbolic quasi-Fuchsian we have included the conditions that such a representation should be both discrete and totally loxodromic.

Definition 3.6 *The space of all marked complex hyperbolic quasi-Fuchsian representations, up to conjugacy, will be called complex hyperbolic quasi-Fuchsian space*

$$\mathcal{Q}_{\mathbb{C}} = \mathcal{Q}_{\mathbb{C}}(\Sigma) \subset \mathrm{Hom}\big(\pi_1, \mathrm{SU}(2, 1)\big)/\mathrm{SU}(2, 1).$$

3.3 Fuchsian representations

It is reasonable to start our study of complex hyperbolic quasi-Fuchsian space by considering the Fuchsian representations. There are two ways to make a Fuchsian representation act on $\mathbf{H}_{\mathbb{C}}^2$. These correspond to the two types of totally geodesic, isometric embeddings of the hyperbolic plane into $\mathbf{H}_{\mathbb{C}}^2$ as we have seen in Section 2.5, namely, totally real Lagrangian planes and complex lines.

(i) If a discrete, faithful representation ρ is conjugate to a representation of π_1 into SO(2, 1) < SU(2, 1) then it preserves a Lagrangian plane and is called $\mathbb{R}$-*Fuchsian*.

(ii) If a discrete, faithful representation ρ is conjugate to a representation of π_1 into $\mathrm{S}\big(\mathrm{U}(1) \times \mathrm{U}(1, 1)\big)$ < SU(2, 1) then it preserves a complex line and is called $\mathbb{C}$-*Fuchsian*.

We shall see later that $\mathbb{C}$-Fuchsian representations are distinguished in the space $\mathcal{Q}_{\mathbb{C}}(\Sigma)$.

3.4 Toledo invariant

Let G be PU(2, 1) or SU(2, 1) and $\rho : \pi_1 \longrightarrow G$ be a representation. There is an important invariant associated to ρ called the *Toledo invariant* [T2], which is defined as follows. Let $\widetilde{f} : \widetilde{\Sigma} \to \mathbf{H}_{\mathbb{C}}^2$ be a ρ-equivariant smooth mapping of the universal cover $\widetilde{\Sigma}$ of Σ. Then the Toledo invariant $\tau(\rho, \widetilde{f})$ is defined as the (normalised) integral of the pull-back of the Kähler form ω on $\mathbf{H}^2_{\mathbb{C}}$. Namely,

$$\tau(\rho, \widetilde{f}) = \frac{1}{2\pi} \int_{\Omega} \widetilde{f}^* \omega$$

where Ω is a fundamental domain for the action of π_1 on $\widetilde{\Sigma}$. The main properties of the Toledo invariant are the following.

Proposition 3.7

   (i) τ is independent of $\widetilde{f}$ and varies continuously with ρ,
   (ii) $\chi \leq \tau(\rho) \leq -\chi$, see [DT], [GuP2].

Geometrically, one may think of the Toledo invariant as follows; see Section 7.1.4 of [Gol4]. Take a triangulation of Σ and consider the lift of this triangulation to the equivariant embedding $\widetilde{f}(\widetilde{\Sigma})$ of the universal cover of Σ constructed above. The result is a triangle Δ in $\mathbf{H}^2_{\mathbb{C}}$ with vertices z_1, z_2, z_3. Suppose that the edges of Δ are the oriented geodesic arcs from z_j to z_{j+1} (with indices taken mod 3). Since ω is an exact form on $\mathbf{H}^2_{\mathbb{C}}$ its integral over Δ only depends on the boundary, therefore the value of the integral is independent of the choice of triangle filling the edges. Let L_{12} be the complex line containing z_1 and z_2 and let Π_{12} denote the orthogonal projection to L_{12}. Then the integral of ω over Δ is the signed area (with the Poincaré metric) of the triangle in L_{12} with vertices z_1, z_2 and $\Pi_{12}(z_3)$. This is called *Toledo's cocycle* [T2]. It is easy to see that this is an extension of Cartan's angular invariant to the case of triangles whose vertices do not necessarily lie on $\partial \mathbf{H}^2_{\mathbb{C}}$; compare Section 2.6. In order to find the Toledo invariant, one repeats this construction over all triangles in the triangulation of Σ and takes the sum. The result is $2\pi\tau(\rho)$. For each triangle, the maximum value of the Toledo cocycle is the hyperbolic area and this occurs if the boundary lies in a complex line. The only way for the Toledo invariant to take its maximal value is if all the triangles lie in the same complex line and are consistently oriented, that is ρ is $\mathbb{C}$-Fuchsian. In which case (when the triangles are all positively oriented) we have $\tau(\rho) = \mathbb{A}/2\pi = -\chi(\Sigma)$, where $\mathbb{A}$ is the area of Σ with a Poincaré metric. Likewise, applying an anti-holomorphic isometry of $\mathbf{H}^2_{\mathbb{C}}$ changes the orientation of the triangle Δ and so changes the sign of the Toledo cocycle. Thus, if all triangles are negatively oriented we get $\tau(\rho) = -\mathbb{A}/2\pi = \chi(\Sigma)$.

If ρ is $\mathbb{R}$-Fuchsian, then it is invariant under $\iota_{\mathbb{R}}$ and so $\tau(\rho) = -\tau(\rho)$. Hence $\tau(\rho) = 0$. This gives a sketch proof of the following result. Note that it may be the case that $\tau(\rho) = 0$ and for ρ to not be $\mathbb{R}$-Fuchsian.

Proposition 3.8 *Suppose that $\rho \in \mathcal{Q}_{\mathbb{C}}(\Sigma)$. Then,*

   (i) ρ is $\mathbb{C}$-Fuchsian if and only if $|\tau(\rho)| = -\chi$, see [T2],
   (ii) if ρ is $\mathbb{R}$-Fuchsian then $\tau(\rho) = 0$, see [GKL].

3.5 Fundamental domains

We have already mentioned that unlike the case of constant curvature there exist no totally geodesic hypersurfaces in complex hyperbolic space. Before attempting to construct a fundamental domain in $\mathbf{H}^2_{\mathbb{C}}$ we must choose the class of real hypersurfaces containing its faces. Moreover, since these faces are not totally geodesic they may intersect in complicated ways. Therefore, constructing fundamental domains in $\mathbf{H}^2_{\mathbb{C}}$ is quite a challenge. In Sections 3.6 and 3.7 below we discuss two classes of hypersurfaces, namely bisectors and packs. Fundamental domains whose faces are contained in bisectors have been widely studied, in particular, this is the case for the construction of Dirichlet domains. The idea goes back to Giraud and was developed further by Mostow and Goldman (see [Gol4] and the references therein). In Section 4 we shall see how such domains were constructed by Goldman and Parker in [GP] and Gusevskii and Parker in [GuP2]. On the other hand, in order to build fundamental domains, Falbel and Zocca used $\mathbb{C}$-spheres [FZ] and Schwartz used $\mathbb{R}$-spheres in [S2] (for the relationship between $\mathcal{C}$-spheres and $\mathbb{R}$-spheres see [FPa]). Packs have been used by Will [W2] and by Parker and Platis [PP1].

3.6 Bisectors

A bisector $\mathcal{B}$ is the locus of points equidistant from a given pair of points z_1 and z_2 in $\mathbf{H}^2_{\mathbb{C}}$. In other words:

$$\mathcal{B} = \mathcal{B}(z_1, z_2) = \left\{ z \in \mathbf{H}^2_{\mathbb{C}} \ : \ \rho(z, z_1) = \rho(z, z_2) \right\}.$$

The corresponding construction in real hyperbolic geometry gives a hyperbolic hyperplane, which is totally geodesic. We have already seen that there are no totally geodesic real hypersurfaces in complex hyperbolic space. Therefore $\mathcal{B}$ is not totally geodesic. However, it is as close as it can be to being totally geodesic, in that it is foliated by totally geodesic subspaces in two distinct ways. For an extensive treatment of bisectors readers should see [Gol4].

Let z_1 and z_2 be any two points of $\mathbf{H}^2_{\mathbb{C}}$. Then z_1 and z_2 lie in a unique complex line L, which we call the *complex spine* of $\mathcal{B}$. (Note that the complex spine is often denoted Σ, but this conflicts with our use of Σ to denote a surface.) The restriction of the Bergman metric to L is the Poincaré metric. Thus the points in L equidistant from z_1 and z_2 lie on a Poincaré geodesic in L, which we call the *spine* of $\mathcal{B}$ and denote by σ. The endpoints of the spine are called the *vertices* of $\mathcal{B}$. Goldman shows (page 154 of [Gol4]) that a bisector is completely determined by its vertices and so $SU(2,1)$ acts transitively on the set of all bisectors.

Following earlier work of Giraud, Mostow describes a foliation of bisectors by complex lines as follows (see also Section 5.1.2 of [Gol4]):

Theorem 3.9 (Mostow [Mos2]) *Let $\mathcal{B}$ be a bisector and let σ and L denote its spine and complex spine respectively. Let Π_L denote orthogonal projection onto L and for each $s \in \sigma$ let L_s be the complex line so that $\Pi_L(L_s) = s$. Then*

$$\mathcal{B} = \bigcup_{s \in \sigma} L_s = \bigcup_{s \in \sigma} \Pi_L^{-1}(s).$$

Moreover, Goldman describes a second foliation of bisectors, this time by Lagrangian planes.

Theorem 3.10 (Goldman [Gol4]) *Let $\mathcal{B}$ be a bisector with spine σ. Then $\mathcal{B}$ is the union of all Lagrangian planes containing σ.*

The complex lines L_s defined in Theorem 3.9 are called the *slices* of $\mathcal{B}$. The Lagrangian planes defined in Theorem 3.10 are called the *meridians* of $\mathcal{B}$. Together the slices and the meridians of a bisector give *geographical coordinates*. The reason for this name is that the boundary of a bisector in $\partial \mathbf{H}^2_{\mathbb{C}}$ is topologically a sphere, called by Mostow a *spinal sphere*. The boundaries of the slices and meridians are $\mathbb{C}$-circles and $\mathbb{R}$-circles that foliate the spinal sphere in a manner analogous to lines of latitude and lines of longitude, respectively, on the earth.

3.7 Packs

Bisectors are rather badly adapted to $\mathbb{R}$-Fuchsian representations. When considering representations close to an $\mathbb{R}$-Fuchsian representation it is more convenient to use a different class of hypersurfaces, called *packs*. There is a natural partial duality, resembling mirror symmetry, in complex hyperbolic space between complex objects (such as complex lines) and totally real objects (such as Lagrangian planes), see the discussion in the introduction to [FPa]. From this point of view, packs are the dual objects to bisectors. Packs were introduced by Will in [W2] and in their general form by Parker and Platis in [PP1].

Let A be a loxodromic map in $SU(2,1)$. Then, its two fixed points in $\partial \mathbf{H}^2_{\mathbb{C}}$ correspond to eigenvectors in $\mathbb{C}^{2,1}$. Their common orthogonal is also an eigenvector of A. Moreover, there exists a complex number $\lambda = l + i\theta \in \mathbb{R}_+ \times (-\pi, \pi]$ (the *complex length* of A) such that the eigenvalues of A are $e^{\lambda}, e^{\bar{\lambda}-\lambda}, e^{-\bar{\lambda}}$. For any $x \in \mathbb{R}$ define A^x to be the element of $SU(2,1)$ which has the same eigenvectors as A, but its eigenvalues are the eigenvalues of A raised to the xth power. Hence we immediately see that A^x is a loxodromic element of $SU(2,1)$ for all $x \in \mathbb{R} - \{0\}$ and $A^0 = I$. The following Proposition, see [PP1] holds.

Proposition 3.11 *Let R_0 and R_1 be disjoint Lagrangian planes in $\mathbf{H}^2_{\mathbb{C}}$ and let ι_0 and ι_1 be the respective inversions. Consider $A = \iota_1 \iota_0$ (which is a loxodromic map by Theorem 2.4) and its powers A^x for each $x \in \mathbb{R}$. Then:*

(i) $\iota_x = A^x \iota_0$ is inversion in a Lagrangian plane $R_x = A^{x/2}(R_0)$.

(ii) R_x intersects the complex axis L_A of A orthogonally in a geodesic denoted γ_x.

(iii) The geodesics γ_x are the leaves of a foliation of L_A.

(iv) For each $x \neq y \in \mathbb{R}$, R_x and R_y are disjoint.

Definition 3.12 *Given disjoint Lagrangian planes R_0 and R_1, then for each $x \in \mathbb{R}$ let R_x be the Lagrangian plane constructed in Proposition 3.11. Define*

$$\mathcal{P} = \mathcal{P}(R_0, R_1) = \bigcup_{x \in \mathbb{R}} R_x = \bigcup_{x \in \mathbb{R}} A^{x/2}(R_0).$$

Then $\mathcal{P}$ is a real analytic 3-submanifold which we call the pack determined by R_0 and R_1. We call the axis of $A = \iota_1 \iota_0$ the spine of $\mathcal{P}$ and the Lagrangian planes R_x for $x \in \mathbb{R}$ the slices of $\mathcal{P}$.

Observe that $\mathcal{P}$ contains L, the complex line containing γ, the spine of $\mathcal{P}$. The following Proposition is obvious from the construction and emphasises the similarity between bisectors and packs (compare it with Section 5.1.2 of [Gol4]).

Proposition 3.13 *Let* $\mathcal{P}$ *be a pack. Then* $\mathcal{P}$ *is homeomorphic to a 3-ball whose boundary lies in* $\partial \mathbf{H}^2_{\mathbb{C}}$. *Moreover,* $\mathbf{H}^2_{\mathbb{C}} - \mathcal{P}$, *the complement of* $\mathcal{P}$, *has two components, each homeomorphic to a 4-ball.*

We remark that the boundary of $\mathcal{P}$ contains the boundary of the complex line L and is foliated by the boundaries of the Lagrangian planes R_x. Since it is also homeomorphic to a sphere, it is an example of an $\mathbb{R}$-sphere (hybrid sphere), see [FPa], [S2].

The definition of packs associated to loxodromic maps A that preserve a Lagrangian plane (that is with $\theta = \mathrm{Im}\,(\lambda) = 0$) was given by Will. We call the resulting pack *flat*.

Proposition 3.14 (Will, Section 6.1.1 of [W2]) *Suppose that the geodesic* γ *lies on a totally real plane* R. *Then the set*

$$\mathcal{P}(\gamma) = \Pi_R^{-1}(\gamma) = \bigcup_{z \in \gamma} \Pi_R^{-1}(z)$$

is the flat pack determined by the Lagrangian planes $R_0 = \Pi_R^{-1}(z_0)$ *and* $R_1 = \Pi_R^{-1}(z_1)$ *for any distinct points* $z_0, z_1 \in \gamma$. *Moreover, for each* $z \in \gamma$, *the Lagrangian plane* $\Pi_R^{-1}(z)$ *is a slice of* $\mathcal{P}(\gamma)$.

Let $A \in \mathrm{SU}(2,1)$ be a loxodromic map with eigenvalues e^{λ}, $e^{\overline{\lambda}-\lambda}$, $e^{-\overline{\lambda}}$ where $\lambda = l + i\theta \in \mathbb{R}_+ \times (-\pi, \pi]$. Suppose that $\mathcal{P}$ is a pack determined by A, as in Definition 3.12. We define the *curl factor* of $\mathcal{P}$ to be $\kappa(\mathcal{P}) = \theta/l = \tan\big(\arg(\lambda)\big)$. Note that flat packs have curl factor 0. Platis [Pla5] proves that two packs $\mathcal{P}_1$ and $\mathcal{P}_2$ are isometric if and only if $\kappa(\mathcal{P}_1) = \kappa(\mathcal{P}_2)$. In particular, $\mathrm{SU}(2,1)$ acts transitively on the set of flat packs.

4 Punctured surfaces

In the study of complex hyperbolic quasi-Fuchsian representations of a surface group, there are qualitatively different conclusions, depending whether $p = 0$ or $p > 0$. Thus we examine each case separately. We will suppose in this section that Σ is a surface with genus g and $p > 0$ punctures, having Euler characteristic $\chi = 2 - 2g - p < 0$.

4.1 Ideal triangle groups and the three times punctured sphere

The first complex hyperbolic deformation space to be completely described is that of the group generated by reflections in the sides of an ideal triangle [GP], [S1]. This group has an index two subgroup corresponding to the thrice punctured sphere. In the case of constant negative curvature these groups are rigid. However, variable negative curvature allows us to deform the group.

Let Σ be a three times punctured sphere and $\pi_1 = \pi_1(\Sigma)$ its fundamental group. Abstractly π_1 is a free group on two generators and we may choose the generators α and β so that loops around the three punctures are represented by α, β and $\alpha\beta$. If $\rho : \pi_1 \longrightarrow \mathrm{SU}(2,1)$ is a type-preserving representation of π_1 then $\rho(\alpha)$, $\rho(\beta)$ and $\rho(\alpha\beta)$ are all parabolic. We consider the special case where each of these three classes is represented by a unipotent parabolic map (that is a map conjugate to a Heisenberg translation). Let u_1, u_2 and u_0 be the fixed points of the unipotent parabolic maps $A = \rho(\alpha)$, $B = \rho(\beta)$ and $AB = \rho(\alpha\beta)$. Let L_j be the complex line spanned by u_{j-1} and u_{j+1} where the indices are taken mod 3. Let $I_j \in \mathrm{SU}(2,1)$ be the complex reflection of order 2 fixing L_j. A consequence of our hypothesis that A, B and AB are unipotent is that $A = I_2 I_0$ and $B = I_0 I_1$. Thus $\rho(\pi_1) = \langle A, B \rangle$ has index 2 in $\langle I_0, I_1, I_2 \rangle$. There is a one (real) dimensional space of representations ρ of π_1 with the above properties, the parameter being the angular invariant of the three fixed points $\mathbb{A}(u_0, u_1, u_2) \in [-\pi/2, \pi/2]$.

It remains to decide which of these groups are quasi-Fuchsian. This question was considered by Goldman and Parker in [GP]. They partially solved the problem and gave a conjectural picture of the complete solution. This conjecture was proved by Schwartz in [S1], who also gave a more conceptual proof in [S4]. We restate the main result in terms of the three times punctured sphere:

Theorem 4.1 (Goldman and Parker [GP], Schwartz [S1]) *Let Σ be a three times punctured sphere with fundamental group π_1. Let $\rho : \pi_1 \longrightarrow \mathrm{SU}(2,1)$ be a representation of π_1 so that the three boundary components are represented by unipotent parabolic maps with fixed points u_0, u_1 and u_2. Let $\mathbb{A} = \mathbb{A}(u_0, u_1, u_2)$ be the angular invariant of these fixed points.*

    (i) If $-\sqrt{125/3} < \tan(\mathbb{A}) < \sqrt{125/3}$ then ρ is complex hyperbolic quasi-Fuchsian.

    (ii) If $\tan(\mathbb{A}) = \pm\sqrt{125/3}$ then ρ is discrete, faithful and geometrically finite and has accidental parabolics.

    (iii) If $\sqrt{125/3} < |\tan(\mathbb{A})| < \infty$ then ρ is not discrete.

    (iv) If $\mathbb{A} = \pm\pi/2$ then ρ is the trivial representation.

Outline of proof. The proof of (i) and (ii) involves constructing a fundamental domain for $\Gamma = \rho(\pi_1)$. The proof of (iii) follows by showing that $I_0 I_1 I_2$ (or equivalently $A^{-1}BAB^{-1} = (I_0 I_1 I_2)^2$) is elliptic of infinite order. The proof of (iv) is trivial since in this case all three lines coincide.

Proposition 4.2 (Gusevskii and Parker [GuP1]) *Let ρ be as in Theorem 4.1. Then the Toledo invariant $\tau(\rho)$ is zero.*

4.2 The modular group and punctured surface groups

Let $\Gamma = \mathrm{PSL}(2, \mathbb{Z})$ be the classical modular group. It is well known that Γ is generated by an element of order two and an element of order three whose product is parabolic. There are six possibilities for the representation depending on the types of the generators of orders 2 and 3.

If A is a complex reflection in a point p then A preserves all complex lines through p; if A is a complex reflection in a complex line L then A preserves all complex lines orthogonal to L. Given two complex reflections there is a unique complex line through their fixed point(s) or orthogonal to their fixed line(s) respectively. This line is preserved by both reflections and hence by the group they generate. For any representation of the modular group in $\mathrm{PU}(2, 1)$, the order 2 generator of the modular group must be a complex reflection. Therefore, if the order 3 generator is also a complex reflection then this representation necessarily preserves a complex line and so is $\mathbb{C}$-Fuchsian. The condition that the product of the two generators is parabolic completely determines the representation up to conjugacy.

In the remaining two cases, we consider representations where the order 3 generator is not a complex reflection, that is it has distinct eigenvalues. There are two cases, namely when the order 2 generator is a complex reflection in a point or in a line. The former case was considered by Gusevskii and Parker [GuP2] and by Falbel and Koseleff [FK] independently. The latter case was considered by Falbel and Parker [FPa]. In both cases the representation may be parametrised as follows. Let u_0 be the fixed point of the product of the generators, which is parabolic by hypothesis. Let u_1 and u_2 be the images of this parabolic fixed point under powers of the order 3 generator. We then define $\mathbb{A} = \mathbb{A}(u_0, u_1, u_2)$ to be the angular invariant of these three fixed points. It turns out that this angular invariant completely parametrises the representation. In particular, in each case the parabolic map is not unipotent, but corresponds to a screw-parabolic map with rotational part of angle $\mathbb{A}$.

The complex hyperbolic quasi-Fuchsian space of the modular group was entirely described by Falbel and Parker in [FPa].

Theorem 4.3 (Falbel and Parker [FPa]) *Let* $PSL(2, \mathbb{Z})$ *be the modular group and consider a representation* $\rho : PSL(2, \mathbb{Z}) \longrightarrow PU(2, 1)$. *Let* $\mathbb{A}$ *be the angular invariant described above. Then:*

(i) *There are four rigid* $\mathbb{C}$-*Fuchsian representations for which the elliptic elements of order 2 and 3 are complex reflections.*

(ii) *If the order 3 generator is not a complex reflection and the order 2 generator is a complex reflection in a point then* ρ *is complex hyperbolic quasi-Fuchsian for all* $\mathbb{A} \in [-\pi/2, \pi/2]$.

(iii) *If the order 3 generator is not a complex reflection and the order 2 generator is a complex reflection in a line then:*

   (a) *If* $\sqrt{15} < |\tan(\mathbb{A})| \leq \infty$ *then* ρ *is complex hyperbolic quasi-Fuchsian.*

   (b) *If* $\tan(\mathbb{A}) = \pm\sqrt{15}$ *then* ρ *is discrete, faithful and geometrically finite and has accidental parabolics.*

   (c) *If* $-\sqrt{15} < \tan(\mathbb{A}) < \sqrt{15}$ *then* ρ *is either not faithful or not discrete.*

In (ii) and (iii) the representation ρ *may be lifted to an isomorphic representation to* $SU(2, 1)$. *In (i) this cannot be done.*

Outline of proof. In (i) the four cases correspond to the different possibilities of reflection in a point or line for each of the generators.

In cases (ii) and (iii)(a) the authors construct a parametrised family of fundamental domains. In case (iii)(c) the commutator of the order 2 and order 3 generators is elliptic. In $PSL(2, \mathbb{Z})$ this element corresponds to

$$\begin{pmatrix} 0 & -1 \\ 1 & 0 \end{pmatrix} \begin{pmatrix} -1 & -1 \\ 1 & 0 \end{pmatrix} \begin{pmatrix} 0 & 1 \\ -1 & 0 \end{pmatrix} \begin{pmatrix} 0 & 1 \\ -1 & -1 \end{pmatrix} = \begin{pmatrix} 1 & 1 \\ 1 & 2 \end{pmatrix}.$$

$\square$

Proposition 4.4 (Gusevskii and Parker [GuP2]) *Let* ρ *and* $\mathbb{A}$ *be as in Theorem 4.3 (ii). Then* $\tau(\rho) = \mathbb{A}/3\pi$.

One remarkable fact is that the groups in Theorem 4.1 (ii) and Theorem 4.3 (iii)(b) are commensurable, and they are each commensurable with an $SU(2,1)$ representation of the Whitehead link complement, as proved by Schwartz [S2].

Let π_1 be the fundamental group of an orbifold Σ and suppose that $\rho : \pi_1 \longrightarrow SU(2,1)$ is a representation with Toledo invariant $\tau = \tau(\rho)$. Let $\widehat{\pi}_1$ be an index d subgroup of π_1 ($\widehat{\pi}_1$ is the fundamental group of an orbifold $\widehat{\Sigma}$ which is a d-fold cover of Σ). The restriction of ρ from π_1 to $\widehat{\pi}_1$ gives a representation $\widehat{\rho} : \widehat{\pi}_1 \longrightarrow SU(2,1)$. In each case the universal cover $\widetilde{\Sigma}$ is the same and so the integral defining $\tau(\widehat{\rho})$ is the same as the integral defining $\tau(\rho)$ but taken over d copies of a fundamental domain. Hence $\tau(\widehat{\rho}) = d\tau(\rho)$. Millington showed that any punctured surface group arises as a finite index subgroup of the modular group:

Proposition 4.5 (Millington [Mil]) *Let $p > 0$, $g \geq 0$, $e_2 \geq 0$ and $e_3 \geq 0$ be integers and write $d = 12(g-1)+6p+3e_2+4e_3$. If $d > 0$ then there is a subgroup of the modular group $PSL(2,\mathbb{Z})$ of index d which is the fundamental group of an orbifold of genus g with p punctures removed, e_2 cone points of angle π and e_3 cone points of angle $2\pi/3$.*

We may combine the above observations to prove that for any punctured surface there exists a representation taking any value of the Toledo invariant in the interval $[\chi, -\chi]$.

Theorem 4.6 (Gusevskii and Parker [GuP2]) *Let Σ be a surface of genus g with p punctures and Euler characteristic $\chi = 2 - 2g - p < 0$. There exists a continuous family of complex hyperbolic quasi-Fuchsian representations ρ_t of $\pi_1(\Sigma)$ into $SU(2,1)$ whose Toledo invariant $\tau(\rho_t)$ varies between χ and $-\chi$.*

Sketch of proof. Let Σ be a surface of genus g with p punctures. Using Proposition 4.5, there is a type-preserving representation of $\pi_1(\Sigma)$ as a subgroup of $PSL(2,\mathbb{Z})$ of index $d = 12g - 12 + 6p$. By restricting the representations constructed in Theorem 4.3 (ii) to this subgroup, we can construct a one parameter family of complex hyperbolic quasi-Fuchsian representations of π_1 to $SU(2,1)$. It remains to compute the Toledo invariants of these representations. Using the discussion above and Proposition 4.4 the Toledo invariant of the representation corresponding to the parameter $\mathbb{A}$ is $\tau = \mathbb{A}d/3\pi = (4g - 4 + 2p)\mathbb{A}/\pi$ where $\mathbb{A}$ varies between $-\pi/2$ and $\pi/2$. Hence τ varies between $-2g + 2 - p = \chi$ and $2g - 2 + p = -\chi$. $\square$

We remark that similar arguments show that the representations constructed in Theorem 4.3 (iii) also have Toledo invariant $\tau = \mathbb{A}/3\pi$. Therefore a similar argument may be used to construct representations of punctured surface groups that interpolate between $\mathbb{C}$-Fuchsian groups and groups with accidental parabolics. Moreover, if τ_1 is the Toledo invariant of these limit groups then $\tan(\pi\tau_1/2\chi) = \mp\sqrt{15}$.

4.3 A special case: the once punctured torus

In his thesis [W1], Will considered the case where $g = p = 1$, that is the once punctured torus. In this section Σ will denote the once punctured torus and π_1 its fundamental group. Consider an $\mathbb{R}$-Fuchsian representation ρ_0 of π_1. This is generated by two loxodromic maps A and B whose axes intersect and whose commutator is parabolic. Suppose $\rho_0(\pi_1)$ fixes the Lagrangian plane R. Let R_0 be the Lagrangian plane orthogonal to R through the intersection of the axes of A and B. Let I_0 denote the (anti-holomorphic) involution fixing R_0. Then I_0 conjugates A to A^{-1} and B to B^{-1}. $I_1 = I_0 A$ and $I_2 = B I_0$ are also anti-holomorphic involutions. We see that $\rho_0(\pi_1) = \langle A, B \rangle$ is an index 2 subgroup of the group $\langle I_0, I_1, I_2 \rangle$. If a representation ρ of π_1 is an index 2 subgroup of a group generated by three anti-holomorphic involutions, then we say ρ is *Lagrangian decomposable*. Will is able to construct a three dimensional family of Lagrangian decomposable punctured torus groups:

Theorem 4.7 (Will [W1]) *Let Σ be the once punctured torus and let $\mathcal{T}(\Sigma)$ denote its Teichmüller space. Then for all $x \in \mathcal{T}(\Sigma)$ and all $\alpha \in [-\pi/2, \pi/2]$ there is a representation $\rho_{x,\alpha} : \pi_1 \longrightarrow \mathrm{SU}(2,1)$ with the following properties:*

    (i) $\rho_{x,\alpha}$ and $\rho_{y,\beta}$ are conjugate if and only if $x = y$ and $\alpha = \beta$.

    (ii) $\rho_{x,\alpha}$ is complex hyperbolic quasi-Fuchsian for all $x \in \mathcal{T}(\Sigma)$ and all $\alpha \in [-\pi/4, \pi/4]$.

    (iii) $\rho_{x,\alpha}$ is $\mathbb{R}$-Fuchsian if and only if $\alpha = 0$.

    (iv) If $\alpha \in [-\pi/2, -\pi/4)$ or $(\pi/4, \pi/2]$ then there exists an $x \in \mathcal{T}(\Sigma)$ so that $\rho(x, \alpha)$ is either not faithful or not discrete.

It is easy to deduce from the details of Will's construction, that $\tau(\rho_{x,\alpha}) = 0$ for all $x \in \mathcal{T}(\Sigma)$ and all $\alpha \in [-\pi/2, \pi/2]$. Recently Will has generalised this result to all oriented punctured surfaces [W3].

Sketch of proof. Will's basic idea is to take an ideal triangle in the hyperbolic plane and to consider points moving along each of the three boundary arcs. These points are parametrised by the signed distance t_0, t_1 and t_2 respectively from a point that is the orthogonal projection of the opposite vertex onto this edge. These three points will be the fixed points of involutions I_0, I_1 and I_2. He shows that the product $(I_0 I_1 I_2)^2$ will be parabolic (fixing one of the vertices of the triangle) if and only if $t_1 + t_2 + t_3 = 0$. If this condition is satisfied then the group generated by $A = I_0 I_1$ and $B = I_2 I_0$ is a representation of a punctured torus group and any two of the lengths parametrise $T(\Sigma)$.

We embed the picture above in a Lagrangian plane R in $\mathbf{H}_{\mathbb{C}}^2$. We construct Lagrangian planes R_0, R_1 and R_2 through the three points on the sides of the triangle and which each make an angle $\pi/2 + \alpha$ with R, in a sense which he makes precise. The (anti-holomorphic) involution I_j fixes R_j for $j = 0, 1, 2$. This then gives a representation of π_1 to $\mathrm{SU}(2,1)$ generated by $A = I_0 I_1$ and $B = I_2 I_0$. The restriction of the action of this group to R is just the action constructed in the previous paragraph and only depends on t_0, t_1 and α. This is the representation $\rho_{x,\alpha}$.

When $\alpha \in [-\pi/4, \pi/4]$ Will constructs a fundamental domain for the action of $\langle I_0, I_1, I_2 \rangle$ on $\mathbf{H}_{\mathbb{C}}^2$. This domain is bounded by three packs each of whose slices makes an angle $\pi/2 + \alpha$ with R. The intersection of this domain with R is just the triangle with which we began.

When $\alpha \in [-\pi/2, -\pi/4)$ or $\alpha \in (\pi/4, \pi/2]$ Will considers groups with $t_0 = t_1$ and he shows that for large $t_0 = t_1$ then $A = I_0 I_1$ is elliptic. Indeed, writing $t = t_0 = t_1$ he shows that

$$\mathrm{tr}(A) = 3 + e^{-4t} + 4e^{-2t} \cos(2\alpha).$$

For these values of α we have $\cos(2\alpha) < 0$ and so A is parabolic when $e^{-2t} = -\cos(2\alpha)$ and elliptic when $e^{-2t} < -\cos(2\alpha)$. $\qquad\square$

4.4 Disconnectedness of complex hyperbolic quasi-Fuchsian space

Dutenhefner and Gusevskii proved in [DG] that in the non compact case, there exist complex hyperbolic quasi-Fuchsian representations of a fundamental group of a surface Σ which cannot be connected to a Fuchsian representation by a path lying entirely in $\mathcal{Q}_{\mathbb{C}}(\Sigma)$. Therefore, the complex hyperbolic quasi-Fuchsian space $\mathcal{Q}_{\mathbb{C}}(\Sigma)$ is not connected.

Theorem 4.8 (Dutenhefner and Gusevskii [DG]) *There exist complex hyperbolic quasi-Fuchsian representations of π whose limit set is a wild knot.*

Outline of proof. In the first place, the authors construct a complex hyperbolic quasi-Fuchsian group whose limit set is a wild knot. To do so, they consider a non trivial knot K (in fact a granny knot) inside the Heisenberg group $\mathfrak{N}$ and a collection of $2n$ Heisenberg spheres placed along K with the following properties. There exists an enumeration $T_1, \ldots, T_{2n}$ of the spheres of this family such that (taking the indices cyclically)

- each sphere T_k is tangent to T_{k-1} and T_{k+1} and
- T_k is disjoint from T_j when $j \neq k-1, k, k+1$.

They call such a collection a *Heisenberg string of beads*. Furthermore, there exists an enumeration $S_1, \ldots, S_n, S'_1, \ldots, S'_n$ of the spheres and maps $A_1, \ldots, A_n \in \mathrm{SU}(2,1)$ such that

- $A_k(S_k) = S'_k$, $k = 1, \ldots, n$,
- A_k maps the exterior of S_k into the interior of S'_k, $k = 1, \ldots, n$ and
- A_k maps the points of tangency of S_k to the points of tangency of S'_k, $k = 1, \ldots, n$.

Consider the group $\Gamma = \langle A_1, \ldots, A_n \rangle$. By Poincaré's polyhedron theorem, Γ is complex hyperbolic quasi-Fuchsian and its limit set is a wild knot. □

5 Closed surfaces

We now turn our attention to the case of closed surfaces Σ without boundary, that is the case $p = 0$ and $g \geq 2$.

5.1 The representation variety

We begin by discussing the representation variety of π_1, the fundamental group of Σ. In particular, for the moment we do not consider discreteness. There are some differences between the case of representations to $\mathrm{SU}(2,1)$ and to $\mathrm{PU}(2,1)$. That is to say, between the representation varieties $\mathrm{Hom}\big(\pi_1, \mathrm{SU}(2,1)\big)/\mathrm{SU}(2,1)$ and $\mathrm{Hom}\big(\pi_1, \mathrm{PU}(2,1)\big)/\mathrm{PU}(2,1)$. Mostly, we concentrate on the case of representations to $\mathrm{SU}(2,1)$. Many of our results go through in both cases. For our first result, we distinguish between the two cases:

Proposition 5.1 (Goldman, Kapovich, Leeb [GKL]) *Let π_1 be the fundamental group of a closed surface of genus $g \geq 2$.*

(i) *Let $\rho : \pi_1 \longrightarrow \mathrm{PU}(2,1)$ be a representation of π_1 to $\mathrm{PU}(2,1)$. Then $\tau \in \frac{2}{3}\mathbb{Z}$.*

(ii) *Let $\rho : \pi_1 \longrightarrow \mathrm{SU}(2,1)$ be a representation of π_1 to $\mathrm{SU}(2,1)$. Then $\tau \in 2\mathbb{Z}$.*

Because τ is locally constant and varies continuously (with respect to the compact-open topology) with ρ then we immediately have:

Corollary 5.2 *The Toledo invariant is constant on components of $\mathrm{Hom}(\pi_1, \mathrm{PU}(2,1))/\mathrm{PU}(2,1)$ or $\mathrm{Hom}(\pi_1, \mathrm{SU}(2,1))/\mathrm{SU}(2,1)$. That is, if ρ_1 and ρ_2 are representations with $\tau(\rho_1) \neq \tau(\rho_2)$ then ρ_1 and ρ_2 are in different components of the representation variety.*

Now, there is a converse to Corollary 5.2 due to Xia, see [X]. Namely, that the Toledo invariant τ distinguishes the components of the whole representation variety.

Theorem 5.3 (Xia [X]) *If ρ_1 and ρ_2 are representations of π_1 with $\tau(\rho_1) = \tau(\rho_2)$ then ρ_1 and ρ_2 are in the same component of the representation variety $\mathrm{Hom}(\pi_1, \mathrm{SU}(2,1))/\mathrm{SU}(2,1)$.*

We know that $\tau(\rho) \in [\chi, -\chi] = [2 - 2g, 2g - 2]$ and (for the $\mathrm{SU}(2,1)$ representation variety) $\tau(\rho) \in 2\mathbb{Z}$. Therefore $\tau(\rho)$ takes one of the $2g - 1$ values $2 - 2g, 4 - 2g, \ldots, 2g - 4, 2g - 2$. For each one of these values of $\tau(\rho)$ there is a component of $\mathrm{Hom}(\pi_1, \mathrm{SU}(2,1))/\mathrm{SU}(2,1)$.

5.2 ℂ-*Fuchsian representations*

Let us consider the components for which $\tau(\rho) = \mp\chi = \pm(2g - 2)$. Suppose that ρ_0 is a ℂ-Fuchsian representation. Then, using Proposition 3.8 we know that $\tau(\rho_0) = \mp\chi$. Because τ is constant on the components of $\mathrm{Hom}(\pi_1, \mathrm{SU}(2,1))/\mathrm{SU}(2,1)$, we see that any deformation ρ_t of ρ_0 also has $\tau(\rho_t) = \tau(\rho_0) = \mp\chi$. Therefore, ρ_t is also ℂ-Fuchsian, again using Proposition 3.8. Thus we have proved the following result, known as the *Toledo-Goldman rigidity theorem*:

Theorem 5.4 (Goldman [Gol2], Toledo [T1], [T2]) *If $\rho_0 \in \mathcal{Q}_{\mathbb{C}}(\Sigma)$ is a $\mathbb{C}$-Fuchsian representation then any nearby representation ρ_t is also $\mathbb{C}$-Fuchsian.*

In fact, we can give a complete description of this component. Let $\rho : \pi_1 \longrightarrow \mathrm{SU}(2,1)$ be a $\mathbb{C}$-Fuchsian representation of π_1. This means that $\rho(\pi_1)$ preserves a complex line L. We may suppose that L is $\{0\} \times \mathbb{C} \subset \mathbb{C}^2$, that is the z_2 axis, in the unit ball model of $\mathbf{H}^2_{\mathbb{C}}$. This means that ρ is a reducible representation

$$\rho : \pi_1 \longrightarrow \mathrm{S}\big(\mathrm{U}(1) \times \mathrm{U}(1,1)\big) < \mathrm{SU}(2,1).$$

That is, each element $A \in \rho(\pi_1)$ is a block diagonal matrix. The upper left block A_1 is a 1×1 block in $\mathrm{U}(1)$, in other words $A_1 = e^{i\theta}$ for some $\theta \in [0, 2\pi)$. The lower right hand block is a 2×2 block A_2 in $\mathrm{U}(1,1)$ with determinant $e^{-i\theta}$. This means that $e^{i\theta/2}A_2 \in \mathrm{SU}(1,1)$. Hence we can write $\rho = \rho_1 \oplus \rho_2$ where $\rho_1 : \pi_1 \longrightarrow \mathrm{U}(1)$ is given by $\rho(\gamma) = A_1$ and $\rho_2 : \pi_1 \longrightarrow \mathrm{U}(1,1)$ is given by $\rho(\gamma) = A_2$. This is equivalent to $\rho'_2 : \pi_1 \longrightarrow \mathrm{SU}(1,1)$ is given by $\rho(\gamma) = e^{i\theta/2}A_2$. Thus ρ_1 is an abelian representation and ρ_2 is (a lift of) a Fuchsian representation. Note that after applying the canonical projection from $\mathrm{U}(1,1)$ to $\mathrm{PU}(1,1)$ the resulting Möbius transformation is independent of $\det(A_2)$. Thus as Fuchsian representations ρ_2 and ρ'_2 are the same. The representations ρ_1 and ρ'_2 are independent. The space of (irreducible) abelian representations is $2g$ copies of S^1, that is a $2g$-dimensional torus T^{2g} (note that since the only relation is a product of commutators, we have a free choice of points in S^1 for each of the $2g$ generators of π_1). The space of Fuchsian representations up to conjugacy is Teichmüller space $\mathcal{T}(\Sigma)$, which is homeomorphic to $\mathbb{R}^{6g-6}$. Thus we have

Proposition 5.5 (Goldman [Gol3]) *Let Σ be a closed surface of genus $g \geq 2$. The two components of $\mathrm{Hom}\big(\pi_1, \mathrm{SU}(2,1)\big)/\mathrm{SU}(2,1)$ for which $\tau = \mp\chi = \pm(2g-2)$ are made up of discrete faithful Fuchsian representations in $T^{2g} \times \mathcal{T}(\Sigma)$. In particular, these two components have dimension $8g - 6$.*

Every representation with $\tau \neq \mp\chi$ is irreducible. One may compute the dimension using Weyl's formula to obtain:

Proposition 5.6 (Goldman [Gol3]) *Let Σ be a closed surface of genus $g \geq 2$. Each of the $2g-3$ components of $\mathrm{Hom}\big(\pi_1, \mathrm{SU}(2,1)\big)/\mathrm{SU}(2,1)$ for which $\tau \neq \mp\chi = \pm(2g-2)$ has dimension $16g - 16$.*

In contrast to the two components of $\mathrm{Hom}(\pi_1, \mathrm{SU}(2,1))/\mathrm{SU}(2,1)$ with $\tau = \mp\chi$, the other components of the representation variety are not easy to describe. In particular, they contain representations that are not quasi-Fuchsian. We expect the picture is that each of these other components contains (probably infinitely many) islands of quasi-Fuchsian representations; see Problem 6.1. As yet very little is known. In the following sections we will summarise the current state of knowledge.

5.3 Complex hyperbolic quasi-Fuchsian space is open

We use the following theorem of Guichard.

Theorem 5.7 (Guichard [Gui]) *Let G be a semi-simple group with finite centre and let G^* be a subgroup of G of rank 1. If Γ is a convex-cocompact subgroup of G^* then there exists an open neighbourhood U of an injection of Γ in G into the space of representations $\mathrm{Hom}(\Gamma, G)/G$ consisting of discrete, faithful, convex-cocompact representations.*

Guichard proves this result using the theory of Gromov hyperbolic metric spaces. In particular, he shows that there is a neighbourhood of Γ comprising quasi-isometric embeddings of Γ into G. Using a theorem of Bourdon and Gromov [Bou] he then deduces that the groups in this neighbourhood are discrete, faithful and convex-cocompact.

Corollary 5.8 (Guichard [Gui]) *Let Σ be a closed surface of genus g and let $\rho_0 : \pi_1 \longrightarrow \mathrm{SU}(2,1)$ be a complex hyperbolic quasi-Fuchsian representation of π_1. Then there exists an open neighbourhood $U = U(\rho_0)$ so that any representation $\rho \in U$ is complex hyperbolic quasi-Fuchsian.*

Combining this Corollary 5.8 and Proposition 5.6 we obtain

Corollary 5.9 *There are open sets of dimension $16g - 16$ in $\mathcal{Q}_{\mathbb{C}}(\Sigma)$.*

In [PP1] Parker and Platis proved a version of Corollary 5.8 in the case where ρ_0 is $\mathbb{R}$-Fuchsian. They started with a fundamental domain for the group $\rho_0(\pi_1)$ whose faces are contained in packs. This domain is the preimage under orthogonal projection of a fundamental polygon for the action of $\rho_0(\pi_1)$ on its invariant Lagrangian plane. They then showed directly that this fundamental domain may be continuously deformed into a fundamental domain for $\rho(\pi_1)$. Such a domain has faces contained in packs and has the same combinatorial type as the original domain.

5.4 The Euler number

Let $\rho : \pi_1 \longrightarrow SU(2,1)$ be a complex hyperbolic quasi-Fuchsian representation of π_1. Let $M = \widetilde{f}(\widetilde{\Sigma})$ be an equivariant surface defined by $\widetilde{f} : \widetilde{\Sigma} \longrightarrow M \subset \mathbf{H}_{\mathbb{C}}^2$. The quotient of M by $\Gamma = \rho(\pi_1)$ is a surface homeomorphic to Σ and $\mathbf{H}_{\mathbb{C}}^2/\Gamma$ is a disc bundle over this surface. The relation in π_1 corresponds to moving around the boundary of a fundamental domain for the action of Γ on M. In doing so it is easy to check that a tangent vector rotates by a total angle of $2\pi\chi = (4 - 4g)\pi$, that is $\chi = 2 - 2g$ whole turns. As this happens the discs in the normal direction also rotate by a certain number of whole turns. This number is called the *Euler number* $e = e(\rho)$ of ρ. The Euler number measures how far $\mathbf{H}_{\mathbb{C}}^2/\Gamma$ is from being a product of Σ with a disc. We now demonstrate how to find the Euler number for $\mathbb{C}$-Fuchsian and $\mathbb{R}$-Fuchsian representations.

An elliptic map in $SU(1,1)$ fixing the origin in the ball model lifts to $S\big(U(1) \times U(1,1)\big)$ as

$$A_\theta = \begin{bmatrix} 1 & 0 & 0 \\ 0 & e^{i\theta/2} & 0 \\ 0 & 0 & e^{-i\theta/2} \end{bmatrix}.$$

A map A_θ of this form acts on $(z_1, z_2) \in \mathbf{H}_{\mathbb{C}}^2$ via its standard lift as $A_\theta : (z_1, z_2) \longmapsto (e^{i\theta/2} z_1, e^{i\theta} z_2)$. In this case M is the (complex) z_2 axis. We may identify tangent vectors to M at the origin with vectors $(0, z_2)$ and normal vectors to M at the origin with vectors $(z_1, 0)$. Hence A_θ acts as rotation by θ on vectors tangent to M and by rotation through $\theta/2$ on vectors normal to M. Suppose $\rho(\pi_1) < S\big(U(1) \times U(1,1)\big) < SU(2,1)$ is a $\mathbb{C}$-Fuchsian representation. The relation in π_1 corresponds to a rotation of a tangent vector to M by $\chi = 2 - 2g$ whole turns (that is by an angle of $(4 - 4g)\pi$). Hence the normal vector makes half this number of turns. Thus the Euler number is $e(\rho) = 1 - g = \chi/2$.

An elliptic map in $SO(2,1)$ fixing the origin in the ball model embeds in $SO(2,1)$ as

$$B_\theta = \begin{bmatrix} \cos\theta & -\sin\theta & 0 \\ \sin\theta & \cos\theta & 0 \\ 0 & 0 & 1 \end{bmatrix}.$$

A map B_θ of this form acts on $(z_1, z_2) \in \mathbf{H}_{\mathbb{C}}^2$ via its standard lift as

$B_\theta : (z_1, z_2) \longmapsto (\cos\theta z_1 - \sin\theta z_2, \sin\theta z_1 + \cos\theta z_2)$. In this case M is the set where z_1 and z_2 are both real. We may identify tangent vectors to M at the origin with vectors (x_1, x_2) and normal vectors to M at the origin with vectors (iy_1, iy_2) where x_1, x_2, y_1, y_2 are all real. It is clear that tangent vectors and normal vectors are all rotated through angle θ. Suppose $\rho(\pi_1) < SO(2, 1) < SU(2, 1)$ is an $\mathbb{R}$-Fuchsian representation. Since the relation in π_1 corresponds to a rotation of a tangent vector to M of $\chi = 2 - 2g$ whole turns. Hence the normal vector makes the same number of turns. Thus the Euler number is $e(\rho) = 2 - 2g = \chi$.

5.5 Construction of examples

Besides Theorem 5.4 and Proposition 5.5 which describe completely the components of $\mathcal{Q}_{\mathbb{C}}(\Sigma)$ comprising $\mathbb{C}$-Fuchsian representations, none of the above results imply anything about discreteness. The first result towards this direction is due to Goldman, Kapovich and Leeb, see Theorem 1.2 of [GKL]:

Theorem 5.10 (Goldman, Kapovich and Leeb [GKL]) *Let Σ be a closed surface of genus g and let π_1 be its fundamental group. For each even integer t with $2 - 2g \le t \le 2g - 2$ there exists a convex-cocompact discrete and faithful representation ρ of π_1 with $\tau(\rho) = t$. Furthermore, the complex hyperbolic manifold $M = \mathbf{H}^2_{\mathbb{C}}/\rho(\pi_1)$ is diffeomorphic to the total space of an oriented $\mathbb{R}^2$ bundle ξ over Σ with the Euler number*

$$e(\xi) = \chi(\Sigma) + |\tau(\rho)/2|.$$

Outline of proof. We have already seen that $\mathbb{R}$-Fuchsian or $\mathbb{C}$-Fuchsian representations give examples of representations for which $\tau = 0$ or $\tau = \pm(2g - 2)$ and with the correct Euler numbers.

Let t be an even integer with $0 < t < 2g - 2$. Let Σ_1 be a (possibly disconnected) subsurface of Σ with the following properties:

(i) $\chi(\Sigma_1) = -t$,
(ii) each boundary component is an essential simple closed curve in Σ and distinct boundary components are not homotopic,
(iii) each component of Σ_1 has an even number of boundary components.

It is easy to see that for each even integer t with $0 < t < 2g - 2$ we can find such a Σ_1. Let Σ_2 be the interior of $\Sigma - \Sigma_1$. Then Σ_2 is also a (possibly disconnected) subsurface of Σ.

The authors start with a piecewise hyperbolic structure on Σ with the following properties. First, Σ_1 has constant curvature -1 and Σ_2 has constant curvature $-1/4$. Secondly, each boundary component of Σ_1 and Σ_2 is a geodesic with the same length l which is sufficiently short, they take $\sinh(l) < 6/35$. Next, the main step for the proof is the construction of a complete complex hyperbolic manifold M together with a piecewise totally geodesic isometric embedding $f : \Sigma \to M$ such that

(i) f is a homotopy equivalence,

(ii) the Toledo invariant of M equals $-\chi(\Sigma_1) = t$ and

(iii) M is diffeomorphic to the total space of an oriented disc bundle which has the Euler number $\chi(\Sigma_1)/2 + \chi(\Sigma_2)$.

The method they use to construct M is obtained by taking a $\mathbb{C}$-Fuchsian representation of the fundamental group of (each component of) Σ_1 and an $\mathbb{R}$-Fuchsian representation of the fundamental group of (each component of) Σ_2. They then use a version of the Klein-Maskit combination theorem to sew these representations together. The hypothesis that all the boundary components are short guarantees that the resulting representation is discrete.

Finally, for even integers t with $2 - 2g < t < 0$, the required representation is obtained by applying an anti-holomorphic isometry to the representation constructed above with $\tau = -t$. $\square$

The reason that Σ_1 is required to have an even number of components (and hence $\chi(\Sigma_1)$ is even) is rather subtle. In order to be able to use the Klein-Maskit combination theorem, each curve γ in the common boundary component of Σ_1 and Σ_2 must be represented by an element A of $SU(2,1)$ that is (up to conjugation) simultaneously a loxodromic element of $S(U(1) \times U(1,1))$ and of $SO(2,1)$. In particular it has real trace greater than $+3$. Hence, the $U(1)$ part of its representation in $S(U(1) \times U(1,1))$ must be $+1$ and hence it lies in $\{+1\} \times SU(1,1)$ and corresponds to an element of $SU(1,1)$ with trace greater than $+2$.

Suppose that Σ_0 is a three-holed sphere with boundary components γ_1, γ_2 and γ_3. For any choice of hyperbolic metric on Σ_0 we may associate a geometrical representation $\rho_0 : \pi_1(\Sigma_0) \longrightarrow PU(1,1)$. Consider any lift of ρ_0 to $SU(1,1)$. Let A_1, A_2 and A_3 be the three elements of $SU(1,1)$ representing the boundary loops. Then either all three traces are negative or else one is negative and the other two positive (see page 9 of Gilman [Gil] for example). This means that for at least one of the boundary components the corresponding element of $SU(2,1)$ has trace less than -1, and hence it corresponds to a glide reflection in $SO(2,1)$. By studying the decomposition of Σ_1 into three-holed spheres, we can see that each boundary component may be represented by an element of $SU(1,1)$ with positive trace if and only if the number of boundary components is even.

Additional examples have been constructed by Anan'in, Grossi and Gusevskii [AGG], [AG]. In their construction, they consider a group generated by complex involutions $R_1, \ldots, R_n$ fixing complex lines $L_1, \ldots, L_n$ respectively with two properties. First, the complex lines are pairwise ultra-parallel and, secondly, the product $R_1 \ldots R_n$ is the identity. They give conditions that determine whether such a group is discrete and they do so by constructing a fundamental polyhedron whose faces are contained in bisectors. These groups contain a complex hyperbolic quasi-Fuchsian subgroup of index 2 or 4. In particular they give a series of examples where the Euler number takes different values. The main result of [AG] is a complex hyperbolic structure on a trivial bundle over a closed oriented surface. Subsequently, a related construction was given by Gaye [Gay]. He considers groups of the same type as those considered in [AGG] and gives additional examples.

5.6 Complex hyperbolic Fenchel-Nielsen coordinates

It is useful to find ways of putting coordinates in complex hyperbolic quasi-Fuchsian space. One way to do it is to mimic the construction of Fenchel-Nielsen coordinates for Teichmüller space and the related complex Fenchel-Nielsen coordinates for quasi-Fuchsian space.

For clarity, we shall recall in brief how Fenchel-Nielsen coordinates are defined; see [FN] or Wolpert [Wol1], [Wol2]. Let γ_j for $j = 1, \ldots, 3g - 3$ be a *curve system* (some authors call this a *partition*) that is, a maximal collection of disjoint, simple, closed curves on Σ that are neither homotopic to each other nor homotopically trivial. The complement of such a curve system is a collection of $2g - 2$ three-holed spheres. If Σ has a hyperbolic metric then, without loss of generality, we may choose each γ_j in our curve system to be the geodesic in its homotopy class. The hyperbolic metric on each three-holed sphere is completely determined by the hyperbolic length $l_j > 0$ of each of its boundary geodesics. There is a real twist parameter k_j that determines how these three-holed spheres are attached to one another. From an initial configuration, the two three-holed spheres with common boundary component may be rotated relative to one another. The absolute value of the twist $|k_j|$ measures the distance they are twisted and the sign denotes the direction of their relative twist. The theorem of Fenchel and Nielsen states that each $(6g - 6)$-tuple

$$(l_1, \ldots, l_{3g-3}, k_1, \ldots, k_{3g-3}) \in \mathbb{R}_+^{3g-3} \times \mathbb{R}^{3g-3}$$

determines a unique hyperbolic metric on Σ and each hyperbolic metric arises in this way.

Kourouniotis [Kou4] and Tan [Tan] extended the theorem of Fenchel and Nielsen to the case of real hyperbolic quasi-Fuchsian space of Σ. In this extension both the length parameters and the twist parameters become complex. The elements of $\rho(\pi_1) \in \mathrm{SL}(2, \mathbb{C})$ corresponding to γ_j in the curve system are now loxodromic with, in general, non-real trace. Thus the imaginary part of the length parameter represents the holonomy angle when moving around γ_j. Likewise, the imaginary part of the twist parameter becomes the parameter of a bending deformation about γ_j; see also [PS] for more details of this correspondence and how to relate these parameters to traces of matrices. The main difference from the situation with real Fenchel-Nielsen coordinates is that, while distinct quasi-Fuchsian representations determine distinct complex Fenchel-Nielsen coordinates, it is not at all clear which set of coordinates give rise to discrete representations, and hence to a quasi-Fuchsian structure. In fact the boundary of the set of realisable coordinates is fractal.

In [PP2], Parker and Platis define Fenchel-Nielsen coordinates for representations of $\pi_1(\Sigma)$ to $\mathrm{SU}(2, 1)$ for which the $3g - 3$ curves in a curve system are represented by loxodromic maps. It is clear that this is a proper subset of the representation variety and contains complex hyperbolic quasi-Fuchsian space. The coordinates are $16g - 16$ real parameters that distinguish non-conjugate irreducible representations and $8g - 6$ real parameters that distinguish non-conjugate representations that preserve a complex line (compare to Propositions 5.5 and 5.6). As with the complex Fenchel-Nielsen coordinates described by Kourouniotis and Tan it is not clear which coordinates correspond to discrete representations. However, the coordinates in [PP2] determine the group up to conjugacy and distinguish between non-conjugate representations. The major innovation of Parker and Platis in this paper is the use of cross-ratios (recall Section 2.7) in addition to complex length and twist-bend parameters. First, for representations that do not preserve a complex line, and so are irreducible, the following holds.

Theorem 5.11 (Parker and Platis [PP2]) *Let Σ be a closed surface of genus g with a simple curve system $\gamma_1, \ldots, \gamma_{3g-3}$. Suppose that $\rho : \pi_1(\Sigma) \longrightarrow \Gamma < \mathrm{SU}(2,1)$ is an irreducible representation of the fundamental group $\pi_1(\Sigma)$ to $\mathrm{SU}(2,1)$ with the property that $\rho(\gamma_j) = A_j$ is loxodromic for each $j = 1, \ldots, 3g - 3$. Then there exist $4g - 4$ complex parameters and $2g - 2$ points on the cross-ratio variety $\mathfrak{X}$ that completely determine ρ up to conjugation.*

For representations that preserve a complex line several of the previous parameters are real and the following holds.

Theorem 5.12 (Parker and Platis [PP2]) *Let Σ be a closed surface of genus g with a simple curve system $\gamma_1, \ldots, \gamma_{3g-3}$. Suppose that $\rho : \pi_1(\Sigma) \longrightarrow \Gamma < \mathrm{S}\big(\mathrm{U}(1) \times \mathrm{U}(1,1)\big)$ is a reducible representation of the fundamental group $\pi_1(\Sigma)$ to $\mathrm{SU}(2,1)$ that preserves a complex line and which has the property that $\rho(\gamma_j) = A_j$ is loxodromic for each $j = 1, \ldots, 3g-3$. Then there exist $8g-6$ real parameters that completely determine ρ up to conjugation.*

5.7 The boundary of complex hyperbolic quasi-Fuchsian space

We now investigate what happens as we approach the boundary of quasi-Fuchsian space. Cooper, Long and Thistlethwaite [CLT] have proved a complex hyperbolic version of Chuckrow's theorem. They prove it for sequences of representations of groups that are not virtually nilpotent. We state it for the special case we are interested in here, namely the case of surface groups.

Theorem 5.13 (Theorem 2.7 of [CLT]) *Let π_1 be the fundamental group of a hyperbolic surface of finite type. Let $\rho_k : \pi_1 \longrightarrow \mathrm{SU}(2,1)$ be an algebraically convergent sequence of discrete, faithful representations of π_1. Then the limit representation is discrete and faithful.*

The classical version of Chuckrow's theorem was used by Greenberg, in Section 2.7 of [Gre], to construct some of the first examples of geometrically infinite Kleinian groups. Roughly speaking, his argument is the following: Consider a space X of discrete, faithful, geometrically finite, type-preserving $\mathrm{SL}(2, \mathbb{C})$-representations of a given abstract group so that X is open and the boundary of X in the space of all representations contains a continuum. The example Greenberg considers is Riley's example of groups generated by two parabolic maps, now known as the Riley slice of Schottky space. Then Greenberg considers a sequence of representations ρ_k converging to the boundary of X. Because X is open the limit representation ρ_0 is not in X. By Chuckrow's theorem ρ_0 is both discrete and faithful. Therefore ρ_0 must either contain additional parabolics or be geometrically infinite. This reasoning applies in our case. Greenberg goes on to argue that, since for a matrix to be parabolic in $SL(2, \mathbb{C})$ involves a codimension 2 condition (trace equals ± 2), there are only countably many boundary points with additional parabolic elements. Therefore there must be geometrically infinite examples on the boundary.

In the case of $SU(2, 1)$ the condition to be parabolic has codimension 1 ($f(\tau) = 0$ where $f(\tau)$ is Goldman's function Theorem 6.2.4 of [Gol4]). Therefore it could be the case that the boundary of complex hyperbolic quasi-Fuchsian space is made up of geometrically finite representations where (at least) one conjugacy class in π_1 is represented by parabolic elements of $SU(2, 1)$. The locus where a particular conjugacy class is parabolic is a real analytic codimension 1 subvariety (with respect to suitable coordinates, such as the complex hyperbolic Fenchel-Nielsen coordinates described in Section 5.6). Therefore a conjectural picture is that complex hyperbolic quasi-Fuchsian space has a polyhedral or cellular structure. The (real) codimension 1 cells correspond to a single extra class of parabolic maps. These intersect in lower dimensional cells. The cells of codimension k correspond to k classes becoming parabolic.

5.8 Complex hyperbolic quakebending

Let ρ_0 be a representation of the fundamental group π_1 of Σ into some Lie group G. A deformation of ρ_0 is a curve $\rho_t = \rho(t)$ such that $\rho(0) = \rho_0$. Deformations in Teichmüller and real quasi-Fuchsian spaces are very well known and have been studied extensively, at least throughout the last thirty years. Below we state some basic facts about them.

In the case of Teichmüller space $\mathcal{T}(\Sigma)$, the basic deformation is the Fenchel-Nielsen deformation; a thorough study of this has been carried out by Wolpert in [Wol2]. We cut Σ_0 along a simple closed geodesic α, rotate one side of the cut relative to the other and attach the sides in their new position. This deformation involves continuously changing the corresponding Fenchel-Nielsen twist parameter while holding all the others fixed. The hyperbolic metric in the complement of the cut extends to a hyperbolic metric in the new surface. In this way a deformation ρ_t (depending on the free homotopy class of α) is defined and its infinitesimal generator t_α is the Fenchel-Nielsen vector field. Such vector fields are very important: at each point of $\mathcal{T}(\Sigma)$, $6g - 6$ of such fields form a basis of the tangent space of $\mathcal{T}(\Sigma)$. Moreover, the Weil-Petersson Kähler form of $\mathcal{T}(\Sigma)$ may be described completely in terms of the variations of geodesic length of simple closed geodesics under the action of these fields. The basic formula for this is Wolpert's first derivative formula: If α, β are simple closed geodesics in Σ_0, l_α is the geodesic length of α and t_β is the Fenchel-Nielsen vector field associated to β then at the point ρ_0 we have

$$t_\beta l_\alpha = \sum_{p \in \alpha \cap \beta} \cos(\phi_p),$$

where ϕ_p is the oriented angle of intersection of α and β at p. Another basic formula concerns the mixed variations $t_\beta t_\gamma l_\alpha$; the reader should see for instance [Wol3] or [Wol2] for details.

In [Kou1] Kourouniotis, using ideas of Thurston and working in the spirit of Wolpert's construction of the Fenchel-Nielsen deformation, constructs a quasiconformal homeomorphism of the complex plane which he calls the *bending homeomorphism*. Given a hyperbolic structure on Σ, from this homeomorphism he obtains a quasi-Fuchsian structure on Σ.

Epstein and Marden took a different and much more general point of view in [EM]. Given a hyperbolic structure ρ_0 on a closed surface Σ, then for every finite geodesic lamination Λ in Σ with complex transverse measure μ and a simple closed geodesic α in Σ_0, there exists an isometric map h, depending on α, of Σ_0 into a hyperbolic 3-manifold M_h (the *quakebend map*). The image of this map is a *pleated surface* Σ_h, that is a complete hyperbolic surface which may be viewed as the original surface bent along the leaves of the lamination in angles depending on the imaginary part of μ, with its flat pieces translated relative to the leaves in distances depending on the real part of μ. The pleated surface Σ_h is then the boundary of the convex hull of M_h. For small $t \in \mathbb{C}$, quakebending along Λ with transverse measure $t\mu$ produces injective homeomorphisms of $\pi_1(\Sigma)$ into $\mathrm{PSL}(2, \mathbb{C})$ with quasi-Fuchsian image and in this way we obtain a deformation $\rho_{t\mu}$ (*the quakebend curve*) of quasi-Fuchsian space $\mathcal{Q}_\mathbb{R}(\Sigma)$ with initial point our given hyperbolic structure, that is a point in the Teichmüller space $\mathcal{T}(\Sigma)$ of Σ. It is evident that the Fenchel-Nielsen deformation as well as Kourouniotis' bending deformation are special cases of the above construction; the first is induced from the case where μ is real (*pure earthquake*) and the second from the case where μ is imaginary (*pure bending*). Infinitesimal generators of quakebend curves are the holomorphic vector fields T_μ. If α is a simple closed geodesic in Σ_0 and in the case where λ is finite with leaves $\gamma_1, \ldots, \gamma_n$, then at the point ρ_0 we have

$$T_\mu l_\alpha = \frac{dl(\rho_{t\mu})}{dt}(0) = \sum_{k=1}^{n} \Re(\zeta_k) \cdot \cos(\phi_k)$$

where $\zeta_k = \mu(\alpha \cap \gamma_k)$ and ϕ_k are the oriented angles of intersection of α and γ_k. This formula is a generalisation of Kerckhoff's formula when $\mu \in \mathbb{R}$, see [Ker]. Epstein and Marden also give formulae for the second derivative as well as generalisation of these in the case where Λ is infinite.

In [Kou2] Kourouniotis revisits the idea of bending. Based on [EM] and using his bending homeomorphism as in [Kou1], he constructs quakebending curves in $\mathcal{Q}(\Sigma)$ but there, the initial point ρ_0 is a quasi-Fuchsian structure on Σ. Moreover in [Kou3] he goes on to define the variations of the *complex length* λ_α of a simple closed curve under bending along (Λ, μ). Platis used Kourouniotis' results to describe completely the complex symplectic form of $\mathcal{Q}_\mathbb{R}(\Sigma)$ in [Pla1], [Pla2]. This form may be thought of as the complexification of the Weil-Petersson symplectic form of $\mathcal{T}(\Sigma)$. We remark finally that generalisations of the derivative formulae were given for instance by Series in [Ser].

In [Apa], Apanasov took the point view of Kourouniotis in [Kou1] to construct bending curves in $\mathcal{Q}_\mathbb{C}(\Sigma)$. In fact he proved the following

Theorem 5.14 (Apanasov [Apa]) *Let ρ_0 be an $\mathbb{R}$-Fuchsian representation of π_1 and write $\Gamma_0 = \rho_0(\pi_1)$. Then for any simple closed geodesic $\alpha \in \mathbf{H}_\mathbb{R}^2/\Gamma_0$ and for sufficiently small $t \in \mathbb{R}$ there exists a (continuous) bending deformation ρ_t of ρ_0 induced by Γ_0-equivariant quasiconformal homeomorphisms F_t of $\mathbf{H}_\mathbb{C}^2$.*

Platis in [Pla3] followed the strategy suggested by Epstein and Marden in [EM]. Let ρ_0 be an $\mathbb{R}$-Fuchsian representation of π_1 with $\Gamma_0 = \rho_0(\pi_1)$. Then, $M_0 = \mathbf{H}_\mathbb{C}^2/\Gamma_0$ is a complex hyperbolic manifold and embedded in M_0 there is a hyperbolic surface $\Sigma_0 = \mathbf{H}_\mathbb{R}^2/\Gamma_0$. For every finite geodesic lamination Λ in Σ with complex transverse measure μ and a simple closed geodesic α in Σ_0, there is an isometric map $B_\mathbb{C}$, depending on α, of M_0 into a complex hyperbolic manifold M_h (the *complex hyperbolic quakebend map*). Restricted to Σ_0 the image of this map is a pleated surface Σ_h, something which is entirely analogous to the classical case.

The pleated surface Σ_h is naturally embedded in M_h. By Corollary 5.8, for small $t \in \mathbb{R}$, complex hyperbolic quakebending along Λ with transverse measure $t\mu$ produces complex hyperbolic quasi-Fuchsian groups. That is,

Theorem 5.15 (Platis [Pla3]) *There is an $\epsilon > 0$ such that for all t with $|t| < \epsilon$ the complex hyperbolic quakebend curve $\rho_{t\mu}$ lies entirely in $\mathcal{Q}_{\mathbb{C}}(\Sigma)$.*

Platis also discusses the variations of the complex hyperbolic length $\lambda = l + i\theta$ of $\rho_{t\mu}$. The induced formulae are natural generalisations of Epstein-Marden's formulae. For instance, for the first derivative of λ at ρ_0 we have the following.

$$\frac{dl}{dt}(0) = \sum_{k=1}^{n} \Re(\zeta_k) \cdot \cos(\phi_k),$$

$$\frac{d\theta}{dt}(0) = \sum_{k=1}^{n} \operatorname{Im}(\zeta_k) \cdot \frac{3\cos^2(\phi_k) - 1}{2}.$$

Remark 5.16 *We remark that in view of Corollary 5.8 and Theorem 5.10, the condition ρ_0 is $\mathbb{R}$-Fuchsian is now quite restrictive. Complex hyperbolic quakebending curves may be constructed in exactly the same way at least when the starting point ρ_0 is an arbitrary point of $\mathcal{Q}_{\mathbb{C}}(\Sigma)$ with even Toledo invariant $\tau(\rho_0)$. This construction is carried out in [Pla4].*

6 Open problems and conjectures

Below we list a number of open problems concerning both compact and non compact cases. We also state a number of conjectures.

Problem 6.1 *Describe the topology of $\mathcal{Q}_{\mathbb{C}}(\Sigma)$. For example:*

   (i) *In the case when $p = 0$ describe the components of $\mathcal{Q}_{\mathbb{C}}(\Sigma)$ with the same Toledo invariant.*

   (ii) *For $|\tau(\rho)| < 2g - 2$, is each of these components homeomorphic to a ball of dimension $16g - 16$?*

   (iii) *In the case when $p > 0$ describe the components of $\mathcal{Q}_{\mathbb{C}}(\Sigma)$.*

   (iv) *For $p > 0$ is $\mathcal{Q}_{\mathbb{C}}(\Sigma)$ an open subset of the representation variety $\operatorname{Hom}(\pi_1, \operatorname{SU}(2,1))/\operatorname{SU}(2,1)$?*

   (v) *Is each component homeomorphic to a ball?*

We have seen that $\mathcal{Q}_\mathbb{C}(\Sigma)$ is disconnected. In the compact case the Toledo invariant distinguishes components of the representation variety $\mathrm{Hom}\big(\pi_1, \mathrm{SU}(2,1)\big)/\mathrm{SU}(2,1)$ [X] and each component is non-empty [GKL]. The Euler number gives a finer classification of $\mathcal{Q}_\mathbb{C}(\Sigma)$. As yet it is not known which values of the Euler number can arise; compare [AGG]. When $p > 0$ there are quasi-Fuchsian representations whose limit set is a wild knot [DG]. It is natural to ask which wild knots arise in this way and whether this construction extends to the case of $p = 0$.

Problem 6.2 *Describe representations in* $\partial\mathcal{Q}_\mathbb{C}(\Sigma)$. *For example:*

 (i) *Does* $\partial\mathcal{Q}_\mathbb{C}(\Sigma)$ *admit a cell structure where cells of codimension* k *correspond to* k *elements of* π_1 *being represented by parabolic maps?*

 (ii) *If the answer to (i) is positive then characterise which elements of* π_1 *correspond to codimension 1 cells and which combinations correspond to cells of higher codimension.*

 (iii) *Is every representation in* $\partial\mathcal{Q}_\mathbb{C}(\Sigma)$ *geometrically finite?*

 (iv) *Does every quasi-Fuchsian representation of a punctured surface arise in the boundary of the quasi-Fuchsian space of a closed surface?*

It is tempting to suggest that the codimension 1 cells correspond exactly to the simple closed curves on Σ. However, the case of the three times punctured sphere, Theorem 4.1, suggests that there are other curves that may be pinched in this way.

A starting point could be to consider surface subgroups of complex hyperbolic triangle groups. Let q_1, q_j, q_3 be integers at least 2 (possibly including ∞) and consider three involutions I_j each fixing a complex line so that $I_{j+1}I_{j-1}$ is elliptic of order q_j when q_j is finite and parabolic when q_j is ∞ (all indices are taken mod 3). Then $\langle I_1, I_2, I_3 \rangle$ is a representation of a (q_1, q_2, q_3) triangle reflection group.

Conjecture 6.3 (Schwartz [S3]) *Let* q_1, q_2 *and* q_3 *be integers at least 2 and let* $\langle I_1, I_2, I_3 \rangle < \mathrm{SU}(2,1)$ *be the corresponding representation of the* (q_1, q_2, q_3) *triangle reflection group. Then* $\langle I_1, I_2, I_3 \rangle$ *is a discrete, faithful, type-preserving, geometrically finite representation if and only if* $I_1 I_2 I_3$ *and* $I_j I_{j+1} I_j I_{j-1}$ *for* $j = 1$, 2, 3 *are all loxodromic.*

Moreover, the values of the q_j *determine which of these four words becomes parabolic first.*

The Coxeter group generated by reflections in the sides of a hyperbolic triangle contains a surface group as a torsion-free, finite index subgroup (indeed it contains infinitely many such surface groups). Therefore every discrete, faithful, type-preserving, geometrically finite representation $\langle I_1, I_2, I_3 \rangle$ contains quasi-Fuchsian subgroups of finite index. If Schwartz's conjecture is true then the subgroups of the corresponding representations where one of $I_1 I_2 I_3$, $I_j I_{j+1} I_j I_{j-1}$ is parabolic will lie in $\partial \mathcal{Q}_{\mathbb{C}}(\Sigma)$.

There are also natural questions about the geometric and analytical structures of $\mathcal{Q}_{\mathbb{C}}(\Sigma)$. Goldman [Gol1] and Hitchin [H] have shown that the representation variety $\mathrm{Hom}(\pi_1, \mathrm{SU}(2,1))/\mathrm{SU}(2,1)$ admits natural symplectic and complex structures. Our next problem concerns these structures.

Problem 6.4 *Let Σ be a closed surface. Examine geometrical and analytical structures of $\mathcal{Q}_{\mathbb{C}}(\Sigma)$.*

  (i) *Do the natural symplectic and complex structures on the representation variety $\mathrm{Hom}(\pi_1, \mathrm{SU}(2,1))/\mathrm{SU}(2,1)$ given by Goldman and Hitchin naturally pass to $\mathcal{Q}_{\mathbb{C}}(\Sigma)$?*

 (ii) *Take a complex hyperbolic quakebending deformation with starting point any irreducible representation. From this, define then geometrically a symplectic structure in the whole $\mathcal{Q}_{\mathbb{C}}(\Sigma)$ such that it agrees with the Weil-Petersson symplectic form of Teichmüller space at $\mathbb{R}$-Fuchsian representations.*

(iii) *What can you say about the complex structure? For instance do complex hyperbolic Fenchel-Nielsen coordinates induce such a complex structure on $\mathcal{Q}_{\mathbb{C}}(\Sigma)$?*

The answer to Problem 6.4 (i) is affirmative in the classical cases of $\mathrm{Hom}(\pi_1, \mathrm{SL}(2,\mathbb{R}))/SL(2,\mathbb{R})$ and $\mathrm{Hom}(\pi_1, \mathrm{SL}(2,\mathbb{C}))/SL(2,\mathbb{C})$. That is, complex and symplectic structures pass naturally to $\mathcal{T}(\Sigma)$ and $\mathcal{Q}_{\mathbb{R}}(\Sigma)$ respectively. We conjecture that the answer is also affirmative in the complex hyperbolic case. We also conjecture that the answer to Problem 6.4 (iii) is negative.

Conjecture 6.5 *For $p = 0$ there is a hyper-Kähler structure in $\mathcal{Q}_{\mathbb{C}}(\Sigma)$.*

Quasi conformal maps are the major tool used to define Teichmüller space. These have been generalised to complex hyperbolic space by Mostow, Chapter 21 of [Mos1]. Also, Korányi and Reimann [KR2] have developed an extensive theory of quasiconformal mappings on the

Heisenberg group. These may be extended to $\mathbf{H}^2_{\mathbb{C}}$ [KR3]. However, it is not known whether these quasiconformal mappings are strong enough to describe the whole of $\mathcal{Q}_{\mathbb{C}}(\Sigma)$.

Conjecture 6.6 *For $p = 0$ any two representations in the same component of $\mathcal{Q}_{\mathbb{C}}(\Sigma)$ are quasiconformally conjugate.*

In the non compact case, by a theorem of Miner, see [Min], type-preserving $\mathbb{C}$-Fuchsian and $\mathbb{R}$-Fuchsian representations are never quasiconformally conjugate. In the compact case, the authors believe that there is strong evidence that this conjecture is true. For instance Aebischer and Miner showed in [AM] that the complex quasi-Fuchsian space of a classical Schottky group has this property.

Notes

1 Supported by a Marie Curie Reintegration Grant fellowship (Contract No. MERG-CT-2005-028371) within the 6th Community Framework Programme.

Bibliography

[AM] Aebischer, B. and Miner, R. Deformation of Schottky groups in complex hyperbolic space. *Conf. Geom. and Dyn.* **3** (1999), 24–36.

[AGG] Anan'in, S., Grossi C. H. and Gusevskii, N. Complex hyperbolic structures on disc bundles over surfaces I: General settings and a series of examples. Preprint.

[AG] Anan'in, S. and Gusevskii, N. Complex hyperbolic structures on disc bundles over surfaces II: Example of a trivial bundle. Preprint.

[Apa] Apanasov, B. Bending deformations of complex hyperbolic surfaces. *J. Reine Angew. Math.* **492** (1997), 75–91.

[Bea] Beardon, A. F. *The Geometry of Discrete Groups*, Springer, 1983.

[Ber] Bers, L. Spaces of Kleinian groups, in *Several Complex Variables I*, Lecture Notes in Mathematics **155** (1970), 9–34.

[Bou] Bourdon, M. Structure conforme au bord et flot géodésique d'un CAT(−1) espace. *Enseign. Math.* **41** (1995), 63–102.

[Bow] Bowditch, B. H. Geometrical finiteness with variable negative curvature. *Duke Math. J.* **77** (1995), 229–274.

[Car] Cartan, É. Sur le groupe de la géométrie hypersphérique. *Comm. Math. Helv.* **4** (1932), 158–171.

[CG] Chen, S. S. and Greenberg, L. Hyperbolic spaces, in *Contributions to Analysis*, ed. L. V. Ahlfors *et al.*, Academic Press, 1974, pp. 49–87.

[CLT] Cooper, D. Long, D. D. and Thistlethwaite, M. Flexing closed hyperbolic manifolds. *Geometry and Topology* **11** (2007), 2413–2440.

[CuGu] Cunha, H. and Gusevskii, N. On the moduli space of quadruples of points on the boundary of complex hyperbolic space, arXiv: 0812.2159v4 [math GT] 33 March 2009.

[DT] Domic, A. and Toledo, D. The Gromov norm of the Kähler class of symmetric domains. *Math. Ann.* **276** (1987), 425–432.

[DG] Dutenhefner, F. and Gusevskii, N. Complex hyperbolic Kleinian groups with limit set a wild knot. *Topology* **43** (2004), 677–696.

[E] Epstein, D. B. A. Complex hyperbolic geometry, in *Analytical and Geometric Aspects of Hyperbolic Space*, ed. D.B.A. Epstein, London Mathematical Society Lecture Notes Series **111** (1987), 93–111.

[EM] Epstein, D. B. A. and Marden, A. Convex hulls in hyperbolic space, a theorem of Sullivan, and measured pleated surfaces, in *Analytical and Geometric Aspects of Hyperbolic Space*, ed. D. B. A. Epstein, London Mathematical Society Lecture Notes Series **111** (1987), 113–253.

[Fal] Falbel, E. Geometric structures associated to triangulations as fixed point sets of involutions. *Topology and its Applications* **154** (2007), 1041–1052.

[FK] Falbel, E. and Koseleff, P.-V. A circle of modular groups in PU(2, 1). *Math. Res. Lett.* **9** (2002), 379–391.

[FPa] Falbel, E. and Parker, J. R. The moduli space of the modular group in complex hyperbolic geometry. *Invent. Math.* **152** (2003), 57–88.

[FPl] Falbel, E. and Platis, I. D. The PU(2, 1)-configuration space of four points in S^3 and the cross-ratio variety. *Math. Annalen* **340** (2008), 935–962.

[FZ] Falbel, E. and Zocca, V. A Poincaré's polyhedron theorem for complex hyperbolic geometry. *J. reine angew. Math.* **516** (1999), 133–158.

[FN] Fenchel, W. and Nielsen, J. *Discontinuous Groups of Isometries in the Hyperbolic Plane.* ed. Asmus L. Schmidt, de Gruyter Studies in Mathematics 29, 2003.

[Gay] Gaye, M. Sous-groupes discrets de PU(2, 1) engendrés par n reflexions complexes et déformation. *Geom. Dedicata* **137** (2008), 27–61.

[Gil] Gilman, J. *Two-Generator Discrete Subgroups of* PSL(2, ℝ), Memoirs of the American Mathematical Society **561**, 1995.

[Gol1] Goldman, W. M. The symplectic nature of fundamental groups of surfaces. *Adv. in Math.* **54** (1984), 200–225.

[Gol2] Goldman, W. M. Representations of fundamental groups of surfaces, in *Geometry and Topology*, ed. J. Alexander and J. Harer, Lecture Notes in Mathematics **1167** (1985), 95–117.

[Gol3] Goldman, W. M. Convex real projective structures on compact surfaces. *J. Diff. Geom.* **31** (1990), 791–845.

[Gol4] Goldman, W. M. *Complex Hyperbolic Geometry*, Oxford University Press, 1999.

[GKL] Goldman, W. M., Kapovich M. E. and Leeb, B. Complex hyperbolic manifolds homotopy equivalent to a Riemann surface. *Comm. Anal. Geom.* **9** (2001), 61–95.

[GP] Goldman, W. M. and Parker, J. R. Complex hyperbolic ideal triangle groups. *J. reine angew. Math.* **425** (1992), 71–86.

[Gre] Greenberg, L. Finiteness theorems for Fuchsian and Kleinian groups, in *Discrete Groups and Automorphic Functions*, ed. W. J. Harvey, Academic Press, 1977, pp. 199–257.

[Gui] Guichard, O. Groupes plongés quasi isométriquement dans un groupe de Lie. *Math. Ann.* **330** (2004), 331–351.

[GuP1] Gusevskii, N. and Parker, J. R. Representations of free Fuchsian groups in complex hyperbolic space. *Topology* **39** (2000), 33–60.

[GuP2] Gusevskii, N. and Parker, J. R. Complex hyperbolic quasi-Fuchsian groups and Toledo's invariant. *Geometriae Dedicata* **97** (2003), 151–185.

[H] Hitchin, N. J. Hyper-Kähler manifolds. Séminaire Bourbaki, Vol. 1991/92. *Astérisque* **206** (1992), Exp. No. 748, 3, 137–166.

[Ker] Kerckhoff, S. P. The Nielsen realization problem. *Annals of Maths.* **117** (1983), 235–265.

[KR1] Korányi, A. and Reimann, H. M. The complex cross-ratio on the Heisenberg group. *L'Enseign. Math.* **33** (1987), 291–300.

[KR2] Korányi, A. and Reimann, H. M. Foundations for the theory of quasiconformal mappings on the Heisenberg group. *Adv. Math.* **111** (1995), 1–87.

[KR3] Korányi, A. and Reimann, H. M. Equivariant extension of quasiconformal deformations into the complex unit ball. *Indiana Univ. Math. J.* **47** (1998), 153–176.

[Kou4] Kourouniotis, C. Complex length coordinates for quasi-Fuchsian groups. *Mathematika* **41** (1994), 173–188.

[Kou2] Kourouniotis, C. Bending in the space of quasi-Fuchsian structures. *Glasgow Math. J.* **33**(1) (1991), 41–49.

[Kou1] Kourouniotis, C. Deformations of hyperbolic structures. *Math. Proc. Camb. Phil. Soc.* **98** (1985), 247–261.

[Kou3] Kourouniotis, C. The geometry of bending quasi-Fuchsian groups, in *Discrete Groups and Geometry*, ed. W. J. Harvey and C. Maclachlan, London Mathematical Society Lecture Notes Series, **173** (1992), 148–164.

[Mil] Millington, M. H. Subgroups of the classical modular group. *J. Lon. Math. Soc.* **1** (1969), 351–357.

[Min] Miner, R. R. Quasiconformal equivalence of spherical CR manifolds. *Ann. Acad. Sci. Fenn. Ser. A I Math.* **19** (1994), 83–93.

[Mos1] Mostow, G. D. *Strong Rigidity of Locally Symmetric Spaces*, Annals of Maths. Studies **78**, 1973.

[Mos2] Mostow, G. D. On a remarkable class of polyhedra in complex hyperbolic space. *Pacific Journal of Maths.* **86** (1980), 171–276.

[P] Parker, J. R. *Complex Hyperbolic Kleinian Groups*, Cambridge University Press (to appear).

[PP1] Parker, J. R. and Platis, I. D. Open sets of maximal dimension in complex hyperbolic quasi-Fuchsian space. *J. Diff. Geom.* **73** (2006), 319–350.

[PP2] Parker, J. R. and Platis, I. D. Complex hyperbolic Fenchel-Nielsen coordinates. *Topology* **47** (2008), 101–135.

[PP3] Parker, J. R. and Platis, I. D. Global, geometrical coordinates for Falbel's cross-ratio variety. *Canadian Mathematical Bulletin* **52** (2009), no. 2, 285–294.

[PS] Parker, J. R. and Series, C. Bending formulae for convex hull boundaries. *J. d'Analyse Math.* **67** (1995), 165–198.

[Pla1] Platis, I. D. Περί της Γεωμετρίας του Χώρου των Quasi-Fuchsian Παραμορφώσεων Μιας Υπερβολικής Επιφάνειας, PhD Thesis, University of Crete, 1999, http://webserver.math.uoc.gr:1080/erevna/didaktorikes/Platis_Phd.pdf

[Pla2] Platis, I. D. Complex symplectic geometry of quasifuchsian space. *Geometriae Dedicata* **87** (2001), 17–34.

[Pla3] Platis, I. D. Quakebend deformations in complex hyperbolic quasi-Fuchsian space. *Geometry and Topology* **12** (2008), 431–459.

[Pla4] Platis, I. D. Quakebend deformations in complex hyperbolic quasi-Fuchsian space II. Preprint.

[Pla5] Platis, I. D. The geometry of complex hyperbolic packs. *Math. Proc.* Cambridge Philos. Soc. **147** (2009), no. 1, 205–234.

[S1] Schwartz, R. E. Ideal triangle groups, dented tori and numerical analysis. *Annals of Maths.* **153** (2001), 533–598.

[S2] Schwartz, R. E. Degenerating the complex hyperbolic ideal triangle groups. *Acta Math.* **186** (2001), 105–154.

[S3] Schwartz, R. E. Complex hyperbolic triangle groups, in *Proceedings of the International Congress of Mathematicians*, Vol. II, ed. T. Li, Beijing, 2002, 339–349.

[S4] Schwartz, R. E. A better proof of the Goldman-Parker conjecture. *Geom. Topol.* **9** (2005), 1539–1601.

[S5] Schwartz, R. E. *Spherical CR Geometry and Dehn Surgery*, Annals of Maths. Studies **165**, 2007.

[Ser] Series, C. An extension of Wolpert's derivative formula. *Pacific J. Math.* **197**(1) (2001), 223–239.

[Tan] Tan, S. P. Complex Fenchel-Nielsen coordinates for quasi-Fuchsian structures. *Internat. J. Math.* **5** (1994), 239–251.

[T1] Toledo, D. Harmonic maps from surfaces to certain Kähler manifolds. *Math. Scand.* **45** (1979), 12–26.

[T2] Toledo, D. Representations of surface groups in complex hyperbolic space. *J. Differential Geometry* **29** (1989), 125–133.

[W1] Will, P. Groupes libres, groupes triangulaires et tore épointé dans PU(2, 1), PhD Thesis, University of Paris VI, 2006.

[W2] Will, P. The punctured torus and Lagrangian triangle groups in PU(2, 1). *J. Reine Angew. Math.* **602** (2007), 95–121.

[W3] Will, P. Bending Fuchsian representations of fundamental groups of cusped surfaces in PU(2, 1). Preprint.

[Wol1] Wolpert, S. A. The length spectra as moduli for compact Riemann surfaces. *Annals of Maths.* **109** (1979), 323–351.

[Wol2] Wolpert, S. A. The Fenchel–Nielsen deformation. *Annals of Maths.* **115** (1982), 501–528.

[Wol3] Wolpert, S. A. On the symplectic geometry of deformations of a hyperbolic surface. Annals of Maths. **117** (1983), 207–234.

[X] Xia, E. Z. The moduli of flat PU(2, 1) structures on Riemann surfaces. *Pacific J. Maths.* **195** (2000), 231–256.

Geometry of optimal trajectories

Mauro Pontani

Scuola di Ingegneria Aerospaziale, University of Rome "La Sapienza"
mauro.pontani@uniroma1.it

Paolo Teofilatto

Scuola di Ingegneria Aerospaziale, University of Rome "La Sapienza"
paolo.teofilatto@uniroma1.it

Abstract

The optimization of orbital manoeuvres and lunar or interplanetary transfer paths is based on the use of numerical algorithms aimed at minimizing a specific cost functional. Despite their versatility, numerical algorithms usually generate results which are local in character. Geometrical methods can be used to drive the numerical algorithms towards the global optimal solution of the problems of interest. In the present paper, Morse inequalities and Conley's topological methods are applied in the context of some trajectory optimization problems.

1 Introduction

Geometrical methods and techniques of differential topology have been useful in the study of dynamical systems for a long time. Classical results are provided by Morse theory and in particular Morse inequalities. These relate the number of critical points of index k of a function $f : M \to \mathbf{R}$, defined on a manifold M, to the k-homology groups of M. The manifold M can be a finite dimensional manifold [Mor1], the infinite dimensional manifold of paths in a variational problem [Mor2], [PS] or the manifold of control functions in an optimal control problem [AV], [V2].

The gradient flow of a Morse function f defines a retracting deformation that maps M into neighborhoods of its critical points of index k. These neighborhoods are identified with cells of dimension k, then a cell decomposition of M is determined through the function f [Mil].
Such handlebody decomposition of a manifold has been generalized to

356

Conley's fundamental results on dynamical systems [Con2]. The critical points of the Morse function f are generalized to isolated invariant sets for the flow of the dynamical system. The gradient flow of the Morse function f is generalized to the existence of a Lyapunov function which is strictly decreasing on the phase space (position and velocity variables) external to the invariant sets. Then the phase space can be decomposed into invariant sets and a gradient-like part.

Morse and Conley geometrical methods have been successfully applied to different branches of science. In the present paper it is stressed that the importance of these methods does not stem from the mathematical elegance of the theory alone, but also from the possibility of using them as fundamental tools in space mission analysis. In particular, we shall deal with trajectory optimization of: i) orbital transfer manoeuvres, ii) transfer orbits to the Moon or to other planets of the solar system.

Some different techniques must be introduced in the case of lunar and interplanetary transfers. An important point consists in searching conditions for lunar or target planet ballistic capture. In other words, the energy of the spacecraft at planet arrival must be such that the spacecraft is captured by the gravitational field of the target planet with no manoeuvre. It will be shown that such capture orbits are close in phase space to orbits asymptotic to invariant sets, and that the topological methods introduced by Conley come into play.

The paper is organized as follows: in Section 2 the numerical methods to find optimal trajectories are recalled and the use of Morse inequalities for mission analysis is shown in a simple example. In Section 3 a method aimed at achieving ballistic capture in lunar transfer missions is analyzed. The geometry of the phase space is described and the capture orbits are characterized from the topological point of view. Final comments conclude the paper.

2 Optimization of orbital transfer manoeuvres

Trajectory optimization consists in the definition of a thrust strategy (control variables) to achieve the final orbit conditions, either starting from the Earth surface or from a given orbit, while minimizing a cost function and satisfying specified constraints along the trajectory. These optimization problems can be solved introducing the Lagrangian multipliers λ and the calculus of variation (or the Pontryagin Principle) to determine the necessary conditions that must be satisfied by an optimal solution [Ces]. The determination of the control variables from the necessary conditions results in a Two-Point Boundary-Value Problem (TPBVP).

The TPBVP usually is not amenable to an analytic solution so numerical methods are to be employed. Shooting methods represent a class of numerical methods for solving such problems. Starting guess values are given to the unknown variables, generally the initial values of the Lagrangian multipliers λ_0, then the final conditions are checked. If they are not satisfied to the prefixed accuracy, the initial guess is improved after evaluating the dependence of the boundary condition violation on the starting guess. A well known drawback characterizes the shooting methods: the region of convergence can be small so an accurate guess must be provided.

However, even providing a reasonable guess, an additional inconvenience cannot be avoided, i.e. the locality of the final result. In fact the solution found by the numerical method corresponds to a single critical point of the cost function, more specifically the critical point "closest" to the starting guess. A possible approach to find other (locally) optimal solutions is based on the use of Genetic Algorithms. They employ an effective search technique that discretizes the space of the initial conditions λ_0 thus reducing the number of trials while guessing the values of the unknown λ_0. However examples show that the number of critical points found by a GA may depend on the discretization in the space of λ_0 [TP]. Then it would be useful to have an a priori estimate on the number of critical points of the cost function J. Morse inequalities can be applied to trajectory optimization problems with this intent.

An example follows: the orbital manoeuvre of a spacecraft from a circular orbit to a coplanar nearby orbit will be discussed. Let $r(t)$ be the position vector of the spacecraft and $r_C(t)$ be the position vector related to the initial circular orbit. The equations of the relative motion $\rho(t) = r(t) - r_C(t)$ in the radial and tangential directions are (Euler–Hill equations):

$$\ddot{x}_1 - 2\dot{x}_2 - 3x_1 = 0 \tag{2.1}$$

$$\ddot{x}_2 + 2\dot{x}_1 = u \tag{2.2}$$

where a tangential thrust u is assumed to perform a rendez-vous manoeuvre, with terminal relative velocity equal to zero, and terminal relative distance equal to ρ_f (specified). The initial conditions are

$$x_1(0) = x_2(0) = \dot{x}_1(0) = \dot{x}_2(0) = 0$$

and the final conditions are

$$\dot{x}_1(t_f) = 0 \qquad \text{and} \qquad \dot{x}_2(t_f) = 0 \tag{2.3}$$

$$x_1^2(t_f) + x_2^2(t_f) = \rho_f^2 \ (= \text{ given}) \tag{2.4}$$

The thrust u is chosen in order to perform the above manoeuvre within a specified terminal time t_f while minimizing the following objective function:

$$J = \frac{1}{2} \int_0^{t_f} u^2 \, dt \tag{2.5}$$

The final condition (2.4) can be replaced with the following pair of boundary conditions:

$$x_1(t_f) = \rho_f \cos\theta \qquad \text{and} \qquad x_2(t_f) = \rho_f \sin\theta \tag{2.6}$$

where the parameter θ will be determined at the end of the optimization process. Applying the necessary conditions for optimality four solutions are found: these correspond to two pairs of symmetrical trajectories, portrayed in Fig. 2.1.

Symmetry with respect to the origin is due to the fact that if u is an admissible control also $-u$ is admissible with the same cost function value. Such a symmetry shows that the problem can split into two problems considering the target region $N = \left\{ x_f^2 + y_f^2 = \rho_f^2 \right\}$ as the disjoint union of the left and right semi-circles: $N = S_- \bigcup S_+$. The problem with target region $N_- = S_-$ is equivalent to the problem with target region $N_+ = S_+$. By considering the derivative $d^2 J / dt^2$, it turns out that two solutions (those associated to the shortest trajectories) are minimizing (optimal) solutions, two solutions (the ones corresponding to the longest trajectories) are maximizing (stationary, non-optimal) solutions. Definitely, the optimal control law turns out to have the following form:

$$u^* = b_0 + b_1 t + c_1 \cos t + c_2 \sin t \tag{2.7}$$

i.e., it is composed of a periodic plus a linear term in time.

It is important to note that the multiplicity of solutions to the above problem depends on the topology of the final target region. Namely, N is composed of two sets that are not simply connected. A known result is that if the target region is convex the above problem admits a unique solution [N], so there exists a relationship between the topology of the target region N and the number of the critical points of J. The target space $N = S_- \bigcup S_+$ has the following homology groups (the double of

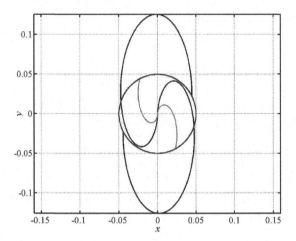

Fig. 2.1. The four solutions corresponding to the critical points of J ($t_f = 2$, $\rho_f = 0.05$)

the homology groups of a circle): $\dim(H_0(N)) = 2$, $\dim(H_1(N)) = 2$, $\dim(H_k(N)) = 0$, $k \geq 2$. On the other hand, the cost function J exhibits two maxima and two minima when the control belongs to the class of periodic plus linear control functions (the class selected by the Euler-Lagrange conditions). However, regarding J as function of the whole control space of functions that are square integrable and compatible with the required final conditions, the two maxima correspond to saddle points, whereas the two minima are again minima. These properties are evident just adding a quadratic term to the control u^*:

$$u = u^* + at^2 = b_0 + b_1 t + c_1 \cos t + c_2 \sin t + at^2 \qquad (2.8)$$

The state equations can be again integrated by using this new (non-optimal) control law. The four boundary conditions on the final values of the state variables ((2.3) and (2.6)) can be employed to express the four (generic) constants b_0, b_1, c_1, and c_2 as functions of θ and a. Hence, the control law introduced through (2.8) becomes a function of θ, a, and t: $u = u(t, \theta, a)$. It follows that the cost functional can be integrated again, to yield (formally):

$$J = \frac{1}{2} \int_0^{t_f} u^2(t, \theta, a) \, dt = J(\theta, a) \qquad (2.9)$$

The two-parameter function $J(\theta, a)$ is associated to the surface portrayed in Fig. 2.2, which exhibits two saddle points and two minima. The saddle points correspond to maximum points over the curve associated to $a = 0$, which identifies the case of the optimal control functions. Therefore, there are two critical points of J with index 0 (minima), two critical points of index 1 (saddle points) and there are no other critical points. In conclusion, $c_k(J) = \dim(H_k(N))$, i.e. the number $c_k(J)$ of critical points of J of index k is equal to the dimension of the k-homology groups of the target space N.

The above result is an example of a more general theory that extends Morse inequalities to the space of control functions. The following theorem holds [AV], [V1]:

Theorem 2.1 *Given the control problem of the form:*

$$\dot{\boldsymbol{x}} = \boldsymbol{f}(\boldsymbol{x}) + \boldsymbol{b}(\boldsymbol{x})\boldsymbol{u} \qquad (2.10)$$

with cost function

$$J(u) = \int_0^T L(\boldsymbol{x}, \boldsymbol{u})\, dt$$

where $\boldsymbol{x}$ is an n-dimensional vector and $\boldsymbol{u}$ is composed of m square integrable functions, let the following two conditions be satisfied
  i) $L(\boldsymbol{x}, \boldsymbol{u}) \geq k \, \|u\|_{L_2}$
  ii) $F_{\boldsymbol{x}_0 T} : L_2^m[0,T] \longrightarrow M$ *has constant rank for any $\boldsymbol{x}_0$ on M and final time T*
(where the input-output map $F_{\boldsymbol{x}_0 T} = \boldsymbol{x}(\boldsymbol{x}_0, \boldsymbol{u}, T)$ is the solution at the time T of (2.10) with initial condition $\boldsymbol{x}_0$, and control function $\boldsymbol{u}$). Then for any submanifold $N \subset M$, let $\mathbf{N}$ be the subspace of the control functions $\mathbf{N} = F_{\boldsymbol{x}_0 T}^{-1}(N)$. Let J^a be the subspace of $\mathbf{N}$ of the control functions $\boldsymbol{u}$ such that $J(\boldsymbol{u}) \leq a$. Let $c_k(J^a)$ be the number of critical points of degree k of J defined on the space J^a. Then the following Morse inequalities are verified:

$$c_k(J^a) \geq \dim(H_k(J^a))$$

Theorem 76 provides a lower bound estimate on the number of critical points of the cost function of an optimal control problem. From the practical point of view there is the problem of computing the homology groups of infinite dimensional subspaces J^a in the space of the control functions $\mathbf{N}$. In fact $\mathbf{N}$ has the structure of a fibred space over the target space N: at any point $\boldsymbol{y}$ of N there is a fiber $\pi^{-1}(\boldsymbol{y})$ consisting of those control functions that generate flows connecting $\boldsymbol{x}_0$ and $\boldsymbol{y}$ in the time T, i.e. $\pi^{-1}(\boldsymbol{y}) = F_{\boldsymbol{x}_0 T}^{-1}(\boldsymbol{y})$. Then the space $\mathbf{N}$ is obtained considering all these fibers $\pi^{-1}(\boldsymbol{y})$ together in a bundle.

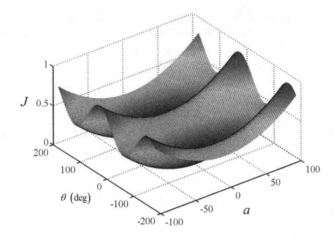

Fig. 2.2. The two minima and the two saddle points of $J(\theta, a)$ ($t_f = 2$, $\rho_f = 0.05$)

It is possible to show that in some circumstances the bundle **N** has the same topological properties of the base space N. For instance if all the fibers $\pi^{-1}(\boldsymbol{y})$ are contractible, i.e. homotopically equivalent to a point, then the topology of the bundle coincides with the topology of the base space [V2].

Of course the above observation applies to linear control systems such as the one represented by equations (7)-(11). In fact it is straightforward to show that the control function space $\pi^{-1}(\boldsymbol{y})$ is convex so $H_k(\mathbf{N}) = H_k(N)$.

Therefore, if the homology groups $H_k(N)$ are nontrivial, the Morse inequalities imply that the optimal control problem has a multiplicity of critical points and one can estimate their number. The joint use of such existence results together with a global search numerical algorithm (e.g. Genetic Algorithms) seems a promising method to determine globally optimal solutions.

3 Optimization of lunar transfer orbits

During the Apollo 11 mission to the Moon the Saturn V launcher left the Apollo spacecraft into an orbit of perigee altitude of 190 km. Then the third stage engine was ignited for 347 seconds to put the spacecraft into a translunar trajectory. At Moon arrival a second (braking) manoeuvre was performed for 357 seconds. Since the total flight time was 100 hours, the thrust could be regarded as impulsive. Basically, two impulses were performed: the first produced a velocity variation $\Delta V_1 = 3.182\,\text{km/sec}$ to reach the Moon orbit, the second impulse of $\Delta V_2 = 0.889\,\text{km/sec}$ decreased the orbital energy so that the spacecraft became a lunar satellite. A current trend in lunar and interplanetary mission analysis consists in reducing to zero the second velocity variation ΔV_2. This is possible if the spacecraft at arrival has the right energy to be captured by the target planet with almost no manoeuvre (ballistic capture). For lunar ballistic capture two different kinds of missions have been proposed: inner missions [Con1] and outer missions [B]. Both kinds of missions produce substantial savings on ΔV_2 and have been carried out. In 1990 the Japanese mission Hiten was sent to the Moon via an outer mission, and in 2003 the European mission Smart reached the Moon performing an inner mission.

An example of inner mission is portrayed in Fig. 3.1, where the spacecraft trajectory is represented in a dimensionless coordinate frame rotating with the Earth–Moon system.

Both the Earth and the Moon are placed on the x axis and their respective coordinates are $-\mu$ and $1 - \mu$ (where μ is the Moon to Earth mass ratio, $\mu \simeq 1/81$). After some rotations around the Earth along elliptic orbits, the spacecraft leaves the elliptic orbit without any burn. The trajectory presents a cusp at coordinate $x \simeq 0.8$, then the spacecraft is captured by the Moon.

The dynamics of inner missions is well modelled as a restricted three-body problem (Earth–Moon and spacecraft) [Sze]. In the rotating (dimensionless) frame the spacecraft equations of motion admit five known equilibrium points: three of them are unstable, i.e. the collinear Lagrangian points L_1, L_2, L_3. These points are on the x axis and for the lunar case they have coordinates $x_{L_1} = 0.837$, $x_{L_2} = 1.155$, $x_{L_3} = -1.005$. A first integral of motion exists, the Jacobi integral:

$$C = 2\Omega(x, y, z) - (u^2 + v^2 + w^2)$$

where Ω is the gravitational field potential. The region where the motion is allowed is referred to as Hill's region H_C, and it is identified by the inequality:

$$2\Omega(x, y, z) - C \geq 0$$

In Fig. 3.2 the boundary of the Hill region, the so called Zero Velocity

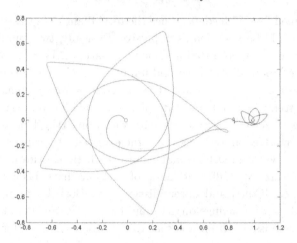

Fig. 3.1. An example of lunar inner mission

Curves (ZVC)

$$C = 2\Omega(x, y)$$

are shown for the planar case ($z = w = 0$) and for three different values
of the Jacobi constant C.

In the left figure, the Jacobi constant has value $C = 3.252$ and the
motion is allowed inside the oval surrounding the Earth, the oval surrounding the Moon or outside the external oval. Increasing the kinetic
energy (thus decreasing C) a small neck opens near the Lagrangian
point L_1 and an inner transit from Earth to Moon becomes feasible.
This occurs in the second picture where the ZVC corresponding to the
Jacobi constant $C = 3.192$ is shown. Such a value of C is less than the
Jacobi constant at the L_1 point: $C_{L_1} = C_1 = 3.200$ but greater than
the Jacobi constant at the L_2 point: $C_{L_2} = C_2 = 3.184$. With a higher
energy a passage through the Lagrangian point L_2 opens and the Moon
can be reached through an outer transfer orbit. This is shown in the third
picture where the ZVC for $C = 3.172 < C_2$ is portrayed. With reference
to three dimensional trajectories, Fig. 3.3 illustrates the Zero Velocity
Surfaces (ZVS) for three different values of the Jacobi constant and the
corresponding deformation of the region of feasible motion H_C as the
value C decreases. The homology groups of these regions for both the
spatial and planar restricted three-body problem have been computed
in [MM]. From the geometrical point of view the state space (position

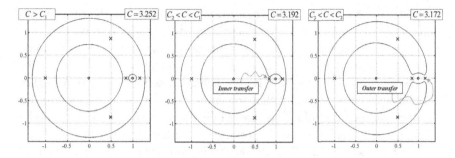

Fig. 3.2. Zero Velocity Curves for three different values of the Jacobi constant C

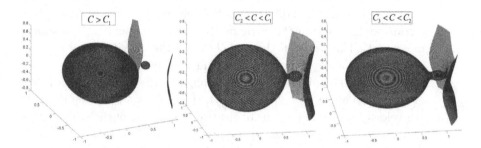

Fig. 3.3. Zero Velocity Surfaces for three different values of the Jacobi constant C

and velocity) of the planar restricted three-body problem is a (trivial) bundle

$$\pi : B \to H_C \qquad (3.1)$$

In the planar case the bundle B is an S^1 bundle, which is singular on the ZVC. Namely the fibers $\pi^{-1}(p)$ are the circles of velocity $u^2 + v^2 = 2\Omega(p) - C$ with vanishing radius at the ZVC. In the spatial case the fibers $\pi^{-1}(p)$ are spheres of velocity $u^2 + v^2 + w^2 = 2\Omega(p) - C$, so B is an S^2 fiber bundle singular at the ZVS.

For lunar capture the spacecraft has to possess an energy as small as possible while going from Earth to Moon. Hence a value of the Jacobi constant close to C_1 must be assumed. The trajectories around L_1 can be classified easily if the equation of motion linearized about L_1 is considered [Con1]. The general solution is

$$U = \alpha_1 e^{\alpha t} v_1 + \alpha_2 e^{-\alpha t} v_2 + 2Re(\beta e^{i\omega_{xy}(t)}) v_3 + 2Re(\gamma e^{i\omega_z(t)}) v_5 \qquad (3.2)$$

where v_1, v_2, v_3, $v_4 = conj(v_3)$, v_5, $v_6 = conj(v_5)$ are the eigenvectors and α, $-\alpha$, $\pm i\omega_{xy}$, $\pm i\omega_z$ the eigenvalues of the linear system.

According to formula (3.2) the orbits are classified as: i) Quasi periodic orbits, for $\alpha_1 = \alpha_2 = 0$; ii) Asymptotic orbits, for $\alpha_1 = 0$ or $\alpha_2 = 0$; iii) Transit orbits, for $\alpha_1\alpha_2 < 0$; iv) Bouncing orbits, for $\alpha_1\alpha_2 > 0$. A further analysis [Con1] points out that transit orbits project themselves into two strips on the (x, y) plane, as shown in Fig. 3.4. In this figure the periodic orbit ($\alpha_1 = \alpha_2 = \gamma = 0$) is portrayed together with the two strips S_1 and S_2 as well as the ZVC and one asymptotic orbit. Transit orbits from Earth to Moon must project themselves into the strip S_1 to the left of the L_1 point and into the strip S_2 to the right of L_1. There is a second condition that must be fulfilled in order to have a transit orbit: at any point p of the strip, the direction of the velocity vector must be inside a certain wedge angle in the circle $\pi^{-1}(p)$ of the S^1 bundle B (as shown in Fig. 3.4). That is, if θ is the angle between the velocity vector and the x axis ("firing angle"), a transit orbit must verify: $\theta_1 < \theta < \theta_2$, where θ_1 and θ_2 are the boundaries of the wedge angle. If the velocity vector is out of the wedge angle ($\theta < \theta_1$ or $\theta > \theta_2$), the orbit is a bouncing orbit. If the velocity vector lies on one of the two boundaries of the wedge angle ($\theta = \theta_1$ or $\theta = \theta_2$) one has an asymptotic orbit. The values θ_1 and θ_2 depend on the point p in the strip. In fact the width of the wedge angle decreases to zero as the point p approaches the boundary of the strip. Hence, at the boundary of the strip there is just one asymptotic solution and all the other solutions are bouncing orbits (no transit orbit). From a geometrical point of view the asymptotic solutions are then identified by two sections σ_1, σ_2 of the S^1 bundle B: $\sigma_i(p) = \theta_i$, $i = 1, 2$. These sections separate B in two regions: transit orbits are in one region and bouncing orbits are in the other.

In the spatial case the out-of-plane motion is a harmonic motion and there is a quasi periodic orbit (for the Earth–Moon system the in-plane frequency ω_{xy} is not commensurable with the out-of-plane frequency ω_z). In the (x, y, z) space, transit orbits belong to a corridor whose upper and lower boundaries are the ZVS, see Fig. 3.5. These orbits project into the strips S_1, S_2 in the (x, y) plane. The width of these strips and the width of the wedge angle of velocity depend also on the out-of-plane component of velocity. Namely, if δ is the angle between the velocity vector and the (x, y) plane, the width of the strips and the width of the wedge angle decrease with δ, and go to zero as $| \delta |$ tends to $\frac{\pi}{2}$. The above analysis provides a complete classification of the orbits in the linearized case.

Now, if the nonlinear dynamics is considered, the following questions

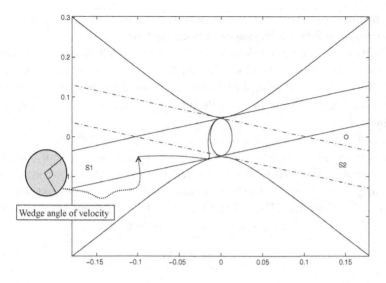

Fig. 3.4. The neck region about L_1, the periodic orbit and the two strips S_1, S_2

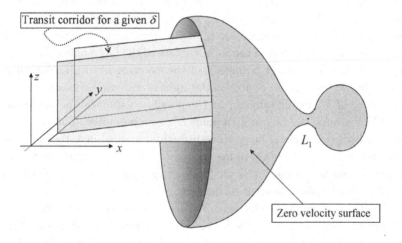

Fig. 3.5. 3 dimensional corridor of transit and asymptotic orbits about L_1

arise: does there still exist an invariant set (such as the quasi periodic orbit of the linearized dynamics)? Are there asymptotic orbits and do they still separate transit orbits from bouncing orbits? In particular how can capture orbits be determined?

The answer to the first two questions is given by Conley analysis. The starting point is the definition of an isolating block: this is a region whose boundary points do not belong to an invariant set. Then, if existent, the invariant set is inside the block. In fact the existence of an invariant set and the existence of asymptotic orbits are proved by looking at the dynamics at the boundary of the isolating block and applying homological arguments.

The technical definition of an isolating block is the following [Moe1]: it is a subbundle, $\pi : M(a,b) \to H(a,b)$, of the bundle B, where $H(a,b)$ is a subspace of H_C defined by the condition $a \leq \xi \leq b$. The boundaries $\xi = a$, $\xi = b$ are such that any orbit that reaches the boundaries tangentially will lie outside $H(a,b)$ both before and after the encounter. More specifically, the following conditions must be satisfied:

$$\xi = a, \quad \dot{\xi} = 0 \Rightarrow \ddot{\xi} = \Omega_x + 2v < 0 \tag{3.3}$$

$$\xi = b, \quad \dot{\xi} = 0 \Rightarrow \ddot{\xi} = \Omega_x + 2v > 0 \tag{3.4}$$

The conditions (3.3)–(3.4) imply that the points in the boundary of $M(a,b)$, $m(a,b) = \partial M(a,b)$, cannot belong to an invariant set. Figure 3.6 shows the possible values a, b for the boundaries of M for different values of the Jacobi constant C. For a specified C, any value in the shaded region on the left of the x-coordinate of the Lagrangian point L_1, x_{L_1}, can be taken as left boundary of the isolating block M. Similarly, any value on the shaded region on the right of x_{L_1} can be taken as right boundary of M. In the same figure the possible boundaries of the isolating block around the Lagrangian point L_2 are represented. The following interesting property can be proved through numerical investigation: *for each value of C in the range of the admissible values (i.e. those values for which a right boundary exists) the furthest right boundary of the isolating block is on the plane tangent to the zero velocity curve.*

Using the isolating blocks $M(a,b)$ it is possible to classify the orbits in the same way as it was done for the linearized case. Let $\boldsymbol{x}$ be the state variable and ϕ_t be the flow generated by the equations of motion. The subset

$$S = \{\boldsymbol{x} : \phi_t(\boldsymbol{x}) \in M(a,b), \forall t \in \Re\}$$

is the invariant set inside $M(a,b)$. The set S generalizes the periodic or quasi periodic solution of the linear case. The subsets

$$A^+ = \{\boldsymbol{x} : \phi_t(\boldsymbol{x}) \in M(a,b), \forall t > 0\}$$

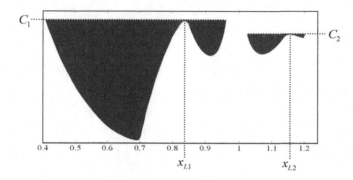

Fig. 3.6. Boundaries of the isolating blocks around L_1 and L_2

$$A^- = \{\boldsymbol{x} : \phi_t(\boldsymbol{x}) \in M(a,b), \forall t < 0\}$$

are the sets of the asymptotic incoming $(+)$ and outgoing $(-)$ orbits. The projections of A^+ and A^- onto the spatial coordinates generalize the strips S_1, S_2 of the linear case; it results in: $S = A^+ \bigcap A^-$. The orbits in the rest of $M(a,b)$ must leave the isolating block and they can be classified according to which boundary ($\xi = a$ or $\xi = b$) they cross when entering or leaving. Let U_{aa} and U_{bb} be the set of orbits which enter and leave $M(a,b)$ through the same boundary ($\xi = a$ and $\xi = b$ respectively). These orbits generalize the bouncing orbits of the linear case. The sets U_{aa} and U_{bb} are not empty, because near any orbit that meets the boundaries $\xi = a$ or $\xi = b$ tangentially there is an orbit that crosses the boundary and then immediately leaves through the same boundary. Finally, let U_{ab} (U_{ba}) be the set of orbits crossing the boundary $\xi = a$ ($\xi = b$) and leaving through the boundary $\xi = b$ ($\xi = a$). The orbits in such a set correspond to the transit orbits incoming along the strip S_1 and outgoing through the strip S_2 from the left to the right (from the right to the left). An analytic proof of the existence of such orbits is given in [Moe2]. Now, for any pair of values (a,b) let us consider the inclusion map:

$$m - a^+ \hookrightarrow m \tag{3.5}$$

where a^+ is the intersection of the asymptotic set A^+ with the boundary of M: $a^+ = A^+ \bigcap m$.

Note that the boundary $m = \partial M$ has two connected components, one on the left, the other on the right of L_1. In the linearized theory, each component of $m(a,b)$ is topologically equivalent to a cylinder that

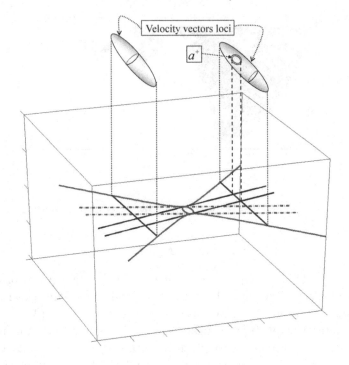

Fig. 3.7. Boundaries of the planar isolating blocks and related phase space

is singular at the extreme points of the segment $\xi = a$ (or $\xi = b$), see Fig. 3.7.

The set a^+ of asymptotic orbits with respect to the linear flow is given by two closed curves, one on each connected component of m.

The pair $(m, m - a^+)$ induces the following exact sequence of relative homology groups [Spa]:

$$\ldots \to H_1(m, m - a^+) \to^\partial H_0(m - a^+) \to^{i^*} H_0(m) \to 0 \qquad (3.6)$$

Take a 0-cycle $z = P_1 - P_0$ on m, where the two points P_0, P_1 are in the same connected component of m, for instance the left component of m. Then $z = 0$ on $H_0(m)$. For the existence of transit and bouncing orbits (in the nonlinear sense) P_0, P_1 can be chosen so that the flow ϕ leads P_0 back to the left component of m ($P_0 \in U_{aa}$) and P_1 to the right component of m ($P_1 \in U_{ab}$). That is, the flow generates a 0-cycle $\phi^*(z)$ that is nontrivial in $H_0(m)$. Due to exactness of the sequence (3.6) the group $H_1(m, m - a^+)$ is nontrivial [E]. The duality theorem [Spa] establishes an isomorphism between homology and Cech cohomology groups:

$$H_1(m, m - a^+) \simeq \breve{H}^{n-1}(a^+) \tag{3.7}$$

where $n = \dim(m)$. In the planar case $n = 2$, so the above isomorphism implies that the first Cech cohomology group of a^+ is nontrivial. In the spatial case $n = 4$, so $\breve{H}^3(a^+)$ is nontrivial. Applying the same arguments [E] to the inclusion maps

$$m \hookrightarrow M - S \hookrightarrow M \tag{3.8}$$

and to the long exact sequence of relative homology groups:

$$\dots H_2(M, M - S) \to^{\partial} H_1(M - S, m) \to^{i^*} H_1(M, m) \dots \tag{3.9}$$

it follows that $H_2(M, M - S)$ is nontrivial and (duality theorem)

$$H_2(M, M - S) \simeq \breve{H}^{N-2}(S)$$

where $N = \dim(M)$. In the planar case $(n = 2)$ $\breve{H}^1(S)$ is nontrivial and in the spatial case $(n = 4)$ $\breve{H}^3(S)$ is nontrivial. That is, the invariant set S and the asymptotic solutions a^+ exist. They have the topological character of a circle in the planar case (note that for the linear case the invariant set is a circle), and of a three sphere in the spatial case.

The above nonlinear analysis shows that invariant and asymptotic sets exist in the nonlinear sense and a classification of orbits is determined according to their behavior on the isolating block. Moreover it turns out that the manifold of asymptotic orbits A^+ separates the isolating block $M(a, b)$ and this holds for any pair (a, b) defining the isolating block. The transit orbits are located at one side with respect to A^+. Numerical evidence shows that the capture orbits of low periselenium are transit orbits which are located close to the asymptotic manifold A^+. Figure 3.8 shows two trajectories starting from the same initial state. The first trajectory follows the linearized flow and is asymptotic to the periodic solution. The second trajectory follows the nonlinear flow and is a capture orbit.

Figure 3.9 illustrates the perigees and periselenia of orbits obtained through backward and forward (numerical) propagation, starting from specified initial conditions on ξ and η. One hundred equally-spaced values are taken on the $\xi = -0.01$ segment belonging to the S_1 strip. The energy h was set to 0.001. Numerical propagations were performed for different firing angles: $\theta = s\theta_1 + (1 - s)\theta_2$; $(s = 0.1, \dots, 0.9)$. Note that lower altitude periselenia and perigees are achieved for values of θ close to θ_1 (see the case $s = 0.9$ in Fig. 3.9).

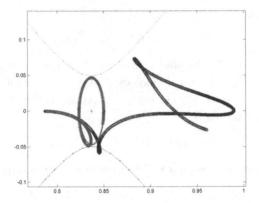

Fig. 3.8. Linear and nonlinear planar flows

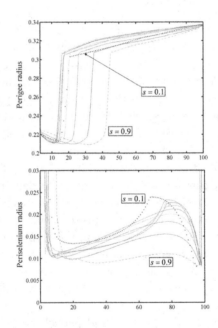

Fig. 3.9. Perigees and periselenia for different conditions at the boundary of the isolating block

Similarly, Fig. 3.10 shows a three dimensional capture orbit (nonlinear flow) together with an orbit asymptotic to the quasi periodic solution about L_2 (linearized flow). Both orbits start from the same initial state.

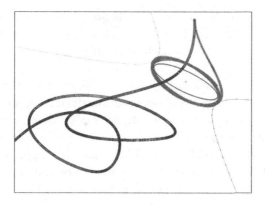

Fig. 3.10. Linear and nonlinear spatial flows about the L_2 point

4 Final comments

In his valuable paper [Con1], Conley wrote *"a scheme for designing periodic Earth–Moon orbits is outlined with the idea in mind of satisfying the following three criteria: i) cost for cycle should be as small as is practical, ii) control and stability problems should be as easy as possible, iii) as much flexibility should be built into the scheme as possible. Unfortunately orbits such as these require a long time to complete a cycle. On the other hand one cannot predict how knowledge will be applied, only that it often is"*.

After forty years the Conley classification of orbits around Lagrangian points is an important tool in the mission analysis of lunar and interplanetary transfers. Transfer orbits with ballistic capture can be generated through the backward or forward propagation of specific initial conditions close to the manifold of asymptotic orbits. Of course for a real mission several perturbations must be taken into account, in particular the eccentricity of the Moon orbit in the context of lunar transfer orbits. This requires some modifications [CT2], however the central idea remains unchanged. The same analysis could be applied to outer missions too, where the L_2 region has to be considered, taking into account the gravitational perturbation of the Sun at Moon arrival [CT1].

In many space missions the propulsion unit has to operate for a significant part of the trajectory, so the trajectory optimization can be formulated as an optimum control problem with continuous thrust. In the search of the global optimal solution, numerical algorithms can be supported by Morse inequalities applied to the infinite dimensional space of control functions **N**. This is a bundle over the manifold N of the final conditions. The difficulty in computing homology groups of the infinite dimensional manifold **N** can be circumvented when the homology groups of **N** coincide with the homology groups of the base space N. For instance this occurs when the dynamics is linear as in the rendez-vous problem. In more general problems the nontrivial topology of the target space implies the existence of a multiplicity of solutions, but an estimate on the number of critical solutions can be given only in particular cases.

In conclusion, the application of the topological approach to the optimization of space trajectories represents a promising field of research, albeit further theoretical developments are needed to obtain improved results in aerospace mission analysis.

Paolo Teofilatto's Acknowledgement. *I am indebted to Bill Harvey for his advice and all his help during my PhD in Mathematics at the King's College London. During my current activity I feel that his influence and his enthusiasm for geometry are still driving me.*

References

[AV] Agrachev, A. A. and Vakhrammev, S. A. Morse theory and optimal control problems, in *Progress in System Control Theory*, Birkhauser, Boston, 1991, pp. 1–11.

[B] Belbruno, E. *Capture Dynamics and Chaotic Motions in Celestial Mechanics*, Wiley, Chichester, 2004.

[Ces] Cesari, L. *Optimization – Theory and Applications*, Springer Verlag, 1983.

[CT1] Circi, C. and Teofilatto, P. On the dynamics of weak stability boundary lunar transfers. *Celestial Mechanics and Dynamical Astronomy* **79** (2001), 41–72.

[CT2] Circi, C. and Teofilatto, P. Effect of planetary eccentricity on ballistic capture in the solar system. *Celestial Mechanics and Dynamical Astronomy* **93** (2005), 69–86.

[Con1] Conley, C. Low energy orbits in the restricted three body problem. *SIAM Journal of Applied Mathematics* **16** (1968), 732–746.

[Con2] Conley, C. *Isolated Invariant Sets and the Morse Index*, Conference Board in Mathematical Science, **38**, AMS, Providence, 1978.

[E] Easton, R. Existence of invariant sets inside a manifold convex to the flow. *Journal of Differential Equations* **7** (1970), 54–68.

[MM] McCord, C. and Meyer, K. Integral manifolds of the restricted three body problem. *Ergod. Th. and Dynam. Sys.* **21** (2001), 885–914.

[Moe1] Moeckel, R. Isolating blocks near the collinear relative equilibria of the three body problem. *Trans. Am. Math. Soc.* **356** (2004), 4395–4425.

[Moe2] Moeckel, R. A variational proof of existence of transit orbits in the restricted three body problem. *Dynamical System* **20** (2005), 45–58.

[Mil] Milnor, J. *Morse Theory*, Princeton University Press, Princeton, 1969.

[Mor1] Morse, M. Relations between the critical points of a real function of n independent variables. *Transactions of the American Mathematical Society* **27** (1925), 345–396.

[Mor2] Morse, M. *The Calculus of Variations in the Large*, American Mathematical Society Press, New York, 1934.

[N] Neustadt, L. Minimum effort control systems. *SIAM Journal of Control* **1** (1962), 16–31.

[PS] Palis, R. and Smale, S. A generalized Morse theory. *Bulletin of the American Mathematical Society* **70** (1964), 165–172.

[Spa] Spanier, E. H. *Algebraic Topology*, McGraw-Hill, New York, 1966.

[Sze] Szebehely, V. *Theory of Orbits*, Academic Press, New York, 1967.

[TP] Teofilatto, P. and Pontani, M. "Numerical and Analytical Methods for Global Optimization", in *Variational Analysis and Aerospace Engineering*, edited by A. Frediani, Springer, New York, 2009, 461–475.

[V1] Vakhrammev, S. A. Hilbert manifolds with corners of finite codimension and the theory of optimal control. *Journal of Mathematical Sciences* **53** (1991), 176–223.

[V2] Vakhrammev, S. A. Morse theory and the Lyusternik–Shnirelman theory in geometric control theory. *Journal of Mathematical Sciences* **71** (1994), 2434–2485.

Actions of fractional Dehn twists on moduli spaces

Robert Silhol

Université Montpellier II, Département de Mathématiques
rs@math.univ-montp2.fr

Abstract

We attempt to give a unified treatment of certain hyperbolic transformations, "fractional Dehn twists", that induce "algebraic actions" on certain subspaces of moduli spaces.

This is done by considering the Teichmüller group $\text{Mod}_{0,\{n\}}$ of the sphere with n *unordered* marked points and an action of this group on $\mathcal{M}_{0,n}$ the moduli space of the sphere with n *ordered* marked points. We show that for covers of the sphere ramified over n points the above action lifts to the action of fractional Dehn twists.

1 Introduction

This paper is an attempt to give a unified treatment of scattered results accumulated over the years on "algebraic actions" of fractional Dehn twists, where by algebraic actions we mean that they induce algebraic transformations on moduli.

The first evidence of the existence of such actions is probably in [BuSi] where it was in fact not realized that they were fractional Dehn twists. They were only identified as such later by A. Aigon in [Ai].

Other examples, in genus 2 and 3, were found in [AiSi] and [Si1]. All these examples have in common the fact that they concerned families of surfaces with non-trivial automorphisms, either hyperelliptic surfaces with an extra non-hyperelliptic automorphism or non-hyperelliptic surfaces with a non-trivial automorphism.

The first example of a family of surfaces without automorphisms exhibiting algebraic actions of fractional Dehn twists can be found in [Si2] (see also [LeSi] for other examples and a more systematic account).

The point of view taken here is to relate all these examples to a known case, albeit not generally recognized as such, the case of the sphere with n punctures on which half-twists act as permutations of the punctures.

Consider a family of surfaces with a ramified covering map of the Riemann sphere of fixed topological type. If we have n branch points we can, in some cases, lift the action of $\mathrm{Mod}_{0,\{n\}}$ to the family. Since the above examples are all of this type this approach allows to bring these under a unified scheme and explains some of the properties they have in common, notably the actions of the permutation groups.

One should however be aware that this is probably not the complete story since [Si2] and [LeSi] give examples of fractional Dehn twists that exchange families and these are not accounted for in this context.

The project for this paper grew out of discussions with Bill Harvey, notably after an illuminating talk he gave on the moduli spaces of spheres with punctures, and I wish to thank him here.

2 The groups $\mathrm{Mod}_{0,n}$ and $\mathrm{Mod}_{0,\{n\}}$

We will recall here the well-known description of the Teichmüller groups for genus 0 surfaces with n marked points. We will need to distinguish the two cases of ordered or unordered marked points.

- We denote by $\mathrm{Mod}_{0,n}$ the Teichmüller group for genus 0 surfaces with n *ordered* marked points.
- We denote by $\mathrm{Mod}_{0,\{n\}}$ the Teichmüller group for genus 0 surfaces with n *unordered* marked points.

For $n \geqslant 5$ we have the classical presentation given by J. Birman as quotients of braid groups and pure braid groups [Bir].

Namely $\mathrm{Mod}_{0,\{n\}}$ is generated by $n-1$ elements

$$\sigma_1, \ldots, \sigma_{n-1}$$

with relations

$$\sigma_i \sigma_j = \sigma_j \sigma_i \quad \text{if } |j - i| \geqslant 2$$
$$\sigma_i \sigma_{i+1} \sigma_i = \sigma_{i+1} \sigma_i \sigma_{i+1}$$
$$\sigma_1 \cdots \sigma_{n-2} \sigma_{n-1}^2 \sigma_{n-2} \cdots \sigma_1 = 1$$
$$(\sigma_1 \cdots \sigma_{n-1})^n = 1. \qquad (2.1)$$

The group $\mathrm{Mod}_{0,n}$ is then identified with the subgroup of $\mathrm{Mod}_{0,\{n\}}$ generated by the squares $\sigma_i{}^2$.

We also note for further use that the quotient $\mathrm{Mod}_{0,\{n\}} / \mathrm{Mod}_{0,n}$ is isomorphic to the symmetric group $\mathfrak{S}_n$.

The σ_i of (2.1) have an easy geometric description. Label the marked points on the sphere $p_1, \ldots, p_n$. Let γ_i be the simple closed geodesic separating the points $\{p_i, p_{i+1}\}$ from the rest of the sphere, $1 \leqslant i \leqslant n-1$. Then σ_i can be identified with a half Dehn twist along γ_i (turning counter-clockwise when seen from the two cusps so that the square is a left twist). Note that for the sphere with n unordered marked points these half-twists are homeomorphisms.

For $n = 4$ the situation is a little different (see for example S. Nag [Nag], p. 129). Considering the geometric description above we immediately note that $\gamma_3 = \gamma_1$ and hence $\sigma_3 = \sigma_1$.

Plugging this in the relations (2.1) one finds that they reduce to

$$\sigma_1\sigma_2\sigma_1 = \sigma_2\sigma_1\sigma_2 \quad \text{is of order } 2$$

$$\sigma_1\sigma_2 \quad \text{and} \quad \sigma_2\sigma_1 \quad \text{are of order } 3.$$

From this we conclude that $\mathrm{Mod}_{0,\{4\}}$ is isomorphic to $\mathrm{PSL}_2(\mathbb{Z})$ with σ_1 and σ_2 identified to

$$A = \begin{pmatrix} 1 & 1 \\ 0 & 1 \end{pmatrix} \quad \text{and} \quad B = \begin{pmatrix} 1 & 0 \\ -1 & 1 \end{pmatrix}$$

respectively.

This also implies that $\mathrm{Mod}_{0,4}$ is isomorphic to the congruence subgroup $\Gamma(2)$ or rather its image in $\mathrm{PSL}_2(\mathbb{Z})$. Hence the quotient of the two groups $\mathrm{Mod}_{0,\{4\}} / \mathrm{Mod}_{0,4}$ is the symmetric group $\mathfrak{S}_3$.

We end this section by recalling that the moduli space $\mathcal{M}_{0,n}$ of genus 0 surfaces with n *ordered* marked points is isomorphic to

$$\left(\mathbb{P}^1(\mathbb{C}) \smallsetminus \{\infty, 0, 1\} \right)^{n-3} \smallsetminus \bigcup \{\text{diagonals}\}$$

where a point $(x_1, \ldots, x_{n-3})$ is identified to the Riemann sphere $\widehat{\mathbb{C}}$ with marked points $(\infty, 0, 1, x_1, \ldots, x_{n-3})$. The action of the symmetric group on $\mathcal{M}_{0,n}$ is generated by

$$f_1 : (x_1, x_2, \ldots, x_{n-3}) \mapsto \left(\frac{x_{n-3}}{x_{n-3}-1}, \frac{x_{n-3}}{x_{n-3}-x_1}, \ldots, \frac{x_{n-3}}{x_{n-3}-x_{n-4}} \right)$$

$$f_2 : (x_1, x_2, \ldots, x_{n-3}) \mapsto \left(\frac{1}{x_1}, \frac{1}{x_2}, \ldots, \frac{1}{x_{n-3}} \right) \tag{2.2}$$

the first corresponding to the circular permutation, the second to the permutation of the first two elements, at least for $n > 4$. For $n = 4$, f_1 is of order 2 and the group generated by f_1 and f_2 is isomorphic to the symmetric group $\mathfrak{S}_3$.

With this description of $\mathcal{M}_{0,n}$, the moduli space $\mathcal{M}_{0,\{n\}}$ of genus 0 surfaces with n *unordered* points is the quotient of $\mathcal{M}_{0,n}$ by the group generated by f_1 and f_2.

3 Actions of symmetric groups on certain moduli spaces

Let S be a genus g hyperelliptic surface with an additional non-hyperelliptic involution. Such a surface always has an equation of the form

$$y^2 = (x^2 - 1)\prod_{i=1}^{g}(x^2 - a_i), \quad \text{the } a_i \text{ distinct and } \neq 0 \text{ and } 1. \quad (3.1)$$

The additional involution is simply induced by $x \mapsto -x$. If g is even, then the two involutions $\varphi_1 : (x,y) \mapsto (-x,y)$ and $\varphi_2 : (x,y) \mapsto (-x,-y)$ both have two fixed points, the two points above $x = 0$ for φ_1 and the two points at infinity for φ_2. If g is odd, then φ_1 has four fixed points, the two points above $x = 0$ and the two points at infinity, while φ_2 is fixed point free.

Call h the hyperelliptic involution. Since h commutes with all other automorphisms the group G generated by φ_1 and h is isomorphic to $\mathbb{Z}/2 \times \mathbb{Z}/2$. Moreover by the above remarks the quotient S/G is the sphere $\widehat{\mathbb{C}}$ with $g + 3$ marked points $\infty, 0, 1, a_1, \ldots, a_g$.

We will be interested in the pairs (S, G), where S and G are as above. In particular we will say that (S, G) and (S', G') are isomorphic if there is an isomorphism $f : S \to S'$ such that $f^{-1}G'f = G$. Denote by $\mathcal{M}_g^h(G)$ the moduli space of such pairs.

With the description of $\mathcal{M}_{0,n}$ given in Section 2 we have an obvious map

$$\mathcal{M}_{0,g+3} \to \mathcal{M}_g^h(G) \quad (3.2)$$

induced by $(a_1, \ldots, a_g) \mapsto$ the curve defined by (3.1).

Two algebraic curves C_1 and C_2 with respective equations $y^2 = (x^2 - 1)(x^2 - a_1)\cdots(x^2 - a_g)$ and $y^2 = (x^2 - 1)(x^2 - b_1)\cdots(x^2 - b_g)$ are isomorphic if and only if there is a Möbius transformation sending the set $\{\pm 1, \pm\sqrt{a_1}, \ldots, \pm\sqrt{a_g}\}$ onto $\{\pm 1, \pm\sqrt{b_1}, \ldots, \pm\sqrt{b_g}\}$. For C_1 and C_2, both with the non-hyperelliptic involutions induced by $x \mapsto -x$, to define the same element in $\mathcal{M}_g^h(G)$ the Möbius transformation must fix globally $\{\infty, 0\}$. Hence be of the form $z \mapsto z/\alpha$ or $z \mapsto \alpha/z$. Finally since 1 must be among the images of $\{\pm 1, \pm\sqrt{a_1}, \ldots, \pm\sqrt{a_g}\}$, the only possible choices are $\alpha = \pm 1$ or $\alpha = \pm\sqrt{a_i}$ for some i, $1 \leqslant i \leqslant g$. In conclusion we have

Lemma 3.1 *Two algebraic curves C_1 and C_2 with respective equations $y^2 = (x^2-1)(x^2-a_1)\cdots(x^2-a_g)$ and $y^2 = (x^2-1)(x^2-b_1)\cdots(x^2-b_g)$, both with the non-hyperelliptic involutions induced by $x \mapsto -x$, define the same point in $\mathcal{M}_g^h(G)$ if and only if the set $\{1, b_1, \ldots, b_g\}$ is the image of $\{1, a_1, \ldots, a_g\}$ under a map of the form $z \mapsto z/\alpha$ or $z \mapsto \alpha/z$ with $\alpha = 1$ or $\alpha = a_i$ for some i, $1 \leqslant i \leqslant g$.*

The important consequence of this is

Corollary 3.2 *The canonical map $\mathcal{M}_{0,g+3} \to \mathcal{M}_{0,\{g+3\}}$ factors through the map $\mathcal{M}_{0,g+3} \to \mathcal{M}_g^h(G)$ of (3.2).*

PROOF. Express the permutations of $\{1, a_1, \ldots, a_g\}$ on the model of (2.2) e.g. permuting 1 and a_1 is $\{1, a_1, \ldots, a_g\} \mapsto \{1, 1/a_1, \ldots, a_g/a_1\}$. Then the Corollary follows from the fact that the transformations described in Lemma 3.1 together with permutations of $\{1, a_1, \ldots, a_g\}$ form a subgroup of the group generated by f_1 and f_2 of (2.2). This subgroup is easily seen to be isomorphic to $\mathfrak{S}_2 \times \mathfrak{S}_{g+1}$.

From Corollary 3.2 the space $\mathcal{M}_g^h(G)$ inherits an action of the symmetric group $\mathfrak{S}_{g+3}$. This action is not faithful of course but we want nevertheless to consider it as an action of $\mathfrak{S}_{g+3}$. In the next section we proceed to give a geometric interpretation of this action.

4 Half Dehn twist actions

Let, as in Section 3, S be a genus g hyperelliptic surface with an additional non-hyperelliptic involution. Let S be defined by the equation $y^2 = (x^2 - a_1)\cdots(x^2 - a_{g+1})$, where we write a_1 for 1 to unify notations in the sequel. Let also h be the hyperelliptic involution, $\varphi_1 : (x,y) \mapsto (-x,y)$ and $G = \langle h, \varphi_1 \rangle$.

Let $S_0 = S/G$ be the quotient. This surface is a sphere with $g + 3$ elliptic points of order 2 at the points ∞, 0 and $a_1, \ldots, a_{g+1}$. Let h_∞, h_0 and $h_1, \ldots, h_{g+1}$ be order 2 elliptics in $\mathrm{PSL}_2(\mathbb{R})$, satisfying $h_\infty h_0 h_1 \cdots h_{g+1} = 1$, that generate a Fuchsian group Γ_0 for S_0, so that

$S_0 = \mathcal{H}/\Gamma_0$. Let p_∞, p_0 and $p_1, \ldots, p_{g+1}$ be the fixed points in $\mathcal{H}$ of the h_i. We assume that these p_i project in S_0 to ∞, 0 and the a_i respectively.

In this context the group Γ generated by the products $h_1 h_j$ and $h_0 h_1 h_j h_0$, $2 \leqslant j \leqslant g+1$ is a Fuchsian group for S. We will denote by w_i^+ the Weierstrass point in S image of p_i, $1 \leqslant i \leqslant g+1$, by w_i^- the image of $h_0(p_i)$ and finally by $q_0^\pm$ (resp. $q_\infty^\pm$) the images of q_0 and $h_1(q_0)$ (resp. q_∞ and $h_1(q_\infty)$).

A half-twist permuting a_j and a_k in S_0 can be written, in terms of the elliptic transformations above, as replacing h_k by $h_j = (h_j h_k) h_k$ and h_j by $h_j h_k h_j = (h_j h_k) h_j$, the other h_i remaining unchanged.

Transposing this to the generators of Γ we obtain different actions depending on the indices j and k. If they are both different from ∞, 0 and 1, the induced transformations are

$$h_1 h_k \mapsto h_1 h_j = (h_1 h_j h_k h_1) h_1 h_k$$
$$h_1 h_j \mapsto h_1 h_j h_k h_j = (h_1 h_j h_k h_1) h_1 h_j$$
$$h_0 h_1 h_k h_0 \mapsto h_0 h_1 h_j h_0 = (h_0 h_1 h_j h_k h_1 h_0) h_0 h_1 h_k h_0$$
$$h_0 h_1 h_j h_0 \mapsto h_0 h_1 h_j h_k h_j h_0 = (h_0 h_1 h_j h_k h_1 h_0) h_0 h_1 h_j h_0 \quad (4.1)$$

the other generators remaining unchanged. This is clearly the composition of two full Dehn twists along the disjoint simple closed geodesics images of the axes of $h_1 h_j h_k h_1$ and $h_0 h_1 h_j h_k h_1 h_0$. Since the images in S of $h_1(p_j)$ and $h_1(p_k)$ are the Weierstrass points w_j^+ and w_k^+ the first is a simple closed geodesic through these two points and, since h_0 induces φ_1 in S, the second is the image of the first under φ_1.

If one of the indices is equal to 1, for example if $h_1 \mapsto h_j$ and $h_j \mapsto h_j h_1 h_j$, the transformations are

$$h_1 h_j \mapsto h_j h_j h_1 h_j = h_1 h_j$$
$$h_1 h_k \mapsto h_j h_k = (h_j h_1) h_1 h_k \quad \text{for } k \neq j$$
$$h_0 h_1 h_j h_0 \mapsto h_0 h_1 h_j h_0$$
$$h_0 h_1 h_k h_0 \mapsto h_0 h_j h_k h_0 = (h_0 h_j h_1 h_0) h_0 h_1 h_k h_0 \quad \text{for } k \neq j \quad (4.2)$$

and again this is the composition of two full Dehn twists along a pair of disjoint simple closed geodesics, one through w_1^+ and w_j^+ the other through w_1^- and w_j^-.

If the two indices are 0 and ∞ we note the following. A simple geodesic arc in S_0 with endpoints 0 and ∞ and not passing through any of the a_i lifts in S to a separating simple closed geodesic if the genus g is even or to a pair of simple closed geodesics the union of which is separating if the genus is odd. Hence if the genus is even the half-twist around 0 and ∞ lifts either to a half-twist along a separating curve, which is a homeomorphism, or to two full Dehn twists if the genus is odd.

The most interesting case is of course the case when only one of the indices is 0 or ∞. We assume it is ∞. If it is 0 the arguments are exactly the same, we only need to replace the generators $h_0 h_1 h_k h_0$ by $h_\infty h_1 h_k h_\infty$. For simplicity we also assume $k \neq 1$ (passing from this to the case $k = 1$, is similar to passing from (4.1) to (4.2)).

To understand better this case we need to take a closer look at the geodesic image $\gamma_{\infty,k}$ in S of the axis $A_{\infty,k}$ of $h_\infty h_k$. The axis contains of course p_∞ and p_k but also $h_k(p_\infty)$, $h_\infty(p_k)$ and $h_\infty h_k h_\infty(p_\infty)$. In S these points get mapped to q_∞^+, w_k^+, q_∞^-, w_k^- and q_∞^-. Moreover p_k is the hyperbolic midpoint of p_∞ and $h_k(p_\infty)$, p_∞ is the midpoint of h_k and $h_\infty(p_k)$, $h_\infty(p_k)$ is the midpoint of p_∞ and $h_\infty h_k h_\infty(p_\infty)$.

Now h_k induces the hyperelliptic involution and h_∞ the involution φ_2, hence $\gamma_{\infty,k}$, the image $A_{\infty,k}$, is a simple closed geodesic through q_∞^+, w_k^+, q_∞^- and w_k^-, globally stable under the action of G.

Finally since any order 2 elliptic centered at a point in the orbit of p_j, $1 \leqslant j \leqslant g + 1$, also induces the hyperelliptic involution, the above arguments show that $\gamma_{\infty,k}$ cannot contain any of the Weierstrass points $w_j^\pm$ for $j \neq k$. The same argument with φ_1 in place of h shows that it cannot pass through $q_\infty^\pm$.

With these notations fixed, the transformations $h_k \mapsto h_\infty$ and $h_\infty \mapsto h_\infty h_k h_\infty$, act on the generators of Γ by

$$h_1 h_k \mapsto h_1 h_\infty = (h_1 h_\infty h_k h_1) h_1 h_k$$
$$h_0 h_1 h_k h_0 \mapsto h_0 h_1 h_\infty h_0 = (h_0 h_1 h_\infty h_k h_1 h_0) h_0 h_1 h_k h_0 \qquad (4.3)$$

leaving the other generators unchanged.

Since h_1 induces the hyperelliptic involution on S and h_0 induces φ_1, the image of the axes of both $h_1 h_\infty h_k h_1$ and $h_0 h_1 h_\infty h_k h_1 h_0$ is $\gamma_{\infty,k}$. On the other hand both transform a preimage of w_k^+ (resp. q_∞^+) into a preimage of w_k^- (resp. q_∞^-), so they do not induce a homeomorphism in general, but correspond to applying a half Dehn twist to $\gamma_{\infty,k}$. Hence we have obtained a new surface S' with an action of the group G.

But by construction the quotient S'/G is S_0 and since we have lifted the half-twist exchanging ∞ and a_k we conclude that S' has equation

$y^2 = (x^2 - a_1') \cdots (x^2 - a_{g+1}')$ where

$$a_k' = 1 - a_k, \quad a_j' = \frac{a_j(1 - a_k)}{a_j - a_k} \quad \text{if } j \neq k . \tag{4.4}$$

To summarize our results let S be a hyperelliptic genus g surface with a non-hyperelliptic involution φ_1. Call G the group generated by the hyperelliptic and non-hyperelliptic involutions. From a topological point of view G is a group generated by a pair of commuting involutions h and φ_1, h with $2g + 2$ fixed points, φ_1 with 2, if the genus is even, or 4 if the genus is odd.

This fixed we consider the Teichmüller space $T_g^h(G)$ of hyperelliptic genus g surfaces with a non-hyperelliptic involution.

Proposition 4.1 *The modular group* $\mathrm{Mod}_{0,\{g+3\}}$ *acts on* $T_g^h(G)$.

More precisely if S is a hyperelliptic genus g surface with a non-hyperelliptic involution φ_1 with $G = \langle h, \varphi_1 \rangle$, h the hyperelliptic involution, then

(i) *If σ is a half-twist on $S_0 = S/G$ exchanging the images of two Weierstrass points, then σ lifts to the composition of two full Dehn twists along a pair of disjoint simple closed geodesics, exchanged by φ_1.*

(ii) *If σ is a half-twist on S_0 exchanging the image of the fixed points of φ_1 and the image of those of $\varphi_2 = h \circ \varphi_1$, then σ lifts to a half-twist along a separating geodesic, if the genus is even, or the composition of two full Dehn twists, if the genus is odd.*

(iii) *If σ is a half-twist on S_0 exchanging the image p of a Weierstrass point and the image q of fixed points of φ_1 (or φ_2), then σ transforms S into a surface S' obtained from S by applying a half-twist along a simple closed geodesic passing through the preimages of p and q.*

The induced action on the moduli space $\mathcal{M}_g^h(G)$ is the action of the symmetric group $\mathfrak{S}_{g+3}$ described in Section 3.

Remark. The methods developed in this section easily generalize to the case of surfaces of genus g admitting a quotient of genus 0 with $g + 3$ elliptic points of order 2. The difficulty is however to recover the induced action on the corresponding moduli space. For the family of genus 3 surfaces with two commuting non-hyperelliptic involutions this is done in [AiSi].

5 Surfaces defined by rectangles

In this section we summarize some of the results of [LeSi].

Let R be a Euclidean rectangle in the complex plane. To simplify we assume the edges of R are horizontal or vertical. Assemble r copies $R_1, \ldots, R_r$ of R by pasting along sides of same length so that not only the resulting polygon is simply connected but it also remains simply connected when one removes the vertices of the rectangles. We will say that such an arrangement of rectangles is an *admissible arrangement*. This restriction on the arrangement is here for purely technical reasons and could be in fact dispensed with.

Identify the remaining edges by pairs, using translations or rotations of angle π (half-turns), in a way that is compatible with the orientation of the rectangles, i.e. the identifications by translation can be top edge to bottom edge and right edge to left edge, and the identifications by half-turns can be top edge to top edge, left edge to left edge, right edge to right edge and bottom edge to bottom edge. We will call such surfaces *surfaces obtained from rectangles*.

Let S be obtained in this way. The differential dz^2 being invariant under the identifications, S comes naturally equipped with a quadratic differential q. In general this could be a meromorphic quadratic differential, but since, except in the case of the quotient spheres, we will not identify edges to themselves or adjacent edges, the differential will in fact be holomorphic and its zeros will necessarily be located at images of vertices of the rectangles.

We will need another technical restriction,

Definition 5.1 *Let S be a surface obtained from rectangles and q the quadratic differential induced by dz^2. We will say that (S, q), or simply S, is balanced if all vertices correspond to zeros of q of the same order.*

Our next aim is to replace the natural locally flat metric on surfaces obtained from rectangles by a hyperbolic metric. To do this we start with a more general construction that will be used later.

Let G_1 be a discrete subgroup of $\mathrm{Aut}(\mathbb{D}) \cong \mathrm{PSU}(1,1)$ of genus 1 and generated by two hyperbolic elements A and B with elliptic commutator $[A, B]$ of order n. In other words the signature of the group is $(1; n)$.

The quotient $S_1 = \mathbb{D}/G_1$ is a hyperbolic genus 1 surface with a cone point of total angle $2\pi/n$. The surface has also a distinguished homology basis (α, β) given by the images of the (oriented) axes of the transformations A and B. Replacing B by B^{-1} if necessary we may assume that this is a canonical basis. In this situation there is a unique τ such that we have a conformal equivalence from S_1 to $\mathcal{C}/\Lambda$, where Λ is the lattice generated by 2 and 2τ, and the image of $[-1, 1]$ (resp. $[-\tau, \tau]$) under the canonical projection $\pi_2 : \mathcal{C} \to \mathcal{C}/\Lambda = S_1$ is in the same homology class as α (resp. β). The conditions define the equivalence up to a translation in $\mathcal{C}/\Lambda$. We make it unique by requiring that the intersection of the axes maps to $\pi_2(0)$ and write $S_1 = \mathcal{C}/\Lambda$.

Call π_1 the covering map $\mathbb{D} \to S_1$. Since $\pi_2 : \mathcal{C} \to S_1$ is the universal cover of S_1 and $\mathbb{D}$ is simply connected, π_1 lifts to a map $\varphi : \mathbb{D} \to \mathcal{C}$, and this lifting is unique if we impose $\varphi(0) = 0$.

If the axes of A and B are orthogonal we may always assume that up to conjugation the axis of A is the real axis and the axis of B is the pure imaginary axis. The reflections with respect to the real axis and the pure imaginary axis induce anti-holomorphic involutions on S_1 or in other words define real structures on S_1. See left part of figure 1 where a fundamental domain for G_1 is represented. We will call such a domain an *equiquadrangle*.

These real structures obviously have two real components, one is the image of the real (resp. pure imaginary) axis, the other is the image of two identified opposite sides of the fundamental domain. Looking at the lattice $\Lambda = \langle 2, 2\tau \rangle$ introduced above, this means that τ is pure imaginary, or in other words that the natural fundamental domain for Λ is a rectangle. By the uniqueness of the map φ, this map commutes with complex conjugation on both sides, i.e. $\overline{\varphi(\bar{z})} = \varphi(z)$ and hence maps $\mathbb{D} \cap \mathbb{R}$ to $\mathbb{R}$ and $\mathbb{D} \cap i\mathbb{R}$ to $i\mathbb{R}$.

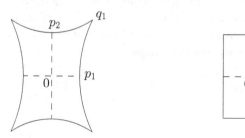

Fig. 1

Lemma 5.2 *Let G_1 be a discrete subgroup of $\mathrm{Aut}(\mathbb{D})$ of signature $(1;n)$ and generated by two hyperbolic elements A and B with respective axes the real axis and the pure imaginary axis. Let A_1 (resp. B_1) be the unique hyperbolic element such that $A_1^2 = A$ (resp. $B_1^2 = B$).*

Let c_m be the image of the pure imaginary axis under A_1^m and d_m the image of the real axis under $B_1{}^m$. Let $S_1 = \mathbb{D}/G_1$ and π_1 the natural projection $\mathbb{D} \to S_1$. Then for all m, $\pi_1(c_m)$ and $\pi_1(d_m)$ are geodesic arcs in S_1 for both the hyperbolic metric induced by that of $\mathbb{D}$ and for the natural conformal flat metric on the torus S_1.

PROOF: The real structure on S_1 induced by complex conjugation in $\mathbb{D}$ has two real connected components $\pi_1(\mathbb{R} \cap \mathbb{D})$ and $\pi_1(d_1)$. Since their union is the fixed locus of an anti-conformal reflection they are geodesic for any metric compatible with the conformal structure and for which the reflection is anti-conformal. The same is true for $\pi_1(i\mathbb{R} \cap \mathbb{D})$ and $\pi_1(c_1)$. Since the c_m and d_m are just the images of c_0, c_1, d_0 and d_1 under G_1 we are done.

Corollary 5.3 *Let S_1 be as in Lemma 5.2 and let $\varphi : \mathbb{D} \to C$ be the lift of the projection map $\pi_1 : \mathbb{D} \to S_1$ normalized as above. Let G be the group generated by A_1 and B_1 — see Lemma 5.2. Then the images of $\mathbb{D} \cap \mathbb{R}$ and $\mathbb{D} \cap i\mathbb{R}$ under G are mapped by φ onto vertical lines through the integers and horizontal lines through $m\tau$, $m \in \mathbb{Z}$.*

In the sequel we will need to use a genus 0 variant of Lemma 5.2. Let A be as above a hyperbolic element with axis the real line. Let e_1 be the elliptic element of order 2 with center 0 and let e_2 be a second elliptic element of order 2 with center in a point τ on the pure imaginary axis. This data being subject to the condition that the group $G_0 = \langle A, e_1, e_2 \rangle$ is a discrete subgroup of $\mathrm{Aut}(\mathbb{D})$ with signature of the form $(0; 2, 2, 2, n)$. By construction the quotient $\mathbb{D}/G_0$ is of genus 0 and is naturally equipped with a singular hyperbolic metric with 4 cone points, 3 with total angle π and one with angle $2\pi/n$.

Fig. 2

The fundamental domain described on the left of figure 2 is conformally equivalent to a rectangle with vertices ± 1 and $\pm 1 + \tau$. Using this rectangle we can also equip S_0 with a singular flat metric (see right of figure 2). For the same reasons as those indicated for S_1 the sides of the fundamental domain coincide with the sides of the rectangle, and in particular are geodesic for both metrics. For further use we write this formally.

Lemma 5.4 *Let G_0 be a discrete subgroup of* $\mathrm{Aut}(\mathbb{D})$ *with signature* $(0; 2, 2, 2, n)$ *and generated by an elliptic element e_1 of order 2 centered at 0 and two hyperbolic elements A and B with respective axes the real axis and the pure imaginary axis.*

Let $S_0 = \mathbb{D}/G_0$. Then S_0 decomposes into two copies of a geodesic hyperbolic trirectangular quadrangle. The sides of this quadrangle are also geodesic for a natural singular flat metric.

We have just taken $B = e_1 e_2$.

Let S be a balanced surface obtained from r copies $R_1, \ldots, R_r$ of a rectangle R. Let d be the common order of the zeros of the associated quadratic differential. Note that in this case the total angle at the image of vertex will be $(2 + d)\pi$ in the locally flat metric.

We first consider the case when all edge identifications are by translations. Let E be the genus 1 surface corresponding to the rectangle R that to fix notations we identify with R_1 and let E^* be the surface obtained from E by removing the image of the vertices of the rectangle. Let p be the center of R_1 and let h and v be the elements of $\pi_1(E^*, p)$, the fundamental group of E^*, corresponding respectively to the horizontal median and the vertical median of the rectangle oriented in the natural way. Since $\pi_1(E^*, p)$ is isomorphic to the free group in two generators, h and v can be thought of as generating a free group.

Let S be as above and let S^* be the surface obtained by removing the vertices of the rectangles. Since the arrangement of rectangles we are starting with is admissible and in particular remains simply connected when the vertices are removed the identifications of edges define a set of generators of the fundamental group of S^*. On the other hand S^* is an unramified covering of E^*. Hence we can consider the fundamental group of S^* as a subgroup of $\pi_1(E^*, p)$. More precisely taking as base point the center of the rectangle R_1 the fundamental group of S^* is generated by words $w_1(h,v),\ldots,w_s(h,v)$ in h and v. For example if we consider the surface obtained from three rectangles (see upper left of figure 3) with the usual identifications (by horizontal and vertical translations) then these words are $h^2,v^2,hvh^{-1},vhv^{-1}$ (see [Sch] and [LeSi] Section 4 for other explicit computations of this type).

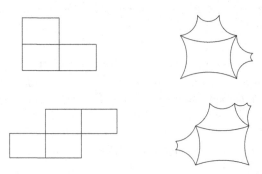

Fig. 3

Let S and q be as above q with zeros of order d. Then there exist hyperbolic elements A and B satisfying the conditions of Lemma 5.2 generating a group G_1 of signature $(1;n)$ and such that $\mathbb{D}/G_1 = E$.

Proposition 5.5 *Let S, the words $w_1,\ldots,w_s$ and A and B be as above. Let G be the group generated by $w_1(A,B),\ldots,w_s(A,B)$. Then*

  (i) *G is a discrete subgroup of $\mathrm{Aut}(\mathbb{D})$;*
 (ii) *$\mathbb{D}/G \cong S$ as Riemann surfaces;*
(iii) *the edges of the rectangles are geodesic arcs for the hyperbolic metric induced by the Poincaré metric on $\mathbb{D}$;*
(iv) *the horizontal and vertical medians of each rectangle are geodesic for the hyperbolic metric and extend to simple closed geodesics on the surface.*

PROOF. Call $\mathscr{R}$ the equiquadrangle defined by A and B. We build a domain in $\mathbb{D}$ starting with $\mathscr{R}$ and using the same combinatorics as the one that defines the arrangements of rectangles, replacing everywhere the copies of the Euclidean rectangle R by copies of $\mathscr{R}$ (see figure 3 for two examples). By construction G identifies in pairs the remaining edges of the equiquadrangles.

If x_0 is a point of S where q has a zero, the total angle at that point will be $(d+2)\pi$ and hence corresponds to the identification of $2(d+2)$ vertices. Since the interior angles of the equiquadrangle are $\pi/(d+2)$ the sum of the hyperbolic interior angles of the vertices identified by G with x_0 will be 2π.

This together with the pairing convention means that we can apply Poincaré's theorem (see for example [Bea] section 9.8) and conclude that the group G is discrete and that the arrangement of equiquadrangles we have constructed is a fundamental domain for this group.

Next we consider the conformal equivalence f from the interior of the equiquadrangle to the rectangle, extended by continuity to the boundary minus the vertices. Using Schwarz's reflection principle we can extend this conformal map to the full arrangement of equiquadrangles. This yields a conformal equivalence from the hyperbolic arrangement to the Euclidean arrangement. By construction of G this conformal equivalence induces a conformal equivalence from $\mathbb{D}/G - \{$images of vertices$\}$ to S^* which extends naturally to a biholomorphic map from $\mathbb{D}/G$ to S. This proves (ii) and (iii).

The median (horizontal or vertical) of a rectangle in S extends to a simple closed curve which corresponds to the decomposition of S into cylinders (horizontal or vertical). By Lemma 5.2 these are locally geodesic and since they intersect orthogonally at midpoints the sides of the equiquadrangles we obtain (iv).

If some of the identifications are by half-turns we have a very similar statement only the initial setup is different.

By changing the arrangement if necessary we may always assume that the identifications of the form $z \mapsto -z + c$ are between horizontal sides.

Let h be the translation that maps the left side of the rectangle R onto the right side. Let r_1 be the rotation of angle π centered at the center of R and let r_2 be the rotation of angle π centered at the middle of the upper side of R. Let S_0^* be the quotient of $R - \{$vertices$\}$ under the identification induced by h, r_1 and r_2. We consider S_0^* as an orbifold with three cone points of order 2 and a cusp. The orbifold fundamental group of S_0^* is generated by three elements, two of order 2 and one of infinite order. We may consider these generators as being r_1, r_2 and h. Let S^* be the surface obtained from S by removing the images of the vertices of the rectangles. Identifying R with R_1 we can, proceeding as in the previous case, write generators for the orbifold fundamental group of S^* as words in h, r_1 and r_2 or better as words in h, $v = r_1 r_2$ and

r_1. We choose these words $w_1(h, v, r_1), \ldots, w_s(h, v, r_1)$ to correspond to side pairings of the arrangement defining S.

Proposition 5.6 *Let* S, *the quadratic differential* q, *the words* $w_1, \ldots, w_s$, A, e_1 *and* e_2 *be as above.*

Let $B = e_1 e_2$ *and let* G *be the group generated by* $w_1(A, B, e_1), \ldots,$ $w_s(A, B, e_1)$. *Then properties* (i) *to* (iv) *of Proposition 5.5 hold also in this case.*

The proof follows exactly the same lines as the proof of Proposition 5.5.

6 Fractional Dehn twists on surfaces obtained from rectangles

Let S be a surface obtained from rectangles and more precisely obtained from copies of a rectangle R. Let P be a parallelogram then P can be seen as the image of R under the (Euclidean) action of an element g of $\mathrm{SL}_2(\mathbb{R})$. Replacing R by P in the construction of S we obtain a new surface that we will denote by $g(S)$. The total orbit of S under $\mathrm{SL}_2(\mathbb{R})$ is what is known as a Teichmüller disk. We will also consider the associated Veech curve which is the quotient of $\mathcal{H}$ by the stabilizer Γ of S in $\mathrm{SL}_2(\mathbb{R})$.

On such Teichmüller disks we of course have an action of $\mathrm{PSL}_2(\mathbb{Z})$. The object of this section is to show that this action is again by fractional Dehn twists and that the induced action on the associated Veech curve is algebraic.

Let S be a balanced surface obtained from rectangles. Then the surface has a natural decomposition into horizontal cylinders $\mathscr{C}_1, \ldots, \mathscr{C}_p$ (see for example [HuLe]). If a cylinder $\mathscr{C}_i$ is formed of n_i rectangles, we will say that it is of width n_i. By Proposition 5.5 or Proposition 5.6 the horizontal medians of these cylinders are disjoint simple closed geodesics γ_i.

We can also consider a decomposition into vertical cylinders $\mathscr{C}_1', \ldots,$ $\mathscr{C}_{p'}'$ of widths n_i' and this defines simple closed geodesics γ_i'.

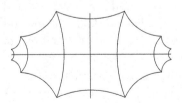 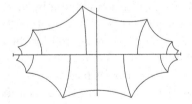

Fig. 4

Proposition 6.1 *Let S be a balanced surface obtained from rectangles and let S' be a surface in the $\mathrm{PSL}_2(\mathbb{R})$ orbit of S. Then the image of S' by an element of $\mathrm{PSL}_2(\mathbb{Z})$ is obtained by a composition of fractional Dehn twists.*

PROOF. We do this first for the surface obtained from rectangles. Let $\mathscr{R}_0$ be an equiquadrangle and let, as in Section 5, A and B be the hyperbolic left-right and bottom-top side pairings. For each integer j denote by $\mathscr{R}_j$ the equiquadrangle $A^j(\mathscr{R}_0)$.

A cylinder $\mathscr{C}$ of width n is the quotient of the infinite union of the $\mathscr{R}_j$ by the identification of each $\mathscr{R}_j$ to $\mathscr{R}_{j+n+1}$ by A^{n+1} (see left of figure 4). We assume that the horizontal middle geodesic is the real axis in the unit disk. In the Fuchsian group the element corresponding to this geodesic is A^n.

Now the action of $\mathrm{PSL}_2(\mathbb{Z})$ on parallelograms is generated by the transformations that transform

(i) the parallelogram with vertices $(0, 1, 1 + \tau, \tau)$ into the one with vertices $(0, 1, 2 + \tau, 1 + \tau)$;

(ii) the parallelogram with vertices $(0, 1, 1 + \tau, \tau)$ into the one with vertices $(0, 1 + \tau, 1 + 2\tau, \tau)$.

In terms of the generators A and B of the Fuchsian group for the genus 1 surface, (i) is the same as replacing (A, B) by (A, AB) while (ii) is the same as replacing (A, B) by (BA, B).

If the surface decomposes into horizontal cylinders $\mathscr{C}_1, \ldots, \mathscr{C}_p$ (respectively vertical cylinders $\mathscr{C}_1', \ldots, \mathscr{C}_{p'}'$) of respective widths n_i (resp. n_i'), then (i) corresponds to applying, for each i, a $1/n_i$-twist along the median of $\mathscr{C}_i$ and (ii) is the same with n_i' and $\mathscr{C}_i'$ in place of n_i and $\mathscr{C}_i$.

Since the transform under (i) or (ii) of a surface obtained from rectangles is again a surface obtained from rectangles this proves the proposition for such surfaces.

Let S be a surface obtained from rectangles and S' a surface in the $\mathrm{PSL}_2(\mathbb{R})$ orbit of S. Such a surface S' is obtained by replacing the rectangles by copies of a parallelogram in the combinatorial construction. Up to multiplication of the quadratic differential by a scalar, in other words by rotation and scaling, we can consider that the vertices of the parallelogram are $0, 1, 1 + \tau, \tau$.

We reformulate this in the hyperbolic context. We first do the case when the identifications are by translations only. Let

$$L' = \sqrt{\frac{\cos(\pi/n)^2 + L^2 - 1}{L^2 - 1}} \tag{6.1}$$

and let A_L and B_L be in $\mathrm{SU}(1,1)$, A_L with axis the real line and trace $2L$, B_L with axis the pure imaginary axis and trace $2L'$. Then a fundamental domain for the group $G_1(L)$ generated by A_L and B_L will be an equiquadrangle with interior angle π/n.

Let $T_{t,L}$ have axis the real line and half-trace $Tw = \cosh(t \operatorname{arccosh}(L))$ and let $B_{t,L} = T_{t,L} B_L$. Let $G_1(t, L)$ be the group generated by A_L and $B_{t,L}$. Then, letting L vary between 1 and $+\infty$ and t vary between $-\infty$ and $+\infty$, the surfaces $\mathbb{D}/G_1(t, L)$ will cover the full set of surfaces of genus 1 with one cone point of total angle $4\pi/n$. In other words replacing A and B by A_L and $B_{t,L}$ in Proposition 5.5 we will recover the full $\mathrm{PSL}_2(\mathbb{R})$ orbit of S.

Lemma 6.2 *Let τ be the normalized period of $\mathbb{D}/G_1(t, L)$ defined by the homology basis induced by the axes of A_L and $B_{t,L}$. Then $1 + \tau$ is the normalized period for $\mathbb{D}/G_1(1 + t, L)$.*

PROOF. By direct computation we have that $T_{1+t,L} = A_L T_{t,L}$ and hence $B_{1+t,L} = A_L B_{t,L}$. Call α and β_t the homology classes of the images of the axes of A_L and $B_{t,L}$, then $\beta_{t+1,L} = \alpha + \beta_t$. The conclusion follows immediately.

If some of the identifications are by half-turns we have a very similar construction, we only need in addition to replace e_1 in Proposition 5.6 by a conjugate by an element of trace $\sqrt{(Tw + 1)/2}$, with Tw as above, and axis the real axis.

Returning to the horizontal cylinders we can view them as on the right of figure 4 where the upper half has been shifted by $T_{t,L}$.

Replacing the parallelogram defined by τ by the one defined by $1 + \tau$ corresponds in the above construction to replacing $T_{t,L}$ by $T_{t+1,L}$ but this is exactly doing a $1/k$-twist, where k is the width of the cylinder, along the axis of the cylinder.

The construction we have made for horizontal cylinders can obviously be transposed to vertical cylinders, and this ends the proof.

We conclude by noting that the induced action on moduli is again algebraic. To see this note that a surface obtained from rectangles is a ramified cover of the genus 1 surface obtained from one rectangle, or in the presence of half-turns the genus 0 surface with four cone points. Hence an equation for the algebraic curve will be obtained algebraically from an equation of the quotient and the branch point (or branch points).

Assume the quotient is of genus 1. Then it has an equation of the form $y^2 = x(x-1)(x-\lambda)$ and we may choose this λ so that the vertices of the rectangle are mapped to $x = 0$, the midpoints of the horizontal edges are mapped to $x = 1$, the center point to $x = \lambda$ and finally the midpoints of the vertical edges to ∞. We can obviously generalize this to the case of parallelograms. With this fixed an equation of the surface will be obtained algebraically from λ.

But now replacing τ by $1 + \tau$ corresponds to exchanging the center point and the midpoint of a vertical edge or in other words exchanging λ and ∞ while leaving 0 and 1 fixed. Hence if τ corresponds to λ then $1 + \tau$ will correspond to $1 - \lambda$.

For the same reason replacing the parallelogram with vertices $(0, 1, 1 + \tau, \tau)$ by the one with vertices $(0, 1 + \tau, 1 + 2\tau, \tau)$ will correspond to replacing λ by $1/\lambda$.

We recover in this way the action of the permutation group $\mathfrak{S}_3$. On the other hand the situation is more involved here than in the previous cases since to one λ may correspond several surfaces and the dependence in λ may be a dependence in the roots of some covering equation with coefficients depending on λ. Hence there is in general an interplay between the above action of $\mathfrak{S}_3$ and the Galois group of the covering equation (see [LeSi] and [Si2] for some examples).

Examples. 1) Consider the surface obtained from 3 squares with the obvious identifications by horizontal and vertical translations. Then the surfaces in the $\mathrm{PSL}_2(\mathbb{R})$-orbit of this surface will have equations of the form

$$y^2 = x(x-1)\left(x^3 + ax^2 - \frac{8}{3}ax + \frac{16}{9}a\right) \qquad (6.2)$$

(see [Si2]). Moreover if E is the elliptic curve corresponding to one parallelogram and λ is the Legendre coefficient defined as above, then

$$a = -9\lambda .$$

The surface has two horizontal cylinders, one of width 1 the other of width 2. Our first elementary move is to apply a full Dehn twist along the cylinder of length 1 and half-twist along the other. As explained above this corresponds to replacing λ by $1 - \lambda$. Hence replacing a by $-9 - a$.

We also have two vertical cylinders of width 1 and 2. Applying a twist and a half-twist along these corresponds to replacing λ by $1/\lambda$ and hence replacing a by $81/a$.

The action of $\mathfrak{S}_3$ on the family of curves defined by the equations (6.2) is generated by

$$a \mapsto -9 - a \quad \text{and} \quad a \mapsto \frac{81}{a} \; .$$

This action is not faithful since rotating the rectangle or the parallelogram yields an isomorphic algebraic curve. In other words replacing a by

$$\frac{-9\,a}{9 + a}$$

in Equation (6.2) yields an isomorphic algebraic curve (see [Si2]).

2) Consider the surface obtained from four squares as on the lower left of figure 3, and identifications by horizontal and vertical translations. Then the surfaces in the $\mathrm{PSL}_2(\mathbb{R})$-orbit of this surface will have equations of the form

$$y^2 = (x^2 - a)(x^2 - 1)(x^2 - a - 1) \tag{6.3}$$

(see [LeSi] or [Si2]). The relation between the coefficient a of Equation (6.3) and the Legendre invariant is this time

$$\lambda = \frac{(a + 1)^2}{(a - 1)^2}$$

(see [LeSi]).

We have two horizontal cylinders, both of width 2 and three vertical cylinders, two of width 1 and one of width 2. The action on the equations of the fractional Dehn twists is best expressed in terms of

$$\nu = a + \frac{1}{a}$$

which is a modular invariant for algebraic curves with equations of the form (6.3) i.e. they are isomorphic if and only if they have the same ν (see [Si2]).

In terms of ν the transformations are

$$\nu \mapsto \frac{2(6 - \nu)}{2 + \nu} \quad \text{and} \quad \nu \mapsto -\nu \tag{6.4}$$

where the first is induced by applying two half-twists along the horizontal cylinders and the second by applying two full Dehn twists and a half-twist along the vertical cylinders (see [LeSi] or [Si2]). The group generated by the transformations of (6.4) is the permutation group $\mathfrak{S}_3$ and in this case the action is faithful.

For other examples see [LeSi] and [Si2].

References

[Ai]   Aigon-Dupuy, A. Half-twists and equations in genus 2. *Annales Academiæ Scientiarum Fennicæ Math.* **29** (2004), 307–328.

[AiSi] Aigon, A. and Silhol, R. Hyperbolic hexagons and algebraic curves in genus 3. *J. London Math. Soc.* **66** (2002), 671–690.

[Bea] Beardon, A. *The Geometry of Discrete Groups*, Springer G.T.M. 91, Berlin, 1991.

[Bir] Birman, J. *Braids, Links and Mapping Class Groups*, Ann. Math. Studies 82, Princeton Univ. Press, Princeton, 1974.

[Bu]   Buser, P. *Geometry and Spectra of Compact Riemann Surfaces*, Birkhäuser, Boston, 1992.

[BuSi] Buser, P. and Silhol, R. Geodesics, periods and equations of real hyperelliptic curves. *Duke Math. J.* **108** (2001), 211–250.

[HuLe] Hubert, P. and Lelièvre, S. Prime arithmetic Teichmüller discs in $\mathcal{H}(2)$. *Israel J. of Math.* **151** (2006), 281–321.

[LeSi] Lelièvre, S. and Silhol, R. Multi-geodesic tessellations, fractional Dehn twists and uniformization of algebraic curves. Preprint Hal-00129643 (2007).

[Nag] Nag, S. *The Complex Analytic Theory of Teichmüller Spaces*, John Wiley, New York, 1988.

[Sch] Schmithüsen, G. Examples for Veech groups of origamis. *Contemporary Math.* **397** (2006), 193–206.

[Si1] Silhol, R. Hyperbolic Lego and equations of algebraic curves. *Contemporary Math.* **311** (2002), 313–334.

[Si2] Silhol, R. On some one parameter families of genus 2 algebraic curves and half-twists. *Commentarii Math. Helvetici* **82** (2007), 413–449.